Deep Learning for Earth Observation and Climate Monitoring

Earth Observation Series

Deep Learning for Earth Observation and Climate Monitoring

Edited by

Uzair Aslam Bhatti
School of Information and Communication Engineering, Hainan University, Haikou, P.R. China

Mir Muhammad Nizamani
Department of Plant Pathology, College of Agriculture, Guizhou University, Guiyang, Guizhou, P.R. China

Yong Wang
Department of Plant Pathology, College of Agriculture, Guizhou University, Guiyang, Guizhou, P.R. China

Hao Tang
School of Information and Communication Engineering, Hainan University, Haikou, P.R. China

ELSEVIER

Elsevier
Radarweg 29, PO Box 211, 1000 AE Amsterdam, Netherlands
125 London Wall, London EC2Y 5AS, United Kingdom
50 Hampshire Street, 5th Floor, Cambridge, MA 02139, United States

ISBN: 978-0-443-24712-5

For Information on all Elsevier publications
visit our website at https://www.elsevier.com/books-and-journals

Publisher: Candice Janco
Acquisitions Editor: Peter Llewellyn
Editorial Project Manager: Teddy Lewis
Production Project Manager: Sruthi Satheesh
Cover Designer: Vicky Pearson

Idea by Prem Pandey and Design by Mr. Kumar Krishna Mohan - Sr. Graphic Designer, Shiv Nadar University

Typeset by MPS Limited, Chennai, India

Contents

Section I
Deep learning for climate change

Section III
Deep learning for GIS

Section IV
Deep learning for lulc

11. Climate change air quality monitoring using sentimental 2 dataset

Mughair Aslam Bhatti, Hao Tang, Uzair Aslam Bhatti, Dekai Li and Mingtian Yu

12. Latest trends in land use and land cover monitoring using deep learning

Ahsan Ahmed Nizamani, Yonis Gulzar, Hao Tang, Uzair Aslam Bhatti and Abdul Haseeb Nizamani

13. Oceanic biometric recognition algorithm based on generalized zero shot learning

Hao Tang, Dekai Li, Ahsan Ahmed Nizamani, Mughair Aslam Bhatti, Jinru Liu, Yonghui Zhang and Uzair Aslam Bhatti

Section V
Deep learning for oceans

14. Remote sensing image fusion based on deep learning and convolutional neural network technique

Emadalden Alhatami, Mengxing Huang and Uzair Aslam Bhatti

List of contributors

Emadalden Alhatami School of Information and Communication Engineering, Hainan University, Haikou, P.R. China

Muhammad Awais Institute of Plant Sciences, Faculty of Natural Sciences, University of Sindh, Jamshoro, Sindh, Pakistan

Shahnawaz Ayoub Glocal School of Science and Technology, Glocal University, Saharanpur, Uttar Pradesh, India

Mughair Aslam Bhatti School of Information and Communication Engineering, Hainan University, Haikou, P.R. China

Uzair Aslam Bhatti School of Information and Communication Engineering, Hainan University, Haikou, P.R. China

Yonis Gulzar Department of Management Information Systems, College of Business Administration, King Faisal University, Al-Ahsa, Saudi Arabia

Mengxing Huang School of Information and Communication Engineering, Hainan University, Haikou, P.R. China

Fahad Ali Kalhoro Department of Plant Breeding and Genetics, Sindh Agriculture University, Tando Jam, Sindh, Pakistan

Anab Khanzada State Key Laboratory of Crop Biology, Agronomy College, Shandong Agricultural University, Taian, Shandong, P.R. China

Maqsood Ahmed Khaskheli Department of Plant Pathology, College of Agriculture, Guizhou University, Guiyang, Guizhou, P.R. China

Raheel Ahmed Khaskheli Department of Plant Pathology, Faculty of Crop Protection, Sindh Agriculture University, Tando Jam, Sindh, Pakistan

Rashid Ali Khaskheli Department of Crop Physiology, Faculty of Crop Production, Sindh Agriculture University, Tando Jam, Sindh, Pakistan

Sharmili Khatun Department of Geography, Ranaghat College, Kalyani University, Ranaghat, West Bengal, India

Vinit Kumar National Key Laboratory of Green Pesticide, Key Laboratory of Green Pesticide and Agricultural Bioengineering, Ministry of Education, Center for Research and Development of Fine Chemicals, Guizhou University, Guizhou, Guiyang, P.R. China; Department of Medical Lab Technology, Markham College of Commerce, Vinoba Bhave University, Hazaribagh, Jharkhand, India

Abdul Hafeez Laghari Department of Agronomy, Sindh Agriculture University Sub Campus Umerkot, Umerkot, Sindh, Pakistan

Umed Ali Laghari Department of Agriculture, Mir Chakar Khan Rind University, Sibi, Balochistan, Pakistan

Dekai Li School of Information and Communication Engineering, Hainan University, Haikou, P.R. China

Dai Lisi School of Information and Communication Engineering, Hainan University, Haikou, P.R. China

Jinru Liu School of Information and Communication Engineering, Hainan University, Haikou, P.R. China

Abdul Haseeb Nizamani Central South University, Computer Science and Engineering, Changsha, P.R. China

Ahsan Ahmed Nizamani School of Information and Communication Engineering, Hainan University, Haikou, P.R. China; Central South University, Computer Science and Engineering, Changsha, P.R. China

Mir Muhammad Nizamani Department of Plant Pathology, College of Agriculture, Guizhou University, Guiyang, Guizhou, P.R. China

Muhammad Rashid Nizamani Crop Science Research Institute, Agriculture Research Centre, Tando Jam, Sindh, Pakistan

Rafeeque Ahmed Nizamani Department of Mechanical Engineering, Wuhan University of Technology, Wuhan, Hubei, P.R. China

Sher Muhammad Nizamani Department of Engineering and Architecture, University of Parma, Parma, Italy

Muhammad Qayyum School of Economics and Statistics, Guangzhou University, Guangzhou, P.R. China

Faheem Ahmad Reegu Department of Electrical and Electronics Engineering, College of Engineering and Computer Science, Jazan, Arabia Arabia

Xin Sun School of Information and Communication Engineering, Hainan University, Haikou, P.R. China

Faria Sundas Department of Plant Pathology, College of Agriculture, Guizhou University, Guiyang, Guizhou, P.R. China

Hao Tang School of Information and Communication Engineering, Hainan University, Haikou, P.R. China

Entaj Tarafder Department of Plant Pathology, College of Agriculture, Guizhou University, Guiyang, Guizhou, P.R. China

Yong Wang Department of Plant Pathology, College of Agriculture, Guizhou University, Guiyang, Guizhou, P.R. China

Mingtian Yu School of Information and Communication Engineering, Hainan University, Haikou, P.R. China

Yonghui Zhang School of Information and Communication Engineering, Hainan University, Haikou, P.R. China

Chapter 1

Introduction: advancing ecological protection through integrated geographic information system-enabled environmental monitoring: a holistic approach to addressing environmental pollution

Hao Tang, Uzair Aslam Bhatti, Jinru Liu, Dai Lisi and Mughair Aslam Bhatti
School of Information and Communication Engineering, Hainan University, Haikou, P.R. China

1.1 Introduction

Environmental monitoring is one of the important means in current ecological environmental protection (Bhatti et al., 2021). Emergency monitoring serves as an effective means to understand the development of the ecological environment (Bhatti et al., 2021; Parra, 2022; Lovett et al., 2007). By monitoring and analyzing pollutants, it can provide support and assistance for environmental management and decision-making (Han et al., 2024; Ullo & Sinha, 2020; Li et al., 2020). It can collect and analyze data related to the ecological environment in real-time and with high precision, swiftly detecting harmful substances in the air, water, soil, and other environments (Bhatti et al., 2022b; Awewomom et al., 2024). Environmental monitoring not only prevents potential environmental pollution but also analyzes existing environmental pollution and seeks solutions, playing an important role in ecological environmental protection work (Bhatti et al., 2022a; Young & Stokke, 2020; Forgács, 2024). This provides valuable references for relevant departments to respond to sudden environmental issues. However, due to the limited traditional level of monitoring technology for the ecological environment, monitoring personnel can only deploy monitoring points based on their work experience (Antunes, Santos & Jordão, 2001; Wang & Stauffer, 1995). Moreover, with numerous external interference factors, it is difficult to ensure the accuracy and timeliness of monitoring results, thereby greatly hindering the implementation of environmental governance work (Gunasekera, 2004). Therefore it is of great significance to study the role and development of environmental monitoring in ecological environmental protection (Bhatti, 2022c; Kolios & Vorobeva, 2017). By using geographic information system (GIS) technology, geographical spatial data can be utilized as a foundation, and spatial data can be acquired, stored, managed, and analyzed, thereby providing monitoring personnel with more accurate and timely information for emergency decision-making (González, 2012).

1.2 Overview of environmental monitoring

1.2.1 Connotation of environmental monitoring

Environmental monitoring mainly refers to the detection of the environment using existing professional monitoring techniques and analyzing the monitoring results to fully understand the surrounding ecological environment and determine whether pollution has occurred (Gavrilaş et al., 2022). Fast and accurate monitoring data are crucial for decision-makers

Deep Learning for Earth Observation and Climate Monitoring. DOI: https://doi.org/10.1016/B978-0-443-24712-5.00002-6

when facing environmental emergencies (Biber, 2013). Environmental emergency monitoring, as a rapidly developing technology, provides strong support for addressing environmental issues. Currently, as China's economy gradually optimizes and scientific and technological development advances, environmental pollution problems are becoming increasingly severe, directly impacting the living environment (Biber, 2013). Its monitoring content involves real-time monitoring of the environment, such as water, soil, and atmosphere, collecting and analyzing data related to the ecological environment, promptly identifying problems in the environment, and providing a basis for emergency response work by monitoring departments (Kumar et al., 2021). Therefore it is necessary to protect the environment, analyze environmental problems, propose corresponding solutions, analyze the details that may cause environmental pollution, and finally protect the environment, playing the role of environmental monitoring (Ye, 2022).

1.2.2 Classification of environmental monitoring

Environmental monitoring is mainly divided into three types: surveillance monitoring, special-purpose monitoring, and research monitoring (Messer, Zinevich & Alpert, 2006). First, regarding surveillance monitoring. Surveillance monitoring belongs to routine monitoring, also known as routine monitoring (Cyranoski & Brumfiel, 2011). It mainly refers to monitoring specific ecological environments. Surveillance monitoring has the characteristics of periodicity and long time, which can evaluate the monitored environment, judge environmental quality, and determine whether pollution exists, pollution sources (Gilbert, 2011; Harlan, Terrill & Burnett, 2009; Williams et al., 2007), etc. With the help of surveillance monitoring, monitoring personnel can measure environmental protection and standard execution and conduct detailed analysis of the environment. Secondly, concerning special-purpose monitoring. Special-purpose monitoring mainly includes special case monitoring and emergency monitoring (Williams et al., 2009, 2011). When encountering sudden environmental pollution incidents, monitoring personnel should conduct special case monitoring for the current pollution event. If an assessment of the environment is needed, special-purpose monitoring is also required. In addition, special-purpose monitoring is also needed in the completion acceptance process (Skön, Kauhanen & Kolehmainen, 2011). Finally, regarding research monitoring. Research monitoring mainly refers to scientific research monitoring. When conducting scientific research for special targets, monitoring personnel need to conduct high-level monitoring (Berglund, Kotovirta & Seinä, 2007). For example, in environmental background monitoring, research on standard analysis methods, material development, etc., monitoring personnel need to use research monitoring to meet high requirements (Kotovirta et al., 2012).

1.2.3 Overview of geographic information system technology

GIS can perform real-time and accurate analysis of data information related to ecological environments and various resources (Chawla, Karthikeyan & Mishra, 2020; Wei et al., 2020; Babaeian et al., 2019; Sishodia, Ray & Singh, 2020). Based on the processing and analysis of data, GIS can build models and form a complete geographical spatial system. Geographic spatial data refer to data related to the Earth's surface, including terrain, landforms, buildings, and roads. GIS technology can collect, process, and manage these data. Collection methods include satellite remote sensing, map scanning, and field surveys (Pérez-Castro et al., 2022).

1.2.4 Advantages of GIS technology in ecological environment emergency monitoring work

1. Strong spatial data analysis capability: GIS technology can integrate various spatial data sources such as topographic maps, satellite images, and meteorological data. By analyzing spatial data, it can more accurately reflect the spatial distribution and change trends of environmental problems (Lechner, Foody & Boyd, 2020; Noor, Abdullah & Hashim, 2018; Estoque, 2020). For example, GIS technology can quickly analyze the diffusion range and impact of pollutants, providing scientific basis for emergency monitoring.
2. Real-time monitoring and early warning: GIS technology can be combined with hardware facilities such as sensors and monitoring devices to achieve real-time environmental monitoring. When environmental indicators exceed threshold values, the GIS system can issue timely warnings, notifying relevant departments to take response measures, which helps reduce the occurrence of environmental pollution accidents and minimize losses (Trofymchuk et al., 2020; Manfreda et al., 2018; Mahrad et al., 2020).
3. Visualization display and decision support: GIS technology can visually display complex environmental data in graphics, images, etc., helping decision-makers better understand environmental conditions and formulate response

strategies (Reinermann, Asam & Kuenzer, 2020). For example, through GIS visualization, the effects of different pollution control measures can be simulated, providing decision-makers with reference basis.

1.3 Role of environmental monitoring in ecological conservation

1.3.1 Environmental safety maintenance

With the assistance of environmental monitoring, the ecological environment can be further protected and put on the path of sustainable development (Lechner, Foody & Boyd, 2020). Therefore environmental monitoring plays an important role in ecological environmental protection. As environmental awareness gradually increases among people and the importance of the environment is continuously emphasized, the government also needs to strengthen its attention to ecological environmental protection (Ahmed et al., 2021; Jamil et al., 2015). It should construct sound, complete, and reliable laws and regulations on environmental protection, strictly control enterprises in urban areas, perform supervisory functions, and increase punishment for enterprises to prevent environmental pollution (Wu & Tham, 2023). At the same time, monitoring personnel should continuously strengthen their abilities, promote the development of environmental monitoring technology through technological updates, and further improve the quality of work.

1.3.2 Strengthening pollution management

For pollution management, environmental monitoring can conduct in-depth analysis of pollution types, pollution levels, etc., to provide data support for management work, help personnel achieve effective control, and fully play the role of ecological environmental protection work, providing guarantees for pollution management. Environmental monitoring departments need to analyze the emission of pollutants according to current standards in China, conduct in-depth research on the quantity and types of pollutants, propose targeted solutions, and reduce the degree of pollution (Siddiqua, Hahladakis & Al-Attiya, 2022; Sohoo et al., 2022; Gao et al., 2017). On this basis, if there is a problem of excessive emission of pollutants, the environmental protection department can further promote emission reduction work with the help of monitoring data and information to ensure that the amount of pollutant emissions meets current requirements (Colpaert, 2022). If the emission of pollutants in a certain area exceeds the standard severely, the environmental protection department needs to conduct further analysis and monitoring, and adjust the existing rectification plan according to the actual situation (Lin et al., 2020). Therefore environmental monitoring can further strengthen the effect of pollution management in ecological conservation.

1.3.3 Guiding environmental protection work

With the implementation of the concept of harmonious development between humans and nature, in order to achieve the goal of ecological environmental protection, effective supervision of current ecological environmental protection work should be carried out. With the help of environmental monitoring, exploration of the ecological environmental quality within the region can be conducted to clarify the environmental situation and lay a good foundation for subsequent environmental protection work (Cicala et al., 2018; Gaglio et al., 2014; Zhang et al., 2015; Tadejko, 2017). At the same time, a comparison can be made between existing environmental monitoring data and historical data, and the comparative results can be analyzed to ensure the objectivity of ecological environmental assessment (Kulkarni et al., 2016; Kamal, 2020).

1.3.4 Promoting social development

Social development cannot be separated from people, and normal human development cannot be separated from ecological environmental protection. Environmental monitoring can lay a good foundation for ecological environmental protection work, accurately grasp the development trends and changes of the ecological environment, and take timely countermeasures. Through data analysis, ecological environmental protection can obtain reliable data support. After analyzing and processing the data, the inevitable connection between human activities and environmental changes can be further strengthened, providing people with a better ecological environment. At the same time, during environmental monitoring, the impact of air pollution, land pollution, and water pollution on people's quality of life can be accurately analyzed, enhancing people's environmental awareness (Li et al., 2017). In addition, environmental monitoring work can further promote environmental law enforcement, ensure the smooth progress of environmental protection work, systematically analyze existing ecological environmental protection problems, clarify the main causes, and find corresponding measures, thereby effectively improving the overall level of environmental protection and contributing to social development (Zhang, Tan & Chan, 2021).

1.3.5 Common issues in ecological environment emergency monitoring

With the rapid development of technology, emergency monitoring of the ecological environment in China is also constantly evolving. However, due to its late start, there are still some problems in the monitoring process that urgently need to be addressed, among which the low level of informatization is a major factor affecting the effectiveness of monitoring. Common problems can be classified as follows:

1. Field investigations pose risks. In-depth field investigations are the key foundation of emergency monitoring of the ecological environment. Monitoring personnel need to understand the environmental pollution status, composition, development trends, etc., through methods such as visits and sampling (Narayana et al., 2024). However, it is important to note that emergency monitoring faces generally high risks, such as fires, explosions, and other emergencies, which can pose a certain threat to the life and health safety of monitoring personnel (Sattar et al., 2019).
2. Point layout relies on experience. In the implementation of emergency monitoring of the ecological environment, only by reasonably arranging monitoring points can the scientificity and accuracy of monitoring results be improved. This requires monitoring personnel to comprehensively consider factors such as the type of emergency events and the surrounding environment (Bedi et al., 2021). Since most emergencies are characterized by randomness and unpredictability, monitoring personnel need to rely on their own practical experience to complete the layout of monitoring points. Failure to consider comprehensively by monitoring personnel can affect the effectiveness of ecological environment monitoring.
3. Monitoring data is overly abstract. After ecological environmental emergencies occur, monitoring personnel need to use professional monitoring instruments to carry out various tasks at monitoring points and collect data related to the ecological environment (Lindenmayer et al., 2022). However, in practice, the data obtained through this monitoring method tends to be abstract, severely lacking in intuitiveness and readability, and poses certain difficulties in analysis and exploration, requiring the use of specialized software for analysis and processing (Shan Neo & Tan, 2017).

1.4 Measures for the development of environmental monitoring in ecological conservation

1.4.1 Enhancing team literacy

The overall literacy of the environmental monitoring team determines the reliability and effectiveness of monitoring results. With the continuous development and progress of science and technology, environmental monitoring personnel should continuously improve their abilities through diverse methods, learn advanced knowledge and content, and enhance their professional levels (Filipe et al., 2014). Specifically, improvement is needed in four aspects. Firstly, environmental monitoring personnel need to deeply study and proficiently master China's environmental monitoring laws, regulations, rules, standards, monitoring methods, etc., to enhance work efficiency and quality and play their roles effectively (Beer & Mulder, 2020). Secondly, regular ideological and political education should be conducted for personnel to integrate environmental protection work into their thinking, continuously strengthen their moral standards and ideological awareness, and build correct values based on this foundation to improve the ideological level of the team (Beer & Mulder, 2020). Thirdly, systematic training should be conducted regularly to provide professional competence training for monitoring personnel, improve their technical skills, and strengthen their professional literacy to ensure that their professional level meets the requirements of environmental monitoring, thereby playing a role in environmental monitoring and making contributions to environmental protection. Finally, discipline and awareness of monitoring personnel should be cultivated regularly (Yang, Duan & Jianhong, 2022). In terms of discipline, it mainly refers to establishing correct, good, and positive work styles to avoid personnel developing feelings of difficulty or other problems that may affect work progress. In terms of awareness, personnel need to cultivate safety awareness and other awareness regularly, remember safety content, adhere to the principle of self-safe development, and establish correct concepts.

1.4.2 Introduction of advanced equipment

Currently, China's environmental monitoring technology is continuously developing and has gradually reached the forefront internationally. However, there is still a gap compared to Western countries, especially in terms of instrument equipment. At present, there are only over 200 types of environmental monitoring instruments and equipment in China, and some equipment are imported from abroad, resulting in problems of insufficient adaptability of equipment in the environmental monitoring process in China, thus failing to fully utilize the functions of the instruments and equipment. Therefore the focus of optimizing environmental monitoring work at present is to introduce advanced equipment. In the actual use process, monitoring personnel need to optimize and improve the equipment according to the actual environmental conditions in China, fully utilize the functions of the equipment, improve the equipment system, and lay a good foundation for future development (Jose et al., 2023; Gałuszka, Migaszewski & Namieśnik, 2015; Ekins & Zenghelis, 2021). At the same time, rational application of intelligent monitoring methods can be applied during the environmental monitoring process to further simplify the existing processes, fully play the role of intelligence, improve work efficiency, and meet the monitoring requirements in different environments.

1.4.3 Analysis and evaluation

In environmental monitoring work, environmental analysis and evaluation are indispensable important links. Monitoring personnel can use current diversified methods to analyze existing environmental pollution situations, such as using remote sensing technology or GISs for analysis work, and then conduct evaluations to clarify the land conditions and vegetation planting situations in different city areas, and finally integrate and process the information to assist in environmental protection work (Goodchild, 2003; Kerry et al., 2022; Li et al., 1982; Carreño-Conde et al., 2021). In this process, monitoring personnel first need to do a good job of data collection and data analysis, with data accuracy as the core to ensure the effectiveness of the data. Secondly, monitoring personnel should try to integrate environmental monitoring with pollution prevention and control work, further streamline processes while ensuring the accuracy of monitoring results, and conduct scientific analysis and evaluation. Finally, based on ecological environmental detection analysis and results, protective measures that meet the requirements of regional ecological environments should be established, and ecological environmental protection work should be implemented to play an important role in environmental monitoring (Bennett et al., 2018).

1.4.4 Improving the early warning system

Environmental monitoring itself serves ecological environmental protection. Therefore during the daily environmental monitoring work, monitoring personnel need to optimize the current ecological environmental monitoring network, handle sudden pollution events promptly, and prevent potential pollution events on this basis to strengthen ecological civilization construction. In this process, data from different levels of environmental monitoring need to be centralized in the same system, and communication and sharing among all parties should be promoted to ensure the authenticity of data information. Different regions need to be closely linked, and technical requirements need to be clarified based on the actual ecological pollution environment conditions to build an environmental monitoring technology system on this basis (Mauri et al., 2021; Domazetović et al., 2021). If a sudden event occurs in a certain area, it needs to be resolved promptly, and early warning methods need to be applied reasonably to build a comprehensive and reliable ecological environmental monitoring network and reduce the probability of risk occurrence. In addition, with the assistance of Internet technology and digital technology, existing pollution problems need to be controlled, monitoring equipment needs to be correctly utilized, and data information needs to be accurately grasped to provide data support for environmental protection.

1.4.5 Advancing standardized management

Incorporating standardized management into the environmental monitoring process can effectively improve the effectiveness of monitoring, play the role of monitoring, and enhance monitoring results. Therefore standardized management needs to be implemented for environmental monitoring work. Management personnel should establish corresponding institutional systems based on China's laws, regulations, and current standards, combined with actual conditions, and establish rules and regulations on this basis to ensure the standardization of management. For example, in the construction and improvement of the environmental quality management system, the current environmental quality and management system should be optimized to ensure its completeness and standardization (Yang et al., 2021). Then, corresponding standardized standards should be established to ensure that different departments and different links can smoothly advance the work according to the standards, providing strong assistance for environmental monitoring (Scerri et al., 2022). At the same time, when facing the data collection work of monitoring, standardized management needs to be carried out for different links, and the existing data information needs to be summarized and reported. When reporting, unified standards and methods should be adopted to ensure further improvement of work quality under the premise of meeting the requirements of environmental monitoring (Zhou, 2020).

1.4.6 Application of information technology

Firstly, a data network platform should be constructed for environmental protection and connected to ecological environmental monitoring. Based on ecological environmental monitoring data, a database should be constructed, and the platform should be used to facilitate communication between different departments and units, transmit information, fully utilize the role of information technology, and lay a good foundation for other work (Johnson et al., 2021). Secondly, further research and development of environmental monitoring database systems should be conducted, and sharing functions should be strengthened to reduce the possibility of monitoring duplication. In this process, technical personnel should do a good job of classification, accurately classify existing environmental data, give full play to the functions of the database, provide convenient conditions for data queries, and reduce the overall difficulty of work (Chen, Zhang & Rajakani, 2022). Finally, construct an ecological environmental monitoring early warning system. This mainly refers to the application of advanced technology by technical personnel to optimize existing ecological environmental monitoring, effectively improve monitoring efficiency and quality, ensure the reliability and accuracy of ecological environmental data collection, fully utilize the role of data, issue early warnings for ecological pollution, and facilitate environmental protection personnel to take corresponding measures.

1.5 Practical application of GIS technology in ecological environment emergency monitoring

1.5.1 Application of GIS in ecological environment simulation

In the field investigation work of ecological environment emergency monitoring, GIS technology can be fully utilized to improve the efficiency and quality of emergency monitoring work in terms of data collection, analysis, and processing. Specifically, monitoring departments can analyze the results of ecological environment investigations by establishing databases (Chen et al., 2022). By simulating the evolution process of the ecological environment, abnormal situations can be promptly identified and reported, enabling relevant departments to respond promptly. It is well known that the evolution of the ecological environment is a dynamic development process (Davenhall & Kinabrew, 2011). If the evolution process is unbalanced, it will increase the difficulty of restoration and cause serious disasters. However, by using GIS technology, the diffusion of pollutants in the natural environment over a period of time can be simulated through the analysis of remote sensing images, and the adverse effects of pollutants on the ecological environment can be predicted, which is conducive to providing reliable basis for relevant departments to formulate emergency measures.

1.5.2 Application of GIS in regional environmental governance

GIS, as a comprehensive information management tool based on geographical spatial data, provides powerful support for ecological environment background investigations. By utilizing GIS technology, data collected from remote sensing technology and global positioning systems can be effectively integrated, and a virtual model consistent with the regional ecological environment can be established. This model has the characteristics of real-time and thoroughness. Monitoring personnel can analyze remote sensing images to visually express complex ecological environment data in the form of graphics and images, facilitating monitoring personnel to quickly understand the situation and make scientific decisions (Firouraghi et al., 2022). It can timely discover vulnerable areas of the ecological environment. Based on this, real-time routes for emergency measures can be optimized, effectively improving emergency response efficiency and environmental governance capabilities. In environmental governance, GIS technology mainly plays the following roles:

In environmental monitoring, GIS collects and analyzes spatial data to monitor environmental pollution conditions in real-time, evaluate environmental quality, and provide a basis for formulating governance measures. For example, GIS technology can analyze the spatial distribution characteristics of environmental elements such as atmosphere, water quality, and soil, discover pollution sources and potential pollution areas, and support the formulation of targeted governance measures. In environmental pollution prevention and control, GIS can provide comprehensive technical support (Nuckols, Ward & Jarup, 2004; Chen et al., 2022). For example, using GIS's spatial analysis capabilities, pollution sources can be located and distributed for the formulation of pollution prevention and control strategies. At the same time, through GIS's data visualization function, environmental pollution situations can be intuitively presented to the public, increasing public awareness of environmental protection. In environmental risk management, GIS can help monitoring personnel better manage environmental risks. By establishing an environmental risk database, monitoring personnel can monitor and evaluate various environmental risk factors in real-time, discover and resolve potential environmental risk issues. At the same time, using GIS's simulation prediction function, environmental risks can be predicted and evaluated to provide a basis for formulating response measures.

1.5.3 Application of GIS in ecological environment background investigation

The application of GIS in ecological environment background investigations is very extensive and has significant practical value. By collecting, organizing, and storing ecological environment-related data, GIS has established a complete spatial database. These data cover various elements such as vegetation coverage, land use, and water resources. Using GIS technology, these elements can be subjected to spatial overlay analysis, distance calculation, and other operations to reveal the spatial distribution laws of the ecological environment and provide scientific basis for ecological environment protection (Xie, Yao & Liu, 2015). Based on GIS technology, monitoring personnel can also establish an environmental spatial database, which includes all relevant information about water, air, noise, etc. This technology can integrate all collected information and geographical locations and adopt a classification management and analysis approach (Faisal & Shaker, 2017). By simply entering keywords, direct queries can be made, and various types of spatial data can be visualized. Currently, some cities have established environmental pollution source monitoring databases and spatial databases based on GIS technology. The specific functions of GIS can be summarized as follows:

Firstly, data collection. GIS can collect and process large amounts of geographical spatial data, including remote sensing images, topographic maps, and meteorological data. These data can be obtained through digitization or manual collection methods and then imported into GIS for processing and analysis. Secondly, data processing. GIS can preprocess, correct, and format data collected to ensure accuracy and consistency. At the same time, GIS can also perform data mining and analysis to extract useful information and features (Faisal & Shaker, 2017). Thirdly, spatial analysis. GIS can conduct spatial analysis of ecological environment data, including terrain analysis, hydrological analysis, and vegetation analysis. These analyses can help researchers understand the spatial distribution characteristics and change rules of the ecological environment, providing scientific basis for ecological environment management and protection. Fourthly, results display. GIS can display research results in the form of graphics and charts, facilitating visualization and interpretation by researchers. At the same time, GIS can generate thematic maps and statistical reports as output forms, providing decision-makers with more intuitive and reliable information support.

1.5.4 Application of GIS in ecological environment risk assessment and early warning

Applying GIS technology to ecological environment emergency management can also play a role in risk assessment and early warning. Specifically, firstly, analyzing risk sources. Using GIS technology, spatial analysis of environmental risk sources can be conducted, such as spatial positioning of factories, landfill sites, and other pollution sources, analyzing their distribution characteristics and diffusion trends to provide basic data for environmental risk assessment. Secondly, exposure assessment. GIS can combine environmental monitoring data to assess the degree of exposure of populations and ecosystems to pollution sources. Through spatial analysis methods, the potential impact range and degree of pollution sources on the surrounding environment can be determined. Thirdly, risk assessment and early warning. Based on GIS mathematical models and algorithms, environmental pollution, ecological damage, and other risks can be assessed and warned against. For example, using GIS technology to establish a water environment risk assessment model for water pollution early warning; through GIS spatial analysis functions, pollution diffusion trends can be predicted, providing decision-makers with timely and accurate environmental risk information. From extensive practice, using GIS technology in ecological environment emergency management has the following advantages: firstly, it can achieve visual analysis of spatial data, intuitively displaying the distribution characteristics and diffusion trends of environmental risks; secondly, it can combine multiple data sources to improve the accuracy and comprehensiveness of environmental risk assessment; finally, it can achieve real-time monitoring and early warning of environmental risks, providing decision-makers with timely and accurate environmental risk information.

1.6 Practical implications of advancing ecological protection through integrated GIS-enabled environmental monitoring

- Enhanced decision-making: Policymakers and environmental agencies can leverage GIS-enabled environmental monitoring systems to make informed decisions based on real-time data and comprehensive spatial analysis. This enables the identification of pollution hotspots and the prioritization of remediation efforts.
- Improved resource allocation: By providing detailed insights into the geographic distribution of environmental pollution, resources can be allocated more efficiently. This ensures that areas with the highest levels of pollution receive the necessary attention and funding for cleanup and preventive measures.
- Public health protection: Integrating GIS with environmental monitoring helps in tracking pollution sources that impact public health. This facilitates timely interventions to reduce exposure to harmful pollutants, thus safeguarding community health.
- Early warning systems: The system can be used to develop early warning mechanisms for potential environmental hazards. For instance, detecting changes in air or water quality can trigger alerts, allowing for prompt responses to mitigate adverse effects.
- Community engagement: The accessibility of GIS data to the public fosters greater community involvement in environmental protection efforts. Citizens can participate in monitoring activities and report pollution incidents, thereby enhancing collective responsibility and action.
- Sustainable development planning: Urban planners and developers can utilize the data from GIS-enabled monitoring systems to design sustainable projects that minimize environmental impact. This ensures that development activities are conducted in harmony with ecological preservation.
- Regulatory compliance: Industries and businesses can use the insights gained from GIS monitoring to comply with environmental regulations. Continuous monitoring helps in maintaining adherence to standards and avoiding potential fines and legal actions.
- Educational and research opportunities: The integration of GIS in environmental monitoring provides a valuable tool for educational institutions and researchers. It supports academic studies and fosters innovation in developing new methods for environmental protection.
- Climate change mitigation: By monitoring environmental pollution on a granular level, this approach contributes to broader climate change mitigation efforts. Identifying and controlling pollution sources helps reduce greenhouse gas emissions and other pollutants that exacerbate climate change.
- Long-term environmental sustainability: The holistic approach of integrating GIS with environmental monitoring supports long-term sustainability goals. It promotes the continuous assessment and improvement of environmental health, ensuring a better quality of life for future generations.

By addressing these practical implications, the study underscores the significant benefits of utilizing integrated GIS-enabled environmental monitoring systems to advance ecological protection and manage environmental pollution effectively.

1.7 Role of deep learning in advancing ecological protection through integrated GIS-enabled environmental monitoring

1.7.1 Enhanced data analysis and pattern recognition

- Analyze large volumes of complex environmental data efficiently.
- Identify hidden patterns and correlations in GIS data.
- Provide more accurate and insightful environmental assessments.

1.7.2 Automated feature extraction

- Use convolutional neural networks (CNNs) for automatic feature extraction.
- Reduce the need for manual data processing.
- Rapidly identify pollution sources and land cover changes.

1.7.3 Improved prediction and forecasting

- Train models on historical environmental data for accurate predictions.

- Forecast future pollution trends and ecological changes.
- Enable early intervention and better environmental planning.

1.7.4 Real-time monitoring and anomaly detection

- Process real-time data streams from environmental sensors.
- Detect anomalies and potential environmental threats promptly.
- Monitor air and water quality continuously.

1.7.5 Image classification and land use mapping

- Classify satellite images and create detailed land use maps.
- Monitor deforestation, urban expansion, and habitat destruction.
- Provide essential data for conservation efforts and land management.

1.7.6 Species identification and biodiversity monitoring

- Identify species from images and audio recordings.
- Monitor biodiversity and track endangered species.
- Assess ecosystem health for informed conservation strategies.

1.7.7 Optimized resource management

- Predict areas at risk of water scarcity or soil degradation.
- Enable targeted conservation practices.
- Promote sustainable resource utilization.

1.7.8 Pollution source attribution

- Trace pollution back to its sources using spatial and temporal data.
- Enforce environmental regulations effectively.
- Implement effective pollution control measures.

1.7.9 Environmental impact assessment

- Integrate various data types for comprehensive assessments.
- Provide a holistic view of environmental interactions.
- Support the development of mitigation strategies and policies.

1.7.10 Enhanced decision support systems

- Equip decision-makers with advanced tools for environmental management.
- Improve the accuracy and efficiency of decision-making processes.
- Lead to more effective ecological protection initiatives.

1.8 Conclusion

With the advancement of science and technology, environmental monitoring has become one of the important means to protect the ecological environment, and it can play a significant role. Under the concept of harmonious development between humans and nature, monitoring personnel need to further research existing environmental monitoring technologies, strengthen their application scope, improve the accuracy and feasibility of data monitoring, and then carry out pollution monitoring, pollution control, and other work to lay a solid foundation for ecological environmental protection. This will meet the requirements of the new era's development while also meeting the requirements of ecological environment, providing people with a good living environment.

In summary, natural environmental emergencies themselves have dynamic and dangerous characteristics. If not handled in a timely manner, they can cause unpredictable negative impacts on ecological, social, and human development. Emergency monitoring, as an effective way for monitoring departments to respond quickly to emergencies, directly affects the effectiveness of emergency response. Using GIS technology for ecological environment emergency monitoring can not only improve monitoring quality and efficiency but also effectively reduce the workload of monitoring personnel. Moreover, against the backdrop of continuous scientific and technological advancements, GIS technology will further optimize and improve, laying a solid foundation for ecological environment protection.

References

Ahmed, S., Nicholson, C. E., Muto, P., Perry, J. J., Dean, J. R., & da Silva, C. R. (2021). Applied aerial spectroscopy: A case study on remote sensing of an ancient and semi-natural woodland. *PLoS ONE*, *16*(11), e0260056. Available from https://doi.org/10.1371/journal.pone.0260056.

Antunes, P., Santos, R., & Jordão, L. (2001). The application of geographical information systems to determine environmental impact significance. *Environmental Impact Assessment Review*, *21*(6), 511–535. Available from https://doi.org/10.1016/s0195-9255(01)00090-7.

Awewomom, J., Dzeble, F., Takyi, Y. D., Ashie, W. B., Ettey, E. N. Y. O., Afua, P. E., Sackey, L. N. A., Opoku, F., & Akoto, O. (2024). Addressing global environmental pollution using environmental control techniques: A focus on environmental policy and preventive environmental management. *Discover Environment*, *2*(1). Available from https://doi.org/10.1007/s44274-024-00033-5.

Babaeian, E., Sadeghi, M., Jones, S. B., Montzka, C., Vereecken, H., & Tuller, M. (2019). Ground, proximal, and satellite remote sensing of soil moisture. *Reviews of Geophysics*, *57*(2), 530–616. Available from https://doi.org/10.1029/2018RG000618, http://agupubs.onlinelibrary.wiley.com/agu/journal/10.1002/(ISSN)1944-9208/.

Bedi, J., Vijay, D., Dhaka, P., Singh Gill, J., & Barbuddhe, S. B. (2021). Emergency preparedness for public health threats, surveillance, modelling & forecasting. *Indian Journal of Medical Research*, *153*(3), 287–298. Available from https://doi.org/10.4103/ijmr.IJMR-653-21, http://www.ijmr.org.in/.

Beer, P., & Mulder, R. H. (2020). The effects of technological developments on work and their implications for continuous vocational education and training: A systematic review. *Frontiers in Psychology*, *11*. Available from https://doi.org/10.3389/fpsyg.2020.00918, http://www.frontiersin.org/Psychology.

Bennett, N. J., Whitty, T. S., Finkbeiner, E., Pittman, J., Bassett, H., Gelcich, S., & Allison, E. H. (2018). Environmental stewardship: A conceptual review and analytical framework. *Environmental Management*, *61*(4), 597–614. Available from https://doi.org/10.1007/s00267-017-0993-2, https://www.springer.com/journal/267.

Berglund, R., Kotovirta, V., & Seinä, A. (2007). A system for icebreaker navigation and assistance planning using spaceborne sar information in the baltic sea. *Canadian Journal of Remote Sensing*, *33*(5), 378–387. Available from https://doi.org/10.5589/m07-042.

Bhatti, U. A., Nizamani, M. M., & Mengxing, H. (2022a). Climate change threatens Pakistan's snow leopards. *Science*, *377*(6606), 585–586.

Bhatti, U. A., Wu, G., Bazai, S. U., Nawaz, S. A., & Baryalai, M. (2022b). A pre- to post-COVID-19 change of air quality patterns in anhui province using path analysis and regression. *Polish Journal of Environmental Studies*.

Bhatti, U. A., Yan, Y., Zhou, M., Ali, S., Hussain, A., Qingsong, H., Yu, Z., & Yuan, L. (2021). Time series analysis and forecasting of air pollution particulate matter (PM2.5): An SARIMA and factor analysis approach. *IEEE Access*, *9*, 41019–41031. Available from https://doi.org/10.1109/ACCESS.2021.3060744, http://ieeexplore.ieee.org/xpl/RecentIssue.jsp?punumber = 6287639.

Bhatti, U. A., Zeeshan, Z., Nizamani, M. M., & Bazai, S. (2022c). Assessing the change of ambient air quality patterns in Jiangsu Province of China pre- to post-COVID-19. *Chemosphere*, *288*.

Biber, E. (2013). The challenge of collecting and using environmental monitoring data. *Ecology and Society*, *18*(4). Available from https://doi.org/10.5751/ES-06117-180468.

Carreño-Conde, F., Sipols, A. E., Blas., & Mostaza-Colado, D. (2021). A forecast model applied to monitor crops dynamics using vegetation indices (Ndvi). *Applied Sciences*.

Chawla, I., Karthikeyan, L., & Mishra, A. K. (2020). A review of remote sensing applications for water security: Quantity, quality, and extremes. *Journal of Hydrology*, *585*, 124826. Available from https://doi.org/10.1016/j.jhydrol.2020.124826.

Chen, L., Mao, Y., & Zhao, R. (2022). GIS application in environmental monitoring and risk assessment. *3rd International Conference on Geology, Mapping and Remote Sensing, ICGMRS*, Institute of Electrical and Electronics Engineers Inc., China. http://ieeexplore.ieee.org/xpl/mostRecentIssue.jsp?punumber = 9849165.

Chen, L., Zhang, L., & Rajakani, K. (2022). Design of ecological environment monitoring system based on internet of things technology. *Wireless Communications and Mobile Computing*, *2022*, 1–7. Available from https://doi.org/10.1155/2022/2531425.

Cicala, L., Angelino, C. V., Parrilli, S., Fiscante, N., Liberata, S., & Addabbo, P. (2018). Unsupervised post-fire assessment of burned areas with free and open multispectral data using OBIA. *Conference Paper*, pp. 1–21.

Colpaert, A. (2022). Satellite and UAV platforms, remote sensing for geographic information systems. *Sensors*.

Cyranoski, D., & Brumfiel, G. (2011). Fukushima impact is still hazy. *Nature*, *477*(7363), 139–140. Available from https://doi.org/10.1038/477139a.

Davenhall, W. F., & Kinabrew, C. (2011). *GIS in health and human services* (pp. 557–578). Springer Science and Business Media LLC. Available from http://doi.org/10.1007/978-3-540-72680-7_29.

Domazetović, F., Šiljeg, A., Marić, I., Faričić, J., Vassilakis, E., & Panđa, L. (2021). Automated coastline extraction using the very high resolution worldview (WV) satellite imagery and developed coastline extraction tool (CET). *Applied Sciences*.

Ekins, P., & Zenghelis, D. (2021). The costs and benefits of environmental sustainability. *Sustainability Science*, *16*(3), 949–965. Available from https://doi.org/10.1007/s11625-021-00910-5.

Estoque, R. C. (2020). A review of the sustainability concept and the state of SDG monitoring using remote sensing. *Remote Sensing*.

Faisal, K., & Shaker, A. (2017). An investigation of GIS overlay and PCA techniques for urban environmental quality assessment: A case study in Toronto, Ontario, Canada. *Sustainability*, *9*(3), 380. Available from https://doi.org/10.3390/su9030380.

Filipe, H. P., Silva, E. D., Stulting, A. A., & Golnik, K. C. (2014). Continuing professional development: Best practices. *Journal of Ophthalmology*, *21*(2), 134–141,. Available from https://doi.org/10.4103/0974-9233.129760, http://www.meajo.org.

Firouraghi, N., Kiani, B., Jafari, H. T., Learnihan, V., Salinas-Perez, J. A., Raeesi, A., Furst, M. A., Salvador-Carulla, L., & Bagheri, N. (2022). The role of geographic information system and global positioning system in dementia care and research: A scoping review. *International Journal of Health Geographics*, *21*(1). Available from https://doi.org/10.1186/s12942-022-00308-1, http://www.ij-healthgeographics.com/start.asp.

Forgács, B. (2024). A medical language for climate discourse. *Frontiers in Climate*, *6*. Available from https://doi.org/10.3389/fclim.2024.1384753.

Gaglio, S.G. L., Martorella, G., Peri, D., & Vassallo, S.D. (2014). Development of an IoT environmental monitoring application with a novel middleware for resource constrained devices. *Proceedings of the 2nd Conference on Mobile and Information Technologies in Medicine*.

Gao, W., Xu, W., Bian, X., & Chen, Y. (2017). A practical approach for calculating the settlement and storage capacity of landfills based on the space and time discretization of the landfilling process. *Waste Management*, *69*, 202–214. Available from https://doi.org/10.1016/j.wasman.2017.07.048.

Gavrilaş, S., Ursachi, C. Ş., Perţa-Crişan, S., & Munteanu, F. D. (2022). Recent trends in biosensors for environmental quality monitoring. *Sensors*, *22*(4). Available from https://doi.org/10.3390/s22041513, https://www.mdpi.com/1424-8220/22/4/1513/pdf.

Gałuszka, A., Migaszewski, Z. M., & Namieśnik, J. (2015). Moving your laboratories to the field – advantages and limitations of the use of field portable instruments in environmental sample analysis. *Environmental Research*, *140*, 593–603. Available from https://doi.org/10.1016/j.envres.2015.05.017.

Gilbert, N. (2011). Drug waste harms fish. *Nature*, *476*(7360), 265. Available from https://doi.org/10.1038/476265a.

González, A. (2012). GIS in environmental assessment: A review of current issues and future needs. *Journal of Environmental Assessment Policy and Management*. Available from http://doi.org/10.1142/S146433321250007X2012.

Goodchild, M. F. (2003). Geographic information science and systems for environmental management. *Annual Review of Environment and Resources*, *28*, 493–519. Available from https://doi.org/10.1146/annurev.energy.28.050302.105521.

Gunasekera, R. (2004). Use of GIS for environmental impact assessment: An interdisciplinary approach. *Interdisciplinary Science Reviews*, *29*(1), 37–48. Available from https://doi.org/10.1179/030801804225012473.

Han, H., Hasnain, A., Bhatti, U. A., & Yue, Y. (2024). Assessing the spatio-temporal impact of the COVID-19 pandemic lockdown on air quality in Jiangsu province, China. *Environment, Development and Sustainability*, 1–21.

Harlan, J., Terrill, E., & Burnett, B. (2009). National high frequency radar network: Update. MTS/IEEE Biloxi - Marine technology for our future: Global and local challenges, OCEANS 2009, United States.

Jamil, M.S., Jamil, M.A., Mazhar, A., Ikram, A., Ahmed, A., & Munawar, U. (2015). Smart environment monitoring system by employing wireless sensor networks on vehicles for pollution free smart cities. *Procedia Engineering*, Elsevier Ltd., United States. http://www.sciencedirect.com/science/journal/18777058.

Johnson, N., Druckenmiller, M. L., Danielsen, F., & Pulsifer, P. L. (2021). The use of digital platforms for community-based monitoring. *BioScience*, *71*(5), 452–466. Available from https://doi.org/10.1093/biosci/biaa162.

Jose, A., Tortorella, G. L., Vassolo, R., Kumar, M., & Mac Cawley, A. F. (2023). Professional competence and its effect on the implementation of healthcare 4.0 technologies: Scoping review and future research directions. *International Journal of Environmental Research and Public Health*, *20*(1). Available from https://doi.org/10.3390/ijerph20010478, http://www.mdpi.com/journal/ijerph.

Kamal, R. (2020). Lesson 11: Internet connected environment (weather, air pollution and forest fire) monitoring. 2017, 1–41.

Kerry, R. G., Montalbo, F. J. P., Das, R., Patra, S., Mahapatra, G. P., Maurya, G. K., Nayak, V., Jena, A. B., Ukhurebor, K. E., Jena, R. C., Gouda, S., Majhi, S., & Rout, J. R. (2022). An overview of remote monitoring methods in biodiversity conservation. *Environmental Science and Pollution Research*, *29*(53), 80179–80221. Available from https://doi.org/10.1007/s11356-022-23242-y, https://www.springer.com/journal/11356.

Kolios, V., Vorobeva, S., GIS and environmental monitoring. (2017). Available from http://doi.org/10.1007/978-3-319-53086-42017.

Kotovirta, V., Toivanen, T., Tergujeff, R., & Huttunen, M. (2012). Finland participatory sensing in environmental monitoring – experiences. *Proceedings - 6th International Conference on Innovative Mobile and Internet Services in Ubiquitous Computing*, IMIS.

Kulkarni, P. H., & Kute, P. (2016). Internet of things based system for remote monitoring of weather parameters and applications. *International Journal of Advances in Electronics and Computer Science*, *3*, 68–73.

Kumar, P., Debele, S. E., Sahani, J., Rawat, N., Marti-Cardona, B., Alfieri, S. M., Basu, B., Basu, A. S., Bowyer, P., Charizopoulos, N., Jaakko, J., Loupis, M., Menenti, M., Mickovski, S. B., Pfeiffer, J., Pilla, F., Pröll, J., Pulvirenti, B., Rutzinger, M., … Zieher, T. (2021). An overview of monitoring methods for assessing the performance of nature-based solutions against natural hazards. *Earth-Science Reviews*, *217*, 103603. Available from https://doi.org/10.1016/j.earscirev.2021.103603.

Lechner, A. M., Foody, G. M., & Boyd, D. S. (2020). Applications in remote sensing to forest ecology and management. *One Earth*, *2*(5), 405–412. Available from https://doi.org/10.1016/j.oneear.2020.05.001.

Li, J., Pei, Y., Zhao, S., Xiao, R., Sang, X., & Zhang, C. (2020). A review of remote sensing for environmental monitoring in China. *Remote Sensing*, *12*(7), 1130. Available from https://doi.org/10.3390/rs12071130.

Li, J., Zhang, Y., Qin, Q., & Yan, Y. (2017). Investigating the impact of human activity on land use/cover change in China's Lijiang River Basin from the perspective of flow and type of population. *Sustainability*, *9*(3), 383. Available from https://doi.org/10.3390/su9030383.

Li, Y., Wu, D., Yang, L., & Zhou, T. (1982). Declining effect of precipitation on the normalized difference vegetation index of grasslands in the inner Mongolian plateau. *Applied Sciences*, *2021*, 1982.

Lin, L., Hao, Z., Post, C. J., Mikhailova, E. A., Yu, K., Yang, L., & Liu, J. (2020). Monitoring land cover change on a rapidly urbanizing island using Google Earth Engine. *Applied Sciences*, *10*, 2020.

Lindenmayer, D. B., Woinarski, J., Legge, S., Maron, M., Garnett, S. T., Lavery, T., Dielenberg, J., & Wintle, B. A. (2022). Eight things you should never do in a monitoring program: An Australian perspective. *Environmental Monitoring and Assessment*, *194*(10). Available from https://doi.org/10.1007/s10661-022-10348-6.

Lovett, G. M., Burns, D. A., Driscoll, C. T., Jenkins, J. C., Mitchell, M. J., Rustad, L., Shanley, J. B., Likens, G. E., & Haeuber, R. (2007). Who needs environmental monitoring? *Frontiers in Ecology and the Environment*, *5*(5), 253–260. Available from https://doi.org/10.1890/1540-9295(2007)5[253:wnem]2.0.co;2.

Mahrad, B. E., Newton, A., Icely, J. D., Kacimi, I., Abalansa, S., & Snoussi, M. (2020). Contribution of remote sensing technologies to a holistic coastal and marine environmental management framework: A review. *Remote Sensing*.

Manfreda, S., McCabe, M. F., Miller, P. E., Lucas, R., Madrigal, V. P., Mallinis, G., & Toth, B. (2018). On the use of unmanned aerial systems for environmental monitoring. *Remote Sensing*.

Mauri, P. V., Parra, L., Mostaza-Colado, D., Garcia, L., Lloret, J., & Marin, J. F. (2021). The combined use of remote sensing and wireless sensor network to estimate soil moisture in golf course. *Applied Sciences*.

Messer, H., Zinevich, A., & Alpert, P. (2006). Environmental monitoring by wireless communication networks. *Science*, *312*(5774), 713. Available from https://doi.org/10.1126/science.1120034.

Narayana, T. L., Venkatesh, C., Kiran, A., J, C. B., Kumar, A., Khan, S. B., Almusharraf, A., & Quasim, M. T. (2024). Advances in real time smart monitoring of environmental parameters using IoT and sensors. *Heliyon*, *10*(7). Available from https://doi.org/10.1016/j.heliyon.2024.e28195, https://www.sciencedirect.com/science/journal/24058440.

Noor, N. M., Abdullah, A., & Hashim, M. (2018). Remote sensing UAV/drones and its applications for urban areas: A review. *IOP Conference Series: Earth and Environmental Science*, *169*(1), 012003. Available from https://doi.org/10.1088/1755-1315/169/1/012003.

Nuckols, J. R., Ward, M. H., & Jarup, L. (2004). Using geographic information systems for exposure assessment in environmental epidemiology studies. *Public Health Services, Environmental Health Perspectives*, *112*(9), 1007–1015. Available from https://doi.org/10.1289/ehp.6738, http://ehp.niehs.nih.gov/.

Parra, L. (2022). Remote sensing and GIS in environmental monitoring. *Applied Sciences*, *12*(16), 8045. Available from https://doi.org/10.3390/app12168045.

Pérez-Castro, M. Á., Schubert, N., Ang-Montes de Oca, G., Leyte-Morales, G. E., Eyal, G., & Hinojosa-Arango, G. (2022). Mesophotic coral ecosystems in the Eastern Tropical Pacific: The current state of knowledge and the spatial variability of their depth boundaries. *Science of the Total Environment*, *806*. Available from https://doi.org/10.1016/j.scitotenv.2021.150576, http://www.elsevier.com/locate/scitotenv.

Reinermann, S., Asam, S., & Kuenzer, C. (2020). Remote sensing of grassland production and management—a review. *Remote Sensing*.

Sattar, H., Bajwa, I. S., Ul-Amin, R., Mahmood, A., Anwar, W., Kasi, B., Kazmi, R., & Farooq, U. (2019). An intelligent and smart environment monitoring system for healthcare. *Applied Sciences*, *9*(19), 4172. Available from https://doi.org/10.3390/app9194172.

Scerri, S., Tuikka, T., Vallejo, I. L. de, & Curry, E. (2022). *Common European data spaces: Challenges and opportunities* (pp. 337–357). Springer Science and Business Media LLC. Available from http://doi.org/10.1007/978-3-030-98636-0_16.

Shan Neo, J. P., & Tan, B. H. (2017). The use of animals as a surveillance tool for monitoring environmental health hazards, human health hazards and bioterrorism. *Veterinary Microbiology*, *203*, 40–48. Available from https://doi.org/10.1016/j.vetmic.2017.02.007, http://www.elsevier.com/locate/vetmic.

Siddiqua, A., Hahladakis, J. N., & Al-Attiya, W. A. K. A. (2022). An overview of the environmental pollution and health effects associated with waste landfilling and open dumping. *Environmental Science and Pollution Research*, *29*(39), 58514–58536. Available from https://doi.org/10.1007/s11356-022-21578-z, https://link.springer.com/journal/11356.

Sishodia, R. P., Ray, R. L., & Singh, S. K. (2020). Applications of remote sensing in precision agriculture: A review. *Remote Sensing*, *12*(19), 1–31. Available from https://doi.org/10.3390/rs12193136, https://www.mdpi.com/2072-4292/12/19/3136/pdf.

Skön, J.P., Kauhanen, O., & Kolehmainen, M. (2011). Finland energy consumption and air quality monitoring system. *Proceedings of the 2011 7th International Conference on Intelligent Sensors, Sensor Networks and Information Processing*, ISSNIP.

Sohoo, I., Ritzkowski, M., Guo, J., Sohoo, K., & Kuchta, K. (2022). Municipal solid waste management through sustainable landfilling: In view of the situation in Karachi, Pakistan. *International Journal of Environmental Research and Public Health*, *19*(2), 773. Available from https://doi.org/10.3390/ijerph19020773.

Tadejko, P. (2017). Environmental monitoring systems using internet of things-standards and protocols Pawel Tadejko Environmental policy and management. *Ekonomia i Srodowisko*, *4*, 2017.

Trofymchuk, O., Klymenko, V., Anpilova, Y., Sheviakina, N., & Zagorodnia, S. (2020). The aspects of using GIS in monitoring of environmental components. *International Multidisciplinary Scientific GeoConference Surveying Geology and Mining Ecology Management*, SGEM, Ukraine. http://www.sgem.org/

Ullo, S. L., & Sinha, G. R. (2020). Advances in smart environment monitoring systems using IoT and sensors. *Sensors*, *20*(11), 3113. Available from https://doi.org/10.3390/s20113113.

Wang, X., & Stauffer, B. (1995). Application of GIS for environmental impact analysis in a traffic relief study. *Computers, Environment and Urban Systems*, *19*(4), 275–286. Available from https://doi.org/10.1016/0198-9715(95)00029-1.

Wei, X., Chang, N. B., Bai, K., & Gao, W. (2020). Satellite remote sensing of aerosol optical depth: Advances, challenges, and perspectives. *Critical Reviews in Environmental Science and Technology*, *50*(16), 1640–1725. Available from https://doi.org/10.1080/10643389.2019.1665944, http://www.tandf.co.uk/journals/titles/10643389.asp.

Williams, M., Cornford, D., Bastin, L., Jones, R., & Parker, S. (2011). Automatic processing, quality assurance and serving of real-time weather data. *Computers and Geosciences*, *37*(3), 353–362. Available from https://doi.org/10.1016/j.cageo.2010.05.010.

Williams, M., Cornford, D., Bastin, L., & Pebesma, E. (2009). *Uncertainty markup language (UncertML)* (2009). Open Geospatial Consortium Inc.

Williams, W., Cornford Ingram, B., Bastin, L., Beaumont, T., Pebesma, E., & Dubois, G. (2007). Supporting interoperable interpolation: The intamap approach. *Proceedings of International Symposium on Environmental Software Systems*.

Wu, Y., & Tham, J. (2023). The impact of environmental regulation, environment, social and government performance, and technological innovation on enterprise resilience under a green recovery. *Heliyon*, *9*(10), e20278. Available from https://doi.org/10.1016/j.heliyon.2023.e20278.

Xie, H., Yao, G., & Liu, G. (2015). Spatial evaluation of the ecological importance based on GIS for environmental management: A case study in Xingguo county of China. *Ecological Indicators*, *51*, 3–12. Available from https://doi.org/10.1016/j.ecolind.2014.08.042.

Yang, T., Duan, Z., & Jianhong, H. (2022). Design of personalized ideological and political education resource recommendation system in the online education platform in the network environment. *Journal of Environmental and Public Health*, *2022*, 1–12. Available from https://doi.org/10.1155/2022/3845749.

Yang, X., Xi, W., Chen, A., Wang, C., & Vijayakumar, P. (2021). An environmental monitoring data sharing scheme based on attribute encryption in cloud-fog computing. *PLoS ONE*, *16*(9), e0258062. Available from https://doi.org/10.1371/journal.pone.0258062.

Ye, P. (2022). Remote sensing approaches for meteorological disaster monitoring: Recent achievements and new challenges. *International Journal of Environmental Research and Public Health*, *19*(6), 3701. Available from https://doi.org/10.3390/ijerph19063701.

Young, O. R., & Stokke, O. S. (2020). Why is it hard to solve environmental problems? The perils of institutional reductionism and institutional overload. *International Environmental Agreements: Politics, Law and Economics*, *20*(1), 5–19. Available from https://doi.org/10.1007/s10784-020-09468-6, http://ww.kluweronline.com/issn/1567-9764.

Zhang, D., Eng, B., Connor, N. E. O., & Regan, P. F. (2015). *Multi-modal smart sensing network for school of electronic engineering*. Dublin, Ireland: Dublin City University.

Zhang, X., Tan, J., & Chan, K. C. (2021). Environmental law enforcement as external monitoring: Evidence from the impact of an environmental inspection program on firm-level stock price crash risk. *International Review of Economics and Finance*, *71*, 21–31. Available from https://doi.org/10.1016/j.iref.2020.08.020.

Zhou, D. (2020). China's environmental vertical management reform: An effective and sustainable way forward or trouble in itself? *Laws*, *9*(4), 25. Available from https://doi.org/10.3390/laws9040025.

Section I

Deep learning for climate change

Chapter 2

Secure data storage and processing architectures for climate Internet of Things systems

Hao Tang, Uzair Aslam Bhatti, Dai Lisi, Jinru Liu, Dekai Li and Mughair Aslam Bhatti
School of Information and Communication Engineering, Hainan University, Haikou, P.R. China

2.1 Introduction

In recent years, the proliferation of big data technologies has catalyzed a paradigm shift in environmental research, particularly in the analysis and understanding of site pollution dynamics (Lucivero, 2020). This introduction provides a comprehensive overview of the integration of big data acquisition techniques and machine learning algorithms for intelligent pollution identification at both local and regional scales (El Khediri et al., 2024). Traditional methodologies for acquiring and processing site environmental data are confronted with significant challenges in accommodating the burgeoning volumes and complexities of big data, necessitating novel approaches and technologies to meet the demands of storage, processing, and analysis (Bibri et al., 2024; Liang et al., 2024). Consequently, recent research efforts have been directed towards harnessing advanced technologies such as 5G connectivity, Internet of Things (IoT) devices, and natural language processing (NLP) techniques to gather, refine, and analyze site environmental data in real-time. Numerous data processing architectures tailored for big data have emerged to tackle its diverse characteristics (Alwahedi et al., 2024). Data acquisition, encompassing the gathering, filtering, and cleansing of data before its storage in a data warehouse or other storage solutions, stands as a foundational step in this process (Fadlallah et al., 2023) (Maria; Hadi). Typically governed by the four Vs—volume, velocity, variety, and value—data acquisition scenarios often grapple with high-volume, high-velocity, and high-variety data, albeit low in value (Hulsen et al., 2019; Mallappallil et al., 2020). Thus, the development of adaptable and time-efficient gathering, filtering, and cleaning algorithms is paramount to ensuring that only high-value data fragments undergo processing in data-warehouse analyses.

With the acceleration of urbanization and industrial relocation processes, a large number of industrial enterprises have been closed or relocated, leaving behind numerous polluted sites. Amidst the backdrop of burgeoning urbanization and industrial advancement, contemporary societies grapple with two salient challenges in urban development (Luo et al., 2020). First, the daunting task of remediating large-scale environmental contamination stemming from industrial activities looms large. Simultaneously, the rejuvenation of urban brownfield sites emerges as a pressing concern. Brownfields, characterized by abandoned or underutilized industrial and commercial land, represent a substantial inventory within urban landscapes, laden with potential for redevelopment (Zhao et al., 2022). However, the specter of environmental degradation casts a shadow over these sites, diminishing their value and impeding local development prospects. The genesis of brownfield site analysis traces back to the early 1970s, primarily spurred by initiatives in Western Europe and the United States (Yang, Guo, et al., 2023). Presently, the European Union grapples with an estimated 340,000 contaminated brownfield sites necessitating environmental rehabilitation prior to redevelopment. Similarly, the United States contends with over 450,000 brownfield sites ripe for revitalization (Berman et al., 2022). Notably, China, with its 120 aging industrial cities and a staggering 2,615,400 hectares of abandoned industrial land, confronts a monumental task in repurposing these sites (Lai et al., 2020). The strategic redevelopment of brownfield sites holds promise in assuaging urban land shortages, mitigating environmental pressures, and catalyzing economic resurgence postrehabilitation (Berman et al., 2022; Jian et al., 2022).

Deep Learning for Earth Observation and Climate Monitoring. DOI: https://doi.org/10.1016/B978-0-443-24712-5.00003-8

The remediation of brownfield sites poses complex challenges due to the dynamic interplay of industrial activities, human interventions, and transportation networks, resulting in fluid and variable soil contamination patterns (Grifoni et al., 2022; Hou et al., 2023). Tailored ecological remediation strategies are imperative, informed by a nuanced understanding of site-specific contamination dynamics. Concurrently, effective pollution risk management necessitates a comprehensive grasp of soil and groundwater environments. Conventional methods falter in delineating pollution sites amidst intricate pollutant interactions and spatial variations (Mystrioti & Papassiopi, 2024). In response, researchers turn to machine learning to construct integrated pollution identification frameworks, leveraging diverse data facets to develop robust prediction models (Shah et al., 2020). This holistic approach promises to streamline pollution site identification and risk assessment, potentially reducing associated costs and advancing sustainable environmental stewardship.

The ecological environment serves as the crucible for human survival and societal progress, embodying the intricate interplay of water, land, biology, and climate (Lawrance et al., 2022). As human activities burgeon, environmental degradation intensifies, necessitating urgent attention from both the public and governmental bodies (Balsari et al., 2020). Mitigating environmental pollution stands as a linchpin for fostering socioeconomic development and ensuring human safety. Leveraging big data holds immense promise in bolstering environmental assessment, monitoring, and intervention efforts (Selvam & Al-Humairi, 2023). An integrated spatial big data platform tailored for the ecological environment is pivotal, streamlining data aggregation and sharing while enhancing decision-making within regulatory frameworks (Li & Huang, 2023). However, fragmented environmental data repositories pose a challenge, necessitating robust governance frameworks and stringent data security measures (Lu et al., 2023). With initiatives like the "Special Project on Soil Pollution Causes and Control Technologies," utilizing big data and machine learning methods emerges as a research frontier, enabling the identification and mitigation of site pollution with unprecedented precision and efficiency.

2.1.1 Novel research focus

In the view of ecological data resources, the burgeoning availability of big data technology has sparked a revolution in understanding and analyzing site pollution dynamics. This study embarks on a comprehensive exploration of the integration of big data acquisition techniques and machine learning algorithms for intelligent pollution identification at both local and regional scales. Traditional methodologies for acquiring and processing site environmental data face significant hurdles in accommodating the burgeoning volumes and complexities of big data. Consequently, recent research efforts have pivoted towards harnessing advanced technologies such as 5G connectivity, Internet of Things (IoT) devices, and natural language processing (NLP) techniques to gather, refine, and analyze site environmental data in real-time. This research seeks to address the following points:

- Investigate the synergistic potential of advanced technologies like 5G, IoT, and NLP in revolutionizing data acquisition and processing methodologies for environmental monitoring and pollution identification.
- Explore the application of machine learning algorithms for uncovering intricate patterns and correlations within multisource environmental datasets, data security, and enabling more precise and efficient pollution identification.
- Develop integrated pollution identification frameworks that leverage machine learning insights to streamline pollution site identification, risk assessment, and remediation efforts, thus advancing sustainable environmental stewardship.
- Address the challenges posed by fragmented environmental data repositories by proposing robust governance frameworks and stringent data security measures to ensure the integrity and confidentiality of shared environmental datasets.

Through this exploration, we aim to shed light on the evolving landscape of data acquisition in the era of big data and elucidate avenues for further research and innovation. The advent of big data has enabled researchers to explore multisource heterogeneous datasets to unveil intricate patterns and correlations, thereby facilitating a deeper understanding of pollution characteristics and formation mechanisms. Machine learning algorithms play a pivotal role in this endeavor, allowing for the extraction of actionable insights from vast and diverse datasets. By leveraging supervised, unsupervised, and semisupervised learning techniques, researchers can effectively mine site environmental data to identify pollution sources, characterize pollution profiles, and predict pollution trends (Fadhel et al., 2024; Guerri et al., 2024). Furthermore, the integration of machine learning with big data analytics enables the development of predictive models for assessing the environmental impact of industrial activities, infrastructure projects, and other human interventions. Evaluate the impact of big data technologies on environmental governance initiatives, highlighting the potential for informed decision-making and preemptive intervention in mitigating environmental pollution and safeguarding ecological integrity.

2.2 Related work

Ecological environment big data, constituting a subset of big data, encapsulates the quintessential characteristics of voluminous, varied, and dynamic datasets (Bibri, 2021; McCleery et al., 2023). It epitomizes the convergence of big data concepts, technologies, and methodologies, marshaled to address multifaceted challenges spanning data collection, storage, analysis, and application within the realm of eco-environmental sciences. In the epoch of incessant information proliferation, the imperative to foster wide-ranging data sharing assumes paramount importance, engendering a democratized access to data resources and catalyzing innovation in eco-environmental research and management endeavors (Alam & Mohanty, 2023; Bibri & Krogstie, 2020). The gamut of data resources within the purview of the ecological environment spans ground monitoring data, remote sensing imagery, geospatial information, social surveys, and scientific research datasets. However, the dispersion of these datasets across disparate government departments engenders a fragmented landscape characterized by isolated data silos. Consequently, interoperability and data sharing remain formidable challenges, necessitating concerted efforts towards integration and harmonization of disparate datasets (Knoble & Yu, 2023). The trajectory of scholarly inquiry into big data, whether theoretical or empirical, underscores discernible trends (Górniak-Zimroz et al., 2024). Firstly, while there exists a plethora of literature expounding upon the conceptual underpinnings of big data, there remains a paucity of discourse surrounding the practical application of specialized technical tools. Secondly, the commercial sector has emerged as a crucible for big data innovation, with industry giants like IBM and Google spearheading transformative initiatives propelled by their substantial economic prowess. Concurrently, governments worldwide evince a palpable interest in harnessing big data for governance and policy formulation across diverse domains, ranging from environmental management to healthcare and urban planning.

Scholars have delved into the symbiotic relationship between environmental public services and big data, unraveling intricate dynamics and advancing novel frameworks for environmental monitoring and governance. Jiang et al. advocate for the establishment of a comprehensive "three-dimensional monitoring system of ecological environment," while highlighting challenges stemming from no standardized monitoring data and inadequate analytical depth. Wang elucidates the burgeoning prospects of Guizhou's big data industry, citing the establishment of a hierarchical environmental protection network and the integration of emission reduction services through digital environmental initiatives (Heping, 2020). Conversely, Hua and Zhou pivot towards harnessing internet-derived information to gage public sentiment towards environmental issues, thereby accentuating the multifaceted dimensions of big data utilization in environmental governance. The confluence of big data and the ecological environment heralds a new era of data-driven decision-making and informed environmental stewardship. However, realizing the full potential of ecological environment big data necessitates concerted efforts towards data integration, technological innovation, and interdisciplinary collaboration, thereby paving the path towards sustainable development and environmental resilience.

2.3 Application of big data in China's ecological environment

Amidst the burgeoning landscape of information technology, the realm of ecological environment data has undergone a profound transformation. China's ecological environment digitization has embarked on a journey towards big data utilization. Unlike conventional data paradigms, big data presents a paradigm shift in terms of volume, diversity, acquisition modalities, storage mechanisms, and analytical approaches (refer to Table 2.1) (Wang, Wu, et al., 2022). As data volumes swell, the efficacy of traditional data processing algorithms faces significant constraints. Outmoded techniques such as environmental data encoding, linear-surface categorization, and multidimensional tree classification struggle to cope with the demands of handling diverse big data sets sourced from multiple origins (Vogelstein et al., 2021). In contrast to conventional ecological environment digitization endeavors, the realm of ecological environment big data is characterized by the heterogeneity of multisourced data. Leveraging mathematical statistical analyses, information retrieval methodologies, machine learning algorithms, expert judgments, and pattern recognition techniques, modern processing systems delve deep into ostensibly disparate ecological data sets, unlocking latent value and furnishing invaluable insights for decision-making processes (de Campos et al., 2024). This fusion of methodologies stands at the crux of contemporary big data technology, empowering stakeholders with the means to navigate the complexities of ecological data landscapes. Shown in Table 2.1.

In the wake of China's "Twelfth Five-Year Plan" and subsequent legislative advancements such as the updated "Environmental Protection Law," "Action Plan for Land Pollution Prevention and Control, Air Pollution Prevention and Control," and "Action Plan for Water Pollution Prevention and Control," the trajectory of big data utilization in ecological environment protection has witnessed a meteoric rise. These legislative milestones have spurred various provinces, autonomous regions, and municipalities to prioritize critical endeavors including environmental quality monitoring,

TABLE 2.1 Ecological environment big data applications versus traditional data applications.

Aspect	Traditional data applications	Big data applications
Data source and type	Singular, centrally distributed sampling monitoring data; aggregated structured data after processing and summarization.	Diverse data sources, including massive heterogeneous data directly obtained and unrelated to the ecological environment.
Data volume	Typically in the GB range; simple data integration.	Ranges from TB to PB; continuous updates; difficult data integration.
Data storage method	Relational databases (RDBMS).	Distributed file storage systems (HDFS), NoSQL databases.
Application method	Data statistical analysis, building numerical models.	Deeply mining data value, verifying numerical models with big data models.
Results presentation	Data analysis results and chart displays.	Various intuitive forms of visualization, virtual reality, etc.
Decision support	Based on data analysis results, decisions are formed according to human experience.	Directly obtaining effective decisions through machine learning and deep learning methods, reducing manual business operations.

pollution source supervision, mobile law enforcement, and pollution discharge fee management. Embracing cutting-edge technologies like cloud computing, big data analytics, remote sensing, geographic information systems, the Internet of Things (IoT), and video surveillance, these entities have embarked on establishing ecological environment information centers. These centers serve as nerve centers for aggregating and coordinating environmental data resources, thereby facilitating the continual advancement of environmental protection informatization (Allioui and Mourdi, 2023). While strides have been made in harnessing big data within atmospheric and water environments, the exploration of soil environment big data for site pollution identification in China has been comparatively nascent. Present research endeavors often grapple with nebulous definitions of soil environment big data concepts, indicative of a broader ambiguity surrounding the interplay between informatization development and big data application (Martínez-Peláez et al., 2023). Particularly, the domain lacks comprehensive investigations into the design of big data architectures tailored for site soil environment data collection, processing, analysis, access, and application. This research lacuna underscores the pressing need for concerted efforts to elucidate the intricacies of soil environment big data utilization, thereby laying the foundation for informed decision-making and sustainable environmental stewardship.

2.4 Application of site environmental big data collection technology

2.4.1 Mobile internet and handheld terminal data collection technology

In the realm of site pollution investigation, extant data predominantly comprises textual records gleaned from investigation reports, acquired through a medley of data collection methodologies, interviews, and on-site surveys. Guided by the site pollution investigation and risk assessment protocol, comprehensive scrutiny entails an analysis of the site and its environs, the status of facilities and structures housing hazardous waste, the nature of site utilization and its vicinities, and the delineation of site pollution characteristics. Some scholarly endeavors have delved into pertinent literature pertaining to site pollution identification, culminating in the compilation and synthesis of enterprise products, raw materials, production processes, and discharge patterns across diverse regions and industries. These efforts have led to the development of an index system tailored for site pollution identification (Zhu et al., 2021). However, the heterogeneous nature of data sources and structures requisite for site pollution identification engenders challenges. Conventional manual retrieval and excerpting methods fall short in furnishing structured data for subsequent analysis and mining, characterized by inefficiencies and a dearth of standardization. The Internet of Things (IoT) has transformed almost every aspect of human life by enabling intelligent and efficient work. IoT equipment such as sensors, controllers, Wi-Fi modules, and cloud computing are critical components of integrated agriculture, which boosts agricultural productivity while decreasing waste. Podlasek et al. proposed smart agricultural research that uses the Internet of Things to measure soil pH, the amount of moisture, climate, humidity, and nutrient levels. Nowadays, smart agriculture solves a variety of farming-related problems. Farmers can use the "INTERNET OF THINGS (IOT)" to connect numerous sensors, actuators, and other embedded devices, supplying them with required information and data. To grow high-quality crops, keep the soil's pH, moisture content, and nutrient level stable (Podlasek et al., 2024).

In the realm of site pollution investigation, crucial data points such as enterprise production years, product specifics, raw materials, pollutant origins, migration pathways, and sensitive receptor locations necessitate meticulous acquisition through on-site inquiries, meticulous data collection, and insightful interviews. Leveraging handheld terminal information collection technology, underpinned by mobile Internet connectivity and global positioning systems (GPS), offers a robust solution for the accurate and efficient retrieval of site-specific information. Contrasted with antiquated paper-based and manual recording methods, the integration of handheld terminal technology enables heightened automation, facilitating the creation of comprehensive electronic records. Moreover, the utilization of handheld terminals for entering preconfigured data, superseding traditional sampling and personnel interview records, embodies a paradigm shift toward enhanced data collection efficiency, standardized record-keeping, and heightened accuracy. This approach significantly streamlines the formation of structured site investigation datasets, laying a robust foundation for subsequent analyses.

Recent years have witnessed a rapid surge in the development of site investigation information management technology. China has spearheaded the adoption of handheld terminals and information management platforms for data collection, storage, and management initiatives. Notably, during initiatives such as the "Detailed Investigation of Soil Pollution of Agricultural Land" and the "Investigation of Soil Pollution of Key Industry Enterprise Land," handheld terminal technology has been instrumental in facilitating data gathering endeavors. Several ecological and environmental departments, alongside prominent enterprises across various provinces, have pioneered the design and development of site investigation and information management platforms. These platforms have been pivotal in supporting soil pollution investigations across closed and relocated enterprise sites, chemical industrial parks, and other industrial enclaves and their surrounding areas. This concerted effort has undoubtedly catalyzed the evolution of soil environmental management informatization in China, fostering a more adept and proactive approach toward environmental stewardship. Shown in Fig. 2.1.

2.4.2 Data collection technology based on web crawling and text classification

The conventional manual approaches to collecting and organizing public data pertaining to sites are fraught with inefficiencies and fail to meet the exigencies of big data analysis and mining. However, the advent of web crawling technology has ushered in a new era of data acquisition, significantly enhancing efficiency and timeliness. Web crawlers, through their emulation of human interaction with web browsers, adeptly scour the internet, extracting relevant text, images, videos, and other content from web pages with precision and speed. This process, facilitated by specialized programs, enables the seamless parsing and extraction of data from diverse online sources. A myriad of web crawler toolkits have emerged, each tailored to suit specific programming languages and user preferences. Notable examples include Nutch and Heritrix, both based on the Java language, as well as Scrapy, Crawley, and PySpider, which find their foundations in Python. Additionally, frameworks like Phpspider and Beanbun, rooted in the PHP scripting language, offer alternative solutions for web data retrieval. Among these, the Scrapy framework stands out as a mature and versatile open-source web crawling framework. Leveraging the efficient and intuitive features of the Python language, Scrapy has garnered widespread adoption in the realm of big data mining research, owing to its robust functionality and ease of use. Previous research endeavors have underscored the pivotal role of web crawling technology in enabling large-scale data acquisition and analysis. Studies have demonstrated the efficacy of web crawlers in gathering diverse datasets from online repositories, ranging from textual information to multimedia content. Moreover, the integration of web crawling techniques with advanced data processing algorithms has facilitated the extraction of actionable insights from vast and heterogeneous datasets. By harnessing the power of web crawling technology, researchers have unlocked new avenues for knowledge discovery and decision-making in various domains, including environmental science, business intelligence, and social research. Shown in Fig. 2.2.

Natural language processing (NLP) stands as a cornerstone in the realm of data mining, wielding significant influence and boasting a plethora of applications ranging from text processing to sentiment analysis. The evolution of NLP has seen the emergence of sophisticated methodologies and tools, rendering it indispensable in contemporary data analysis workflows. The processing pipeline of NLP typically encompasses text acquisition, corpus preprocessing, feature engineering, model training, and evaluation, each constituting a crucial step in the journey towards actionable insights.

Prior to feeding text corpora into language models, preprocessing steps such as word segmentation, part-of-speech tagging, and named entity recognition are imperative to ensure optimal performance. A plethora of open-source tools such as Jieba, ANSJ, THULAC, and LTP have streamlined these preprocessing tasks, empowering researchers with

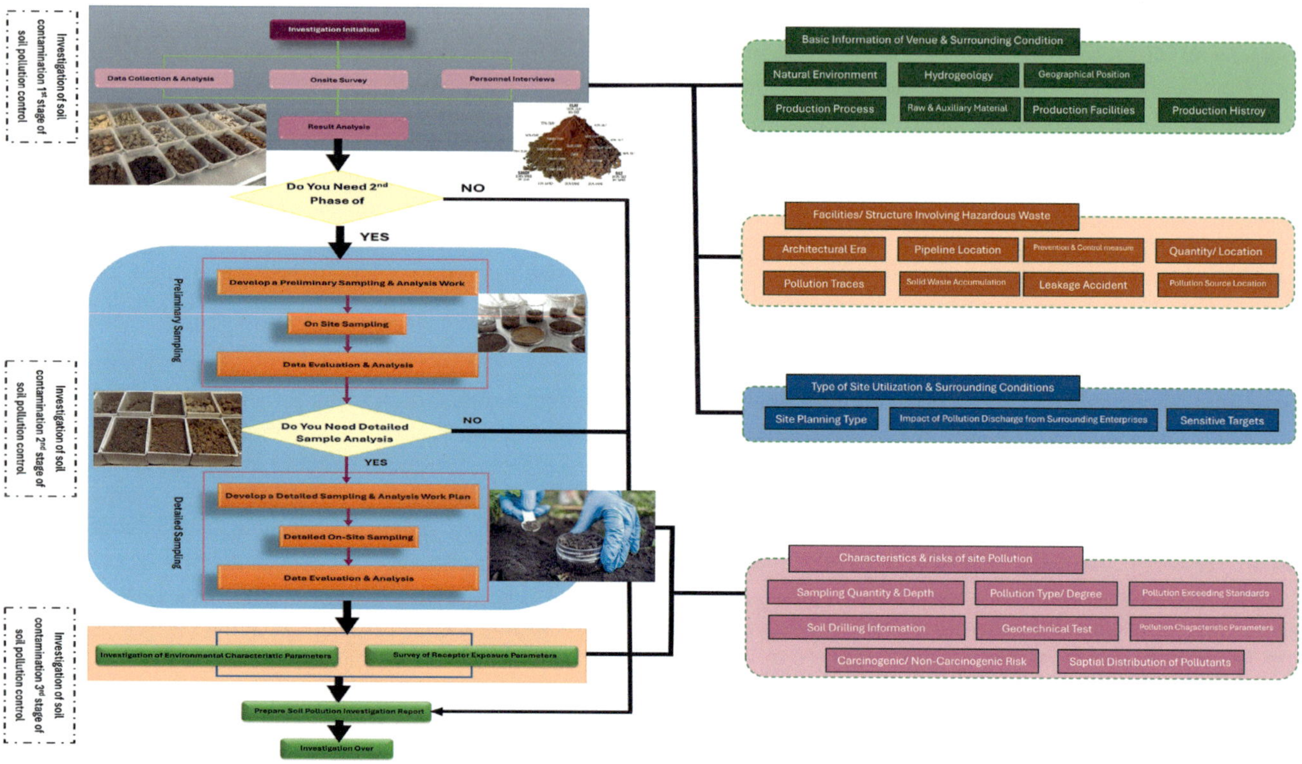

FIGURE 2.1 Site ecological analysis and threat assessment procedure.

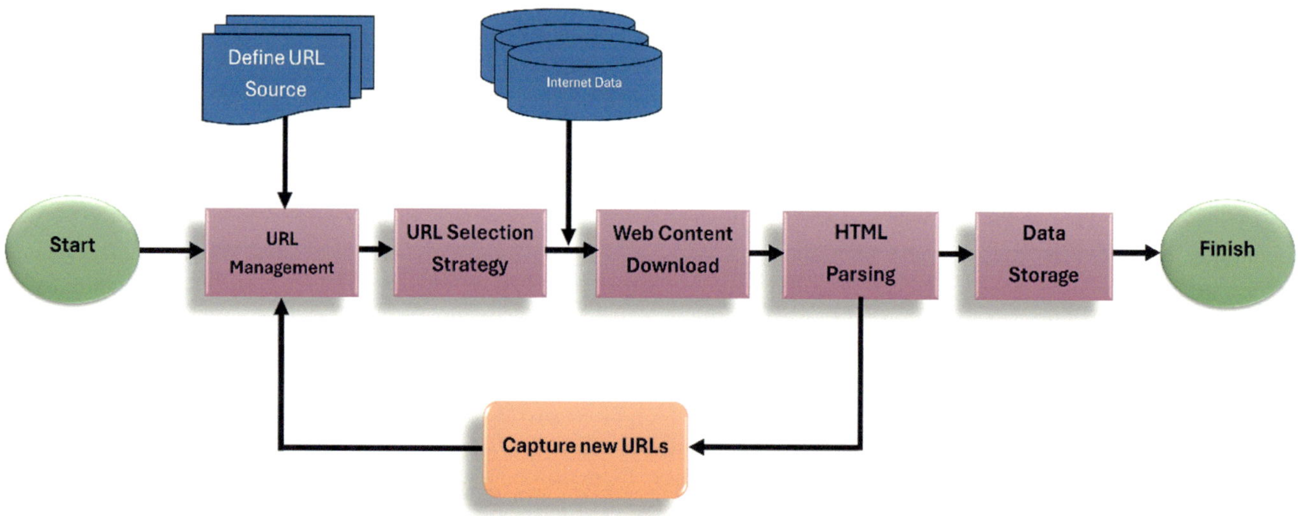

FIGURE 2.2 Diagram for web miner architecture.

efficient and scalable solutions. Traditional NLP models, rooted in rule-based and statistical frameworks, have long served as the bedrock of text analysis methodologies. Rule-based approaches leverage regular expressions to identify patterns in text, offering simplicity and flexibility, albeit with limitations in handling nonstandardized texts. Conversely, statistical methods like Hidden Markov Models (HMM) and Conditional Random Fields (CRF) rely on probability calculations to partition words in input samples, yielding predictions based on the most probable outcomes. However, the advent of deep learning algorithms has revolutionized the landscape of NLP, propelling deep neural network models to the forefront of research endeavors. Renowned for their prowess in text representation and learning, deep neural networks have garnered widespread attention across various NLP domains. The inherent ability of these models to discern intricate

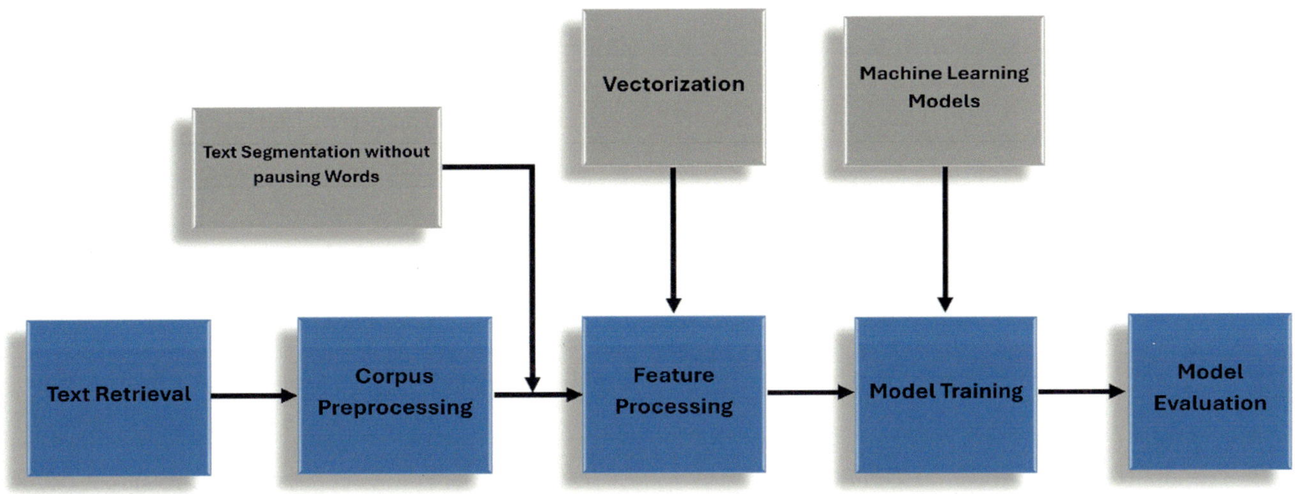

FIGURE 2.3 Illustration of data extraction using machine learning for natural language processing.

patterns and relationships within textual data has propelled them to the forefront of NLP research, offering unparalleled capabilities in tasks such as sentiment analysis, machine translation, and text summarization.

Previous research endeavors have underscored the transformative impact of NLP on diverse domains, showcasing its efficacy in extracting meaningful insights from textual data. Leveraging a combination of rule-based and statistical approaches, researchers have developed sophisticated NLP models capable of handling diverse text formats and languages. Moreover, the integration of deep learning techniques has unlocked new avenues for knowledge discovery, enabling researchers to delve deeper into textual data and uncover hidden patterns with unprecedented accuracy and efficiency. As NLP continues to evolve, it remains poised to shape the future of data mining and analysis, driving innovation and discovery across a myriad of fields. Shown in Fig. 2.3.

Leveraging diverse data sources such as images, text, and other unstructured data holds immense promise in acquiring valuable insights for site pollution identification. Contemporary research endeavors underscore the efficacy of natural language processing (NLP) and machine learning methodologies in extracting features and converting data types, thereby enriching the pool of site environmental big data. Wang et al. exemplify this approach by employing a blend of rule-based and statistical techniques to extract pertinent soil environmental fields and information from textual data sources like soil investigation reports and on-site surveys (Yang et al., 2022). This process facilitates the conversion of unstructured textual data into structured formats, enhancing data accessibility and usability. Integration of web crawlers with NLP-based text extraction techniques offers a streamlined approach to data acquisition. By efficiently retrieving public internet data and extracting relevant information using NLP algorithms, researchers can simplify the data collection process. This methodological synergy not only enhances the efficiency of data acquisition but also renders it more accessible and user-friendly. Overall, the fusion of diverse data acquisition strategies underpinned by NLP and machine learning principles represents a promising avenue for expanding the scope and utility of site environmental big data.

2.4.3 Construction of site pollution intelligent identification information system based on big data framework

The relentless evolution of Hadoop and Spark, the paradigm of distributed frameworks, emerges as a defining hallmark in the realm of big data applications. This transformative shift capitalizes on parallel processing to overcome the limitations of single-machine algorithms, thereby amplifying the efficiency of handling vast, intricate datasets across storage, analysis, and computation domains. Hadoop, hailed as the cornerstone of distributed big data processing, boasts a robust ecosystem comprising core components like HDFS, Yarn, and MapReduce. These components orchestrate distributed storage, resource allocation, and computation with unparalleled reliability, efficiency, and scalability. The ideal location for fruits and vegetables growing is 500 m above sea level. Pacheco Rocha et al., development on fruit-producing technology proposed for desert places. An important fruit crop in dry and semiarid regions. Arid environments are characterized by high temperatures, irregular, low rainfall, and frequent droughts, as well as soils that lack both water-holding capacity and nutrient availability (Pacheco Rocha et al., 2022). The selection of a fruit crop for a dry region is critical for economic output because the environmental conditions are unfavorable for plant survival. Chang et al. planned

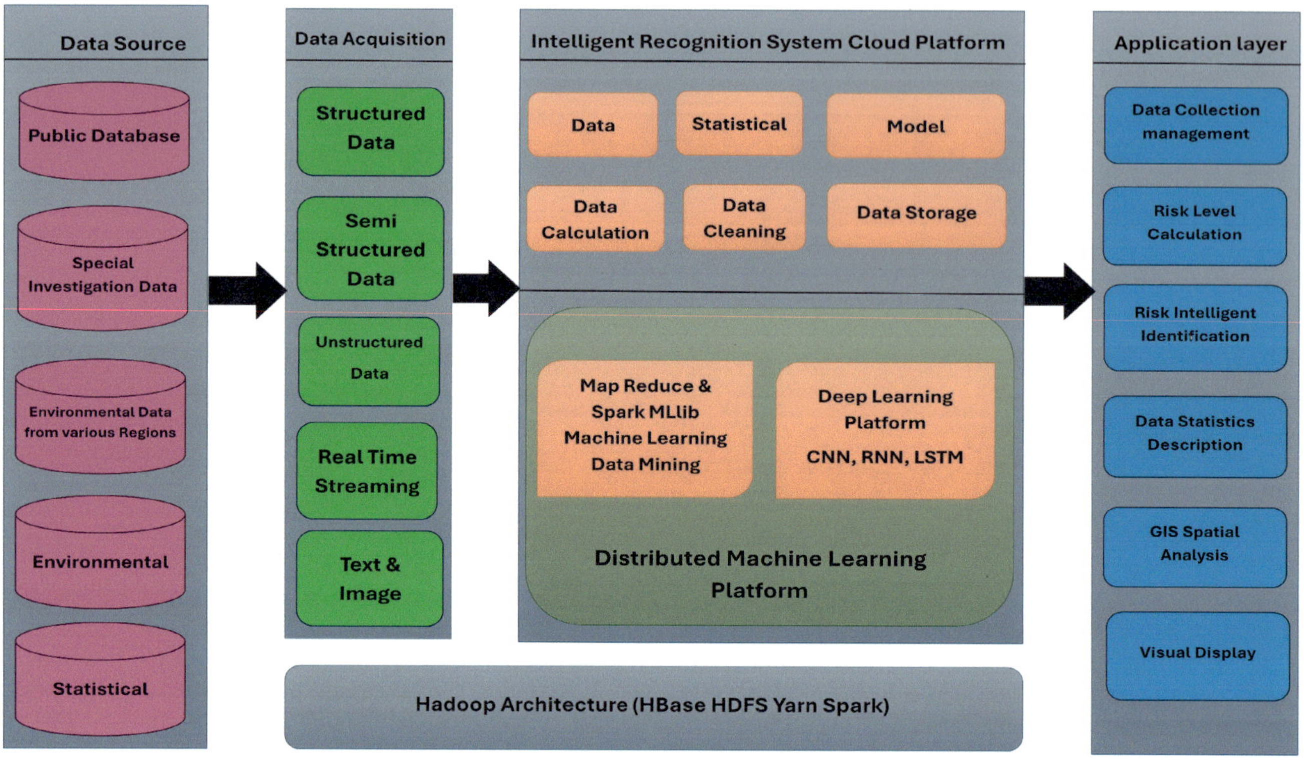

FIGURE 2.4 Schematic diagram of a big data infrastructure that enables advanced recognition via a cloud-based system.

upgrades to the soil moisture monitoring system. The Internet of Things (IoT) refers to the linking of multiple devices via the internet (Chang et al., 2020).

In tandem, Spark complements Hadoop's prowess by prioritizing in-memory storage and processing, thereby unlocking higher computational efficiency, broader compatibility, and enhanced fault tolerance. Leveraging Hadoop's distributed file system HDFS, Spark harnesses the power of intermediate data storage, underpinning its superiority in performance and resilience. Both MapReduce and Spark MLlib stand as formidable pillars in the realm of distributed data processing, offering a rich repository of algorithms for machine learning tasks within distributed architectures. Researchers have spearheaded investigations into ecological environment big data, leveraging Hadoop's robust storage and computation capabilities to analyze multisource remote sensing data and continuous environmental monitoring datasets. In China, researchers have begun exploring the deployment of site environmental big data platforms tailored to the exigencies of pollution identification. Looking ahead, the envisaged trajectory entails the design and implementation of intelligent pollution identification systems anchored on big data frameworks. These systems aim to harness enhanced computing power to decipher vast swathes of unstructured data, facilitating data cleansing, integration, mining, and decision analysis across heterogeneous data sources. This visionary pursuit holds the promise of revolutionizing pollution identification efforts, ushering in an era of data-driven environmental stewardship and sustainability. Shown in Fig. 2.4.

2.5 Research progress of machine learning methods in site pollution identification in China

2.5.1 Common machine learning algorithms for site pollution identification

Machine learning stands as a transformative active discipline wherein computers harness data to construct probabilistic statistical models, wielding these models to predict and analyze intricate datasets. This interdisciplinary field, amalgamating data mining, artificial intelligence, and association analysis, emerges as a potent tool set for unraveling complex environmental conundrums. From atmospheric dynamics to oceanic phenomena and mining activities, machine learning demonstrates remarkable efficacy in addressing a myriad of environmental challenges. Machine learning algorithms, categorized into regression, classification, and clustering paradigms, cater to diverse learning tasks. Supervised learning, unsupervised learning, semisupervised learning, and reinforcement learning further delineate the algorithmic landscape based on the presence or absence of labels. Notably, supervised learning algorithms like artificial neural networks (ANN), support vector machines (SVM), and random forests reign supreme in big data analysis and mining endeavors. These algorithms excel in function approximation, pattern recognition, and regression tasks, finding extensive application in ecological environments.

In environmental domains, like soil, water, and air machine learning algorithms serve multifaceted roles, from predicting environmental pollution to recognizing salient features crucial for ecosystem health. Techniques such as ANN, SVM, and random forests furnish invaluable insights into environmental quality changes, enabling timely warnings and comprehensive evaluations. Leveraging these algorithms, researchers have garnered commendable success in mitigating environmental risks and enhancing ecosystem resilience. The Naive Bayes algorithm, rooted in probability theory, emerges as a cornerstone in text classification tasks owing to its simplicity and efficiency. Widely employed in environmental research, Naive Bayes facilitates the categorization of textual data, streamlining processes such as document classification and sentiment analysis. As machine learning continues to evolve, its integration with big data analytics promises to unlock new frontiers in environmental research, empowering stakeholders with actionable insights to address pressing ecological challenges.

2.5.2 Research on site pollution identification at the plot scale

In the complex field of plot-scale pollution identification, researchers have used a variety of approaches, developing linear and nonlinear connections between site feature indicators and pollutant concentrations. These endeavors, often complemented by advanced methodologies like three-dimensional modeling, aim to predict pollutant concentrations and their spatial distribution with precision and foresight. Yang et al. pioneered the application of multivariate statistical methods to dissect the correlation between soil pollutants in a complex site contaminated with heavy metals and PAHs (Yang et al., 2022). Leveraging a sparse dataset, they ingeniously constructed a BP neural network model to prognosticate heavy metal and PAH content in missing soil samples, showcasing the potential of predictive modeling in pollution assessment. Han and Suh blazed trails by introducing the RFOK model, a hybrid fusion of Random Forest and ordinary kriging techniques (Nawaz et al., 2021; Tang et al., 2023), to the spatial distribution of arsenic in soil across a vast arsenic slag site (Han and Suh, 2024). By discerning the nonlinear relationship between pollutant concentrations and environmental factors, including terrain elements and remote sensing data, their innovative approach yielded insights crucial for targeted remediation efforts.

Mathematical simulation and machine learning methods have been harnessed to unveil the physical and chemical properties of site soil and groundwater. Sharifrazi et al. pioneered the integration of the Sobel operator and convolutional neural network (CNN) algorithm for resistivity tomography imaging inversion in contaminated sites (Sharifrazi et al., 2021). Their groundbreaking work markedly enhanced the accuracy of pollution area and location identification, heralding a new era in pollution source delineation. Herui et al. introduced a trifecta of clustering algorithms—K-means, fuzzy C-means (FCM), and Gaussian mixture model (GMM)—into the realm of electrical resistivity tomography (ERT) monitoring systems (Herui et al., 2022). Their innovative amalgamation empowered the identification of leachate pollution ranges from landfill sites, underscoring the pivotal role of machine learning in pollution source characterization. These pioneering studies exemplify the convergence of advanced methodologies and cutting-edge technologies in pollution identification at the plot scale, offering promising avenues for mitigating environmental risks and fostering sustainable land management practices.

The aforementioned research predominantly employs machine learning regression and clustering algorithms to investigate the source-sink dynamics of site pollution. It delves into unraveling the intricate "cause-and-effect" relationship between feature indicators and various facets of pollution, including types, severity, and spatial extents. However, owing to constraints such as the scarcity of site samples, diverse data acquisition methodologies, and varied approaches, notable disparities exist in the driving factors of pollution and the sensitivity of indicators across different regions and industries. As a consequence, findings from pollution identification studies conducted on individual polluted sites often lack generalizability when extrapolated to characterize pollution patterns across regions or different industrial sectors.

2.5.3 Research on regional scale site pollution identification with big data

Since the 1980s, China has undertaken numerous national-scale soil environmental surveys, spanning investigations into soil environmental background values, pollution statuses, regional geochemical compositions, and more. These surveys yield a wealth of data, ranging from structured datasets to unstructured textual reports and graphical representations. However, these datasets are often fragmented across various government departments, posing challenges in comprehensive integration and application, particularly in the realm of site pollution identification and causative analysis. Researchers have endeavored to bridge this gap by extracting pertinent information from sources like business registration websites, industry databases, and literature repositories. Sharifrazi et al. recommended developing an Internet of Things-based smart temperature and humidity monitor. During the COVID-19 pandemic (Aamir et al., 2021; Bhatti et al., 2022), the coronavirus spreads swiftly in cold, humid environments. Many people are aware of the settings in which they live (Sharifrazi et al., 2021). A smart environment monitoring system is needed to support those individuals. A temperature sensor and a relative humidity sensor are valuable components of a smart Internet of Things (IoT) temperature and humidity real-time monitoring and reporting system. Muhammad Adnan et al. proposed work for Farming Ultimate Guide. Dry, semiarid weather is required for optimal pomegranate growing, with cold winters and extremely dry summers boosting fruit yield. Fruit plants may be considered drought-tolerant and frost-resistant. The optimal temperature for fruit development is 35°C−38°C (Adnan et al., 2024).

For instance, Muhammad Adnan et al. tapped into the China Pesticide Information Network to swiftly identify soil pollution at pesticide production sites, leveraging data on production activities. Meanwhile, Choi et al. conducted a comprehensive analysis of potential pollutants and associated risks at smelting facilities across 13 provinces, drawing from a diverse array of domestic and international literature sources (Choi et al., 2020). Furthermore, researchers have gleaned valuable insights from platforms like the National Pollutant Discharge Permit Management Information Platform and Green Web public databases, extracting details on production processes, raw materials, and pollution discharge. These data serve as crucial indicators for analyzing and understanding site pollution characteristics and driving factors at both regional and industry-specific scales. Despite these efforts, challenges persist in aggregating and harmonizing heterogeneous datasets to unlock deeper insights into soil pollution dynamics and inform targeted mitigation strategies.

Basic data such as soil type, physicochemical properties, and land use serve as crucial inputs for understanding site pollution dynamics. Singh et al. employed a metaanalysis approach to quantify the impact of metal mine sites on surrounding soil pollution, leveraging literature retrieval and land use information under varying scenarios (Singh et al., 2023). Meanwhile, Chen and Ma delved into the temporal and spatial transformations of energy extraction and processing industry sites in China, harnessing high-resolution remote sensing imagery, land cover data, soil types, and environmental thematic data. In exploring the drivers of soil heavy metal accumulation (Chen and Ma, 2023), Yang et al. integrated factors like land cover, proximity to mines or smelters, road distances, and terrain elevation, utilizing the geographic detector model to assess the intensity of these influences (Yang, Wang, et al., 2023). Their study unveiled the principal sources and contribution rates of heavy metal pollution in a prefecture-level city in Hubei. Furthermore, leveraging data sharing methods facilitated by search engine APIs like Google and Baidu, as well as OSM electronic maps, researchers access Point of Interest (POI) data through web crawling. Bratanova et al. tapped into the Google search engine API to gather geographic coordinates and names of over 7000 enterprises in the Yangtze River Delta region, employing machine learning algorithms to predict industry categories (Bratanova et al., 2022). They further examined the relationship between different industry categories and soil Cd and Hg concentrations using bivariate local Moran's index analysis. Similarly, Zhang et al. utilized NLP and machine learning techniques to predict industry categories and map pollution hotspots in a southern Chinese prefecture-level city, showcasing the efficacy of their approach in enhancing pollution identification efforts (Zhang et al., 2023).

2.6 Site pollution identification data mining and analysis method improvement

2.6.1 Construction and optimization of site pollution identification index system

The diversity in methodologies and data interpretation across different studies poses a challenge in constructing effective feature index systems for site pollution identification. While existing research endeavors have developed various index systems, they often prioritize quantity over quality, overlooking the crucial "causal relationships" inherent in pollution processes. This oversight contributes to low precision and scientific inadequacy, particularly when identifying site pollution across regional scales and diverse industry categories. Addressing this challenge necessitates a holistic approach that encompasses pollution processes and production emission characteristics. By comprehensively considering these factors, establishing a unified and comprehensive index system becomes imperative for robust pollution identification across various industry enterprises and regional scales. This entails integrating insights from diverse research perspectives and data mining analyses to formulate a refined index system that captures the nuanced causal relationships driving site pollution dynamics. Only through such an approach can we achieve the precision and scientific rigor required to effectively address pollution challenges across different contexts.

The inventory method of site pollution identification stands as a pivotal approach for discerning pollution across diverse industry sites and regional scales. By establishing correlations between site characteristics and soil pollution, this method offers a cost-effective means of pollution identification, minimizing sampling and analysis costs while expediting the process. In 2019, the Chinese Ministry of Ecology and Environment introduced technical regulations centered on risk screening and classification, leveraging the pollution source-pathway-receptor model. These regulations encompass key categories such as soil and groundwater environmental risk management levels for operational enterprises, site pollution status, pollutant migration pathways, and pollution receptors. While providing essential guidance for pollution identification in China, these regulations are not devoid of subjectivity, primarily due to expert scoring in index establishment.

To address this, Mehmood et al. conducted comprehensive evaluations of risk screening methods for closed and relocated enterprise sites. Employing uncertainty analysis, sensitivity analysis, and empirical case studies, they proposed adjustments to internal grading and score allocation within the indicators, enhancing objectivity (Adnan et al., 2024). Traditional methods for determining index weights and scores, including the analytic hierarchy process and fuzzy membership analysis, are supplemented by feature selection methods based on machine learning. These methods, such as filtering, wrapping, and embedding, play a crucial role in selecting sensitive indicators and determining their importance in evaluation. The embedding method, which derives the contribution rate or importance of feature indicators through model training results like random forests, proved adept at optimizing indicator weights. Leveraging machine learning methods to refine the pollution identification index system and weight allocation using extensive data from soil pollution investigations of key industry enterprises in China holds promise. This approach is poised to bolster the accuracy and applicability of site pollution identification, marking a significant advancement in environmental management practices.

2.6.2 Preprocessing and mining methods for small sample and imbalanced data

Navigating the challenges of obtaining comprehensive site investigation data poses a significant hurdle in pollution identification efforts. Whether due to data scarcity, historical data inadequacy, or system instability, issues like missing or abnormal data plague model training, limiting the dataset's effectiveness. To mitigate these constraints, various strategies emerge. One such tactic involves leveraging methods like nearest neighbors, clustering, and regression analysis to fill missing values and weed out outliers, thereby expanding the pool of viable training data.

Moreover, in the realm of small sample data, where supervised learning suffers, alternative approaches gain prominence. Weakly supervised learning, transfer learning, and metalearning harness prior knowledge to compensate for the dearth of supervised learning information, offering pathways to surmount data scarcity challenges. Yet, even when data is available, imbalances often skew the landscape. For instance, data sourced from key industry enterprise investigations may tilt heavily towards samples exceeding pollution standards, while clean samples are scant. Learning from imbalanced datasets presents its own set of hurdles, with classifiers prone to favoring majority classes and discarding minority ones as noise. Addressing this imbalance necessitates a multifaceted approach, blending preprocessing and mining algorithms. Preprocessing techniques like sampling and feature selection, including PCA and SVD, strive to rebalance the sample space and enhance model performance. Mining algorithms further augment model efficacy, employing semi-supervised clustering, weight-enhanced boosting, and cost-sensitive models to counteract imbalance and bolster generalization. Thus, alongside ensuring data quality meets identification standards, delving into the numerical mechanisms of data processing and model application emerges as pivotal for enhancing the accuracy and applicability of pollution identification efforts.

2.7 Prospects for site environmental big data and pollution intelligent identification

2.7.1 Intelligent soil pollution information collection technology

In the domain of ecological environmental research, big data is gradually finding its foothold, with applications spanning climate change prediction, ecological monitoring networks, and regional atmospheric pollution control. China's strategic blueprints, including the "Overall Plan for Construction of Ecological Environment Big Data," "Plan for Construction of Ecological Environment Monitoring Network," and "Action Plan for Soil Pollution Prevention and Control," underscore the imperative of robust ecological and soil environmental monitoring networks. While China has made significant strides in atmospheric monitoring, soil environmental monitoring remains in its infancy, necessitating concerted efforts to bolster foundational capabilities, technical infrastructure, information management, and institutional frameworks. Future research endeavors must harness the potential of big data technology across various fronts, from information collection and preprocessing to storage and management. Emphasizing the transformative role of technologies like 5G, the Internet of Things (IoT), and global positioning systems (GPS) in soil environmental monitoring networks is paramount for advancing intelligent soil pollution identification and facilitating comprehensive decision-making processes.

During the "Thirteenth Five-Year Plan" period, China transitioned from traditional data collection methods to modern approaches reliant on mobile terminals and information platforms. This shift has enabled efficient, standardized, and accurate acquisition of fundamental site information and soil pollution investigation data. Furthermore, the maturation of rapid detection technologies, such as optical ion, X-ray fluorescence spectroscopy, and pollution sensors, has laid the groundwork for developing instrument systems capable of simultaneous detection of heavy metals and organic pollutants in soil. Additionally, integrating soil pollution remote sensing technology into research and applications promises to enhance three-dimensional monitoring of soil environments. This integration offers insights into the spatial and temporal distribution of pollutants, empowering rapid comprehension of soil pollution dynamics. As China steers ahead, leveraging big data and cutting-edge technologies will be pivotal in fortifying its ecological and soil environmental monitoring endeavors.

2.7.2 Pollution intelligent identification methods based on deep learning

As the triumvirate of artificial intelligence—comprising computing power, algorithms, and data—continues its iterative evolution, computer vision (CV) technology has transcended traditional boundaries, venturing into increasingly intricate domains. From optical character recognition (OCR) to edge detection and handwritten digit recognition, CV now encompasses sophisticated realms like face recognition, dynamic background detection, and image generation. Within the ecological environmental sphere, there's a burgeoning focus on extracting features and mining unstructured image data. Historically, researchers have employed various feature extraction algorithms—such as SIFT (Fang et al., 2022; Liu et al., 2019), Gabor, HOG, and Haar—to analyze image color, texture, and spectral bands. These methods, coupled with PCA for spectral band selection, clustering, and spatial texture and edge feature extraction algorithms, have enabled the identification of pollution sources and ranges. However, recent years have witnessed a paradigm shift, with the ascendancy of CNN-based deep learning models (Bhatti, 2021) like AlexNet, VGG, GoogLeNet, and ResNet significantly enhancing predictive performance in image recognition tasks. Looking ahead, future research endeavors are poised to leverage advancements in technologies like unmanned aerial vehicle (UAV) remote sensing and hyperspectral imaging. These innovations, coupled with deep learning algorithms (Aamir et al., 2022), hold promise for automating abstraction and implicit learning, thereby achieving high-level semantic understanding and analysis of ecological environmental imagery.

2.8 Conclusion

In recent years, the proliferation of big data technologies has ignited a transformative revolution in environmental research, particularly in the realm of site pollution dynamics analysis. This comprehensive exploration has delved into the amalgamation of big data acquisition techniques and machine learning algorithms for intelligent pollution identification across both local and regional scales. However, traditional methodologies for acquiring and processing site environmental data confront substantial challenges in accommodating the burgeoning volumes and complexities of big data, necessitating novel approaches and technologies to meet the demands of storage, processing, and analysis.

Write simple positive argumentative essay "Soil NPK Monitoring System An IoT (Internet of Things)-based soil NPK (nitrogen, phosphorus, and potassium)monitoring system uses contemporary technology to track and control the amount of nutrients in the soil. For maximum agricultural productivity, the levels of NPK, the main nutrients in soil, must be closely monitored. NPK is essential for plant growth. Real-time data from soil sensors can be gathered thanks to Internet of Things technology. This data can then be analyzed and utilized to make knowledgeable decisions about fertilization and other agricultural practices. Here's a breakdown of the key components and functionalities of a Soil NPK Monitoring System using IoT: 1. Soil Sensors: To measure the concentrations of potassium, phosphorus, and nitrogen in the soil, specialized sensors are embedded in the soil. These sensors come in a variety of technological forms, such as spectroscopy-based sensors, electrical conductivity sensors, and others. 2. IoT Devices: The soil sensors are linked to these gadgets, which include microcontrollers and hardware that supports the Internet of Things, such as Arduino or Raspberry Pi. They collect sensor data and send it to a central data storage facility. 3. Connectivity: Data is transmitted by IoT devices via wired or wireless connectivity. Frequently used communication protocols for Internet of Things applications comprise MQTT, HTTP, and other protocols. 4. Centralized Server or Cloud Platform: The data collected from the sensors is sent to a centralized server or a cloud platform for storage and further analysis. Cloud platforms like AWS, Azure, or Google Cloud are commonly used for this purpose.5. Data Processing and Analysis: Algorithms are employed to process and analyze the raw data, providing insights into the nutrient levels of the soil. Machine learning models may be used to make predictions and recommendations based on historical data. 6. User Interface: Farmers or users can access the data through a user-friendly interface, often in the form of a web or mobile application. This interface displays real-time and historical data, making it easier for farmers to understand the soil conditions. 7. Alerts and Notifications: The system can be configured to send alerts or notifications to farmers when nutrient levels fall below or exceed certain thresholds. This proactive approach allows for timely intervention. 8. Integration with Agricultural Practices: The data collected from the sensors is sent to a centralized server or a cloud platform for storage and further analysis. Cloud platforms like AWS, Azure, or To provide comprehensive insights, the system can be integrated with other agricultural management practices. For a more comprehensive approach to crop management, it might be linked to weather information or irrigation systems, for instance. 9. Fertilization Recommendations: The system may produce recommendations for the kind and quantity of fertilizers to be applied in order to maximize crop yield and soil health based on the data that has been analyzed. Water pH Monitoring System An IoT (Internet of Things)-based water pH monitoring system is a technological advancement intended to continuously measure and track the pH levels of water in real-time. In many different applications, such as water treatment, aquaculture, agriculture, and environmental monitoring, it is crucial to keep an eye on the pH of the water. The Internet of Things (IoT) component enables automated and remote monitoring, delivering timely data for improved decision-making. Here are the key components and functionalities of a Water pH Monitoring System using IoT: 1. pH Sensor: The data collected from the sensors is sent to a centralized server or a cloud platform for storage and further analysis. Cloud platforms like AWS, Azure, or To determine how acidic or alkaline the water is, specialized pH sensors are submerged in it. Accurate pH readings are made possible by these sensors, which produce electrical signals proportionate to the concentration of hydrogen ions. 2. IoT Devices: The pH sensors are connected to microcontrollers or Internet of Things(IoT)-capable devices like Arduino or Raspberry Pi. These gadgets gather pH data and send it to a cloud platform or central server. Connectivity: To transfer pH data from the Internet of Things devices to a central server or cloud platform, use wired or wireless communication protocols, such as MQTT or HTTP. 3. Cloud Platform or Centralized Server: The pH data is transferred to a cloud platform or centralized server for storage, analysis, and visualization. 4. Centralized Cloud Platform or Server: After being gathered, the pH data is transferred to a cloud platform or centralized server, where it is stored and available for analysis and visualization. 5. Data Analysis and Processing: The pH data is processed and analyzed by algorithms, which then look for trends, patterns, or departures from the ideal pH range. Predictive analytics may use machine learning models. 6. User Interface: Offer a simple user interface for displaying both historical and real-time pH data via a mobile and web application. This enables users to keep an eye on water conditions and react quickly to changes. 7. Notifications and Alerts: Set up an alert system to inform users when pH levels deviate from predetermined bounds. Prompt notifications allow for prompt action to stop negative impacts on aquatic life, crops, or water quality. 8. Connectivity to Water Treatment Systems: When water treatment systems are connected to the pH monitoring system, pH levels can be automatically adjusted as needed. Temperature, Humidity and Moisture monitoring For pomegranate growers, a temperature, humidity, and moisture monitoring system is intended to give them up-to-date information on the environmental circumstances in their orchard. It is essential to keep an eye on these factors in order to maximize growing conditions, avoid illnesses, and guarantee a good harvest.

Here's an overview of the components and functionalities of such a monitoring system: 1. Temperature Sensors: Place temperature sensors in the orchard to gauge the temperatures of the soil and air. This is necessary to determine temperature variations and evaluate the general climate. 2. Humidity Sensor: Install humidity sensors to keep an eye on the amount of moisture in the air. It's essential to maintain the proper humidity levels to keep diseases at bay and encourage strong plant growth. 3. Moisture Sensors: To determine the amount of water in the soil, soil moisture sensors should be inserted into the root zone. This aids in maximizing irrigation techniques and prevents over- or under watering. 4. Connectivity: To transfer data from IoT devices to a central server or cloud platform, use wireless communication protocols like MQTT or other appropriate protocols. 5. Centralized Server or Cloud Platform: To gather, store, and handle the data, set up a centralized server or make use of a cloud platform. Scalability and accessibility are two reasons why cloud platforms like AWS, Azure, or Google Cloud are frequently used. 6. Data Processing and Analysis: Create algorithms to handle and examine the information gathered from sensors measuring moisture, humidity, and temperature. Trends, patterns, and correlations can be understood through the analysis that follows. 7. User Interface: Create a user-friendly interface that is accessible via a web or mobile application. This interface should provide farmers with real-time and historical data on temperature, humidity, and soil moisture, allowing them to monitor conditions at a glance. 8. Notifications and Alerts: Implement an alert system that notifies farmers when conditions deviate from optimal ranges. Temperature extremes, low humidity, or insufficient soil moisture, for example, can all trigger alerts. 9. Irrigation System Integration: Connect the monitoring system to irrigation systems to enable real-time soil moisture data-driven adjustments. This ensures that water is use deficiently and that overwatering is avoided

"An IoT-based Soil NPK Monitoring System represents a significant leap forward in modern agriculture, offering farmers a sophisticated tool to precisely monitor and manage soil nutrient levels. With nitrogen, phosphorus, and potassium being the key nutrients for plant growth, maintaining their optimal levels is crucial for maximizing agricultural productivity. Thanks to real-time data gathered by specialized soil sensors, farmers can now make informed decisions about fertilization and other agricultural practices, ensuring healthier crops and higher yields.

At the heart of this monitoring system are the soil sensors, strategically embedded in the soil to measure the concentrations of NPK. These sensors, ranging from spectroscopy-based to electrical conductivity sensors, provide accurate readings essential for effective nutrient management. Connected to IoT devices like Arduino or Raspberry Pi, these sensors transmit data to a centralized server or cloud platform using communication protocols like MQTT, ensuring seamless connectivity and data transmission.

Once the data reaches the centralized server or cloud platform, sophisticated algorithms process and analyze it, revealing insights into soil nutrient levels. Farmers can access this valuable information through user-friendly interfaces, accessible via web or mobile applications. Real-time and historical data on NPK levels empower farmers to monitor soil conditions closely and take timely action when necessary.

Furthermore, the system can be configured to send alerts or notifications to farmers when nutrient levels deviate from optimal ranges, enabling proactive intervention to maintain soil health. Integrated with other agricultural management practices like irrigation systems, this monitoring system offers a comprehensive approach to crop management, ensuring efficient water usage and maximizing crop yield.

In conclusion, the IoT-based Soil NPK Monitoring System revolutionizes nutrient management in agriculture, offering farmers a powerful tool to optimize soil health and maximize crop productivity. By harnessing the latest technology and real-time data, this system paves the way for sustainable and efficient farming practices, ensuring a brighter future for agriculture."

Consequently, recent research endeavors have pivoted towards harnessing advanced technologies such as 5 G connectivity, Internet of Things (IoT) devices, and natural language processing (NLP) techniques to gather, refine, and analyze site environmental data in real-time. In light of the evolving landscape of ecological data resources, this exploration embarked on an in-depth investigation of the integration of big data acquisition techniques and machine learning algorithms for intelligent pollution identification. By scrutinizing the synergistic potential of advanced technologies like 5G, IoT, and NLP, this research has illuminated pathways for revolutionizing data acquisition and processing methodologies for environmental monitoring and pollution identification. Furthermore, the exploration into the application of machine learning algorithms has uncovered intricate patterns and correlations within multisource environmental datasets, enabling more precise and efficient pollution identification. The development of integrated pollution identification frameworks leveraging machine learning insights promises to streamline pollution site identification, risk assessment, and remediation efforts, thus advancing sustainable environmental stewardship. Moreover, the exploration addressed the challenges posed by fragmented environmental data repositories, proposing robust governance frameworks and stringent data security measures to ensure the integrity and confidentiality of shared environmental datasets.

Through this comprehensive analysis, the research endeavors have shed light on the evolving landscape of data acquisition in the era of big data, elucidating avenues for further research and innovation. Leveraging big data and deep learning methods for feature extraction and data fusion holds the promise of revolutionizing pollution identification, offering more accurate and efficient methodologies for real-time comprehensive analysis and decision-making. Advent of big data has enabled researchers to explore multisource heterogeneous datasets, unveiling intricate patterns and correlations critical for understanding pollution characteristics and formation mechanisms. By harnessing machine learning algorithms, stakeholders can extract actionable insights from vast and diverse datasets, empowering them to address pressing ecological challenges effectively. Moving forward, the integration of big data technologies with environmental governance initiatives holds the potential for informed decision-making and preemptive intervention, thus mitigating environmental pollution and safeguarding ecological integrity.

References

Aamir, M., Bhatti, M. A., Bhatti, U. A., Bazai, S. U., & Marjan, S. (2022). Predicting the environmental change of carbon emission patterns in South Asia: A deep learning approach using BiLSTM. *Atmosphere*, *13*(12).

Aamir, M., Li, Z., Bhatti, U. A., & Wagan, R. A. (2021). Spatiotemporal change of air-quality patterns in Hubei province—a pre-to post-COVID-19 analysis using path analysis and regression. *Atmosphere*, *12*(10). Available from: https://scholar.google.it/citations?view_op = view_citation&hl = en&user = aQSMyzAAAAAJ&citation_for_view = aQSMyzAAAAAJ:NMxIlDl6LWMC.

Adnan, M., Xiao, B., Ali, M. U., Xiao, P., Zhao, P., Wang, H., & Bibi, S. (2024). Heavy metals pollution from smelting activities: A threat to soil and groundwater. *Ecotoxicology and Environmental Safety*, *274*, 116189. Available from https://doi.org/10.1016/j.ecoenv.2024.116189.

Alam, A., & Mohanty, A. (2023). Educational technology: Exploring the convergence of technology and pedagogy through mobility, interactivity, AI, and learning tools. *Cogent Engineering*, *10*(2). Available from https://doi.org/10.1080/23311916.2023.2283282.

Allioui, H., & Mourdi, Y. (2023). Exploring the full potentials of IoT for better financial growth and stability: A comprehensive survey. *Sensors (Basel)*, *23*(19), 8015. Available from https://doi.org/10.3390/s23198015.

Alwahedi, F., Aldhaheri, A., Ferrag, M. A., Battah, A., & Tihanyi, N. (2024). Machine learning techniques for IoT security: Current research and future vision with generative AI and large language models. *Internet of Things and Cyber-Physical Systems*, *4*, 167–185. Available from https://doi.org/10.1016/j.iotcps.2023.12.003.

Balsari, S., Dresser, C., & Leaning, J. (2020). Climate change, migration, and civil strife. *Current Environmental Health Reports*, *7*(4), 404–414. Available from https://doi.org/10.1007/s40572-020-00291-4, http://link.springer.com/journal/40572.

Berman, L., Morar, C., Unkart, S., & Erdal, S. (2022). An overview of brownfields redevelopment in the United States through regulatory, public health, and sustainability lenses. *Journal of Environmental Health*, *84*(9), 8–15. Available from: https://www.neha.org/publications/journal-environmental-health.

Bhatti, U. A. (2021). Local similarity-based spatial–spectral fusion hyperspectral image classification with deep CNN and Gabor filtering. *IEEE Transactions on Geoscience and Remote Sensing*, *60*, 1–15. Available from: https://scholar.google.it/citations?view_op = view_citation&hl = en&user = aQSMyzAAAAAJ&citation_for_view = aQSMyzAAAAAJ:JV2RwH3_ST0C.

Bhatti, U. A., Zeeshan, Z., Nizamani, M. M., & Bazai, S. (2022). Assessing the change of ambient air quality patterns in Jiangsu Province of China pre-to post-COVID-19. *Chemosphere*, *288*(1). Available from: https://scholar.google.it/citations?view_op = view_citation&hl = en&user = aQSMyzAAAAAJ&citation_for_view = aQSMyzAAAAAJ:hMod-77fHWUC.

Bibri, S. E. (2021). Data-driven smart sustainable cities of the future: An evidence synthesis approach to a comprehensive state-of-the-art literature review. *Sustainable Futures*, *3*, 2021.

Bibri, S. E., & Krogstie, J. (2020). Environmentally data-driven smart sustainable cities: Applied innovative solutions for energy efficiency, pollution reduction, and urban metabolism. *Energy Informatics*, *3*(1). Available from https://doi.org/10.1186/s42162-020-00130-8.

Bibri, S. E., Krogstie, J., Kaboli, A., & Alahi, A. (2024). Smarter eco-cities and their leading-edge artificial intelligence of things solutions for environmental sustainability: A comprehensive systematic review. *Environmental Science and Ecotechnology*, *19*, 100330. Available from https://doi.org/10.1016/j.ese.2023.100330.

Bratanova, A., Pham, H., Mason, C., Hajkowicz, S., Naughtin, C., Schleiger, E., Sanderson, C., Chen, C., & Karimi, S. (2022). Differentiating artificial intelligence activity clusters in Australia. *Technology in Society*, *71*, 102104. Available from https://doi.org/10.1016/j.techsoc.2022.102104.

de Campos, L. M., Fernández-Luna, J. M., Huete, J. F., Ribadas-Pena, F. J., & Bolaños, N. (2024). Information retrieval and machine learning methods for academic expert finding. *Algorithms*, *17*(2). Available from https://doi.org/10.3390/a17020051, http://www.mdpi.com/journal/algorithms/.

Chang, J., Nimer Kadry, S., & Krishnamoorthy, S. (2020). Review and synthesis of big data analytics and computing for smart sustainable cities. *IET Intelligent Transport Systems*, *14*(11), 1363–1370. Available from https://doi.org/10.1049/iet-its.2020.0006.

Chen, L., & Ma, K. (2023). Spatial and temporal distribution and source analysis of heavy metals in agricultural soils of Ningxia, Northwest of China. *Sustainability*, *15*(21), 15360.

Choi, J. Y., Jeong, H., Choi, K.-Y., Hong, G. H., Yang, D. B., Kim, K., & Ra, K. (2020). Source identification and implications of heavy metals in urban roads for the coastal pollution in a beach town, Busan, Korea. *Marine Pollution Bulletin*, *161*, 111724. Available from https://doi.org/10.1016/j.marpolbul.2020.111724.

El Khediri, S., Benfradj, A., Thaljaoui, A., Moulahi, T., Ullah Khan, R., Alabdulatif, A., & Lorenz, P. (2024). Integration of artificial intelligence (AI) with sensor networks: Trends, challenges, and future directions. *Journal of King Saud University – Computer and Information Sciences*, *36*(1), 101892. Available from https://doi.org/10.1016/j.jksuci.2023.101892.

Fadhel, M. A., Duhaim, A. M., Saihood, A., Sewify, A., Al-Hamadani, M. N. A., Albahri, A. S., Alzubaidi, L., Gupta, A., Mirjalili, S., & Gu, Y. (2024). Comprehensive systematic review of information fusion methods in smart cities and urban environments. *Information Fusion*, *107*, 102317. Available from https://doi.org/10.1016/j.inffus.2024.102317.

Fadlallah, H., Kilany, R., Dhayne, H., El Haddad, R., Haque, R., Taher, Y., & Jaber, A. (2023). BIGQA: Declarative big data quality assessment. *Journal of Data and Information Quality*, *15*(3), 1–30. Available from https://doi.org/10.1145/3603706.

Fang, Y., Liu, J., Li, J., & Bhatti, U. A. (2022). Robust zero-watermarking algorithm for medical images based on SIFT and Bandelet-DCT. *Multimedia Tools and Applications*, *81*(12), 16863–16879.

Górniak-Zimroz, J., Romańczukiewicz, K., Sitarska, M., & Szrek, A. (2024). Light-pollution-monitoring method for selected environmental and social elements. *Remote Sensing*, *16*(5), 774.

Grifoni, M., Franchi, E., Fusini, D., Vocciante, M., Barbafieri, M., Pedron, F., Rosellini, I., & Petruzzelli, G. (2022). Soil remediation: Towards a resilient and adaptive approach to deal with the ever-changing environmental challenges. *Environments*, *9*(2), 18.

Guerri, M. F., Distante, C., Spagnolo, P., Bougourzi, F., & Taleb-Ahmed, A. (2024). Deep learning techniques for hyperspectral image analysis in agriculture: A review. *ISPRS Open Journal of Photogrammetry and Remote Sensing*, *12*100062.

Han, H., & Suh, J. (2024). Spatial prediction of soil contaminants using a hybrid random forest−Ordinary Kriging Model. *Applied Sciences*, *14*(4), 1666.

Heping, L. (2020). Establishment of a monitoring system of eco-environment − Suggestions and reflection. *Advances in Social Science, Education and Humanities Research*, *466*. Available from https://doi.org/10.2991/assehr.k.200826.0802020.

Herui, Z., Guolin, W., Xiaozhen, T., & Xiaohui, Z. (2022). Distributed fuzzy clustering analysis of time-lapse electrical resistivity tomography for water inrush monitoring in coal mines. *Sustainability*, *14*(24), 17011. Available from https://doi.org/10.3390/su142417011.

Hou, D., Al-Tabbaa, A., O'Connor, D., Hu, Q., Zhu, Y.-G., Wang, L., Kirkwood, N., Ok, Y. S., Tsang, D. C. W., Bolan, N. S., & Rinklebe, J. (2023). Sustainable remediation and redevelopment of brownfield sites. *Nature Reviews Earth & Environment*, *4*(4), 271−286. Available from https://doi.org/10.1038/s43017-023-00404-1.

Hulsen, T., Jamuar, S. S., Moody, A. R., Karnes, J. H., Varga, O., Hedensted, S., Spreafico, R., Hafler, D. A., & McKinney, E. F. (2019). From big data to precision medicine. *Frontiers in Medicine*, *6*. Available from https://doi.org/10.3389/fmed.2019.00034, https://www.frontiersin.org/articles/10.3389/fmed.2019.00034/full.

Jian, H., Hao, H., Haize, P., Chuan, L., Xiaoqin, L., Yan, W., Haidan, J., Changliang, Z., & Xue, B. (2022). Research on brownfield redevelopment based on Wuli-Shili-Renli system theory and catastrophe progression method. *PLoS ONE*, *17*(11), e0277324. Available from https://doi.org/10.1371/journal.pone.0277324.

Knoble, C., & Yu, D. (2023). Bridging the gap: Analyzing the relationship between environmental justice awareness on twitter and socio-environmental factors using remote sensing and big data. *Remote Sensing*, *15*(23), 5510. Available from https://doi.org/10.3390/rs15235510.

Lai, Y., Chen, K., Zhang, J., & Liu, F. (2020). Transformation of industrial land in urban renewal in Shenzhen, China. *Land*, *9*(10), 371. Available from https://doi.org/10.3390/land9100371.

Lawrance, E. L., Thompson, R., Newberry Le Vay, J., Page, L., & Jennings, N. (2022). The impact of climate change on mental health and emotional wellbeing: A narrative review of current evidence, and its implications. *International Review of Psychiatry*, *34*(5), 443−498. Available from https://doi.org/10.1080/09540261.2022.2128725, http://www.tandfonline.com/loi/iirp20.

Li, C., & Huang, M. (2023). Environmental sustainability in the age of big data: Opportunities and challenges for business and industry. *Environmental Science and Pollution Research International*, *30*(56), 119001−119015. Available from https://doi.org/10.1007/s11356-023-30301-5.

Liang, H., Zhang, Z., Hu, C., Gong, Y., & Cheng, D. (2024). A survey on spatio-temporal big data analytics ecosystem: Resource management, processing platform, and applications. *IEEE Transactions on Big Data*, *10*(2), 174−193. Available from https://doi.org/10.1109/TBDATA.2023.3342619, https://www.ieee.org/membership-catalog/productdetail/showProductDetailPage.html?product = PER472-ELE.

Liu, J., Li, J., Chen, Y., Zou, X., Cheng, J., & Bhatti, U. A. (2019). A robust zero-watermarking based on SIFT-DCT for medical images in the encrypted domain. *Computers, Materials & Continua*, *61*(1), 363−378. Available from https://scholar.google.it/citations?view_op = view_citation&hl = en&user = aQSMyzAAAAAJ&cstart = 20&pagesize = 80&citation_for_view = aQSMyzAAAAAJ:hqOjcs7Dif8C.

Lu, X., Du, J., Zheng, L., Wang, G., Li, X., Sun, L., & Huang, X. (2023). Feature fusion improves performance and interpretability of machine learning models in identifying soil pollution of potentially contaminated sites. *Ecotoxicology and Environmental Safety*, *259*. Available from https://doi.org/10.1016/j.ecoenv.2023.115052, http://www.elsevier.com/inca/publications/store/6/2/2/8/1/9/index.htt.

Lucivero, F. (2020). Big data, big waste? A reflection on the environmental sustainability of big data initiatives. *Science and Engineering Ethics*, *26*(2), 1009−1030. Available from https://doi.org/10.1007/s11948-019-00171-7.

Luo, Y., Xiang, P., & Wang, Y. (2020). Investigate the relationship between urbanization and industrialization using a coordination model: A case study of China. *Sustainability*, *12*(3), 916. Available from https://doi.org/10.3390/su12030916.

Mallappallil, M., Sabu, J., Gruessner, A., & Salifu, M. (2020). A review of big data and medical research. *SAGE Open Medicine*, *8*. Available from https://doi.org/10.1177/2050312120934839, 205031212093483.

Martínez-Peláez, R., Ochoa-Brust, A., Rivera, S., Félix, V. G., Ostos, R., Brito, H., Félix, R. A., & Mena, L. J. (2023). Role of digital transformation for achieving sustainability: Mediated role of stakeholders, key capabilities, and technology. *Sustainability, MDPI*, *15*(14), 1−27.

McCleery, R., Guralnick, R., Beatty, M., Belitz, M., Campbell, C. J., Idec, J., Jones, M., Kang, Y., Potash, A., & Fletcher, R. J. (2023). Uniting experiments and big data to advance ecology and conservation. *Trends in Ecology & Evolution*, *38*(10), 970−979. Available from https://doi.org/10.1016/j.tree.2023.05.010.

Mystrioti, C., & Papassiopi, N. (2024). A comprehensive review of remediation strategies for soil and groundwater contaminated with explosives. *Sustainability*, *16*(3), 961. Available from https://doi.org/10.3390/su16030961.

Nawaz, S. A., Li, J., Bhatti, U. A., Bazai, S. U., & Zafar, A. (2021). A hybrid approach to forecast the COVID-19 epidemic trend. *PLoS ONE*, *16*(10). Available from https://scholar.google.it/citations?view_op = view_citation&hl = en&user = aQSMyzAAAAAJ&cstart = 20&pagesize = 80&citation_for_view = aQSMyzAAAAAJ:blknAaTinKkC.

Pacheco Rocha, N., Dias, A., Santinha, G., Rodrigues, M., Rodrigues, C., Queirós, A., Bastardo, R., & Pavão, J. (2022). Systematic literature review of context-awareness applications supported by smart cities' infrastructures. *SN Applied Sciences*, *4*(4). Available from https://doi.org/10.1007/s42452-022-04979-0.

Podlasek, A., Vaverková, M. D., Jakimiuk, A., & Koda, E. (2024). A comprehensive investigation of geoenvironmental pollution and health effects from municipal solid waste landfills. *Environmental Geochemistry and Health*, *46*(3). Available from https://doi.org/10.1007/s10653-024-01852-4.

Selvam, A.P. & Al-Humairi, S.N. S. *The impact of IoT and sensor integration on real-time weather monitoring systems: A systematic review*. Research Square, Malaysia Research Square. (2023), Available from: https://www.researchsquare.com/browse. doi: 10.21203/rs.3.rs-3579172/v1

Shah, S.K., Tariq, Z., Lee, J., & Lee, Y. (2020). Real-time machine learning for air quality and environmental noise detection, *Proceedings – 2020 IEEE International Conference on Big Data*, Big Data 2020, Institute of Electrical and Electronics Engineers Inc., United States, pp. 3506–3515. Available from http://ieeexplore.ieee.org/xpl/mostRecentIssue.jsp?punumber = 9377717, https://doi.org/10.1109/BigData50022.2020.9377939

Sharifrazi, D., Alizadehsani, R., Roshanzamir, M., Joloudari, J. H., Shoeibi, A., Jafari, M., Hussain, S., Sani, Z. A., Hasanzadeh, F., Khozeimeh, F., Khosravi, A., Nahavandi, S., Panahiazar, M., Zare, A., Islam, S. M. S., & Acharya, U. R. (2021). Fusion of convolution neural network, support vector machine and Sobel filter for accurate detection of COVID-19 patients using X-ray images. *Biomedical Signal Processing and Control*, *68*. Available from https://doi.org/10.1016/j.bspc.2021.102622, http://www.elsevier.com/wps/find/journalbibliographicinfo.cws_home/706718/description#bibliographicinfo.

Singh, S., Maiti, S. K., & Raj, D. (2023). An approach to quantify heavy metals and their source apportionment in coal mine soil: A study through PMF model. *Environmental monitoring and assessment*, *195*(2). Available from https://doi.org/10.1007/s10661-023-10924-4.

Tang, H., Bhatti, U. A., Li, J., Marjan, S., & Baryalai, M. (2023). A new hybrid forecasting model based on dual series decomposition with long-term short-term memory. *International Journal of Intelligent Systems*, *2023*(1). Available from https://scholar.google.it/citations?view_op = view_citation&hl = en&user = aQSMyzAAAAAJ&cstart = 20&pagesize = 80&citation_for_view = aQSMyzAAAAAJ:738O_yMBCRsC.

Vogelstein, J. T., Bridgeford, E. W., Tang, M., Zheng, D., Douville, C., Burns, R., & Maggioni, M. (2021). Supervised dimensionality reduction for big data. *Nature Communications*, *12*(1). Available from https://doi.org/10.1038/s41467-021-23102-2, http://www.nature.com/ncomms/index.html.

Wang, L., Wu, Y., Huang, Z., & Wang, Y. (2022). How big data drives green economic development: Evidence from China. *Frontiers in Environmental Science*, *10*. Available from https://doi.org/10.3389/fenvs.2022.1055162.

Yang, J., Wang, J.-y, Zhao, C., Wang, L., Wan, X., Shi, H., Lei, M., Chen, T., & Liao, X. (2023). Identifying driving factors of arsenic distribution at the mining area scale: Methods and practice. *SSRN Electronic Journal*. Available from https://doi.org/10.2139/ssrn.4352028.

Yang, Y., Guo, T. T., Liu, X. L., & Tie, B. Q. (2023). Spatial distribution characteristics and assessment of heavy metal pollution in cultivated soil of agricultural small watershed in typical mining area in South China. *Environmental Science*, *44*(3), 1602–1610. Available from https://doi.org/10.13227/j.hjkx.202204292, https://www.hjkx.ac.cn/hjkx/ch/reader/view_abstract.aspx?file_no = 20230339&flag = 1.

Yang, Z. N., Liu, Z. S., Wang, K. H., Liang, Z. L., Abdugheni, R., Huang, Y., Wang, R. H., Ma, H. L., Wang, X. K., Yang, M. L., Zhang, B. G., Li, D. F., Jiang, C. Y., Corvini, P. F. X., & Liu, S. J. (2022). Soil microbiomes divergently respond to heavy metals and polycyclic aromatic hydrocarbons in contaminated industrial sites. *Environmental Science and Ecotechnology*, *10*. Available from https://doi.org/10.1016/j.ese.2022.100169, http://www.journals.elsevier.com/environmental-science-and-ecotechnology.

Zhang, B., Duan, M., Sun, Y., Lyu, Y., Hou, Y., & Tan, T. (2023). Air quality index prediction in six major chinese urban agglomerations: A comparative study of single machine learning model. Ensemble model, and hybrid model. *Atmosphere*, *14*.

Zhao, W., Liao, Y., Zhou, S., & Zhou, B. (2022). Ecological remediation strategy for urban brownfield renewal in Sichuan Province, China: A health risk evaluation perspective. *Scientific Reports*, *12*(1). Available from https://doi.org/10.1038/s41598-022-08268-z.

Zhu, W., Yang, X., He, J., Wang, X., Lu, R., & Zhang, Z. (2021). Investigation and systematic risk assessment in a typical contaminated site of hazardous waste treatment and disposal. *Frontiers in Public Health*, *9*. Available from https://doi.org/10.3389/fpubh.2021.764788, http://journal.frontiersin.org/journal/public-health/section/public-health-education-and-promotion#archive.

Further reading

Zhang, H., Yang, Y., Ma, S., Yuan, W., Gao, M., Li, T., Wei, Y., Wang, Y., Xiong, Y., Li, A., & Zhao, B. (2024). Development of a multifaceted perspective for systematic analysis, assessment, and performance for environmental standards of contaminated sites. ACS Omega. Available from https://doi.org/10.1021/acsomega.3c05187, pubs.acs.org/journal/acsodf.

Zhang, L., Lin, X., Qiu, B., Zhang, M., & He, Q. (2022). The industrial sprawl in China from 2010 to 2019: A multi-level spatial analysis based on urban scaling law. *International Journal of Environmental Research and Public Health*, *19*(23). Available from https://doi.org/10.3390/Ijerph192316255, http://www.mdpi.com/journal/ijerph.

Zhao, K., Zhang, L., Dong, J., Wu, J., Ye, Z., Zhao, W., Ding, L., & Fu, W. (2020). Risk assessment, spatial patterns and source apportionment of soil heavy metals in a typical Chinese hickory plantation region of southeastern China. *Geoderma*, *360*, 114011. Available from https://doi.org/10.1016/j.geoderma.2019.114011.

Zhao, S., Yan, Y., & Han, J. (2021). Industrial land change in Chinese Silk Road Cities and its influence on environments. *Land*, *10*(8), 806. Available from https://doi.org/10.3390/land10080806.

Zheng, B., & Masrabaye, F. (2023). Sustainable brownfield redevelopment and planning: Bibliometric and visual analysis. *Heliyon*, *9*(2), e13280. Available from https://doi.org/10.1016/j.heliyon.2023.e13280.

Chapter 3

Artificial intelligence for remote sensing and climate monitoring

Hao Tang, Uzair Aslam Bhatti, Dekai Li, Dai Lisi, Jinru Liu and Mughair Aslam Bhatti
School of Information and Communication Engineering, Hainan University, Haikou, P.R. China

3.1 Introduction

In recent years, big data has emerged as one of the hottest terms in the IT field following cloud computing and the internet of things (Bhatti et al., 2024). It has been widely discussed and researched in various fields such as economy, society, engineering, and science, attracting high attention from academia, industry, and government both domestically and internationally (Earthdata, 2022). International renowned academic journals such as Nature, Science, and ERCIM News have published special issues focusing on big data, discussing key technologies from different perspectives (Bhatti et al., 2024). China has also organized academic forums on big data to delve into theoretical, engineering, and application aspects, as well as organizational methods and resource support for research work (Xu et al., 2021). In March 2012, the US government officially launched the "Big Data Initiative," which was considered by the industry as another significant move in the field of information science following the "Information superhighway initiative." According to this initiative, research institutions such as NASA, NOAA, and USGS have successively conducted research on acquiring, storing, processing, analyzing, and sharing massive remote sensing data under the background of big data (Bhatti et al., 2023; Hong et al., 2021; Liu et al., 2024, n.d.; Yang et al., 2024). Remote sensing is a comprehensive discipline that utilizes noncontact sensors to obtain information about distant objects. It is widely applied in various industries such as resource surveying, environmental protection, agricultural yield estimation, forestry surveying, water conservancy construction, urban management, marine conservation, and disaster prevention, becoming one of the important technical means to support economic and social development decisions (Sun et al., 2022). Remote sensing can be categorized into three types based on the method of data acquisition: spaceborne remote sensing, airborne remote sensing, and ground-based remote sensing. Due to its advantages such as all-weather, wide coverage, short observation cycle, and strong dynamic characteristics, spaceborne remote sensing has become the focus of research and development in the field of remote sensing with the rapid development of space technology (Moigne, 2021).

Spaceborne remote sensing systems are organic entities with functions including acquiring, processing, managing, analyzing, and applying remote sensing data (Bhatti, Bazai, et al., 2023). They consist of six aspects: energy sources, observation targets, data acquisition systems, information inversion systems, information application systems, and users. The working mechanism of spaceborne remote sensing systems involves electromagnetic waves from energy sources interacting with targets to reflect or radiate into space (Yuan et al., 2024). The system captures electromagnetic wave signals carrying target information through satellite sensors and transmits them back to the ground. After data preprocessing, remote sensing data is obtained. Using remote sensing information models, feature information is extracted from remote sensing data. The extracted information is analyzed to form knowledge, which is provided to users to support their application decisions (Yang et al., 2023). The control system is responsible for monitoring the operation status of the spacecraft and issuing system task commands according to user observation requirements to acquire new remote sensing data.

Currently, spaceborne remote sensing systems face typical big data challenges in key areas such as space-to-ground data transmission, data storage management, data preprocessing, data analysis application, and result presentation visualization. Firstly, a global multi-scale, multi-angle, comprehensive and all-weather stereo-geodetic observation network

Deep Learning for Earth Observation and Climate Monitoring. DOI: https://doi.org/10.1016/B978-0-443-24712-5.00004-X

has been formed (Kamarulzaman et al., n.d.). This means that while humans have the ability to obtain detailed information about the Earth, it also brings massive remote sensing data. As of 2006, NASA's nine data archives have accumulated and stored satellite remote sensing data and experimental data totaling 4.9 PB, with an additional 4.5 TB of data added daily and distributing 2.5 TB of data to users daily (Cannistra et al., 2021). China's three major satellite centers, namely the National Satellite Meteorological Center, China Center for Resources Satellite Application, and National Satellite Ocean Application Center, store approximately 1.1 PB of meteorological satellite data, approximately 3.5 PB of land observation (resource) satellite data, and about 25 TB of ocean satellite data, respectively (John et al., n.d.). Secondly, all remote sensing data are unstructured data. From the operation process of spaceborne remote sensing systems, it can be seen that various types of sensors such as visible light, infrared, microwave, laser, gravity, and electromagnetic sensors actively or passively obtain Earth target information (Richards, 2009). They record raw data in imaging and non-imaging forms, and after a series of processing such as preprocessing, inversion (extraction), and analysis application, remote sensing data is obtained in the form of images or signals. The semantic relationships of these data are difficult to express using the two-dimensional logic of traditional relational databases. In addition, there is a large amount of redundancy between the data, which does not meet the atomic consistency requirements of structured data. Thirdly, with the rapid increase in the number of remote sensing satellites and the significant improvement in observation resolution in terms of space, time, and spectrum, the growth rate of remote sensing data volume is very fast (Bhatti, Masud, et al., 2023; Dinh et al., n.d.). For example, the terra and aqua satellites of the earth observation system (EOS) acquire data at the terabyte level daily. The National Satellite Meteorological Center is responsible for archiving 801 types of data from 21 meteorological satellites at home and abroad, with a daily addition of approximately 2 TB of data. The China Center for Resources Satellite Application currently manages eight domestic land observation (resource) satellites, which acquire data at a rate of up to 4 TB per day (Moreira et al., 2013; Oguchi et al., 2022).

In recent years, there has been a surge in the advancement and utilization of artificial intelligence, both domestically and internationally. This trend can be attributed to two key factors: "big data" and "high computing power." This surge is also evident in the realm of geospatial data analysis and application. For instance, initiatives like the National High-Resolution Earth Observation Major Science and Technology Program ("High Score Program") have led to the accumulation of vast amounts of geospatial data. Furthermore, the evolution of high-performance computing frameworks, such as the "cloud + end" architecture, has bolstered computational capabilities. Against this backdrop, leveraging artificial intelligence technology to unlock the profound insights within geospatial big data is poised to become a cornerstone for future developments in geospatial data analysis and application (Dassot et al., 2011; Deems et al., 2013; Devaney et al., 2015; Xu et al., 2024). The goal of developing intelligent analysis technology for remote sensing data is to transform extensive archives of remote sensing data into comprehensive observations, analyses, and interpretations of observed phenomena. This entails extracting rich and precise attribute information and uncovering the evolutionary patterns of target areas. Key tasks include precise data processing, spatiotemporal analysis, target element classification, and association mining. Notably, significant strides have been made in open-source remote sensing datasets and sharing platforms, catering to the growing demand for sophisticated analytical tools (Disney et al., 2010; Nawaz et al., 2024).

In terms of precise data processing, the challenge lies in accurately calibrating sensor-acquired spectral reflection or radar scattering data to produce image products that faithfully represent ground objects (Allison et al., 2016). Traditional methods rely on predefined models and extensive field calibration tests, which are becoming increasingly impractical with the advent of new sensor technologies and higher resolutions. Leveraging deep learning networks to bridge the gap between observation data and accurate image products holds promise for multi-satellite integrated remote sensing image processing. Spatial and temporal analysis of remote sensing data is crucial for capturing dynamic changes and revealing underlying trends. However, challenges such as cloud cover and shadows often obscure ground information, hampering the effectiveness of remote sensing applications (Bhatti, Huang et al., 2023). Addressing these challenges requires innovative approaches to enhance data availability and enable comprehensive time-series analysis. Classification and recognition of remote sensing target elements are hindered by noise and limited computational capabilities (Zhang et al., 2024). Harnessing the self-learning capacity of computers through artificial intelligence methods presents an opportunity to improve accuracy and mitigate false alarms (Manolakis et al., 2016). Developing an automated analysis framework based on deep learning can enhance target element extraction and recognition (Sun & Du, 2019).

As we enter the era of remote sensing big data, there is a growing need for rapid association, organization, and analysis of multidimensional information (Dong & Chen, 2017). Leveraging heterogeneous remote sensing data sources to achieve this goal represents a pivotal direction for future research. Despite progress in remote sensing datasets and platforms, challenges persist regarding data scale and compatibility with deep learning frameworks (Weiss et al., 2020). This paper explores the current landscape, emerging trends, challenges, and future directions of intelligent analysis technology for remote sensing, focusing on five key research areas. Bridging this gap requires the development of open-source datasets and sharing platforms tailored to remote sensing applications, thereby fostering a more intelligent and practical ecosystem (Ghosh et al., 2014). This paper analyzes the technical status of spaceborne remote sensing systems from five key aspects: space-to-ground data transmission, data storage management, data preprocessing, data application, and visualization presentation. It explores how to solve the big data challenges faced by spaceborne remote sensing systems under the background of big data, and meet the needs of rapid acquisition, efficient management, in-depth application, and intuitive presentation of massive remote sensing data by combining key technologies in five hot research areas of big data, including information communication, massive storage, high-performance computing, spatial data mining, and visualization.

3.2 Space-to-ground data transmission

3.2.1 Current technology status

The data transmission system of remote sensing satellites is responsible for sending target information collected by sensors back to the ground and is an important component of spaceborne remote sensing systems. With the advancement of space technology, the detection capabilities of remote sensing satellites have significantly improved, with spatial resolutions reaching submeter levels, spectral resolutions reaching nanometer levels, and radiometric quantization reaching 16 bits (Lary et al., 2016). This has led to continuous increases in the transmission code rate of data transmission systems, with transmission bandwidths approaching the civil X-band radio bandwidth specified by the International Telecommunication Union. For example, the transmission code rate of the French SPOT-5 satellite is 150 Mb/s, that of the US Landsat-7 satellite is 280 Mb/s, and that of China's "ZY-1" satellite is 113 Mb/s. The commercial satellites IKONOS-2 and Quickbird have transmission code rates of 320 Mb/s each. Most high-resolution satellites currently use polarization to increase the amount of transmitted data. For example, the accumulated transmission code rate of the French Pleiades satellite is 600 Mb/s, while that of the US commercial satellites WorldView-1 and GeoEye-1 is 800 and 740 Mb/s, respectively. The Japanese ASNARO-1 satellite has an accumulated transmission code rate of 832 Mb/s, and China's "ZY-3" and "GF-1" satellites both have accumulated transmission code rates of 900 Mb/s (Sarker, 2021; Sun, Cristea, & Rivas, 2023). Shown in Fig. 3.1.

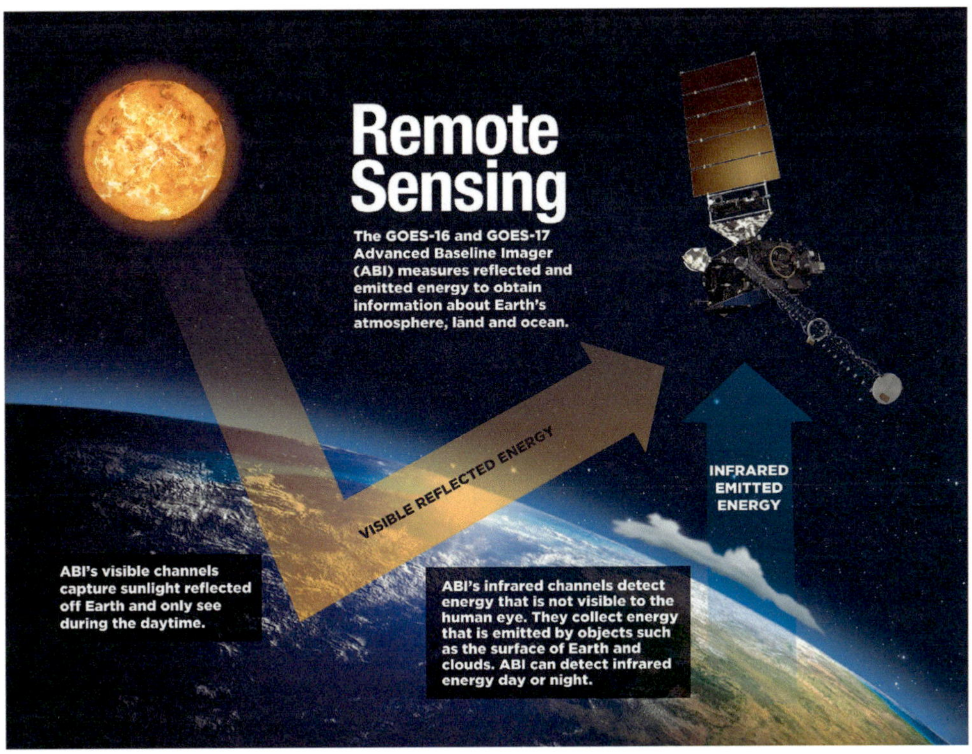

FIGURE 3.1 Remote sensing data transmission.

Due to the complex space-to-ground transmission environment and limitations in onboard antenna power, data transmission systems typically employ techniques such as data compression, channel coding, and modulation to improve transmission efficiency and reduce transmission error rates. Data compression reduces data redundancy and transmission volume while maintaining data quality. Currently, differential pulse code modulation (DPCM), discrete cosine transform (DCT), and wavelet transform are widely used compression coding methods onboard. DPCM, a predictive coding method, achieves lossless compression but with relatively low compression ratios, typically around 2:1 (Bhatti, Hashmi, et al., 2023; Miao et al., 2012; Saini & Ghosh, 2017; Zhang et al., 2022). DCT and wavelet transform, belonging to transform coding methods, offer higher compression ratios of over 8:1 but may result in information loss. Channel coding uses error-control techniques to reduce transmission error rates in adverse communication conditions. NASA's standardized space communication concatenated coding scheme, established in 1984, employs Reed-Solomon linear block codes as outer codes, Viterbi convolutional codes as inner codes, and a certain depth of interleaving (Schapire, 1999). This scheme exhibits strong error correction capabilities for burst errors, random errors, and continuous errors and has been officially adopted by the consultative committee for space data systems as the advanced orbiting system standard, widely used in space communication. Modulation techniques transform compressed and encoded baseband signals into frequency-band signals suitable for transmission in long-distance and complex environments. Currently, remote sensing satellites commonly use quadrature phase shift keying (QPSK) and some derived modulation techniques, achieving transmission bit rates of 2 b/(s Hz) (Bhatti, Tang, Wu, Marjan, et al., 2023; Rodriguez-Galiano et al., 2012).

3.2.2 Compression, coding, and modulation

Compressed sensing theory breaks through the limitations of the Nyquist sampling theorem in traditional signal processing by transforming frequency sampling of signals into sparse information sampling within signals, significantly reducing sampling data volume and hardware costs (Mascaro et al., 2014). If a signal can be sparsely represented in a certain transformation domain, it can be projected onto an orthogonal transformation basis satisfying noncoherence or equidistance constraints through a sensing matrix to obtain sparse or approximately sparse transformation vectors. The original signal can then be reconstructed by solving a nonlinear optimization problem (Phan et al., 2020). Compressed sensing theory can greatly reduce data transmission volume as a source compression method. Additionally, compressed sensing theory enables sampling at frequencies much lower than the Nyquist frequency to obtain high-quality images, providing theoretical support for new remote sensing imaging methods.

Turbo codes and low-density parity-check (LDPC) codes are channel coding technologies approaching the Shannon limit. Turbo codes absorb the advantages of traditional concatenated codes, using recursive systematic convolutional codes as constructing cascaded subcodes and employing random interleavers to cascade various subcodes (Yang et al., n.d.). Based on different cascading structures, Turbo codes are divided into parallel concatenation, serial concatenation, and hybrid concatenation. Turbo codes make full use of additional information between different subcodes, offering excellent coding gains. Furthermore, Turbo codes exhibit good distance properties, maintaining excellent performance in low signal-to-noise ratio environments, and have become the channel coding standard for 3G mobile communication (Rittger et al., 2021). LDPC codes are linear block codes constructed on a sparse matrix and can be categorized into regular codes and irregular codes based on whether the number of rows and columns in the construction matrix is fixed. LDPC decoding complexity is low, suitable for parallel operations, and has minimal delay, making it suitable for real-time requirements in remote sensing satellite data transmission.

Multiple quadrature amplitude modulation is an efficient modulation technique that can increase transmission efficiency to 8 b/(s Hz) or higher. Orthogonal frequency division multiplexing is a new wireless multicarrier transmission technology that divides the channels of effective bandwidth into multiple subchannels, modulates the data on each subchannel, and then transmits them (Belgiu & Drăgu, 2016; Sabat-Tomala et al., 2020). When the number of subchannels is large, high spectrum utilization can be achieved. Additionally, trellis coded modulation and bit interleaving coding technologies combine coding with modulation, changing the previous separate design of coding and modulation and balancing the effectiveness and reliability of data transmission. These techniques can improve error performance without sacrificing bandwidth width.

3.2.3 Atmospheric laser communication

Atmospheric laser communication boasts higher carrier frequencies, with channel capacities exceeding shortwave by hundreds of thousands of times, potentially alleviating the current scarcity of frequency resources and becoming one of the research hotspots in the field of information communication (Behnamian et al., 2019). Current research focuses on optical systems, coding, modulation, and acquisition, pointing, and tracking (APT) technologies for space-to-ground laser communication. Optical systems typically employ adaptive zoom, increased transmission power, and multiple antenna backups to address atmospheric attenuation, refraction, and the effects of weather conditions like rain, fog, and snow on laser transmission stability. Additionally, optical systems require filters with ultra-narrow bandwidth and high transmittance to reduce background radiation interference. Atmospheric laser communication operates in a point-to-point fashion with extremely small emission angles, necessitating constant alignment between the transmitting and receiving ends. Even minor changes in terminal platform status can cause communication link interruptions (Chen & Carlos, 2016; Georganos et al., 2018; Ghatkar et al., 2019; Millard & Richardson, 2015). APT technology dynamically adjusts antenna positions based on interference source characteristics and tracking accuracy requirements to ensure that the laser at the transmitting end remains centered at the receiving end, maintaining a clear communication link. Coding and modulation techniques such as wavelet transform, Turbo codes, LDPC codes, and QPSK, which are commonly used in shortwave communication, are also being researched and applied in atmospheric laser communication. Additionally, researchers have proposed new coding methods like Alamouti codes and new modulation techniques like polarization shift keying (Bhatti, Marjan, et al., 2023; Mountrakis et al., 2011).

3.2.4 Real-time onboard processing

Real-time onboard processing installing high-speed data processing systems onboard satellites to enable real-time data processing capabilities, transmitting only valuable processed data back to the ground, is another technique for improving space-to-ground transmission efficiency. For example, the "Ocean Mapper" hyperspectral imager developed by the United States incorporates a high-speed data processing system capable of identifying ground objects based on spectral image characteristics (Ma et al., 2019; Sheykhmousa et al., 2020; Zhu et al., 2017). This system possesses data analysis, feature extraction and classification, and data compression capabilities. The processed effective data typically only accounts for 1/10 of the original data, significantly reducing transmission data volume. Germany's BIRD fire monitoring microsatellite, launched by the German Aerospace Center in 2001, employs a new type of onboard data processing system that integrates data preprocessing, image feature extraction, and image data identification and classification technologies (Yuan et al., n.d.). It can generate thematic map data containing only key information in real-time, significantly reducing transmission data volume. Shown in Fig. 3.2.

3.3 Accurate interpretation of satellite imagery signals

In the realm of remote sensing data processing, precision is paramount. Leveraging intelligent techniques for data preprocessing is essential for tasks like registration and correction of remote sensing data, whether optical, SAR, or spectral.

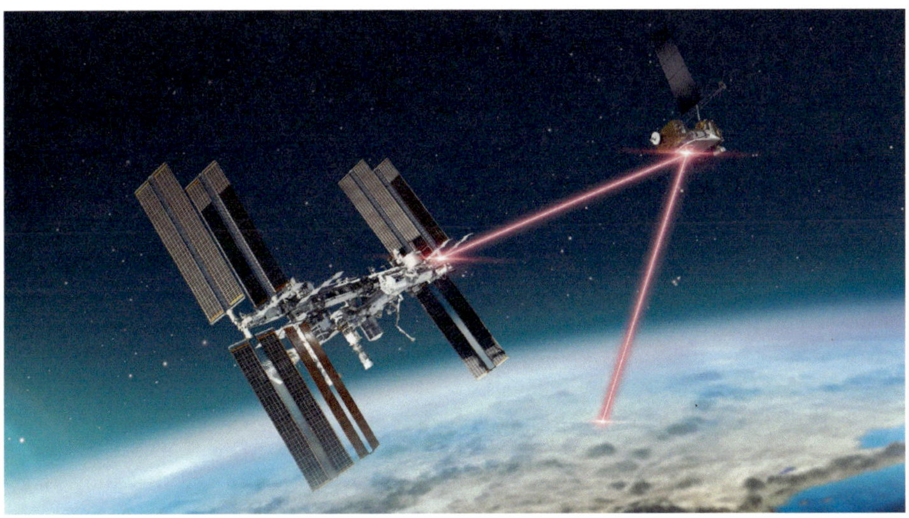

FIGURE 3.2 Laser communication.

3.3.1 Advanced computation

Remote sensing images serve as the foundation for various applications such as agricultural monitoring, urban planning, and military reconnaissance. The quality of these images directly impacts their utility, with factors like image resolution, spatial characteristics, spectral attributes, and geometric accuracy playing crucial roles (Goodfellow et al., 2016). Achieving high-quality remote sensing images involves two main approaches: refining traditional imaging algorithms and enhancing existing low-quality images through denoising, cloud removal, and super-resolution reconstruction.

3.3.1.1 Imaging advancements

In the SAR domain, innovative algorithms based on deep learning have simplified the imaging process. For instance, Zhang et al. (2016) introduced a line spectrum estimation method based on deep networks, significantly expediting 3D SAR imaging (Zhang et al., 2016). Traore proposed a depth SAR imaging algorithm to reduce sampling requirements and mitigate motion errors (Traore et al., 2018). Additionally, super-resolution techniques, such as those proposed by Chen et al., offer a cost-effective means to enhance image resolution without relying solely on satellite capabilities (Chen, Bai, et al., n.d.).

3.3.1.2 Artificial intelligence integration for image enhancement

Artificial intelligence, particularly machine learning methods like GAN networks, has revolutionized remote sensing image processing. By harnessing transfer learning and zero-shot learning, researchers aim to bridge the gap between natural and remote sensing image quality (Agarap, n.d.; Aloysius & Geetha, 2017). Despite the inherent challenges, such as resolution limitations, AI-based approaches hold promise for extracting richer information from remote sensing imagery (Dubey & Jain, 2019; Ronneberger et al., n.d.; Zhang et al., 2018).

3.3.1.3 Spatial feature extraction and visual enhancement

To maximize the insights gleaned from remote sensing images, it's imperative to extract spatial features and enhance visual clarity. Techniques like attention mechanisms and wavelet transforms enable better detection and interpretation of small and weak targets, thus mitigating information loss due to missing details (Badrinarayanan et al., 2017).

3.3.2 Quantifiable enhancement

The advancement of remote sensing imagery quality, encompassing visible light, SAR, and hyperspectral domains, primarily revolves around enhancing geometric, radiometric, spectral, and polarization aspects. Through meticulous identification of stable points, inherent distortions during the imaging process are rectified, leading to quantifiable enhancements in data product fidelity, thus enabling a more precise depiction of ground objects (Alom et al., n.d.; Chen, Papandreou, et al., n.d.; Zhao et al., n.d.). In the global research arena, neural network methodologies are predominantly employed to bolster image matching accuracy, consequently propelling geometric positioning precision. Leveraging machine learning paradigms has ushered in a systematic evolution in remote sensing image matching technologies, with notable strides in matching accuracy. Two prevalent methodologies, namely local feature point matching and region matching, have garnered significant attention.

Local feature point matching methods center around key point detection and descriptor extraction. Pioneering advancements include strategies like coarse-to-fine approaches by He et al. and descriptor training techniques by Li Serra et al. (He et al., 2016; Li et al., 2019). On the other hand, regional feature matching, characterized by its robustness in handling regional deformation, has witnessed innovative adaptations in the realm of deep learning. Noteworthy approaches include classification-based matching by Han et al. and descriptor-driven matching by Redmon et al. (Redmon et al., 2016). The intelligent execution system, orchestrated by the intelligent expert system, orchestrates imaging tasks encompassing large-area low-resolution imaging, local medium-resolution imaging, and high-resolution imaging of pivotal targets (Redmon, 2023). This hierarchical approach enables wide-area perception, medium-resolution search, and high spectral resolution identification, facilitating comprehensive spatial coverage and target identification. Intelligent semantic interpretation, the process involves data characterization, feature extraction, semantic mapping, and model training, culminating in knowledge-based semantic models for image interpretation (Redmon & Farhadi, n.d.; Wu et al., 2018; Xu & Wu, 2020). Deep learning networks have profoundly enriched remote sensing image matching capabilities, with ongoing advancements promising further refinement in accuracy and efficacy.

3.4 Satellite imagery data collection and evaluation at a time and geographic scale

In recent years, significant strides have been made in the collaborative processing and analysis of multisource remote sensing data, aiming to achieve synergistic fusion and reconstruction across temporal and spatial domains, thereby enhancing remote sensing data's time-space analysis capabilities.

3.4.1 Temporal preservation of satellite data

Researchers have developed numerous methods for restoring and reconstructing time series remote sensing images, categorized into three main approaches: self-completion, reference completion, and multi-temporal completion. Self-completion methods utilize data from cloud-free or shadow-free areas within the same image to repair missing data in cloud or shadow-covered regions (Yang, 2021). Conversely, reference completion methods leverage spectral bands insensitive to clouds in multispectral or hyperspectral images to reconstruct missing data. Meanwhile, multi-temporal completion methods capitalize on the availability of multiple images of the same area over time, employing similar pixel or patch matching techniques to predict missing data in cloud or shadow-covered regions (Terven & Cordova-Esparza, n.d.). While these methods have achieved notable success, challenges persist, such as the need for recurrent neural networks to capture temporal trends effectively and the inherent spatial and temporal uncertainties associated with cloud and shadow-covered areas (Jiang et al., 2022). Shown in Fig. 3.3.

3.4.2 Space-temporal fusion of multisource satellite imaging

Remote sensing image fusion research has evolved in two stages, focusing initially on panchromatic enhancement algorithms and subsequently advancing to multisource and multi-resolution spatial-temporal fusion algorithms (Terven & Cordova-Esparza, n.d.). Spatio-temporal fusion algorithms aim to integrate high spatial resolution and high temporal resolution remote sensing data, offering significant potential for remote sensing applications (Ren et al., n.d.). While various fusion principles exist, including unmixing-based, filtering-based, and learning-based methods, recent efforts have emphasized learning-based approaches utilizing deep convolutional neural networks (DCNNs) to overcome spatial disparities between sensor data (Girshick, 2015). While DCNN-based models show promise, challenges remain in determining optimal network architectures and addressing spatial differences between sensor data types (Vaswani et al., 2017). Shown in Fig. 3.4.

3.4.3 Operational method of the smart laser satellite navigation system

Intelligent functions, such as target recognition and automatic tracking, promise to enhance satellite operations' efficiency and timeliness. Illustrated through the observation of river basins, the intelligent satellite remote sensing system orchestrates imaging tasks in a systematic manner, optimizing user satisfaction through targeted data acquisition and interpretation (Aleissaee et al., n.d.).

FIGURE 3.3 Remote sensing images of mountains. A snowy mountain range description automatically generated.

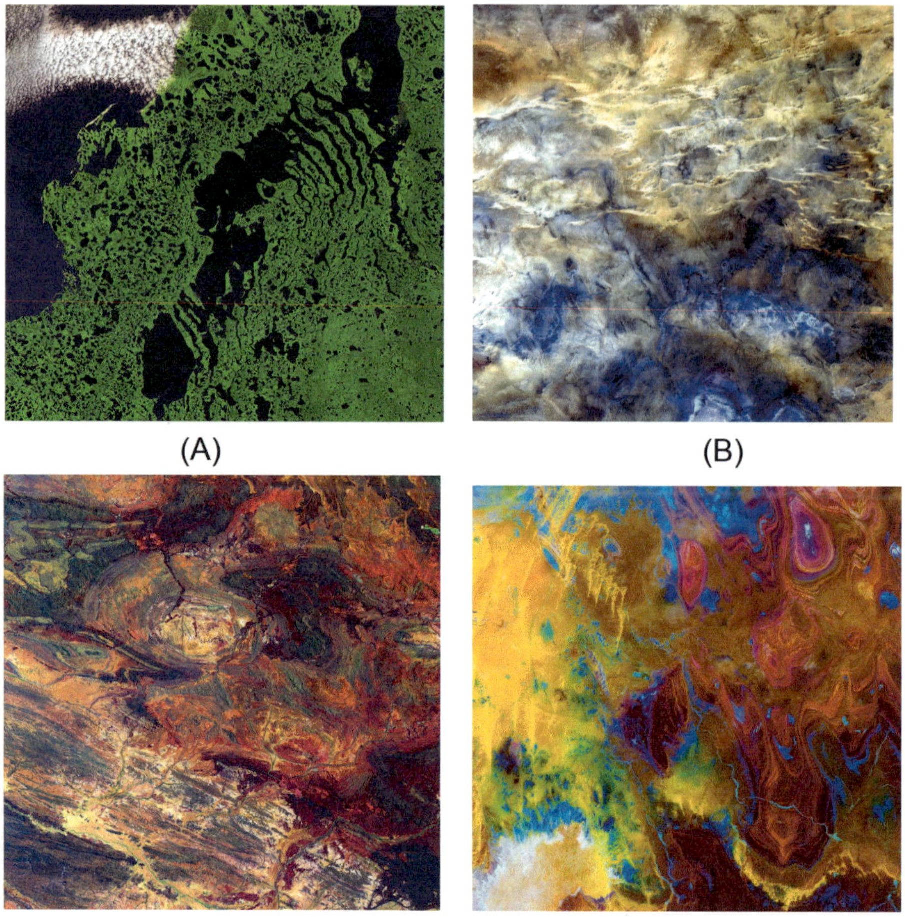

FIGURE 3.4 (A)-(D) shows different multisource remote sensing images.

3.4.4 Potential growth of advanced satellite navigation techniques in China

China's research on intelligent remote sensing systems is nascent, necessitating learning from global advancements. Future developments should focus on sensor autonomy, satellite mission planning, and in-orbit semantic interpretation technologies to bolster the efficiency and effectiveness of remote sensing operations, meeting evolving user demands and enhancing China's leadership in remote sensing technology (Devlin et al., n.d.; He et al., 2020).

3.5 Data storage management

3.5.1 Current technology status

Remote sensing data storage management systems mainly comprise two major functions: efficient data indexing management, enabling rapid location of data storage physical locations; and stable and reliable data storage, as well as fast read/write, movement, and deletion operations (Bhatti et al., 2022). Currently, China's three major remote sensing data centers, namely the National Satellite Meteorological Center, the China Remote Sensing Satellite Application Center, and the National Satellite Ocean Application Center, all use mature commercial relational databases to store metadata information of remote sensing data and perform operations such as insertion, querying, updating, and deletion of metadata using SQL statements (Ankan et al., n.d.; Goodfellow et al., 2014; Graves, 2012; Sun et al., 2019). This approach fully utilizes the stability of commercial relational databases, simplifies the data layer access interface, and facilitates system development and maintenance. However, when facing millions of data records, it may result in performance losses in data operations. Therefore, NASA's data archive center adopts a hierarchical directory organization based on the Linux operating system to manage data files, speeding up data access for users without complex data storage relationship information. Moreover, it exhibits good horizontal scalability, meeting the management needs of rapidly growing data (Jozdani et al., 2022).

In terms of data storage, distributed parallel storage systems based on SAN and cluster storage systems based on NAS are the two most commonly used methods. The former deploys a distributed parallel file system on a disk array constructed based on a SAN network, with clients directly writing data files to the disk array in block form, and the system managing data blocks through metadata. The system's performance mainly depends on parameters such as CPU, cache, and bus bandwidth in the disk array controller (Creswell et al., 2018). This system can only achieve vertical scalability, and as the number of disks increases, performance will decrease. Additionally, it does not support online expansion, and expanded hardware devices must be compatible with the original manufacturer, resulting in high expansion costs (Xu & Zhao, 2018) The latter is composed of multiple storage nodes with relatively small storage capacities under the traditional NAS architecture, forming a complete storage system together. Each node in the cluster storage is both a storage node and an I/O node, and data is distributed and stored in different storage nodes. This system features horizontal scalability and does not reduce I/O performance as storage capacity increases. Furthermore, storage nodes can adopt a common X86 server architecture, reducing expansion costs.

According to the different physical locations of remote sensing data storage, existing storage systems can be divided into distributed storage and centralized storage. The former disperses the storage system construction across different physical locations, only centrally managing it logically, which can solve the unified management problem between different businesses and storage technologies (Isola et al., 2017; Ledig et al., 2017). Typical systems include NASA's EOS system and the European Space Agency's (ESA) G-POD system. The latter concentrates the storage system construction at one physical location, facilitating unified data management and system maintenance. Most of China's remote sensing data centers adopt a centralized storage solution (Sun, Wang, et al., 2020). As remote sensing data processing is computationally intensive, to resolve the contradiction between data access speed and storage capacity, most storage systems adopt a three-tier architecture of online, nearline, and offline storage. Online storage utilizes high-speed storage systems to meet the requirements of rapid read/write during data processing (Lin et al., 2017). Nearline storage uses low-speed tape libraries to meet the long-term storage requirements of large-capacity data. Offline tapes are responsible for permanent data preservation.

3.5.2 Cloud storage

Looking at the development trends of aerospace remote sensing systems both domestically and internationally, ground systems are moving toward the direction of unified management and distribution sharing of data from multiple satellites and sensors. For example, the United States' EOS system integrates ground systems from various remote sensing agencies, not only storing and distributing data from all Landsat satellites but also providing storage and distribution services for subsequent observation satellite data (Hu et al., 2020; Liu et al., 2018). China also plans to integrate domestic satellite ground systems through major technology projects with high-resolution imagery, achieving unified data sharing. Cloud storage technology can integrate storage systems between different units, businesses, and locations, providing unified storage services (Singh & Komodakis, 2018). Currently, typical cloud storage systems include Google's GFS (Google File System) and Hadoop's HDFS (Hadoop Distributed File System), with many internet companies' massive data storage systems based on improvements to these two systems.

Multiple copy management is a key technology for improving the reliability and access efficiency of cloud storage systems, including aspects such as copy creation, copy selection, copy location, copy deletion, and data consistency (Arulkumaran et al., 2017). The best client strategy is a popular copy creation strategy that can improve data access efficiency while reducing network bandwidth consumption. It requires detailed records of user data access frequencies on each storage grid node, creating copies on the node when access frequency exceeds a threshold. Copy selection technology is responsible for selecting the best copy based on user requirements, utilizing storage network performance models or historical records of copy access to predict copy response times (Li et al., 2018). Copy location technology needs to quickly find the physical storage location of data in the cloud storage system based on data access requests. The directory copy service algorithm based on hierarchical directories centrally manages all data directory information, with features of simplicity and flexible configuration, widely applied (Mou et al., 2021). Copy deletion technology dynamically deletes copies with low access rates to optimize storage space, commonly using strategies such as delayed deletion, offline deletion, and deletion based on copy lifecycle. Copy consistency technology ensures consistency in content and logic of data copies during multiple user operations. Eventually consistent strategies do not update data copies in real-time during user operations, only updating and propagating data copies under certain conditions, resulting

FIGURE 3.5 Prediction of cyclone path through analysis of cumulonimbus cloud. Cumulonimbus analysis to predict cyclone paths.

in low hardware resource utilization and widely used in commercial cloud storage systems like Amazon (Fu et al., 1922). Shown in Fig. 3.5.

Storage virtualization technology unifies mapping of heterogeneous storage devices, shielding the physical locations of devices, and is a key technology for cloud storage systems to provide unified storage services externally. Storage virtualization technology is divided into three types: host-based, array-based, and network-based. Host-based storage virtualization technology requires installing dedicated virtualization software on the host, increasing the load on the host and having poor scalability (Filar & Vrieze, 2012; Li et al., 2018; Maxwell et al., 2018). Array-based virtualization technology equips storage device controllers with virtualization control programs, mapping logical storage units to multiple physical disk devices. This technology can provide good storage performance but has poor scalability due to device vendor limitations. Network-based storage virtualization technology neither increases the load on the host nor is compatible with storage devices from different vendors, with strong scalability and is the mainstream technology for current storage virtualization. Common methods include in-band virtualization and out-of-band virtualization (Song et al., 2019). The former requires dedicated virtualization devices to complete all virtualization work. When there are frequent data access requests, a large amount of data flow can bottleneck virtualization devices' I/O performance (Methodology & Accuracy Summary 10m Global Land Use Land Cover Maps, 2022). The latter places control flow in virtualization devices for processing, with data flow transmitted through storage networks, resulting in less virtualization overhead than the former.

3.5.3 NoSQL databases

In the era of big data, over 85% of data consists of unstructured data like images, videos, and semi-structured data like documents. According to the consistency, availability, partition tolerance theorem, no distributed file system can simultaneously meet consistency, availability, and partition tolerance requirements (Gemstone Case Space Know, Study, 2022; AI Enables Rapid Creation of Global Land Cover Map, 2021). In cloud storage systems, NoSQL databases abandon the strong SQL language and transactional consistency and paradigm constraints of relational databases, designing for the availability and partition tolerance of massive unstructured data to meet high-concurrency data access requests and rapid data growth requirements (Qi, 2022). In terms of consistency, many mainstream NoSQL databases adopt weak consistency principles, meaning they do not rigorously control consistency in real-time for multiple user operations (Yang et al., 2022). Eventually consistent principles are a special case of weak consistency, only guaranteeing that user operation results are eventually consistent. Consistent hashing algorithms are widely used for partition tolerance (Schmitt & Zhu, 2016).

Common NoSQL databases are divided into four types: key-value stores, column-oriented stores, document-oriented stores, and graph-oriented stores. Key-value store databases are the simplest NoSQL systems, using a key-value mapping structure, a type of unstructured data storage mode, well-applied in Amazon's Dynamo (Floodly, 2021). Different from relational databases storing data by row, column-oriented database systems store data of the same column together, supporting dynamic column expansion and meeting the storage requirements of rapidly growing data; additionally, they feature data compression, deferred materialization, and block cycling (Paganini et al., 2020). Typical column-oriented storage systems include Google's BigTable and Hadoop's Hbase. Document-oriented storage mode maps keys to documents containing certain format information, with the document format being free, able to store lists and recursively nested data. Graph-oriented storage mode uses concepts like nodes, attributes, and edges, representing data objects stored, object description information, and relationships between objects, suitable for storing spatial vector data (Redins, 2023; Technology, 2022; Woodie, 2020; Zhong et al., 2022).

3.6 Data preprocessing

3.6.1 Current status of technology

The purpose of remote sensing data preprocessing is to maximize the authenticity of the terrain object information. Taking optical imaging remote sensing systems as an example, the data preprocessing system generally generates standard Level-2 products (Skulovich & Gentine, 2023). Its processing workflow includes format decoding/decompression, scene segmentation, radiometric correction, geometric correction, and product generation steps (Esen & Wentworth, 2020). Early preprocessing systems focused too much on improving processing accuracy and program parallelization, lacking consideration for top-level issues such as general task scheduling and workflow management. Additionally, the hardware of these systems typically used expensive and bulky specialized devices such as array processors, support vector machines (SVMs), and symmetric multiprocessing machines (Holland et al., n.d.). This led to early systems usually employing specific job scheduling strategies, with relatively fixed processing functions and workflows unable to handle data from multiple satellite payloads. Currently, the number and variety of remote sensing satellites are rapidly increasing, exposing the drawbacks of designing patterns that target specific satellites, such as long development cycles, high costs, and difficulty in upgrading and expanding (Dörnhöfer & Oppelt, 2016; Verbesselt et al., 2012).

With the development of computer technology, the new generation of preprocessing systems is moving toward multi-satellite, multitask, comprehensive, and universal systems (Bhatti, Wu et al., 2022). They use reconfigurable process technology to customize processing workflows for different satellite data on demand and achieve flexible combinations of multiple data processing functional modules (Wang et al., 2022). Furthermore, the new generation of preprocessing systems is built on ordinary PC server clusters, with hardware resources managed through virtualization and general job scheduling technology, possessing hardware reconfiguration capabilities for convenient hardware device upgrades and expansions as needed (Nizamani et al., 2023).

Remote sensing data processing systems have long used high-performance computing to parallelly process massive remote sensing data. Common distributed parallel strategies include data partitioning-based, function partitioning-based, and task partitioning-based strategies (Alzubaidi et al., 2021). Data processing systems typically select one or more parallel strategies based on processing needs. The data partitioning-based strategy is one of the commonly used strategies in processing systems. It first divides data into multiple blocks in some way, applies the same processing operation to each data block on different processors, and then merges and outputs the results (Soydaner, 2020). This strategy requires a lot of real-time thread communication during processing, so systems usually use high-speed, low-latency. Shown in Fig. 3.6.

networks such as 10 Gigabit Ethernet, fiber networks, and InfiniBand networks. The function partitioning-based strategy divides a complete data processing program into multiple sub-algorithms with different functions. Each processor executes a sub-algorithm, and data is shared between them, with the output of the previous sub-algorithm being the input data for the next sub-algorithm (Sheng et al., 2008). The task partitioning-based strategy utilizes task scheduling technology to parallelly schedule a large number of identical processing tasks to various computing nodes in the cluster based on the workload of computing resources.

3.6.2 Mapreduce programming model

Traditional distributed parallel processing systems usually use message passing models such as message passing interface and Sheng, V.S. parallel virtual machine to develop parallel processing algorithms. Programmers need to consider issues such as data and task partitioning, task communication synchronization, deadlock detection, etc., resulting in heavy programming burden and low programming efficiency (Bhatti, Yu, et al., 2022). As one of the core technologies of cloud computing, MapReduce is a distributed parallel programming model proposed by Google for processing large datasets (Gao et al., 2015). It abstracts away the underlying implementation details, reduces the difficulty of developing parallel programs, and improves programming efficiency, making it one of the most widely used parallel programming models currently.

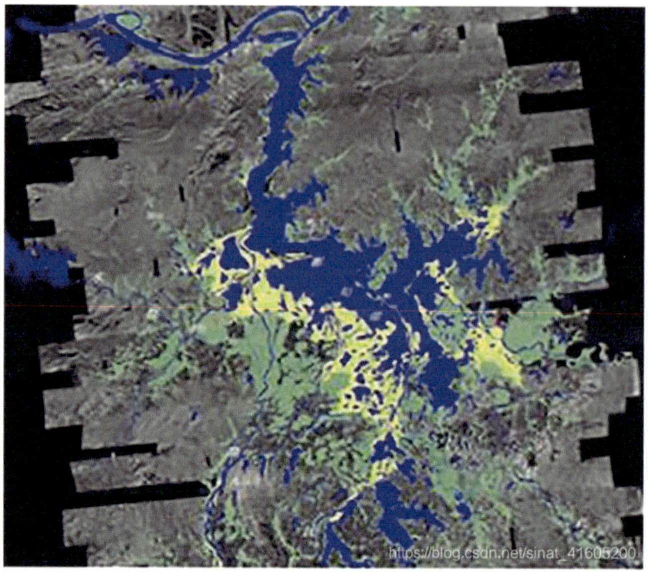

FIGURE 3.6 Flood distribution map of Poyang Lake (Part).

In the MapReduce model, the problem to be processed is decomposed into two stages: mapping (Map) and reducing (Reduce). The data to be processed is decomposed into "<key, value>" key-value pairs and then entered into Map tasks and reduce tasks sequentially (Petitjean et al., 2012). The Map function is responsible for processing these data blocks separately and then mapping the results into result data according to defined rules. After multiple Map tasks are completed, multiple result data blocks organized as key-value pairs are generated. Result data blocks are grouped and sorted based on the key value, and key-value pairs with the same key value are assigned to one Reduce task. According to the defined merging rules, the Reduce function is responsible for merging result sets and outputting the final parallel computation results. Alvaarez-Vanhard et al. integrated ecological remote sensing parameter inversion methods into the MapReduce model parallel framework, using long time series MODIS data to parallelly invert ecological environment parameters in the Sanjiangyuan region, proving to have good parallel acceleration ratio and parallel inversion efficiency (Alvarez-Vanhard et al., n.d.; Griffith & Chun, 2016; Güntner et al., 2007; Tang et al., 2023).Himeur et al. used the MapReduce model to integrate large amounts of remote sensing images from different regions and resolutions into a unified framework for batch parallel construction of tile pyramids, increasing processing efficiency by 39% compared to conventional methods and demonstrating good scalability (Himeur et al., 2022). Shown in Fig. 3.7.

3.6.3 Field programmable gate arrays processing technology

FPGAs have multiple adders and shifters and can set multiple parallel processing channels, suitable for parallel processing algorithms such as image enhancement, noise removal, and edge detection of image pixels, becoming an important research direction in high-performance computing. Kakogeorgiou & Karantzalos designed a remote sensing image parallel processing prototype system based on FPGA, realizing parallel processing of linear stretching, binarization, and other algorithms for remote sensing images (Kakogeorgiou & Karantzalos, 2021; von Eschenbach, 2021). The calculation speed was increased by 3–4 times compared to serial processing on ordinary PCs, and the processing speed of remote sensing image data could reach over 160 MB/s, with a parallel efficiency close to 20%. Belle & Papantonis implemented an image restoration algorithm based on edge extraction and wavelet packet adaptive Wiener filtering using FPGA and applied it to a real-time image restoration processing system on satellites, verifying the possibility of real-time image processing in satellite/airborne scenarios (Belle & Papantonis, n.d.; Shorten & Khoshgoftaar, 2019).

For image advanced processing tasks such as feature extraction, target recognition, and feature information processing, the processing algorithm logic is relatively complex and has some degree of serialization (Torrey & Shavlik, 2009). Simple FPGA processing not only requires a large number of hardware resources but also is difficult to implement. Embedding CPUs on FPGA platforms to form heterogeneous FPGA platforms, separating algorithm control flow and data flow, utilizing FPGA's multi-pipeline parallel structure to accelerate pixel-level processing of images, and using CPUs to perform task scheduling work for complex algorithms is the mainstream technology to solve the parallelization problem of image advanced processing (Ganaie et al., 2022). Xu Jinbo built a hardware acceleration prototype

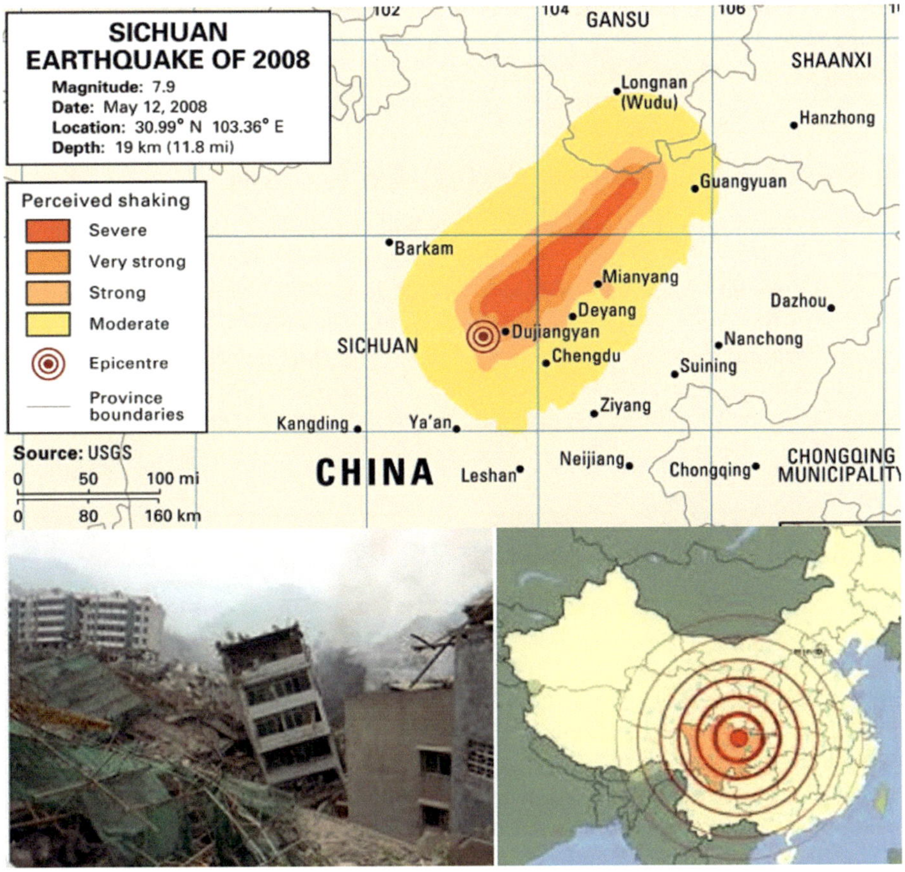

FIGURE 3.7 Map of Dujiangyan during the Wenchuan earthquake.

system on a heterogeneous FPGA platform, realizing pedestrian detection, recognition, and tracking based on active shape model, with significant improvement in the performance of pedestrian target detection and recognition (Chi et al., 2016).

3.6.4 Graphics processing unit processing technology

A graphics processing unit (GPU) can provide more computing units in a limited area, possessing strong parallel computing capabilities, making it a new technique for parallel processing large data blocks. In 2007, NVIDIA introduced the CUDA model, which was the first to use a C-like language for GPU programming development. It also provided a rich function library, allowing developers to develop high-performance parallel processing programs without needing to understand too many details of GPU architecture. This reduced the development difficulty, significantly shortened development time, and promoted the development of GPU processing technology in the field of high-performance computing (Benchaita & Mccarthy, 2023; Wang et al., 2023; Xie et al., 2016).

Currently, research on parallel processing of remote sensing data based on GPUs mainly focuses on some processing algorithms that require a large amount of repetitive computation, such as fast Fourier transform and matrix calculations. The main idea is to first partition the remote sensing data and allocate thread tasks, with each thread responsible for processing one data block (Mehrabi et al., 2021). Through the collaboration of multiple threads, the entire remote sensing image data processing is completed, including algorithms like remote sensing image denoising and digital image orthorectification. For remote sensing image processing algorithms with complex logical judgments, such as geometric correction, it is necessary to first decompose the algorithm logic. In this case, the GPU serves as a coprocessor responsible for computing-intensive data processing, while other task scheduling work is handled by the CPU, achieving CPU-GPU collaboration (Roselli et al., 2019). Raji et al. decomposed the geometric correction algorithm, with the CPU handling satellite attitude and orbit data while the GPU was responsible for establishing grid points and parallel processing of resampling, achieving a processing performance acceleration of around 40 times (Raji et al., 2020). Alkhelaiwi et al. implemented a real-time modulation transfer function (MTF) compensation algorithm based on GPU, where the CPU was responsible for computing MTF curves, constructing point spread functions, and generating two-dimensional MTF matrices, while the GPU parallelized image restoration processing, resulting in a 59.3 times speed improvement for the MTF compensation algorithm (Alkhelaiwi et al., n.d.; Zhang et al., 2012).

3.7 Identification and detection of satellite-borne destination features

Traditionally, remote sensing element classification and recognition rely on "single input and single output" model architectures, which are often insufficient for handling diverse target elements, different modal data, or various classification tasks. In contrast, real-world scenarios demand multi-input and multi-output capabilities akin to human perception systems, integrating multiple inputs to provide comprehensive target information. However, conventional models face challenges in adapting to new scenarios, sharing features across different data types, and managing computational complexity (Poktonjak et al., 2010). To address these issues, current research trends focus on parallel extraction of multifactor target information, leveraging shared learning of multimodal data and multitask features to enhance model complexity and generalization.

3.7.1 Separation of conventional satellite imagery goal features

The extraction of remote sensing target features typically involves two key processes: analyzing data characteristics and designing specialized network models. Understanding the unique characteristics of remote sensing data, such as electromagnetic wave scattering and atmospheric radiation, is crucial for building effective models (Colomina & Molina, 2014). Researchers have developed innovative approaches, including SAR image simulation based on 3D CAD models and generation-countermeasure network models for hyperspectral image classification. Specialized network designs tailored to remote sensing data characteristics, such as rotation convolution-based multisource data extraction networks, have demonstrated significant advancements in semantic segmentation and feature classification tasks. Additionally, efforts to optimize model parameters and streamline operations have led to the development of lightweight networks, ensuring efficient processing of satellite remote sensing data without sacrificing performance.

3.7.2 Concurrent retrieval of multifactor data

Research in parallel extraction methods aims to achieve high-precision acquisition of multi-class remote sensing feature attributes within a unified model framework. Existing methods focus on feature representation of multimodal input data and the sharing and fusion of multitask output features (Jain et al., 2020). Feature sharing techniques, including hard parameter sharing and soft parameter sharing, facilitate the integration of multiple tasks within the same network architecture. These methods enable joint feature representation for diverse tasks, such as flood area detection and forest canopy height estimation, leveraging the complementary information from multi-temporal and multimodal datasets to enhance classification accuracy and performance forest fire in West Mountain of Kunming (local). Shown in Fig. 3.8.

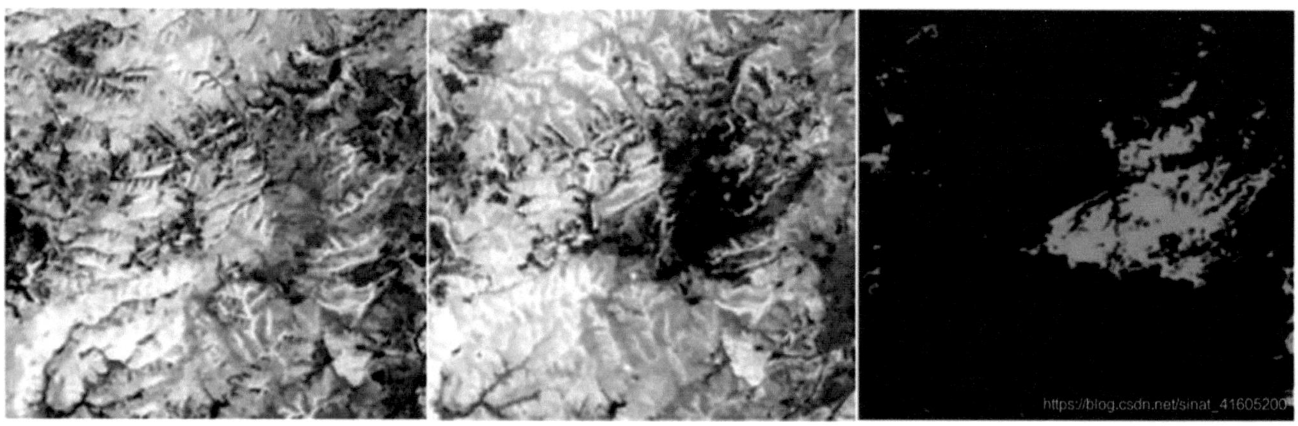

FIGURE 3.8 Remote sensing image and change detection image before and after forest fire in West Mountain of Kunming (local).

3.8 Data analysis applications

3.8.1 Current technological status

Applications are the purpose of obtaining remote sensing data and are a crucial part of demonstrating the value of aerospace remote sensing systems. Remote sensing applications include two major parts: remote sensing information inversion and analysis. The objects processed are preprocessed standard data products. The processing workflow involves interpreting or extracting various features of land objects, such as spatial features, physicochemical features, visual features, and information features, from standard data products, conducting comprehensive analysis, obtaining industry application knowledge, and finally providing users with assistance in decision-making. Analysis and processing methods include data fusion, matching recognition, change detection, and image classification (Bouguettaya et al., 2022).

Data fusion, which integrates multisource remote sensing data, has advantages such as enhancing target features, improving classification accuracy, and enhancing image interpretation capabilities. It is an important method for remote sensing analysis applications. Data fusion is divided into three levels: pixel-level, feature-level, and decision-level. Pixel-level fusion involves fusing information from multiple remote sensing image data pixels. Feature-level fusion considers the correlation between the information of the data itself, extracting feature information from each remote sensing data first, and then analyzing and fusing the feature information to obtain joint feature vectors. Its characteristics are strong pertinence, information compression capabilities, and favorability for real-time data processing. Decision-level fusion first analyzes each remote sensing data using a certain decision rule to obtain preliminary judgment results based on individual data, and then performs global optimization processing on the preliminary judgment results of all data to obtain the final decision (Nguyen et al., 2019). It has the least information content and strong anti-interference capabilities. In addition, pixel-level fusion can only be applied to image-type remote sensing data, while feature-level and decision-level fusion can be applied to all types of remote sensing data.

Matching recognition accurately registers multiple remote sensing image data from different sensors or at different times in spatial positions for target identification and change monitoring. Currently, image matching mainly includes methods based on image grayscale matching and image feature matching (Amani et al., 2020). The former uses the grayscale information of remote sensing data as the measurement basis, employing common measurement criteria such as cross-correlation, sequential similarity, and mutual information. The latter first extracts features from remote sensing data, uses similarity and constraint conditions to determine geometric transformations, and then applies this transformation to the data to be matched. Singular pixel feature, edge feature, and region feature are three commonly used spatial features. In recent years, some nonspatial feature methods have gradually become research hotspots, such as spectral reflectance, radiance or radiosity, absorbance, polarized spectra, contrast, texture, hue, saturation, information density, entropy, vegetation index, land cover index, river turbidity index, and other application features.

Change detection analyzes, processes, and compares remote sensing data of the same area at different times to obtain information on dynamic changes in targets. The processing workflow generally includes image preprocessing, change information discovery, change area extraction, and change type determination. Traditional change detection methods include image algebra, principal component analysis, multiband synthesis, spectral feature variation, and classification detection methods. Garbini studied the fusion of change detection results of remote sensing images at different scales using the Markov random field model, considering the correlation between adjacent pixels and the relationship between different-scale detection results, making the fusion results more detailed and accurate, with good practicality and robustness (Garbini, 2023).

Image classification divides each pixel in a remote sensing image into different categories based on its spectral feature information, spatial structural information, or other information according to certain rules or algorithms. This classification is a process of machine learning based on samples. Therefore, methods such as Bayesian, maximum likelihood, neural networks, fuzzy classification, decision trees, SVMs, and other methods in data mining are widely used in remote sensing image classification (Mujetahid et al., 2023). Object-oriented classification methods have emerged in recent years, where the minimum unit of analysis and processing is no longer individual pixels but objects defined according to certain rules. Since some pixels in remote sensing images usually contain information from multiple land objects, the classification of mixed pixels is an important research content in remote sensing image classification. Linear spectral unmixing is a method of classifying mixed pixels based on the spectral characteristics of materials to determine the relative abundance of materials in multispectral images.

Various types of spatial data managed and analyzed in geographic information systems (GIS) can serve as auxiliary knowledge for remote sensing feature extraction and classification, helping to improve the efficiency and accuracy of remote sensing applications. Deng Shubin et al. constructed a spatial database containing remote sensing images and GIS information based on the GIS platform, reducing the impact of same-material-different-spectrum and different-material-same-spectrum phenomena on the classification of forest-type remote sensing data, and improving classification accuracy. Li Xue et al. used semantic information in GIS-assisted data to integrate and analyze remote sensing data with GIS data, improving the accuracy and reliability of change detection results (González-Rivero et al., n.d.; Lou et al., 2023).

3.8.2 Data mining

Remote sensing data contains rich spectral, texture, shape, structure, topology, orientation, and their combined feature information. Therefore, to some extent, remote sensing applications can be regarded as a type of information application. Data mining aims to achieve the transformation from data to information, from information to knowledge, and from knowledge to wisdom, and it is an important technical means for information application. In recent years, many data mining algorithms, such as decision trees, SVMs, neural networks, etc., have begun to be combined with remote sensing application research, helping to extract more accurate feature information from remote sensing data and conduct comprehensive analysis, understanding, and deeper application. Common data mining algorithms include association rule methods, classification methods, clustering methods, etc.

Association rule methods can discover a large number of existing correlations or rules between datasets. Tang Xiaoping applied the classical Apriori association rule algorithm to mine and analyze meteorological data in the Tibet region, discovering three types of rules related to sand and dust weather: rules consistent with existing meteorological theories; rules that were previously unrecognized or not given attention; and new rules that have not been discovered yet (Ditria et al., 2022).

Classification constructs a classifier based on the characteristics of the dataset to map unknown category data samples into a class. Traditional classifiers include decision trees, Bayesian classifiers, K-nearest neighbors, SVMs, etc. In recent years, algorithms such as neural networks, fuzzy sets, rough sets, cloud theory, genetics, etc., have been proposed to handle large volumes of complex data, such as semantic fuzziness, dynamic changes, and incompleteness, making them hotspots in spatial data mining research. Hasbagan combined neural networks with rough sets, wavelet transforms, fuzzy mean clustering, etc., significantly improving the classification accuracy of traditional neural network classification in remote sensing applications (Shafiq, 2023).

Clustering is an unsupervised learning process that clusters data into classes with different features, minimizing inter-class similarity and maximizing intra-class similarity. Asanobu conducted mining and analysis of global satellite data for typhoon forecasting. They first conducted principal component analysis on the data to reduce its dimensionality, and then used clustering analysis to derive typhoon cloud pattern modes and state transition rules (Silvestro et al., 2022).

3.9 The fourth paradigm

Currently, although remote sensing data has been initially applied in various fields such as economy, society, national defense security, and scientific research, it is still at a relatively low level of data analysis application. For example, in crop monitoring applications, a single type of visible light remote sensing data is first classified to identify crop planting areas, then, based on crop inversion models, vegetation indices reflecting crops are extracted, and finally crop growth is estimated. This application method mainly relies on deterministic causal physical models based on single-source data, ignoring the inherent correlations between multiple data sources and making it difficult to leverage the complementary and verification roles of multiple data sources. In fact, the real world is an open dynamic system, and most phenomena have only a certain degree of correlation with each other, making it difficult to accurately describe them with one or a few deterministic causal relationships. Turing Award winner and renowned database expert Dr. Jim Gray proposed the fourth paradigm, separating big data research from the third paradigm (computational science) as an independent research paradigm. This also prompts us to change traditional causal thinking and re-understand the important role of massive data and explore the mutual correlations between data (Alnuaim et al., 2023; Bragazzi et al., 2020; Toivonen et al., 2019; Tong et al., 2023). Shown in Fig. 3.9.

Currently, Internet companies such as Google, Facebook, Amazon, Baidu, and Alibaba are leaders in big data applications. They only need to discover measures strongly correlated with increased enterprise profits from data mining, without delving into why they can increase profits, let alone discovering the underlying rules and models. Through tracking, analyzing, and mining Internet big data, they can quickly and accurately achieve big data marketing for enterprises and generate commercial value. For example, Google can accurately calculate the return brought by each keyword in advertisements through data mining; eBay reduced advertising expenses for product sales by 99% since 2007 through optimized advertising placements, while the proportion of top sellers in total sales increased to 32%. Therefore, when analyzing massive remote sensing data, there is no need to first establish complex causal relationship models in

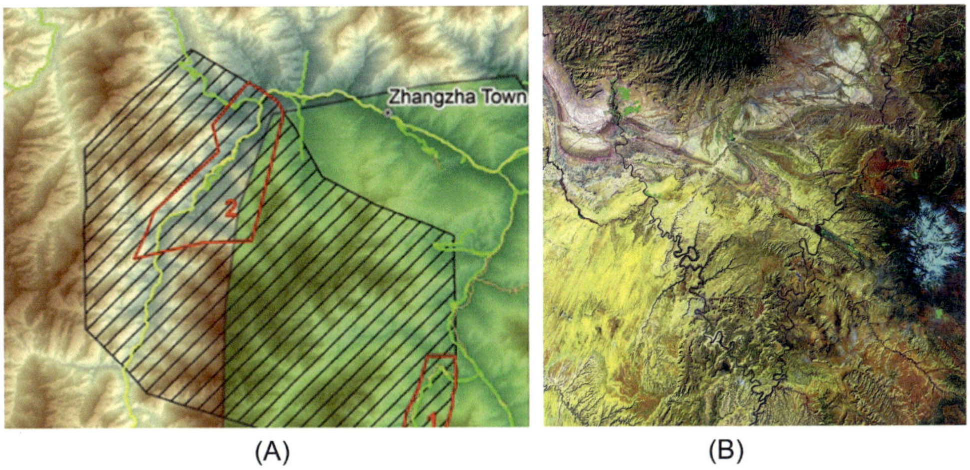

FIGURE 3.9 (A) Satellite view of landslide areas near Zhangzha Town, with green lines marking main roads and red markers indicating landslide locations. (B) Topographic map of the surrounding area, with yellow and brown hues showing elevation changes, and green markers identifying landslide zones.

mathematics. It is only necessary to rely on large amounts of repeated remote sensing information from different data sources, throw massive data into computing clusters with powerful computing capabilities, and extract information and knowledge that traditional remote sensing science methods cannot discover.

3.10 Visualization display

3.10.1 Technical status

In remote sensing application systems, maps with text labels are commonly used to display the results of analysis. However, users often find it difficult to directly, clearly, and effectively understand the meaning of the data results. Three-dimensional visualization technology can display the three-dimensional terrain features of the observed area stereoscopically. It also supports users to rotate, roam, and transform the view from different angles in a three-dimensional scene, providing users with a better experience and enhancing the understandability of the results. Geology is one of the typical application fields of remote sensing three-dimensional visualization. Alnaim et al. combined ASTER multispectral remote sensing information with digital elevation models (DEMs) to achieve three-dimensional visualization of geological bodies such as Earth surface active structures and volcanoes. This not only intuitively displays the three-dimensional spatial terrain and landform features of geological bodies but also enables quantitative analysis related to active structures and volcanoes (Alnaim et al., n.d.; Bhatti, Chen, et al., 2022; Bragazzi et al., 2020; Vaishya et al., 2020).

Extracting three-dimensional information is a prerequisite for three-dimensional visualization. Currently, commonly used multi-angle high-resolution satellite remote sensing stereo image pairs are used to automatically extract the three-dimensional spatial structural information of observed targets. First, the objects in the stereo image pair are matched, and then the three-dimensional spatial features of the objects are extracted. These features are used to describe and manage the three-dimensional entities based on their relationships. The matching of homonymous points in multi-angle remote sensing image pairs is the key to extracting three-dimensional information, and its matching accuracy seriously affects the accuracy of extraction. Liu et al. used the SIFT algorithm to extract feature points from stereo image pairs, selected feature points based on the similarity of feature points, performed region-propagation dense matching on the selected feature points, and then used the symmetric epipolar distance method to eliminate false matches to obtain accurate matching points (Liu et al., n.d.). Sharma et al. used orbit and attitude interpolation methods to determine the initial values of the external orientation elements of multicenter projections, constructed a general stereo measurement mathematical model using forward and backward colinearity equations and error equations, which helps extract three-dimensional DEM data from high-resolution satellite remote sensing stereo image pairs (Sharma et al., 2023). In addition, laser scanning data is also a means of obtaining three-dimensional information about target objects, but it is currently mostly used in aerial remote sensing.

Texture mapping can effectively enhance the detailed features of the surface of objects. Therefore, extracting texture information from remote sensing data and mapping it to the surface of three-dimensional objects are important research topics in remote sensing visualization technology (Stereńczak, 2023). Zhao Wei et al. used the correlation between image data to statistically analyze the grayscale covariation pattern of objects based on the spectral characteristics of objects and recorded them in a grayscale level difference correlation probability matrix (Liu et al., 2022). Finally, they extracted stereo texture features from this matrix.

In addition to displaying the results of remote sensing analysis applications, three-dimensional visualization technology is also commonly used in aerospace remote sensing systems to simulate and simulate the operation orbit, flight direction, areas scanned by sensors on Earth, and working status of remote sensing satellites (Alhatami et al., 2023; Amani et al., 2020). It is widely used in satellite development, satellite mission evaluation, and satellite operation monitoring.

3.10.2 Scientific computing visualization

Scientific computing visualization was the earliest research direction of visualization technology. It uses computer graphics and image processing techniques to display data generated during scientific computing processes and computation results on screens in graphic or image form, supporting interactive processing with users (Lyu et al., 2022). The core of scientific computing visualization technology lies in the visualization of three-dimensional spatial data fields. Mainstream algorithms include surface rendering, volume rendering, and point rendering. Among them, volume rendering algorithms are the most widely used, including indirect rendering methods and direct rendering methods. Direct volume rendering can not only render surfaces but also display internal structures of objects, and it has been widely used. However, direct volume rendering requires a large amount of computational data, and when the viewpoint changes, recalculations are necessary. Hu Huijun proposed a three-dimensional texture volume rendering method based on GPU hardware acceleration characteristics, which improves the drawing ability of the direct volume rendering method based on texture mapping for large-scale data fields (Rahman et al., 2023). Ray casting algorithms have the advantages of high drawing quality, drawing performance independent of the scale of volume data, and no drawing artifacts caused by changes in viewpoints, but they have longer computation complexity and drawing time than other volume rendering algorithms (Alnaim & Sun, 2022). Yang et al. proposed a GPU-based ray casting algorithm, which can complete all resampling operations along the line of sight in a single drawing, avoiding frequent data exchange between the CPU and GPU caused by multi-drawing (Yang, John et al., 2023). By early ray termination and invalid voxel elimination, high-quality real-time drawing of volume data can be achieved.

3.10.3 Information visualization

Unlike scientific computing visualization, information visualization focuses on representing the relationships between data and trends in changes, helping to discover important information hidden behind large amounts of data. Using information visualization technology, GIS can visually and vividly convey the analysis process and results of spatiotemporal data to users, enhancing the understandability of the process and results and providing a basis for decision-making. For example, the United States Geological Survey overlays various disaster information on GIS to simulate and describe the losses that will be suffered under different response measures when disasters such as floods, earthquakes, and forest fires occur, in order to find the best response measures (Al-Bakri et al., n.d.; An & Rui, n.d.; Xiang et al., 2021). China once established a GIS for the SARS epidemic, which used information visualization technology to display the spatial evaluation model and warning map of the SARS epidemic, study the spatiotemporal diffusion patterns of SARS, and simulate the spatiotemporal distribution characteristics and laws of SARS under various possible conditions (Sun & Scanlon, 2019). Cities are one of the typical application areas of GIS information visualization. By overlaying information such as streets, shopping malls, and real estate on GIS, urban information is vividly displayed in a three-dimensional map mode, providing functions such as public transit transfers and driving navigation (Sun, Bocchini, et al., 2020). Maps are another application area of information visualization technology . Compared with paper maps, modern maps overlay various dynamic information such as graphics, images, and sounds, making it easier to express the laws of spatial evolution.

3.10.4 Knowledge visualization

Knowledge visualization is a new research field of visualization developed on the basis of scientific computing visualization and information visualization (Imran et al., 2014). Its purpose is to use visualization technology to promote the dissemination and innovation of knowledge. This also conforms to the development direction of remote sensing technology from data to information and then to knowledge (Gevaert et al., 2021). Since the knowledge extracted from remote sensing data has spatial attributes, the expression process of remote sensing knowledge is more complex, and the expression methods are more diverse. Knowledge graphs have unique advantages in the visualization expression of spatial knowledge. Cao proposed the concept of geographic knowledge visualization and used the visualization of knowledge about dam system planning in the Loess Plateau small watershed as an example to illustrate (Cao, 2023). By studying the forms, types, and relationships between concept proposition graphs and similar graph interpretation graphs in the knowledge graph supporting dam system spatial planning, representing the quantitative relationships between dam height and storage capacity, as well as sedimentation area, it is used to calculate dam height or sedimentation area based on sedimentation years, control area, and soil erosion modulus during dam system planning, or calculate storage capacity based on dam height (Bhatti, Zeeshan, et al., 2022).

3.11 Development trend and prospect

Although AI has achieved good development and application in the field of disaster reduction and remote sensing, there are still many deficiencies and breakthroughs, as follows:

1. At present, the application of AI in disaster reduction and remote sensing mainly depends on the analysis of remote sensing image data. The next step is to fuse and analyze satellite data, ground data and other sensor data, such as UAV data. Using AI methods to analyze meteorological big data and multisource data fusion will obtain more information and better prediction than simply analyzing remote sensing data.
2. Most of the existing intelligent analysis methods still transmit the data to the ground operation center for analysis after obtaining the data. In the future, they will directly conduct rapid analysis on the data acquisition platform, such as direct intelligent analysis on the satellite, and transmit the useful "information" after analysis to the ground, rather than transmitting the original satellite data to the ground for further analysis. This is like installing a "brain" on various sensors to improve the efficiency and response speed of the overall analysis.

Most of the existing artificial intelligence methods are still used to solve the assessment and rescue after disasters, and most of them play a role in assisting the existing disaster prevention, meteorological analysis and other methods. With the continuous development of the field of artificial intelligence, the author suspects that in the future, artificial intelligence will play a leading role in disaster prediction and prevention, and will become the protagonist from a supporting role.

3.12 Conclusion

This paper takes the overall observation, analysis interpretation and rule mining of observation objects in the process of remote sensing data transformation as the main line, and combs the research status and progress in the field of remote sensing data precision processing, remote sensing data spatio-temporal processing and analysis, remote sensing target element classification and recognition, remote sensing data association mining, remote sensing open source datasets and sharing platforms by integrating domestic and foreign literature and relevant reports. First of all, for the task of precise processing of remote sensing data, the research progress of fine processing is reviewed from the aspects of improving the imaging quality of optical, SAR and other remote sensing data and low quality image reconstruction, and the research progress of quantitative improvement is reviewed from the aspects of local feature matching and regional feature matching of remote sensing images. Secondly, for the task of remote sensing data spatio-temporal processing and analysis, the research progress is reviewed from two aspects of remote sensing image time series restoration and multisource remote sensing spatio-temporal fusion. Thirdly, for the classification and recognition task of remote sensing target elements, the research progress is reviewed from two aspects: typical feature extraction and multifeature parallel extraction. Finally, the research progress of remote sensing data association mining is reviewed from two aspects: data organization and association, and the construction of professional knowledge map.

With the rapid increase in the number of remote sensing satellites and the continuous improvement of spatial, temporal, and spectral resolutions, the characteristics of large-scale remote sensing data, such as massive scale, complex structure, and rapid growth rate of data volume, have become increasingly obvious, bringing enormous challenges to key links in aerospace remote sensing systems such as satellite-ground data transmission, data storage management, data preprocessing, data analysis applications, and result visualization. Currently, big data has become a common research hotspot in data-intensive scientific fields, providing solutions to the big data challenges in aerospace remote sensing systems: (1) In the transmission link, the combination of compressed sensing theory, LDPC codes, Turbo codes, and modulation technologies such as frequency division multiplexing can improve the utilization and reliability of large data transmission in limited channels. Atmospheric laser communication has higher channel capacity than wireless short-wave communication as an emerging communication method. On-board processing technology can extract effective information from remote sensing observation data in real time and reduce the amount of data transmission. (2) In the storage link, cloud storage technology and NoSQL database technology can meet the needs of reliable storage, efficient management, and rapid growth of massive unstructured remote sensing data. (3) In the preprocessing link, high-performance computing technologies represented by MapReduce, FPGA, and GPU can provide strong data parallel processing capabilities, accelerate data processing speed, and meet the real-time processing requirements of massive remote sensing data. (4) In the analysis application link, data mining methods and the scientific thinking of the fourth paradigm help to discover more valuable information and knowledge from massive remote sensing data. (5) In the visualization display link, intuitive and vivid visualization technology helps to increase the understandability of remote sensing data processing processes and analysis results, effectively supporting user judgment and decision-making.

However, compared with other types of big data, spatial data also has the characteristics of massive data volume, complex data semantics, obvious scale characteristics, and incomplete data. The massive data volume makes it impossible to implement some conventional big data research algorithms due to excessive computational load. The complex data semantics are manifested in fuzzy attributes, nonlinear relationships between data, etc., reflecting the complexity and openness of the internal mechanisms of the Earth system. Scale characteristics can be used to study the characteristic gradient rules reflected in the process of spatial data generalization and refinement, but it can also lead to an increase in the dimensionality of spatial data. Incomplete data occurs when external forces make spatial data inaccessible or lost, resulting in incomplete spatial data. Therefore, it is necessary to reform or create new big data research methods based on the characteristics of spatial data to overcome the technical difficulties caused by the complexity of spatial data. In summary, research methods in hot areas such as information communication, massive storage, high-performance computing, spatial data mining, and visualization can help solve the big data challenges in aerospace remote sensing systems such as rapid acquisition, efficient storage, in-depth applications, and intuitive and vivid display of data, bringing new opportunities for the development of aerospace remote sensing systems. In addition, this paper also reviews the research progress of remote sensing open source datasets and sharing platforms for the development of large intelligent analysis technology. On this basis, the research on intelligent analysis and interpretation of remote sensing data is sorted out and summarized, and the future development trend and prospect of this field are given.

Compared with traditional imaging methods, satellite remote sensing payload systems based on artificial intelligence technology have the following advantages: (1) on-demand imaging, with stronger pertinence; (2) low-resolution perception capability and high-resolution detailed inspection capability for key targets; (3) reduced downlink data volume, reducing satellite information transmission pressure; (4) high information utilization timeliness; (5) learning and evolutionary capabilities. Satellite remote sensing payload systems based on artificial intelligence technology can systematically design the entire chain of intelligence collection, processing, distribution, and application of information from the source of data acquisition, improving the timeliness and usefulness of information, and providing effective support for users to obtain information on demand and in a timely manner. They can be applied in wide-range remote sensing imaging, automatic search, identification, and positioning of interested targets, dynamic/static target tracking and detailed inspection, with autonomous perception and intelligence processing capabilities. They have significant application value in on-orbit immediate response, rapid location and interpretation of sudden events, rapid capture and tracking of moving targets, and other fields.

References

Agarap, A.F. (n.d.). Deep learning using rectified linear units (relu). *arXiv 2018*.

AI Enables Rapid Creation of Global Land Cover Map. (2021).

Al-Bakri, J.T., Urso, G., Calera, A., Abdalhaq, E., Altarawneh, M., & Margane, A. (n.d.). Remote sensing for agricultural water management in Jordan.

Aleissaee, A.A., Kumar, A., Anwer, R.M., Khan, Cholakkal, H., Xia, G.-S., & Khan, F.S. (n.d.). Transformers in remote sensing: A survey.

Alhatami, E., Bhatti, U. A., Huang, M. X., & SiLing, F. (2023). *Review and enhancement of discrete cosine transform (DCT) for medical image fusion* (614, pp. 89–97). Springer Science and Business Media LLC. Available from 10.1007/978-981-19-9331-2_8.

Alkhelaiwi, M., Boulila, W., Ahmad, J., Koubaa, A., & Driss, M. (n.d.). An efficient approach based on privacy-preserving deep learning for satellite image classification.

Allison, R., Johnston, J., Craig, G., & Jennings, S. (2016). Airborne optical and thermal remote sensing for wildfire detection and monitoring. *Sensors, 16*(8), 1310. Available from https://doi.org/10.3390/s16081310.

Alnaim, A., & Sun, Z. (2022). Using geoweaver to make snow mapping workflow FAIR, 1 2022/01*Proceedings – 2022 IEEE 18th international conference on e-Science, eScience 2022*, 409–410. Available from https://doi.org/10.1109/eScience55777.2022.00062, http://ieeexplore.ieee.org/xpl/mostRecentIssue.jsp?punumber = 9973400.

Alnaim, A., Sun, Z., Tong, D. (n.d.). Evaluating machine learning and remote sensing in monitoring NO_2 emission of power plants.

Alnuaim, A., Sun, Z., & Islam, D. (2023). *AI for improving ozone forecasting. Artificial intelligence in earth science: Best practices and fundamental challenges* (pp. 247–269). Elsevier. Available from https://www.sciencedirect.com/book/9780323917377, https://doi.org/10.1016/B978-0-323-91737-7.00002-5.

Alom, M.Z., Taha, T.M., Yakopcic, C., Westberg, S., Sidike, P., Nasrin, M.S., Esesn, Awwal, A.A. S., Asari, V.K. (n.d.). The history began from alexnet: A comprehensive survey on deep learning approaches. *arXiv 2018.*

Aloysius, N., & Geetha, M. (2017). A review on deep convolutional neural networks. *Proceedings of the 2017 IEEE international conference on communication and signal processing, ICCSP 2017, 2018*, 588–592. Available from https://doi.org/10.1109/ICCSP.2017.8286426.

Alvarez-Vanhard, E., Corpetti, T., & Houet, T. (n.d.). UAV & satellite synergies for optical remote sensing applications: A literature review.

Alzubaidi, L., Zhang, J., Humaidi, A. J., Al-Dujaili, A., Duan, Y., Al-Shamma, O., Santamaría, J., Fadhel, M. A., Al-Amidie, M., & Farhan, L. (2021). Review of deep learning: Concepts, CNN architectures, challenges, applications, future directions. *Journal of Big Data*, *8*(1). Available from https://doi.org/10.1186/s40537-021-00444-8, https://journalofbigdata.springeropen.com.

Amani, M., Ghorbanian, A., Ahmadi, S. A., Kakooei, M., Moghimi, A., Mirmazloumi, S. M., Moghaddam, S. H. A., Mahdavi, S., Ghahremanloo, M., Parsian, S., Wu, Q., & Brisco, B. (2020). Google earth engine cloud computing platform for remote sensing big data applications: A comprehensive review. *IEEE Journal of Selected Topics in Applied Earth Observations and Remote Sensing*, *13*, 5326–5350. Available from https://doi.org/10.1109/JSTARS.2020.3021052, http://ieeexplore.ieee.org/xpl/RecentIssue.jsp?punumber = 4609443.

An, & Rui, X. (n.d.). A high-precision water body extraction method based on improved lightweight U-net.

Ankan, D., Ye, J., & Wang, G. (n.d.). A review of generative adversarial networks (GANs) and its applications in a wide variety of disciplines—From medical to remote sensing.

Arulkumaran, K., Deisenroth, M. P., Brundage, M., & Bharath, A. A. (2017). Deep reinforcement learning: A brief survey. *IEEE Signal Processing Magazine*, *34*(6), 26–38. Available from https://doi.org/10.1109/MSP.2017.2743240, http://ieeexplore.ieee.org/xpl/RecentIssue.jsp?punumber = 79&year = 2008.

Badrinarayanan, V., Kendall, A., & Cipolla, R. (2017). SegNet: A deep convolutional encoder-decoder architecture for image segmentation. *IEEE Transactions on Pattern Analysis and Machine Intelligence*, *39*(12), 2481–2495. Available from https://doi.org/10.1109/TPAMI.2016.2644615.

Behnamian, A., Banks, S., White, L., Millard, K., Pouliot, D., Pasher, J., & Duffe, J. (2019). Dimensionality reduction in the presence of highly correlated variables for random forests: Wetland case study. *International geoscience and remote sensing symposium (IGARSS)*, 9839–9842. Available from https://doi.org/10.1109/IGARSS.2019.8898308.

Belgiu, M., & Drăgu, L. (2016). Random forest in remote sensing: A review of applications and future directions. *ISPRS Journal of Photogrammetry and Remote Sensing*, *114*, 24–31. Available from https://doi.org/10.1016/j.isprsjprs.2016.01.011, http://www.elsevier.com/inca/publications/store/5/0/3/3/4/0.

Belle, V., & Papantonis, I. (n.d.). Principles and practice of explainable machine learning.

Benchaita, S., & Mccarthy, B. H. (2023). *NASA IBM, open source largest geospatial AI foundation model on hugging face*. IBM Newsroom.

Bhatti, M., Huang, N.-M., Harold, M., Shah, B., Mehmood, H., Tang, W., & Sibghat, B. (2023). MFFCG – Multi feature fusion for hyperspectral image classification using graph attention network. *Expert Systems with Applications*. Available from https://doi.org/10.1016/j.eswa.2023.120496.

Bhatti, U., Wu, G., Bazai, S. U., Nawaz, S. A., Baryalai, M., Bhatti, M. A., Hasnain, A., & Nizamani, M. M. (2022). A pre- to post-COVID-19 change of air quality patterns in Anhui province using path analysis and regression. *Polish Journal of Environmental Studies*, *31*(5), 4029–4042. Available from https://doi.org/10.15244/pjoes/148065.

Bhatti, U. A., Bazai, S. U., Hussain, S., Fakhar, S., Ku, C. S., Marjan, S., Yee, P. L., & Jing, L. (2023). Deep learning-based trees disease recognition and classification using hyperspectral data. *Computers, Materials & Continua*, *77*(1), 681–697. Available from https://doi.org/10.32604/cmc.2023.037958.

Bhatti, U. A., Bhatti, M. A., Tang, H., Syam, M. S., Awwad, E. M., Sharaf, M., & Ghadi, Y. Y. (2024). Global production patterns: Understanding the relationship between greenhouse gas emissions, agriculture greening and climate variability. *Environmental Research*, *245*, 118049. Available from https://doi.org/10.1016/j.envres.2023.118049.

Bhatti, U. A., Chen, Y. W., Huang, M., Zheng, J., Li, J., & Liu, J. (2022). A novel convolution kernel-based robust watermarking scheme applied in medical image. *International Journal of Wireless and Mobile Computing*, *22*(3/4), 290. Available from https://doi.org/10.1504/ijwmc.2022.10049464.

Bhatti, U. A., Hashmi, M. Z., Sun, Y., Masud, M., & Nizamani, M. M. (2023). Editorial: Artificial intelligence applications in reduction of carbon emissions: Step towards sustainable environment. *Frontiers in Environmental Science*, *11*. Available from https://doi.org/10.3389/fenvs.2023.1183620.

Bhatti, U. A., Marjan, S., Wahid, A., Syam, M. S., Huang, M., Tang, H., & Hasnain, A. (2023). The effects of socioeconomic factors on particulate matter concentration in China's: New evidence from spatial econometric model. *Journal of Cleaner Production*, *417*, 137969. Available from https://doi.org/10.1016/j.jclepro.2023.137969.

Bhatti, U. A., Masud, M., Bazai, S. U., & Tang, H. (2023). Editorial: Investigating AI-based smart precision agriculture techniques. *Frontiers in Plant Science*, *14*. Available from https://doi.org/10.3389/fpls.2023.1237783.

Bhatti, U. A., Nizamani, M. M., & Mengxing, H. (2022). Climate change threatens Pakistan's snow leopards. *Science (New York, N.Y.)*, *377*(6606), 585–586. Available from https://doi.org/10.1126/science.add9065. Available from, https://www.science.org/doi/10.1126/science.add9065.

Bhatti, U. A., Tang, H., Wu, G., Marjan, S., Hussain, A., & Sarker, S. K. (2023). Deep learning with graph convolutional networks: An overview and latest applications in computational intelligence. *International Journal of Intelligent Systems*, *2023*, 1–28. Available from https://doi.org/10.1155/2023/8342104.

Bhatti, U. A., Tang, H., & Wu, S. (2023). Mangrove decline puts Pakistan's coasts at risk. *Science (New York, N.Y.)*, *382*(6671), 654–655. Available from https://doi.org/10.1126/science.adl3073.

Bhatti, U. A., Yu, Z., Yuan, L., Nawaz, S. A., Aamir, M., & Bhatti, M. A. (2022). A robust remote sensing image watermarking algorithm based on region-specific SURF. *Lecture notes in networks and systems*, *350*, 75–85. Available from https://doi.org/10.1007/978-981-16-7618-5_7, http://springer.com/series/15179.

Bhatti, U. A., Zeeshan, Z., Nizamani, M. M., Bazai, S., Yu, Z., & Yuan, L. (2022). Assessing the change of ambient air quality patterns in Jiangsu Province of China pre-to post-COVID-19. *Chemosphere*, *288*, 132569. Available from https://doi.org/10.1016/j.chemosphere.2021.132569.

Bouguettaya, A., Zarzour, H., Taberkit, A. M., & Kechida, A. (2022). A review on early wildfire detection from unmanned aerial vehicles using deep learning-based computer vision algorithms. *Signal Processing*, *190*, 108309. Available from https://doi.org/10.1016/j.sigpro.2021.108309.

Bragazzi, N. L., Dai, H., Damiani, G., Behzadifar, M., Martini, M., & Wu, J. (2020). How big data and artificial intelligence can help better manage the covid-19 pandemic. *International Journal of Environmental Research and Public Health*, *17*(9). Available from https://doi.org/10.3390/ijerph17093176, https://www.mdpi.com/1660-4601/17/9/3176/pdf.

Cannistra, A. F., Shean, D. E., & Cristea, N. C. (2021). High-resolution CubeSat imagery and machine learning for detailed snow-covered area. *Remote Sensing of Environment*, *258*. Available from https://doi.org/10.1016/j.rse.2021.112399. Available from, http://www.elsevier.com/inca/publications/store/5/0/5/7/3/3.

Cao, L. (2023). AI and data science for smart emergency, crisis and disaster resilience. *International Journal of Data Science and Analytics*, *15*(3), 231–246. Available from https://doi.org/10.1007/s41060-023-00393-w, https://www.springer.com/journal/41060.

Chen, L., Bai, L.Q., Yang, J., Jiang, S., & Miao, Y. (n.d.). Review of image classification algorithms based on convolutional neural networks.

Chen, L.-C., Papandreou, G., Kokkinos, I., Murphy, K., & Yuille, A.L. (n.d.). DeepLab, semantic image segmentation with deep convolutional nets, atrous convolution, and fully connected CRFs. *arXiv 2016*.

Chen, T., & Carlos, G. (2016). XGBoost, a scalable tree boosting system. *arXiv*.

Chi, M., Plaza, A., Benediktsson, J. A., Sun, Z., Shen, J., & Zhu, Y. (2016). Big data for remote sensing: Challenges and opportunities. *Proceedings of the IEEE*, *104*(11), 2207–2219. Available from https://doi.org/10.1109/JPROC.2016.2598228, http://ieeexplore.ieee.org/xpl/RecentIssue.jsp?punumber = 5.

Colomina, I., & Molina, P. (2014). Unmanned aerial systems for photogrammetry and remote sensing: A review. *ISPRS Journal of Photogrammetry and Remote Sensing*, *92*, 79–97. Available from https://doi.org/10.1016/j.isprsjprs.2014.02.013, http://www.elsevier.com/inca/publications/store/5/0/3/3/4/0.

Creswell, A., White, T., Dumoulin, V., Arulkumaran, K., Sengupta, B., & Bharath, A. A. (2018). Generative adversarial networks: An overview. *IEEE Signal Processing Magazine.*, *35*(1), 53–65. Available from https://doi.org/10.1109/MSP.2017.2765202. Available from, http://ieeexplore.ieee.org/xpl/RecentIssue.jsp?punumber = 79&year = 2008.

Dassot, M., Constant, T., & Fournier, M. (2011). The use of terrestrial LiDAR technology in forest science: Application fields, benefits and challenges. *Annals of Forest Science*, *68*(5), 959–974. Available from https://doi.org/10.1007/s13595-011-0102-2.

Deems, J. S., Painter, T. H., & Finnegan, D. C. (2013). Lidar measurement of snow depth: A review. *Journal of Glaciology*, *59*(215), 467–479. Available from https://doi.org/10.3189/2013JoG12J154UnitedStates, http://docserver.ingentaconnect.com/deliver/connect/igsoc/00221430/v59n215/s6.pdf?expires = 1381213168&id = 75755663&titleid = 6497&accname = Elsevier + BV&checksum = 4D668C9947CA5FF7DBAAFAF11A4C2792.

Devaney, J., Barrett, B., Barrett, F., Redmond, J., O'Halloran, J., & Bond-Lamberty, B. (2015). Forest cover estimation in Ireland using radar remote sensing: A comparative analysis of forest cover assessment methodologies. *PLoS One*, *10*(8), e0133583. Available from https://doi.org/10.1371/journal.pone.0133583.

Devlin, J., Chang, M.W., Lee, K., & Toutanova, K. (n.d.). Bert, pre-training of deep bidirectional transformers for language understanding. *arXiv 2018*.

Dinh, H.T. M., Hanssen, R., & Rocca, F. (n.d.). Radar interferometry: 20 years of development in time series techniques and future perspectives.

Disney, M. I., Kalogirou, V., Lewis, P., Prieto-Blanco, A., Hancock, S., & Pfeifer, M. (2010). Simulating the impact of discrete-return lidar system and survey characteristics over young conifer and broadleaf forests. *Remote Sensing of Environment*, *114*(7), 1546–1560. Available from https://doi.org/10.1016/j.rse.2010.02.009.

Ditria, E. M., Buelow, C. A., Gonzalez-Rivero, M., & Connolly, R. M. (2022). Artificial intelligence and automated monitoring for assisting conservation of marine ecosystems: A perspective. *Frontiers in Marine Science*, *9*. Available from https://doi.org/10.3389/fmars.2022.918104, https://www.frontiersin.org/journals/marine-science#.

Dong, P., & Chen, Q. (2017). *LiDAR remote sensing and applications. LiDAR remote sensing and applications* (pp. 1–220). CRC Press. Available from https://www.routledge.com/LiDAR-Remote-Sensing-and-Applications/Dong-Chen/p/book/9781138747241, https://doi.org/10.4324/9781351233354.

Dubey, A. K., & Jain, V. (2019). Comparative study of convolution neural network's relu and leaky-relu activation functions. *Lecture notes in electrical engineering*, 873–880. Available from https://doi.org/10.1007/978-981-13-6772-4_76, http://www.springer.com/series/7818.

Dörnhöfer, K., & Oppelt, N. (2016). Remote sensing for lake research and monitoring – Recent advances. *Ecological Indicators*, *64*, 105–122. Available from https://doi.org/10.1016/j.ecolind.2015.12.009. Available from, http://www.elsevier.com/locate/ecolind.

Earthdata Cloud Evolution. Earthdata. (2022).

von Eschenbach, W. J. (2021). Transparency and the black box problem: Why we do not trust AI. *Philosophy and Technology*, *34*(4), 1607–1622. Available from https://doi.org/10.1007/s13347-021-00477-0, http://www.springer.com/philosophy/epistemology + and + philosophy + of + science/journal/1334.

Esen, B., & Wentworth, J. (2020). Remote sensing and machine learning. *Parliament*.

Filar, J., & Vrieze, K. (2012). *Competitive Markov decision processes*. Springer Science & Business Media.

Floodly, A.I. (2021). Esa, Int.

Fu, K., Li, Y., Sun, H., Yang, X., Xu, G., Li, Y., Sun, X., & Ship, A. (1922). Rotation detection model in remote sensing images based on feature fusion pyramid network and deep reinforcement learning. *Remote Sensing, 10.*

Ganaie, M. A., Hu, M., Malik, A. K., Tanveer, M., & Suganthan, P. N. (2022). Ensemble deep learning: A review. *Engineering Applications of Artificial Intelligence.*

Gao, F., Hilker, T., Zhu, X., Anderson, M., Masek, J., Wang, P., & Yang, Y. (2015). Fusing landsat and MODIS data for vegetation monitoring. *IEEE Geoscience and Remote Sensing Magazine*, *3*(3), 47–60. Available from https://doi.org/10.1109/MGRS.2015.2434351. Available from, http://ieeexplore.ieee.org/xpl/RecentIssue.jsp?punumber = 6245518.

Garbini, S. (2023). How geospatial AI can help you comply with EU's deforestation law—Customers. *Picterra.*

Gemstone Case Space Know, Study. (2022). Global economic monitoring using satellite data and AI/ML technology.

Georganos, S., Grippa, T., Vanhuysse, S., Lennert, M., Shimoni, M., Kalogirou, S., & Wolff, E. (2018). Less is more: Optimizing classification performance through feature selection in a very-high-resolution remote sensing object-based urban application. *GI Science and Remote Sensing*, *55*(2), 221–242. Available from https://doi.org/10.1080/15481603.2017.1408892, http://www.tandfonline.com/loi/tgrs20?open = 50&repitition = 0#vol_50.

Gevaert, C. M., Carman, M., Rosman, B., Georgiadou, Y., & Soden, R. (2021). Fairness and accountability of AI in disaster risk management: Opportunities and challenges. *Patterns*, *2*(11), 100363. Available from https://doi.org/10.1016/j.patter.2021.100363.

Ghatkar, J. G., Singh, R. K., & Shanmugam, P. (2019). Classification of algal bloom species from remote sensing data using an extreme gradient boosted decision tree model. *International Journal of Remote Sensing*, *40*(24), 9412–9438. Available from https://doi.org/10.1080/01431161.j2019.1633696, https://www.tandfonline.com/loi/tres20.

Ghosh, A., Fassnacht, F. E., Joshi, P. K., & Kochb, B. (2014). A framework for mapping tree species combining hyperspectral and LiDAR data: Role of selected classifiers and sensor across three spatial scales. *International Journal of Applied Earth Observation and Geoinformation*, *26*(1), 49–63. Available from https://doi.org/10.1016/j.jag.2013.05.017, http://www.elsevier.com/locate/jag.

Girshick, R. (2015). Fast R-CNN. *Proceedings of the IEEE international conference on computer vision*, 1440–1448. Available from https://doi.org/10.1109/ICCV.2015.169, http://ieeexplore.ieee.org/xpl/conhome.jsp?punumber = 1000149.

González-Rivero, M., Beijbom, O., Rodriguez-Ramirez, A., Bryant, D.E., Ganase, A., Gonzalez-Marrero, Y., Herrera-Reveles, A., Kennedy, E.V., Kim, C.J., Lopez-Marcano, et al. (n.d.). Monitoring of coral reefs using artificial intelligence: A feasible and cost-effective approach.

Goodfellow, I., Bengio, Y., & Courville, A. (2016). *Deep learning.* MIT Press.

Goodfellow, I.J., Pouget-Abadie, J., Mirza, M., Xu, B., Warde-Farley, D., A. Ozair, Courville, & Bengio, Y. (2014). Generative adversarial nets. *Neurips.Cc.*

Graves, A. (2012). *Long short-term memory* (pp. 37–45). Springer Science and Business Media LLC. Available from http://doi.org/10.1007/978-3-642-24797-2_4.

Griffith, D. A., & Chun, Y. (2016). Spatial autocorrelation and uncertainty associated with remotely-sensed data. *Remote Sensing*, *8*(7). Available from https://doi.org/10.3390/rs8070535. Available from, http://www.mdpi.com/journal/remotesensing.

Güntner, A., Stuck, J., Werth, S., Döll, P., Verzano, K., & Merz, B. (2007). A global analysis of temporal and spatial variations in continental water storage. *Water Resources Research*, *43*(5). Available from https://doi.org/10.1029/2006WR005247.

He, J., Zhao, L., Yang, H., Zhang, M., & Li, W. (2020). HSI-BERT: Hyperspectral image classification using the bidirectional encoder representation from transformers. *IEEE Transactions on Geoscience and Remote Sensing*, *58*(1), 165–178. Available from https://doi.org/10.1109/TGRS.2019.2934760, https://ieeexplore.ieee.org/servlet/opac?punumber = 36.

He, K., Zhang, X., Ren, S., & Sun, J. (2016). Deep residual learning for image recognition. *Proceedings of the IEEE computer society conference on computer vision and pattern recognition*, 770–778. Available from https://doi.org/10.1109/CVPR.2016.90.

Himeur, Y., Rimal, B., Tiwary, A., & Amira, A. (2022). Using artificial intelligence and data fusion for environmental monitoring: A review and future perspectives. *Information Fusion*, *86–87*, 44–75. Available from https://doi.org/10.1016/j.inffus.2022.06.003, http://www.elsevier.com/inca/publications/store/6/2/0/8/6/2/index.htt.

Holland, S., Hosny, A., Newman, S., Joseph, J., & Chmielinski, K. (n.d.). The dataset nutrition label.

Hong, D., He, W., Yokoya, N., Yao, J., Gao, L., Zhang, L., Chanussot, J., & Zhu, X. (2021). Interpretable hyperspectral artificial intelligence: When nonconvex modeling meets hyperspectral remote sensing. *IEEE Geoscience and Remote Sensing Magazine*, *9*(2), 52–87. Available from https://doi.org/10.1109/MGRS.2021.3064051, http://ieeexplore.ieee.org/xpl/RecentIssue.jsp?punumber = 6245518.

Hu, A., Xie, Z., Xu, Y., Xie, M., Wu, L., & Qiu, Q. (2020). Unsupervised haze removal for high-resolution optical remote-sensing images based on improved generative adversarial networks. *Remote Sensing*, *12*(24), 4162. Available from https://doi.org/10.3390/rs12244162.

Imran, M., Castillo, C., Lucas, J., Meier, P. & Vieweg, S. (2014). AIDR: Artificial intelligence for disaster response. *WWW 2014 companion – Proceedings of the 23rd international conference on world wide web* (pp. 159–162). Available from https://doi.org/10.1145/2567948.2577034.

Isola, P., Zhu, J.Y., Zhou, T., & Efros, A.A. (2017). Image-to-image translation with conditional adversarial networks 2017. *Proceedings – 30th IEEE conference on computer vision and pattern recognition*. CVPR. Available from https://doi.org/10.1109/CVPR.2017.632 5967 5976.

Jain, P., Coogan, S. C. P., Subramanian, S. G., Crowley, M., Taylor, S., & Flannigan, M. D. (2020). A review of machine learning applications in wildfire science and management. *Environmental Reviews*, *28*(4), 478–505. Available from https://doi.org/10.1139/er-2020-0019, https://www.nrcresearchpress.com/loi/er.

Jiang, P., Ergu, D., Liu, F., Cai, Y., & Ma, B. (2022). A review of yolo algorithm developments. *Procedia Computer Science*, *199*, 1066–1073. Available from https://doi.org/10.1016/j.procs.2022.01.135.

John, A., Cannistra, A.F., Yang, K., Tan, A., Shean, D., Lambers, J.H. R., & Cristea, N. (n.d.). High-resolution snow-covered area mapping in forested mountain ecosystems using planet scope imagery.

Jozdani, S., Chen, D., Pouliot, D., & Johnson, B. A. (2022). A review and meta-analysis of generative adversarial networks and their applications in remote sensing. *International Journal Applied Earth Observation and Geoinformation*.

Kakogeorgiou, I., & Karantzalos, K. (2021). Evaluating explainable artificial intelligence methods for multi-label deep learning classification tasks in remote sensing. *International Journal of Applied Earth Observation and Geoinformation*, *103*, 102520. Available from https://doi.org/10.1016/j.jag.2021.102520.

Kamarulzaman, A.M. M., Jaafar, W.S. W.M., Said, M.N. M., Saad, S.N. M., & Mohan, M. (n.d.). UAV implementations in urban planning and related sectors of rapidly developing nations: A review and future perspectives for Malaysia.

Lary, D. J., Alavi, A. H., Gandomi, A. H., & Walker, A. L. (2016). Machine learning in geosciences and remote sensing. *Geoscience Frontiers*, *7*(1), 3–10. Available from https://doi.org/10.1016/j.gsf.2015.07.003, https://www.sciencedirect.com/journal/geoscience-frontiers.

Ledig, C., Theis, L., Huszár, F., Caballero, J., Cunningham, A., Acosta, A., Aitken, A., Tejani, A., Totz, J., Wang, Z., & Shi, W. (2017). Photo-realistic single image super-resolution using a generative adversarial network. *Proceedings – 30th IEEE conference on computer vision and pattern recognition*. CVPR. Available from https://doi.org/10.1109/CVPR.2017.19 105 114.

Li, H., Wu, Xj, & Durrani, T. S. (2019). Infrared and visible image fusion with ResNet and zero-phase component analysis. *Infrared Physics and Technology*, *102*. Available from https://doi.org/10.1016/j.infrared.2019.103039, https://www.journals.elsevier.com/infrared-physics-and-technology.

Li, Y., Zhang, H., Xue, X., Jiang, Y., & Shen, Q. (2018). Deep learning for remote sensing image classification: A survey. *WIREs Data Mining and Knowledge Discovery*, 1264.

Lin, D., Fu, K., Wang, Y., Xu, G., & Sun, X. (2017). MARTA GANs: Unsupervised representation learning for remote sensing image classification. *IEEE Geoscience and Remote Sensing Letters*, *14*(11), 2092–2096. Available from https://doi.org/10.1109/LGRS.2017.2752750, http://ieeexplore.ieee.org/xpl/RecentIssue.jsp?punumber = 8859.

Liu, L., Zhang, Q., Guo, Y., Chen, E., Li, Z., Li, Y., Wang, B., & Ri, A. (n.d.). Mapping the distribution and dynamics of coniferous forests in large areas from 1985 to 2020 combining deep learning and google earth engine.

Liu, W., Quijano, K., & Crawford, M. M. (2022). YOLOv5-tassel: Detecting tassels in RGB UAV Imagery with improved YOLOv5 based on transfer learning. *IEEE Journal of Selected Topics in Applied Earth Observations and Remote Sensing*, *15*, 8085–8094. Available from https://doi.org/10.1109/JSTARS.2022.3206399, http://ieeexplore.ieee.org/xpl/RecentIssue.jsp?punumber = 4609443.

Liu, X., Wang, Y., & Liu, Q. (2018). Psgan: A generative adversarial network for remote sensing image pan-sharpening. *Proceedings – International conference on image processing* (pp. 873–877). ICIP. Available from https://doi.org/10.1109/ICIP.2018.8451049.

Liu, Z., Li, J., Ashraf, M., Syam, M. S., Asif, M., Awwad, E. M., Al-Razgan, M., & Bhatti, U. A. (2024). Remote sensing-enhanced transfer learning approach for agricultural damage and change detection: A deep learning perspective. *Big Data Research*, *36*, 100449. Available from https://doi.org/10.1016/j.bdr.2024.100449.

Lou, R., Lv, Z., Dang, S., Su, T., & Li, X. (2023). Application of machine learning in ocean data. *Multimedia Systems*, *3*, 1815–1824. Available from https://doi.org/10.1007/s00530-020-00733-x, https://www.springer.com/journal/530.

Lyu, F., Wang, S., Han, S. Y., Catlett, C., & Wang, S. (2022). An integrated cyberGIS and machine learning framework for fine-scale prediction of Urban Heat Island using satellite remote sensing and urban sensor network data. *Urban Inform*, 1–15.

Ma, L., Liu, Y., Zhang, X., Ye, Y., Yin, G., & Johnson, B. A. (2019). Deep learning in remote sensing applications: A meta-analysis and review. *ISPRS Journal of Photogrammetry and Remote Sensing*, *152*, 166–177. Available from https://doi.org/10.1016/j.isprsjprs.2019.04.015, http://www.elsevier.com/inca/publications/store/5/0/3/3/4/0.

Manolakis, D. G., Lockwood, R. B., & Cooley, T. W. (2016). *Hyperspectral imaging remote sensing: Physics, sensors, and algorithms*. Cambridge University Press. Available from http://doi.org/10.1017/CBO9781316017876.

Mascaro, J., Asner, G. P., Knapp, D. E., Kennedy-Bowdoin, T., Martin, R. E., Anderson, C., Higgins, M., Chadwick, K. D., & Bond-Lamberty, B. (2014). A tale of two "forests": Random forest machine learning aids tropical forest carbon mapping. *PLoS One*, *9*(1), e85993. Available from https://doi.org/10.1371/journal.pone.0085993.

Maxwell, A. E., Warner, T. A., & Fang, F. (2018). Implementation of machine-learning classification in remote sensing: An applied review. *International Journal of Remote Sensing*, *39*(9), 2784–2817. Available from https://doi.org/10.1080/01431161.2018.1433343, http://www.tandf.co.uk/journals/titles/01431161.asp.

Mehrabi, N., Morstatter, F., Saxena, N., Lerman, K., & Galstyan, A. (2021). A survey on bias and fairness in machine learning. *ACM Computing Surveys*, *54*(6). Available from https://doi.org/10.1145/3457607. Available from, http://dl.acm.org/citation.cfm?id = J204.

Methodology & Accuracy Summary 10m Global Land Use Land Cover Maps (2022). Impactobservatory.com.

Miao, X., Heaton, J. S., Zheng, S., Charlet, D. A., & Liu, H. (2012). Applying tree-based ensemble algorithms to the classification of ecological zones using multi-temporal multi-source remote-sensing data. *International Journal of Remote Sensing*, *33*(6), 1823–1849. Available from https://doi.org/10.1080/01431161.2011.602651, https://www.tandfonline.com/loi/tres20.

Millard, K., & Richardson, M. (2015). On the importance of training data sample selection in Random Forest image classification: A case study in peatland ecosystem mapping. *Remote Sensing*, *7*(7), 8489–8515. Available from https://doi.org/10.3390/rs70708489, http://www.mdpi.com/2072-4292/7/7/8489/pdf.

Moigne. (2021). *Proceedings of the 2021 international space university (ISU) alumni conference artificial intelligence and machine learning for earth science*.

Moreira, A., Prats-Iraola, P., Younis, M., Krieger, G., Hajnsek, I., & Papathanassiou, K. P. (2013). A tutorial on synthetic aperture radar. *IEEE Geoscience and Remote Sensing Magazine*, *1*(1), 6–43. Available from https://doi.org/10.1109/MGRS.2013.2248301, http://ieeexplore.ieee.org/xpl/RecentIssue.jsp?punumber = 6245518.

Mou, L., Saha, S., Hua, Y., Bovolo, F., Bruzzone, L., & Zhu, X.X. (2021). Deep reinforcement learning for band selection in hyperspectral image classification. *arXiv*, Germany arXiv. 23318422. Available from: https://arxiv.org, https://doi.org/10.48550/arxiv.2103.08741.

Mountrakis, G., Im, J., & Ogole, C. (2011). Support vector machines in remote sensing: A review. *ISPRS Journal of Photogrammetry and Remote Sensing*, *66*(3), 247–259. Available from https://doi.org/10.1016/j.isprsjprs.2010.11.001.

Mujetahid, A., Nursaputra, M., & Soma, A. S. (2023). Monitoring illegal logging using google earth engine in Sulawesi Selatan tropical forest, Indonesia. *Forests*, *14*(3), 652. Available from https://doi.org/10.3390/f14030652.

Nawaz, S. A., Li, J., Bhatti, U. A., Shoukat, M. U., Li, D., & Raza, M. A. (2024). Hybrid watermarking algorithm for medical images based on digital transformation and MobileNetV2. *Information Sciences*, *653*, 119810. Available from https://doi.org/10.1016/j.ins.2023.119810.

Nguyen, G., Dlugolinsky, S., Bobák, M., Tran, V., López García, Á., Heredia, I., Malík, P., & Hluchý, L. (2019). Machine learning and deep learning frameworks and libraries for large-scale data mining: A survey. *Artificial Intelligence Review*, *52*(1), 77–124. Available from https://doi.org/10.1007/s10462-018-09679-z.

Nizamani, A. H., Chen, Z., Nizamani, A. A., & Bhatti, U. A. (2023). Advance brain tumor segmentation using feature fusion methods with deep U-Net model with CNN for MRI data. *Journal of King Saud University – Computer and Information Sciences*, *35*(9), 101793. Available from https://doi.org/10.1016/j.jksuci.2023.101793.

Oguchi, T., Hayakawa, Y. S., & Wasklewicz, T. (2022). *Remote data in fluvial geomorphology: Characteristics and applications. Treatise on geomorphology* (pp. 1116–1142). Elsevier Available from. Available from https://www.sciencedirect.com/book/9780128182352, http://doi.org/10.1016/B978-0-12-818234-5.00103-6.

Paganini, M., Wyniawskyj, N., Talon, P., White, S., Watson, G., & Petit, D. (2020). Total ecosystem management of the InterTidal habitat (TEMITH). esa.int. 12.

Petitjean, F., Inglada, J., & Gancarski, P. (2012). Satellite image time series analysis under time warping. *IEEE Transactions on Geoscience and Remote Sensing*, *50*(8), 3081–3095. Available from https://doi.org/10.1109/tgrs.2011.2179050.

Phan, T. N., Kuch, V., & Lehnert, L. W. (2020). Land, cover classification using google earth engine and random forest classifier—The role of image composition. *Remote Sensors*, *12*.

Poktonjak, M., Meguerdichian, S., & Wong, J. L. (2010). Trusted sensors and remote sensing. *Proceedings of IEEE sensors*, 1104–1107. Available from https://doi.org/10.1109/ICSENS.2010.5690721.

Qi, W. (2022). Object detection in high resolution optical image based on deep learning technique. *Natural Hazards Research*, *2*(4), 384–392. Available from https://doi.org/10.1016/j.nhres.2022.10.002. Available from, https://www.sciencedirect.com/journal/natural-hazards-research.

Rahman, A., Roy, S. S., & Talukdar, S. (2023). *Shahfahad, advancements in urban environmental studies: application of geospatial technology and artificial intelligence in urban studies GI science and geo-environmental modelling*. Springer International Publishing. Available from http://doi.org/10.1007/978-3-031-21587-2.

Raji, I.D., Smart, A., White, R.N., Mitchell, M., Gebru, T., Hutchinson, B., Smith-Loud, J., Theron, D., & Barnes, P. (2020). Closing the AI accountability gap: Defining an end-to-end framework for internal algorithmic auditing. *FAT* 2020 – Proceedings of the 2020 conference on fairness, accountability, and transparency* (pp. 33 – 44). Available from http://doi.org/10.1007/978-3-031-21587-2, http://dl.acm.org/citation.cfm?id = 3351095.

Redins, L. (2023). Metaspectral's AI platform uses hyperspectral imaging, edge computing to transform space, recycling and other industries.

Redmon, J. (2023). Open source neural networks in C. Pjreddie.Com. 2013. Available online.

Redmon, J., Divvala, S., Girshick, R., & Farhadi, A. (2016). You only look once: Unified, real-time object detection. *Proceedings of the IEEE computer society conference on computer vision and pattern recognition* (pp. 779 – 788). Available from http://doi.org/10.1109/CVPR.2016.91.

Redmon, J., & Farhadi, A. (n.d.). Better, faster, stronger. *arXiv 2016*.

Ren, S., He, K., Girshick, R., & Sun, J. (n.d.). Faster R-CNN, towards real-time object detection with region proposal networks. *arXiv 2015*.

Richards, J. A. (2009). *Remote sensing with imaging radar signals and communication technology*. Springer Berlin Heidelberg. Available from http://doi.org/10.1007/978-3-642-02020-9.

Rittger, K., Krock, M., Kleiber, W., Bair, E. H., Brodzik, M. J., Stephenson, T. R., Rajagopalan, B., Bormann, K. J., & Painter, T. H. (2021). Multi-sensor fusion using random forests for daily fractional snow cover at 30 m. *Remote Sensing of Environment*, *264*. Available from https://doi.org/10.1016/j.rse.2021.112608, http://www.elsevier.com/inca/publications/store/5/0/5/7/3/3.

Rodriguez-Galiano, V. F., Ghimire, B., Rogan, J., Chica-Olmo, M., & Rigol-Sanchez, J. P. (2012). An assessment of the effectiveness of a random forest classifier for land-cover classification. *ISPRS Journal of Photogrammetry and Remote Sensing*, *67*(1), 93–104. Available from https://doi.org/10.1016/j.isprsjprs.2011.11.002.

Ronneberger, O., Fischer, P., & Brox, T. (n.d.). U-Net, convolutional networks for biomedical image segmentation. *arXiv 2015*.

Roselli, D., Matthews, J., & Talagala, N. (2019). *Managing bias in AI. The web conference 2019 – Companion of the world wide web conference*. WWW 2019, Available from http://dl.acm.org/citation.cfm?id = 3308560, https://doi.org/10.1145/3308560.3317590 539 544.

Sabat-Tomala, A., Raczko, E., & Zagajewski, B. (2020). Comparison of support vector machine and random forest algorithms for invasive and expansive species classification using airborne hyperspectral data. *Remote Sensing*, *12*(3), 516. Available from https://doi.org/10.3390/rs12030516.

Saini, R., & Ghosh, S.K. (2017). Ensemble classifiers in remote sensing: A review. *Proceeding – IEEE international conference on computing, communication and automation* (pp. 1148 – 1152). ICCCA. Available from https://doi.org/10.1109/CCAA.2017.8229969.

Sarker, I. H. (2021). Deep learning: A comprehensive overview on techniques, taxonomy, applications and research directions. *SN Computer Science*.

Schapire, R.E. (1999). A brief introduction to boosting. *IJCAI international joint conference on artificial intelligence* (pp. 1401–1406), United States, 10450823.

Schmitt, M., & Zhu, X. X. (2016). Data fusion and remote sensing: An ever-growing relationship. *IEEE Geoscience and Remote Sensing Magazine, 4* (4), 6–23. Available from https://doi.org/10.1109/MGRS.2016.2561021, http://ieeexplore.ieee.org/xpl/RecentIssue.jsp?punumber = 6245518.

Shafiq, S. I. (2023). Artificial intelligence and big data science for oceanographic research in Bangladesh: Preparing for the future. *Journal of Data Acquisition Processing, 38.*

Sharma, M. K., Mujawar, R., Mujawar, A., & Dhayalini, K. (2023). Precision forestry: Integration of robotics and sensing technologies for tree measurement and monitoring. *European Chemical Bulletin, 12*, 4747–4764.

Sheng, V.S., Provost, F., & Ipeirotis, P.G. (2008). Get another label? Improving data quality and data mining using multiple, noisy labelers. *Proceedings of the ACM SIGKDD international conference on knowledge discovery and data mining* (pp. 614–622). Available from https://doi.org/10.1145/1401890.1401965.

Sheykhmousa, M., Mahdianpari, M., Ghanbari, H., Mohammadimanesh, F., Ghamisi, P., & Homayouni, S. (2020). Support vector machine versus random forest for remote sensing image classification: A meta-analysis and systematic review. *IEEE Journal of Selected Topics in Applied Earth Observations and Remote Sensing, 13*, 6308–6325. Available from https://doi.org/10.1109/JSTARS.2020.3026724, http://ieeexplore.ieee.org/xpl/RecentIssue.jsp?punumber = 4609443.

Shorten, C., & Khoshgoftaar, T. M. (2019). A survey on image data augmentation for deep learning. *Journal of Big Data, 6*(1). Available from https://doi.org/10.1186/s40537-019-0197-0, https://journalofbigdata.springeropen.com.

Silvestro, D., Goria, S., Sterner, T., & Antonelli, A. (2022). Improving biodiversity protection through artificial intelligence. *Nature Sustainability, 5* (5), 415–424. Available from https://doi.org/10.1038/s41893-022-00851-6, http://www.nature.com/natsustain/.

Singh, P., & Komodakis, N. (2018). Cloud-GAN: Cloud removal for sentinel-2 imagery using a cyclic consistent generative adversarial networks. *International geoscience and remote sensing symposium (IGARSS)* (pp. 1772–1775). Available from https://doi.org/10.1109/IGARSS.2018.8519033.

Skulovich, O., & Gentine, P. (2023). A long-term consistent artificial intelligence and remote sensing-based soil moisture dataset. *Scientific Data, 10*(1). Available from https://doi.org/10.1038/s41597-023-02053-x, https://www.nature.com/sdata/.

Song, J., Gao, S., Zhu, Y., & Ma, C. (2019). A survey of remote sensing image classification based on CNNs. *Big Earth Data, 3*(3), 232–254. Available from https://doi.org/10.1080/20964471.2019.1657720, https://www.tandfonline.com/action/journalInformation?journalCode = tbed20.

Soydaner, D. (2020). A comparison of optimization algorithms for deep learning. *International Journal of Pattern Recognition and Artificial Intelligence* (2052).

Stereńczak, K. (2023). Precision forestry. *IDEAS NCBR—Intelligent algorithms for digital economy.*

Sun, A. Y., & Scanlon, B. R. (2019). How can Big Data and machine learning benefit environment and water management: A survey of methods, applications, and future directions. *Environmental Research Letters, 14*(7). Available from https://doi.org/10.1088/1748-9326/ab1b7d, https://iopscience.iop.org/article/10.1088/1748-9326/ab1b7d/pdf.

Sun, H., Wang, P., Chang, Y., Qi, L., Wang, H., Xiao, D., Zhong, C., Wu, X., Li, W., & Sun, B. (2020). HRPGAN: A GAN-based model to generate high-resolution remote sensing images. *IOP Conference series: Earth and Environment Science.*

Sun, W., Bocchini, P., & Davison, B. D. (2020). Applications of artificial intelligence for disaster management. *Natural Hazards, 103*(3), 2631–2689. Available from https://doi.org/10.1007/s11069-020-04124-3. Available from, http://www.wkap.nl/journalhome.htm/0921-030X.

Sun, W., & Du, Q. (2019). Hyperspectral band selection: A review. *IEEE Geoscience and Remote Sensing Magazine, 7*(2), 118–139. Available from https://doi.org/10.1109/MGRS.2019.2911100, http://ieeexplore.ieee.org/xpl/RecentIssue.jsp?punumber = 6245518.

Sun, Z., Cristea, N., & Rivas, P. (2023). *Artificial intelligence in earth science: Best practices and fundamental challenges. Artificial intelligence in earth science: Best practices and fundamental challenges* (pp. 1–418). Elsevier. Available from https://www.sciencedirect.com/book/9780323917377, 10.1016/C2021-0-01206-2.

Sun, Z., Di, L., & Fang, H. (2019). Using long short-term memory recurrent neural network in land cover classification on landsat and cropland data layer time series. *International Journal of Remote Sensing, 40*(2), 593–614. Available from https://doi.org/10.1080/01431161.2018.1516313, https://www.tandfonline.com/loi/tres20.

Sun, Z., Sandoval, L., Crystal-Ornelas, R., Mousavi, S. M., Wang, J., Lin, C., Cristea, N., Tong, D., Carande, W. H., Ma, X., et al. (2022). A review of earth artificial intelligence. *Computers & Geosciences.*

Tang, H., Bhatti, U. A., Li, J., Marjan, S., Baryalai, M., Assam, M., Ghadi, Y. Y., Mohamed, H. G., & Gastaldo, P. (2023). A new hybrid forecasting model based on dual series decomposition with long-term short-term memory. *International Journal of Intelligent Systems, 2023*, 1–15. Available from https://doi.org/10.1155/2023/9407104.

Technology (2022). Metaspectral.

Terven, J., & Cordova-Esparza, D., Comprehensive, from YOLOv1 and beyond. *arXiv 2023.* n.d.

Toivonen, T., Heikinheimo, V., Fink, C., Hausmann, A., Hiippala, T., Järv, O., Tenkanen, H., & Di Minin, E. (2019). Social media data for conservation science: A methodological overview. *Biological Conservation, 233*, 298–315. Available from https://doi.org/10.1016/j.biocon.2019.01.023, http://www.elsevier.com/inca/publications/store/4/0/5/8/5/3.

Tong, D. Q., Gill, T. E., Sprigg, W. A., Van Pelt, R. S., Baklanov, A. A., Barker, B. M., Bell, J. E., Castillo, J., Gassó, S., Gaston, C. J., Griffin, D. W., Huneeus, N., Kahn, R. A., Kuciauskas, A. P., Ladino, L. A., Li, J., Mayol-Bracero, O. L., McCotter, O. Z., Méndez-Lázaro, P. A., Mudu, P., Nickovic, S., Oyarzun, D., Prospero, J., Raga, G. B., Raysoni, A. U., Ren, L., Sarafoglou, N., Sealy, A., Sun, Z., & Vimic, A. V. (2023). Health and safety effects of airborne soil dust in the Americas and beyond. *Reviews of Geophysics, 61*(2). Available from https://doi.org/10.1029/2021RG000763, http://agupubs.onlinelibrary.wiley.com/agu/journal/10.1002/(ISSN)1944-9208/.

Torrey, L., & Shavlik, J. (2009). *Transfer learning. Handbook of research on machine learning applications and trends: Algorithms, methods, and techniques* (pp. 242–264). IGI Global. Available from http://www.igi-global.com/book/handbook-research-machine-learning-applications/486, http://doi.org/10.4018/978-1-60566-766-9.ch011.

Traore, B. B., Kamsu-Foguem, B., & Tangara, F. (2018). Deep convolution neural network for image recognition. *Ecological Informatics, 48*, 257–268. Available from https://doi.org/10.1016/j.ecoinf.2018.10.002, http://www.elsevier.com/wps/find/journaldescription.cws_home/705192/description#description.

Vaishya, R., Javaid, M., Khan, I. H., & Haleem, A. (2020). Artificial intelligence (AI) applications for COVID-19 pandemic. *Diabetes and Metabolic Syndrome: Clinical Research and Reviews, 14*(4), 337–339. Available from https://doi.org/10.1016/j.dsx.2020.04.012, http://www.journals.elsevier.com/diabetes-and-metabolic-syndrome-clinical-research-and-reviews/.

Vaswani, A., Shazeer, N., Parmar, N., Uszkoreit, J., Jones, L., Gomez, A.N., Kaiser, L., Polosukhin, I. (2017). Polosukhin, I. Attention is all you need.

Verbesselt, J., Zeileis, A., & Herold, M. (2012). Near real-time disturbance detection using satellite image time series. *Remote Sensing of Environment, 123*, 98–108. Available from https://doi.org/10.1016/j.rse.2012.02.022.

Wang, J., Lan, C., Liu, C., Ouyang, Y., Qin, T., Lu, W., Chen, Y., Zeng, W., & Yu, P. S. (2023). Generalizing to unseen domains: A survey on domain generalization. *IEEE Transactions on Knowledge and Data Engineering, 35*(8), 8052–8072. Available from https://doi.org/10.1109/TKDE.2022.3178128, http://ieeexplore.ieee.org/xpl/RecentIssue.jsp?punumber = 69.

Wang, Q., Ma, Y., Zhao, K., & Tian, Y. (2022). A comprehensive survey of loss functions in machine learning. *Annals of Data Science, 9*(2), 187–212. Available from https://doi.org/10.1007/s40745-020-00253-5, http://springer.com/journal/40745.

Weiss, M., Jacob, F., & Duveiller, G. (2020). Remote sensing for agricultural applications: A meta-review. *Remote Sensing of Environment, 236*, 111402. Available from https://doi.org/10.1016/j.rse.2019.111402.

Woodie, A. (2020). AI opens door to expanded use of LIDAR data.

Wu, Z., Chen, X., Gao, Y., & Li, Y. (2018). Rapid target detection in high resolution remote sensing images using YOLO model. *International archives of the photogrammetry, remote sensing and spatial information sciences – ISPRS archives, 42*(3), 1915–1920. Available from https://doi.org/10.5194/isprs-archives-XLII-3-1915-2018, http://www.isprs.org/proceedings/XXXVIII/4-W15/.

Xiang, X., Li, Q., Khan, S., & Khalaf, O. I. (2021). Urban water resource management for sustainable environment planning using artificial intelligence techniques. *Environmental Impact Assessment Review, 86*, 106515. Available from https://doi.org/10.1016/j.eiar.2020.106515.

Xie, M., Jean, N., Burke, M., Lobell, D., & Ermon, S. (2016). Transfer learning from deep features for remote sensing and poverty mapping. *Proceedings of the AAAI Conference on Artificial Intelligence, 30*(1). Available from https://doi.org/10.1609/aaai.v30i1.9906.

Xu, B., Bhatti, U. A., Tang, H., Yan, J., Wu, S., Sarhan, N., Awwad, E. M., S., S. M., & Ghadi, Y. Y. (2024). Towards explainability for AI-based edge wireless signal automatic modulation classification. *Journal of Cloud Computing, 13*(1). Available from https://doi.org/10.1186/s13677-024-00590-3.

Xu, C., & Zhao, B. (2018). Satellite image spoofing: Creating remote sensing dataset with generative adversarial networks. *Leibniz international proceedings in informatics*, LIPIcs Schloss Dagstuhl-Leibniz-Zentrum fur Informatik GmbH, Dagstuhl Publishing, United States. Available from https://doi.org/10.4230/LIPIcs.GIScience.2018.67. http://drops.dagstuhl.de/opus/institut_lipics.php?fakultaet = 04.

Xu, D., & Wu, Y. (2020). Improved YOLO-V3 with densenet for multi-scale remote sensing target detection. *Sensors, 20*(15), 4276. Available from https://doi.org/10.3390/s20154276.

Xu, S., Cheng, J., & Zhang, Q. (2021). A random forest-based data fusion method for obtaining all-weather land surface temperature with high spatial resolution. *Remote Sensing, 13*(11), 2211. Available from https://doi.org/10.3390/rs13112211.

Xu, Y., Liu, Y., Xin, Y., & Zhu (n.d.). Comparison of gap-filling methods for producing all-weather daily remotely sensed near-surface air temperature.

Yang, F. (2021). An improved YOLO v3 algorithm for remote sensing image target detection. *Journal of Physics: Conference Series, 2132*(1), 012028. Available from https://doi.org/10.1088/1742-6596/2132/1/012028.

Yang, H., Nguyen, T. N., & Chuang, T. W. (2023). An integrative explainable artificial intelligence approach to analyze fine-scale land-cover and land-use factors associated with spatial distributions of place of residence of reported dengue cases. *Tropical Medicine and Infectious Disease, 8*(4). Available from https://doi.org/10.3390/tropicalmed8040238, http://www.mdpi.com/journal/tropicalmed.

Yang, J., Sadiq, T., Xiong, J., Awais, M., Bhatti, U. A., Alizadehsani, R., & Gorriz, J. M. (2024). A novel myocarditis detection combining deep reinforcement learning and an improved differential evolution algorithm. *CAAI Transactions on Intelligence Technology*. Available from https://doi.org/10.1049/cit2.12289, 2468-2322.

Yang, K., John, A., Shean, D., Lundquist, J.D., Sun, Z., Yao, F., Todoran, S., & Cristea, N. (n.d.). High-resolution mapping of snow cover in montane meadows and forests using planet imagery and machine learning.

Yang, K., John, A., Sun, Z., & Cristea, N. (2023). *Machine learning for snow cover mapping. Artificial Intelligence in Earth Science: Best Practices and Fundamental Challenges* (pp. 17–39). Elsevier Available from. Available from https://www.sciencedirect.com/book/9780323917377, http://doi.org/10.1016/B978-0-323-91737-7.00004-9.

Yang, W., Song, H., Du, L., Dai, S., & Xu, Y. (2022). A change detection method for remote sensing images based on coupled dictionary and deep learning. *Computational Intelligence and Neuroscience, 2022*. Available from https://doi.org/10.1155/2022/3404858, http://www.hindawi.com/journals/cin.

Yuan, Q., Shen, H., Li, T., Li, Z., Li, Y., Jiang, H., Xu, W., Tan, Q., Yang, Wang, J. (n.d.). Deep learning in environmental remote sensing: Achievements and challenges.

Yuan, Y., Li, J., Liu, J., Bhatti, U. A., Liu, Z., & Chen, Y. -w (2024). Robust zero-watermarking algorithm based on discrete wavelet transform and daisy descriptors for encrypted medical image. *CAAI Transactions on Intelligence Technology, 9*(1), 40–53. Available from https://doi.org/10.1049/cit2.12282.

Zhang, L., Zhang, L., & Du, B. (2016). Deep learning for remote sensing data: A technical tutorial on the state of the art. *IEEE Geoscience and Remote Sensing Magazine*, *4*(2), 22–40. Available from https://doi.org/10.1109/MGRS.2016.2540798, http://ieeexplore.ieee.org/xpl/RecentIssue.jsp?punumber = 6245518.

Zhang, X., Zhu, G., & Ma, S. (2012). Remote-sensing image encryption in hybrid domains. *Optics Communications*, *285*(7), 1736–1743. Available from https://doi.org/10.1016/j.optcom.2011.12.023.

Zhang, Y., Chen, J., Ma, X., Wang, G., Bhatti, U. A., & Huang, M. (2024). Interactive medical image annotation using improved Attention U-net with compound geodesic distance. *Expert Systems with Applications*, *237*, 121282. Available from https://doi.org/10.1016/j.eswa.2023.121282.

Zhang, Y., Liu, J., & Shen, W. (2022). A review of ensemble learning algorithms used in remote sensing applications. *Applied Science*, *12*.

Zhang, Y. D., Pan, C., Sun, J., & Tang, C. (2018). Multiple sclerosis identification by convolutional neural network with dropout and parametric ReLU. *Journal of Computational Science*, *28*, 1–10. Available from https://doi.org/10.1016/j.jocs.2018.07.003, http://www.elsevier.com/wps/find/journaldescription.cws_home/721195/description#description.

Zhao, Y., Zhang, X., Feng, W., & Xu, J. (n.d.). Deep learning classification by ResNet-18 based on the real spectral dataset from multispectral remote sensing images.

Zhong, H., Lin, W., Liu, H., Ma, N., Liu, K., Cao, R., Wang, T., & Ren, Z. (2022). Identification of tree species based on the fusion of UAV hyperspectral image and LiDAR data in a coniferous and broad-leaved mixed forest in Northeast China. *Frontiers in Plant Science*.

Zhu, J.Y., Park, T., Isola, P., Efros, A.A. (2017). Unpaired image-to-image translation using cycle-consistent adversarial networks. *Proceedings of the IEEE international conference on computer vision* (pp. 2242–2251). Available from https://doi.org/10.1109/ICCV.2017.244. http://ieeexplore.ieee.org/xpl/conhome.jsp?punumber = 1000149.

Further reading

Bhatti, M. A., Song, Z., Bhatti, U. A., & Syam, M. S. (2024). AIoT-driven multi-source sensor emission monitoring and forecasting using multi-source sensor integration with reduced noise series decomposition. *Journal of Cloud Computing*, *13*(1). Available from https://doi.org/10.1186/s13677-024-00598-9.

Zi, Y., & Xie, F. (2021). Thin cloud removal for remote sensing images using a physical-model-based cycle GAN with unpaired data. *IEEE Geoscience and Remote Sensing Letters*.

Chapter 4

Carbon emission pattern analysis and its relationship with climate change

Mir Muhammad Nizamani[1], Muhammad Awais[2], Muhammad Qayyum[3], Entaj Tarafder[1], Maqsood Ahmed Khaskheli[1] and Yong Wang[1]

[1]*Department of Plant Pathology, College of Agriculture, Guizhou University, Guiyang, Guizhou, P.R. China,* [2]*Institute of Plant Sciences, Faculty of Natural Sciences, University of Sindh, Jamshoro, Sindh, Pakistan,* [3]*School of Economics and Statistics, Guangzhou University, Guangzhou, P.R. China*

4.1 Introduction to carbon emissions and climate change

Carbon emissions predominantly refer to carbon dioxide (CO_2), a greenhouse gas (GHG) that contributes to global warming by trapping heat in the Earth's atmosphere (Yoro & Daramola, 2020). These emissions can also include other carbon-based gases like methane (CH_4) and carbon monoxide (CO), which have similar warming effects (Zhou et al., 2024). While natural sources of carbon emissions exist, such as volcanoes and respiration, the focus in climate studies typically remains on anthropogenic (human-caused) sources due to their significant impact and potential for mitigation (Sen Gupta & Tiwari, 2021). The primary source of anthropogenic carbon emissions is the combustion of fossil fuels (Huang et al., 2015). Coal, oil, and natural gas burned for electricity generation, heating, and as fuel in transportation release vast amounts of CO_2 (Oberschelp et al., 2019). The industrial sector also significantly contributes through chemical processes necessary for manufacturing goods (Costa & Ribeiro, 2020). For example, the production of cement releases CO_2 when limestone is transformed into lime (Simoni et al., 2022).

Land use changes are another critical source of emissions (Hong et al., 2021). Deforestation, for instance, not only destroys carbon-absorbing trees but also releases the carbon stored in them when they are burned or decompose. Agricultural practices further exacerbate carbon emissions; for example, methane is released from livestock and rice paddies, while nitrous oxide is emitted from the use of nitrogen-based fertilizers (Adegbeye et al., 2020). The management of waste also plays a role, with landfills emitting methane as organic waste breaks down. Similarly, residential and commercial energy use contributes to the carbon footprint, particularly in developed countries where heating, cooking, and electricity demands are high (Kamyab et al., 2024). The transportation sector adds to the problem, with vehicles emitting CO_2 and other pollutants through the combustion of petroleum-based fuels (Aminzadegan et al., 2022; Kalghatgi et al., 2018).

Climate change refers to significant changes in global temperatures and weather patterns over time (Rummukainen, 2013). While climate variability is a natural phenomenon, the scientific community agrees that the rapid climate change observed since the industrial era is predominantly due to human activities, particularly the emission of greenhouse gases like CO_2 and CH_4. This consensus is supported by numerous scientific organizations worldwide, including the Intergovernmental Panel on Climate Change (IPCC) and the National Aeronautics and Space Administration (NASA), which confirm that the warming of the climate system is unequivocal and primarily driven by human-induced emissions (Meyer et al., 2020).

The effects of climate change are diverse, widespread, and of significant concern for both natural environments and human populations. One of the most direct effects is the increase in global average temperatures, leading to longer and more intense heatwaves that are becoming more frequent across the globe (Frölicher et al., 2018). This warming also causes the melting of ice sheets and glaciers, contributing to rising sea levels that threaten coastal communities and ecosystems, resulting in more frequent and severe coastal flooding (Lu et al., 2018). Climate change is linked to the increased frequency and intensity of extreme weather events such as hurricanes, floods, and droughts, which cause

Deep Learning for Earth Observation and Climate Monitoring. DOI: https://doi.org/10.1016/B978-0-443-24712-5.00005-1

immediate human suffering, economic damage, and long-term environmental impacts (Gulzar et al., 2021). Additionally, changing weather patterns are significantly impacting ecosystems by altering migration patterns, breeding seasons, and species distribution, thereby affecting biodiversity. Coral reefs, for instance, are highly sensitive to warming waters, leading to coral bleaching and the decline of reef ecosystems (Hoegh-Guldberg et al., 2017). The agricultural sector is also affected, as changes in temperature and precipitation patterns, along with increased occurrences of pests and diseases, challenge traditional farming practices and threaten food security (Skendžić et al., 2021). Furthermore, climate change poses direct and indirect health risks, including heat-related illnesses and deaths during heatwaves, increased respiratory and cardiovascular diseases due to poorer air quality, and greater risks of vector-borne diseases like malaria and dengue fever (Jacobsen et al., 2022). The economic impacts are significant, ranging from damaged infrastructure due to extreme weather events to increased healthcare and disaster response costs, with economies dependent on agriculture or tourism being particularly vulnerable (Dogru et al., 2019).

Understanding the greenhouse effect is essential for comprehending the relationship between carbon emissions and global warming. This natural process, crucial for life on Earth, involves gases in the Earth's atmosphere—primarily water vapor, but also carbon dioxide, methane, and nitrous oxide—that trap heat from the sun (Kamyab et al., 2024). These greenhouse gases allow sunlight to enter the atmosphere freely, where the Earth absorbs it. Once the Earth radiates this energy back toward space as heat, greenhouse gases absorb and reradiate the heat, distributing it around the planet to maintain a livable temperature (Ming et al., 2016). However, human activities, particularly the burning of fossil fuels and deforestation, have intensified this effect (Yoro & Daramola, 2020). The primary issue is the increased concentration of carbon dioxide in the atmosphere. Preindustrial levels of carbon dioxide were about 280 parts per million (ppm); today, they exceed 420 ppm. This sharp increase, predominantly due to the combustion of coal, oil, and natural gas, enhances the natural greenhouse effect, leading to higher global temperatures, a phenomenon commonly referred to as global warming (Jogdand, 2020). Carbon dioxide plays a crucial role in atmospheric warming because of its ability to trap heat and its longevity in the atmosphere (Kabir et al., 2023). Unlike other greenhouse gases that may stay in the atmosphere for a few years or decades, CO_2 persists for up to a century or more. This means that the carbon dioxide released today continues to impact the climate system for generations. Its accumulation ensures a prolonged effect on atmospheric temperatures, making it a primary target for climate mitigation policies (Subramanian et al., 2023). The warming prompted by increased CO_2 levels also triggers various feedback loops that can exacerbate global warming. For example, melting polar ice reduces the Earth's albedo (reflectivity), causing the planet to absorb more solar energy, which in turn leads to further warming. Similarly, thawing permafrost releases methane, a potent GHG, which significantly accelerates warming in a shorter period, even though its atmospheric lifespan is less than that of CO_2 (Saros et al., 2022). The enhanced greenhouse effect resulting from increased carbon emissions leads to a cascade of global changes (Mora et al., 2018). These include rising sea levels, changing precipitation patterns, more extreme weather events, and shifts in ecosystems and wildlife populations. Each of these changes feeds back into the climate system, often amplifying the original warming effect and making the global climate increasingly unstable and unpredictable (Turner et al., 2020).

4.2 Historical perspective of carbon emissions

4.2.1 Global trends in carbon emissions over the past century

Over the past century, the trajectory of global carbon emissions has shown a marked increase, closely paralleled by industrial, economic, and population growth worldwide (Gomes da Silva et al., 2020). This period, particularly since the mid-20th century, is characterized by an unprecedented rise in emissions due to the expansion of the fossil fuel-based economy, widespread industrialization, and significant changes in land use (Mai et al., 2018). The early 20th century marked the beginning of significant carbon emission increases, primarily in Western Europe and North America, where industrialization first took root (Hegerl et al., 2018). The use of coal as a primary energy source propelled emissions upward. After World War II, the economic boom in these regions, coupled with the expansion of oil and natural gas use, further accelerated carbon emissions. The postwar era also saw a rise in automobile ownership and suburbanization, contributing to higher fossil fuel consumption (Dodson et al., 2018). By the latter half of the 20th century, other regions began to industrialize. Rapid economic growth in Asia, particularly in China and India, from the late 20th century into the 21st century, has been a major driver of recent global emission increases. These countries have relied heavily on coal to fuel their economic growth, mirroring earlier patterns observed in the Western economies (Obschonka et al., 2018).

The last few decades have seen some shifts in global emission trends. While total global emissions have continued to rise, the rate of increase has shown variability due to several factors. Advances in technology and greater awareness of sustainability issues have led to more efficient use of energy, somewhat tempering emissions growth in developed nations (Manigandan et al., 2023). There has also been a significant increase in the deployment of renewable energy sources like wind and solar power, particularly in the EU, USA, and China, which has begun to offset some of the emissions from fossil fuels (Holechek et al., 2022). Additionally, the recent shift toward service-oriented economies in the West has reduced direct industrial emissions, although this shift often displaces emissions to developing countries with less stringent environmental controls (Tian et al., 2019). International climate agreements such as the Kyoto Protocol and the Paris Agreement have aimed to curb global carbon emissions. While these agreements have led to actionable commitments from many countries, global emissions continue to rise, highlighting the challenges of coordinating and implementing effective global climate policies (Adshead et al., 2021).

An unprecedented anomaly occurred in 2020 due to the COVID-19 pandemic, which caused a temporary but significant drop in global emissions due to reduced travel, industrial activity, and economic slowdowns. However, this decrease was short-lived, as emissions rebounded with the resumption of economic activities (Zheng et al., 2021). The current trajectory suggests that without drastic and sustained reductions in carbon emissions, achieving the goals set out in international climate agreements to limit global warming will be increasingly difficult. The future of global emissions depends heavily on policy decisions, technological advancements, and the global community's ability to transition toward sustainable energy solutions while supporting economic growth in developing regions in an environmentally responsible way (Brown et al., 2019).

4.2.2 The role of industrialization in rising emission levels

The role of industrialization in increasing carbon emission levels is profound and multifaceted (Wang et al., 2023). Beginning with the Industrial Revolution in the late 18th century, the shift from agrarian economies to industrial economies marked the first significant increase in carbon emissions. This era saw the introduction of machinery powered by fossil fuels, primarily coal, which replaced human and animal labor in production processes. As factories proliferated, especially in Europe and North America, so did the demand for energy, leading to rising emissions of carbon dioxide and other greenhouse gases (Lamb et al., 2021). As industrialization spread throughout the 19th and 20th centuries, other fossil fuels such as oil and natural gas became integral to industrial processes, further intensifying carbon emissions (Acker et al., 2023). The development of internal combustion engines and the increase in automobile manufacturing and usage during the 20th century contributed significantly to global emission levels. These industries not only increased direct emissions but also facilitated greater urbanization, requiring more buildings heated and powered by fossil fuels, which added to the atmospheric concentration of CO_2 (Yoro et al., 2020).

The latter half of the 20th century saw a shift in industrial activities from developed countries to developing countries. This globalization of manufacturing was driven by cheaper labor costs and less stringent environmental regulations in developing nations (Lall, 2004). Consequently, countries like China and India became major centers of industrial production, using coal-intensive energy processes that mirrored the earlier stages of Western industrialization but on a much larger scale. This shift has led to a significant increase in global emissions in recent decades, as these economies have grown rapidly (Astarita, 2016). Industrial processes themselves, such as the production of cement, steel, and chemicals, are intensive in terms of energy use and carbon emissions (Griffin et al., 2018). These processes often involve the transformation of raw materials into finished goods, releasing significant amounts of CO_2 in the process. For example, cement production involves calcination, during which limestone (calcium carbonate) is heated to produce lime (calcium oxide), releasing CO_2 as a byproduct (Yaashikaa et al., 2019).

In modern times, while industrialized nations have made strides in improving energy efficiency and adopting cleaner technologies, the overall increase in production and consumption has kept emission levels high. Regulatory frameworks like the Paris Agreement aim to mitigate these emissions by setting international targets for reduction and encouraging investment in renewable energy and other green technologies (Lamb et al., 2021).

4.2.3 Key events and their impacts on emission trends (e.g., Industrial Revolutions, oil crises)

The First Industrial Revolution, starting in the late 18th century, marked the beginning of high-scale carbon emissions due to the widespread use of coal to power steam engines and later, the advent of electricity generation (Jacobs, 2023). This period was characterized by a shift from agrarian economies to industrial powerhouses, where manufacturing centered around fossil fuels. The Second Industrial Revolution in the late 19th and early 20th centuries introduced more advanced manufacturing processes and expanded the use of electricity and oil, significantly boosting CO_2 emissions as industries and transportation networks expanded globally (Wang et al., 2017). The economic boom following World War II, particularly in Western countries, led to a sharp increase in industrial activities and urbanization, both of which were heavily reliant on oil (Wang et al., 2017) and natural gas. This period also saw a significant rise in consumerism, with an increase in manufactured goods and automobiles, further driving up fossil fuel use and emissions. The suburban sprawl encouraged by the automobile industry meant more people were driving longer distances, cementing the role of oil in global economies and CO_2 emissions (Glaeser & Kahn, 2004).

The oil crises of the 1970s were pivotal in highlighting the world's dependence on fossil fuels and the vulnerability of this dependence. The first oil crisis in 1973, triggered by an organization of the petroleum exporting countries (OPEC) oil embargo, led to a steep increase in oil prices and prompted some countries to begin exploring alternative energy sources. However, the immediate reaction for many was to secure more stable oil supplies rather than reducing dependency, which temporarily slowed the growth rate of carbon emissions but eventually led to an increase as new sources and supplies were secured (Nacke et al., 2022). The fall of the Soviet Union in 1991 led to a significant but temporary decrease in carbon emissions from the former Soviet bloc countries (Salahodjaev & Isaeva, 2022). The economic downturn that followed the dissolution resulted in reduced industrial output and a decline in oil and coal consumption. This period provides a clear instance of how geopolitical events can significantly impact emission trends, albeit often temporarily as economic recovery tends to restore or even increase previous levels of carbon emissions (Bersalli et al., 2023).

The adoption of the Kyoto Protocol in 1997 and its subsequent entry into force in 2005 marked the first significant international treaty aimed at reducing carbon emissions (Metz, 2013). Countries that ratified the treaty committed to reducing their GHG emissions to levels below their 1990 totals. This agreement impacted emission trends by encouraging participating countries to invest in cleaner technologies and renewable energy sources, though its overall effectiveness was limited by the nonparticipation of key nations like the USA (Larch & Wanner, 2024). China's rise as an industrial power in the early 2000s dramatically altered global emission dynamics (Zheng et al., 2020). As the country industrialized at an unprecedented rate, it became the world's largest emitter of CO_2, primarily due to its heavy reliance on coal for energy. This period correlates with significant global emissions growth, as China's industrial output served both domestic needs and a booming global demand for manufactured goods (Li et al., 2020).

The last two decades have seen a significant increase in the adoption of renewable energy technologies, driven by technological advancements, reduced costs, and policy incentives, such as those encouraged by the Paris Agreement of 2015 (Malik et al., 2019). These developments are starting to influence global emission trends, though the full impact remains to be seen as global emissions continue to rise, underscoring the challenge of offsetting emissions from developing economies and the ongoing global dependence on fossil fuels (Foster et al., 2023).

4.3 Sector-specific emission patterns

4.3.1 Energy sector: dominance in carbon emissions and underlying factors

The energy sector is the largest contributor to global GHG emissions, accounting for approximately three-quarters of all emissions (Olivier et al., 2017). This sector encompasses activities related to energy production, conversion, and consumption, including electricity generation, heat production, and transportation fuels. Its dominance in carbon emissions stems from the continued reliance on fossil fuels—coal, oil, and natural gas—due to their availability, energy density, and historically low economic cost (Cao et al., 2016). Coal, being the most carbon-intensive, remains a major energy source in countries like China and India, while oil fuels the transportation sector and natural gas, although cleaner, still contributes significantly to emissions (Yu et al., 2021). Inefficiencies in energy production and use exacerbate the problem; for instance, traditional coal-fired power plants and internal combustion engines lose substantial energy as waste heat, leading to higher fuel consumption and emissions (Barma et al., 2017; Jung & Pandit, 2018). Furthermore, global economic and population growth drive higher energy demand, often outpacing efficiency improvements (Khan et al., 2021). Transitioning to renewable energy sources like wind, solar, and hydroelectric power presents a significant opportunity to reduce emissions, but challenges such as variability, insufficient storage technologies, and substantial initial investments persist (Cantarero, 2020; Kabeyi & Olanrewaju, 2022). Effective policies and technological innovations are essential for decarbonizing the energy sector, including carbon pricing mechanisms, subsidies for renewable technologies, and stringent emission regulations, alongside advances in energy efficiency, cleaner combustion technologies, carbon capture and storage (CCS), and battery storage (Azam et al., 2022; Lau et al., 2021).

4.3.2 Transportation: trends, challenges, and the shift toward sustainable mobility

The transportation sector is a major contributor to global GHG emissions, responsible for about a quarter of all CO_2 emissions from fuel combustion (Khalili et al., 2019). The sector's emissions have grown at a faster rate than those of any other energy end-use sector, largely due to increased demand for personal mobility and global trade. Historically, the majority of these emissions come from road vehicles, including cars, trucks, and buses, which are predominantly powered by gasoline and diesel fuels. Aviation and maritime transport also contribute significantly, with their emissions projected to increase unless significant mitigation strategies are implemented (Sharmina et al., 2021). One of the primary challenges in reducing transportation emissions is the sector's heavy reliance on petroleum-based fuels (Vimmerstedt et al., 2013). These fuels are energy-dense and convenient, making them hard to replace with current technologies. Another challenge is the increasing demand for transportation linked to global population growth and economic development, particularly in emerging economies where car ownership and air travel are becoming more common (Rode et al., 2017). Infrastructure limitations also pose significant hurdles, especially in densely populated urban areas where traffic congestion can lead to idling and increased per capita emissions (Hui & Chu, 2023). Additionally, the logistics and costs associated with transitioning to low-emission vehicles (such as electric cars or hydrogen fuel cell vehicles) and creating the necessary charging or fueling infrastructure are substantial, requiring coordinated efforts between governments, industry, and consumers (Burnham et al., 2017).

Efforts to shift toward sustainable mobility encompass a range of strategies focused on reducing dependency on fossil fuels and minimizing environmental impacts. Key approaches include electrification of the vehicle fleet, with a strong push to increase the number of electric vehicles (EVs) on the road, including passenger cars, buses, and freight vehicles. The success of this strategy depends heavily on the continued decarbonization of the electricity grid; otherwise, the benefits of electrification in terms of emission reductions will be undermined (Le Treut et al., 2021). Beyond electrification, the development and adoption of alternative fuels such as biofuels, hydrogen, and synthetic fuels are seen as vital for reducing emissions, particularly in aviation and maritime transport where electrification is more challenging (Sandaka & Kumar, 2023). Governments worldwide have implemented regulations that require manufacturers to increase the fuel efficiency of vehicles, leading to significant improvements over the past decades, although the full potential of these technologies needs broader deployment (Oke et al., 2024). Encouraging the use of public transportation, cycling, and walking—along with smarter urban planning that reduces the need for long commutes—can significantly cut emissions. Integrated transport strategies that prioritize public transport and nonmotorized modes are essential in urban settings (Di et al., 2018). Advances in technology, such as autonomous vehicles and better traffic management systems, offer potential to improve efficiency and reduce emissions. Innovative logistics and freight management can also play a role in making goods movement more sustainable (Igliński & Babiak, 2017). Effective policy initiatives are crucial for supporting these shifts, including financial incentives for EVs, investments in public transport infrastructure, and stringent emissions standards (Santarromana et al., 2020). International cooperation is also vital, as emissions from international aviation and maritime transport are particularly challenging to address due to their cross-border nature (Albayrak et al., 2021).

4.3.3 Agriculture: emission profiles, land use changes, and mitigation strategies

Agriculture is a significant source of GHG emissions, contributing approximately 24% of global emissions when including related land use changes. The sector's emissions are diverse, originating from various sources: CH_4, a potent GHG, is released primarily through enteric fermentation in ruminant animals like cows and sheep and during the management of organic waste in livestock operations, including manure (Parihar et al., 2019). Nitrous oxide (N_2O), which has a global warming potential many times that of CO_2, is primarily released from soil management practices, especially the use of synthetic and organic fertilizers (Verma et al., 2020). Significant amounts of CO_2 are released through land use changes, particularly deforestation for agricultural expansion, the burning of agricultural residues, and the energy used in farming operations (Raihan et al., 2022). Land use changes are closely tied to agriculture, often resulting in significant environmental impacts. Deforestation, driven by agricultural expansion, releases the carbon stored in trees and reduces the planet's capacity to absorb future CO_2 emissions (Arneth et al., 2017). Intensive farming practices lead to soil degradation and reduced carbon sequestration in soil, including erosion, compaction, loss of soil structure, and the decline in organic matter content, all of which can release CO_2 previously stored in the soil (Karlen & Cambardella, 2020).

Addressing the high emission profile of agriculture involves a multifaceted approach. Improved animal husbandry can reduce methane emissions from livestock through better dietary regimes, improved genetics, and more efficient manure management, such as anaerobic digestion of manure to capture methane for use as a bioenergy source (Løvendahl et al., 2018). Enhanced soil management practices, such as reduced tillage, cover cropping, and organic farming, enhance soil carbon sequestration (Crystal-Ornelas et al., 2021). Precision agriculture techniques can optimize fertilizer use, reducing nitrous oxide emissions (Besen et al., 2024). Integrating trees into agricultural landscapes (agroforestry) can provide additional carbon sinks and mitigate the impact of deforestation, while sustainable land management practices, such as maintaining wetlands and other natural ecosystems, play a critical role (Chapman et al., 2020). Crop rotation and diversification can improve soil health, reduce the need for chemical inputs, and build resilience to climate change (Cordovil et al., 2020). Using renewable energy sources for agricultural operations, such as solar-powered irrigation systems, can reduce the sector's reliance on fossil fuels (Farrok, et al., 2022; Khan, et al., 2022). Effective mitigation in agriculture will require supportive policies that encourage sustainable practices and provide financial incentives for farmers to adopt these methods. Research and development into new technologies for monitoring and managing agricultural emissions are also crucial, including satellite imaging for land use monitoring, genetic modifications to crops and livestock for greater efficiency, and biotechnological advances in soil management (Piñeiro et al., 2020).

4.3.4 Industrial processes: emission characteristics and opportunities for reduction

Industrial processes contribute significantly to global GHG emissions through a variety of mechanisms distinct from the combustion of fossil fuels for energy (Yoro & Daramola, 2020). These processes involve chemical reactions necessary for producing materials like cement, steel, and chemicals, each of which has its own unique emission profile. Cement production is one of the most carbon-intensive processes, involving the calcination of limestone (calcium carbonate) to produce lime (calcium oxide), releasing CO_2 as a byproduct and accounting for about 5% of global CO_2 emissions (Gür, 2022). Iron and steel production is another major emitter, primarily of CO_2, due to the energy-intensive reduction of iron ore using coke in blast furnaces. Additionally, significant amounts of methane and nitrous oxide are released through various associated industrial processes (Quader et al., 2015). The production of chemicals like ammonia, nitric acid, and other petrochemicals involves emissions of both CO_2 and various other gases such as N_2O, a potent GHG especially associated with the production of nitric acid (Griffin et al., 2018). Various other processes such as paper production, food processing, and textile manufacturing contribute additional emissions, mostly in the form of CO_2, from both energy use and chemical reactions involved in production (Roy Choudhury, 2014).

Reducing emissions from industrial processes requires a combination of technological innovation, process improvements, and policy interventions. Key opportunities include energy efficiency improvements, which can be achieved through modernizing equipment, optimizing processes, and recovering waste heat for other uses, thereby reducing the amount of energy needed and the emissions produced per unit of output (Fais et al., 2016). Developing and adopting alternative materials that require less carbon-intensive production methods, such as different types of cement mixtures that require less lime or incorporating recycled materials into manufacturing processes, can reduce the need for raw input materials and lower emissions (Cadez & Czerny, 2016). Carbon capture, utilization, and storage (CCUS) technology, though still in development for scalable use, offers significant potential for reducing emissions from industrial processes by capturing CO_2 emissions at the source and either storing them underground or utilizing them in other products (Lin et al., 2022). Where feasible, replacing carbon-intensive processes with lower-emission alternatives, such as moving from traditional blast furnaces to electric arc furnaces in steel production, can have a substantial impact, particularly if the electricity used is sourced from renewable energy (Fais et al., 2016). Strong policy measures, including carbon pricing, emissions trading schemes, and direct regulations on maximum allowable emissions, can drive industrial innovation and encourage investment in cleaner technologies. Financial incentives for research and development in new technologies can also accelerate adoption (Zhang et al., 2022).

4.3.5 Residential and commercial: emissions from heating, electricity, and waste management

The residential and commercial sectors are significant contributors to global GHG emissions, primarily through the consumption of energy for heating, cooling, lighting, and operating appliances, as well as through waste management practices. These emissions are mainly in the form of CO_2 from the combustion of fossil fuels, but also include CH_4 from waste decomposition and a variety of other pollutants depending on specific practices and technologies used (Zhou et al., 2024). Heating and cooling are the largest sources of emissions in these sectors, especially in regions with extreme weather conditions requiring significant indoor climate control. Most residential and commercial buildings use fossil fuels such as natural gas, heating oil, or coal for these purposes, directly releasing CO_2 into the atmosphere

(Shahsavari & Akbari, 2018). While the emissions from electricity consumption depend on the energy mix of the grid, many regions still rely heavily on coal and natural gas power plants, making electricity use a major indirect source of GHG emissions. Efficient building design and appliance use can significantly reduce these emissions (Li et al., 2021; Yang, et al., 2021). Methane emissions from the decomposition of organic waste in landfills represent a significant emission source within these sectors. Improper waste management can also lead to the release of other potent greenhouse gases, including hydrofluorocarbons (HFCs) from discarded refrigeration and air-conditioning equipment (Zhou et al., 2024).

Mitigating emissions in the residential and commercial sectors involves a combination of technological upgrades, regulatory changes, and shifts in consumer behavior. Enhancing the energy efficiency of buildings is a critical step, achievable through better insulation, energy-efficient windows, and the use of energy-efficient heating and cooling systems. Building codes and regulations that require or incentivize these improvements can accelerate their adoption (Vijayan et al., 2022). Integrating renewable energy sources such as solar, wind, and geothermal directly into residential and commercial buildings reduces reliance on the grid and its fossil fuel-dominated energy. Solar panels on rooftops, for example, can significantly offset the electricity needs of a building (Ouerghi et al., 2024). Using advanced thermostats, lighting systems, and other automated technologies can optimize energy use and reduce emissions by adjusting energy consumption based on actual needs, thus minimizing waste (Aste et al., 2017). Implementing comprehensive recycling programs and waste reduction strategies can lower the amount of waste sent to landfills, thereby reducing methane emissions. Composting organic waste is another effective strategy to mitigate methane release (Pujara et al., 2019). Replacing traditional heating systems with electric heat pumps can significantly reduce emissions, particularly if the electricity is sourced from renewable energies. Heat pumps are much more efficient than combustion-based systems and can provide both heating and cooling (Walker et al., 2022).

Governments play a crucial role in shaping the emissions trajectory of the residential and commercial sectors through policies that encourage or mandate reductions (Nilsson et al., 2021). These can include subsidies for renewable energy installations, tax rebates for energy-efficient renovations, stricter building codes, and regulations that phase out the use of fossil fuels for heating and cooling. Additionally, public awareness campaigns that educate consumers about their energy use and its impacts can drive more sustainable choices (von Borgstede et al., 2013).

4.4 Geographic disparities in emissions

4.4.1 Comparative analysis of emissions by continent and economic development status

4.4.1.1 Overview

GHG emissions vary significantly across continents and are influenced by factors such as economic development status, industrial activity, energy sources, and population size (Lamb et al., 2021). Developed countries typically have higher per capita emissions due to greater industrial activity and higher standards of living, which involve more energy consumption. Conversely, many developing nations have lower per capita emissions but may experience rapid increases as they industrialize and improve living standards (Schandl et al., 2016).

4.4.1.2 Emissions in developed versus developing continents

North America, primarily the United States and Canada, has high per capita CO_2 emissions driven by industrial activities, car dependency, and fossil fuel use. Europe, with its aggressive sustainability policies and commitment to the Paris Agreement, has made significant strides in reducing emissions, though Eastern Europe still faces challenges. Asia, particularly China and India, is a major emitter due to rapid industrialization, though per capita emissions remain lower than in Western nations. Africa, with the lowest emissions per capita, is experiencing rising emissions due to development pressures but has significant potential for renewable energy. South America's low per capita emissions benefit from hydroelectric power, but deforestation is a major concern. Australia and Oceania, led by Australia, have high per capita emissions largely due to coal dependency and a significant mining sector (Table 4.1).

The link between economic development and emissions is evident when comparing industrialized and developing regions (Wang et al., 2018). Developed economies generally have higher infrastructure and industrial bases that increase emissions, though many are transitioning to greener practices. Developing countries often face a double-edged sword of needing to expand economically while also being encouraged to leapfrog to cleaner technologies, a process that can be hindered by financial and technological constraints (Zhou et al., 2017). Mitigation strategies must be tailored to the specific needs and capabilities of each continent and its countries' development status. Developed nations can aid

TABLE 4.1 Regional CO_2 emissions profile and opportunities.

Region	Emissions profile	Key drivers	Challenges	Opportunities	Reference
North America	High per capita emissions	Industrial activities, car dependency, fossil fuels	High energy use	Renewable energy shift, regulations	Azarpour et al. (2022)
Europe	High energy efficiency, aggressive sustainability policies	Renewable energy, public transport	Emissions variation in East Europe	Invest in renewables, transport	Kuzemko et al. (2022)
Asia (China and India)	Largest absolute emitters, lower per capita emissions	Industrialization, large populations, coal	High absolute emissions, coal use	Renewable energy investments	Meng et al. (2019)
Africa	Lowest per capita emissions, rising due to development	Development, urbanization	Rising emissions with growth	Solar and wind resources	Avila et al. (2017)
South America	Low per capita emissions, hydroelectric power, deforestation	Hydroelectric power, deforestation	Deforestation	Sustainable forestry, renewables	Peng et al. (2024)
Australia and Oceania	High per capita emissions, coal dependency, mining sector	Coal power, mining	Political challenges	Renewable energy transition	Burke et al. (2022)

developing countries by transferring technology and providing financial support for renewable energy projects (Suzuki, 2015). Integrating sustainability into the economic development framework can help manage and mitigate emissions growth in developing regions (Chishti, 2021). Global challenges require global solutions, and international cooperation is essential to achieve emission reduction targets across all continents (Reid et al., 2010).

4.4.2 Case studies of high versus low emitter countries

To illustrate the diverse strategies and challenges countries face in managing carbon emissions, let us explore case studies from two different spectrums: a high emitter (the United States) and a low emitter (Costa Rica). These examples highlight the contrast in emissions profiles, policies, and approaches toward climate change mitigation (Higham et al., 2016). The United States is one of the world's largest GHG emitters both in absolute terms and on a per capita basis, driven primarily by its reliance on fossil fuels for electricity generation and widespread car usage (Zheng et al., 2019), significant outputs from manufacturing, chemicals, and heavy industries (Wu et al., 2023), and substantial emissions from intensive livestock production and agricultural practices (Sakadevan & Nguyen, 2017). The United States has implemented various federal and state-level regulations aimed at reducing carbon emissions, including vehicle emission standards and incentives for renewable energy investments (Schwarzman et al., 2021), a significant shift toward renewable energy sources, particularly wind and solar (Gielen et al., 2019), and technological innovations in CCS, EVs, and energy efficiency solutions (Xin et al., 2023). However, challenges such as political polarization (Egan & Mullin, 2017) and economic dependencies on fossil fuel industries (Nalule, 2019) complicate these efforts.

In contrast, Costa Rica stands out as a low emitter with a strong commitment to sustainability, largely due to its reliance on renewable energy, with over 98% of its electricity coming from sources such as hydropower, wind, geothermal, and solar (Farrok, et al., 2022; Khan, et al., 2022), and the widespread adoption of sustainable agriculture and ecotourism practices (Higham et al., 2016). Costa Rica aims to become carbon neutral by 2050 and has implemented rigorous measures to reach this goal, including significant investments in renewable energy and ecosystem restoration (Higham et al., 2016) and a comprehensive decarbonization plan (Nawaz et al., 2022). Despite its ambitious goals, Costa Rica faces challenges related to funding these initiatives, requiring substantial international financial support (Valenciano-Salazar et al., 2022), and its small-scale emissions present unique challenges in further reducing its footprint (Nwaka et al., 2020).

The United States and Costa Rica's economic structures heavily influence their emissions profiles and mitigation strategies. The US economy is diversified and heavily industrialized, while Costa Rica's is more focused on agriculture and tourism, allowing easier integration of sustainable practices (Peng et al., 2024). The United States tends to have a top-down approach driven by federal and state policies and significant private sector involvement, whereas Costa Rica has adopted a more holistic national strategy integrating environmental sustainability into various facets of its development (Jones & Kammen, 2014). As a major emitter, the United States has a significant impact on global emissions, whereas Costa Rica's role is more symbolic, demonstrating that sustainable development is feasible and beneficial even for smaller nations (Peng et al., 2024). These case studies illustrate that while the challenges and strategies may vary dramatically between high and low emitter countries, effective policy, innovation, and international cooperation are key components in addressing global climate change (Fekete et al., 2021).

4.4.3 Urban versus rural emission dynamics

Understanding the differences in emission sources and characteristics between urban and rural areas is critical for designing effective environmental policies. Urban and rural regions contribute to GHG emissions in distinct ways due to their different economic activities, population densities, and energy uses (Jones & Kammen, 2014). Urban areas are hubs of economic activity, consuming a large portion of a country's energy primarily for transportation, heating, cooling of buildings, and industrial processes (Karner et al., 2016). The concentration of population and businesses leads to significant vehicle emissions, often exacerbated by traffic congestion (Dulal et al., 2011). Additionally, larger cities host more industrial facilities and commercial establishments, contributing to higher emissions (Guttikunda et al., 2019). Urban areas also generate a large amount of waste, leading to emissions from waste processing and disposal activities, including methane from landfills (Ramachandra et al., 2018).

To mitigate urban emissions, enhancing public transportation systems, promoting nonmotorized transport (walking and cycling), and encouraging the use of EVs can significantly reduce emissions (Makarova et al., 2017). Implementing strict building codes that mandate energy efficiency and retrofitting older buildings with modern energy-saving technologies can reduce heating and cooling demands (Han et al., 2021). Smart urban planning that reduces the need for long commutes and encourages green spaces can mitigate the urban heat island effect and improve overall sustainability (Cheshmehzangi et al., 2021).

In contrast, rural areas primarily emit greenhouse gases through agriculture, producing methane from livestock and nitrous oxide from soil management practices (Snyder et al., 2014). Deforestation and other land use changes for agricultural expansion lead to significant CO_2 emissions (Raihan et al., 2022). Rural homes often use fossil fuels for heating and cooking due to less access to centralized heating systems and natural gas networks, and they may also rely on more pollutive energy sources due to limited electricity access (Kerimray et al., 2017).

Mitigation strategies for rural emissions include adopting sustainable agricultural techniques to reduce methane and nitrous oxide emissions, such as precision farming, improved livestock diets, and integrated pest management (Plaza-Bonilla et al., 2020). Encouraging forest growth in rural areas can serve as a carbon sink, absorbing CO_2 (Raihan et al., 2022). Expanding access to renewable energy sources, such as solar panels for homes and biogas for cooking and heating, can reduce reliance on fossil fuels (Suman, 2021).

Urban areas, with their high population density, have better infrastructure to support efficient mass transit and centralized energy solutions, which can reduce per capita emissions compared to rural areas (Cooke & Behrens, 2017). The predominant economic activities in urban and rural areas—industrial and commercial in urban settings versus agricultural in rural—dictate different types of emissions and require tailored mitigation approaches (Vos, 2019). Urban areas generally have better access to newer technologies and more capital for investments in green solutions than rural areas, which can be disadvantaged by a lack of financial resources and technology access (Nerini et al., 2016). Effective climate action requires differentiated strategies that consider the unique characteristics and needs of urban and rural areas, with urban strategies focusing on reducing transportation emissions and improving building efficiencies, while rural strategies prioritize agricultural emissions and land use management (Milojevic-Dupont & Creutzig, 2021).

4.5 Technological innovations and their impact

4.5.1 Advances in renewable energy technologies

The transition to renewable energy is crucial for mitigating climate change, and recent advances in technology have significantly improved the efficiency, affordability, and scalability of these energy sources. Table 4.2 provides an overview of key developments across various types of renewable energy technologies.

TABLE 4.2 Key developments in renewable energy technologies.

Type	Category	Development	Efficiency	Benefits	References
Solar power	PV cells	High-efficiency panels	>30% (Perovskite)	Increased energy capture	Lewis (2016)
		Flexible solar films	Flexible, lightweight	More application surfaces	Li et al. (2021), Li, Yang, et al. (2021)
	Concentrating solar power (CSP)	Advanced CSP technologies	Improved storage (molten salt)	Energy generation without sunlight	Islam et al. (2018)
Wind power	Turbine tech	Larger turbine designs	Offshore >10 MW	Greater energy capture, lower cost per MW	Burnham et al. (2017)
		Floating turbines	Deep water deployment	New offshore locations	Bento and Fontes (2019)
	Efficiency	Advanced materials, aerodynamics	Lighter, durable materials	Enhanced efficiency and lifespan	Tong (2018)
		Smart technologies	Real-time blade adjustment	Optimized wind capture	
Hydroelectric	Pumped storage	Enhanced storage solutions	Modernized turbines	Better grid integration	Hunt et al. (2020)
	Small-scale hydro	Modular hydro systems	Small rivers/ streams	Minimal infrastructure investment	Zhou and Deng (2017)
Geothermal	Enhanced geothermal systems (EGS)	Drilling innovations	Reduced costs, increased feasibility	Accessing nontraditional areas	Song et al. (2023)
	Heat pumps	Ground source heat pumps	Improved extraction efficiency	Heating and cooling with minimal electrical input	Singh et al. (2019)
Bioenergy	Advanced biofuels	Cellulosic ethanol	Nonfood biomass	Sustainable alternative to traditional biofuels	Sharma et al. (2020)
		Algae-based fuels	High-yield, CO_2 capture	Sustainable fuel production	Sahu et al. (2024)
System integration	Smart grid	Grid responsiveness	Advanced battery storage	Better management of energy loads	Tan et al. (2021)
	Energy management	AI and IoT	Real-time optimization	Reduced waste, increased efficiency	Li et al. (2023)

These technological advances are rapidly changing the landscape of renewable energy, making it more efficient, cost-effective, and scalable. This progress is essential for transitioning away from fossil fuels and achieving global sustainability and climate goals (Subramanian et al., 2023).

4.5.2 Carbon capture and storage technologies

CCS is a crucial technology in the efforts to mitigate climate change by capturing CO_2 emissions from sources like power plants and industrial processes, and storing it underground to prevent its release into the atmosphere. This technology is particularly important for enabling the continued use of fossil fuels in a more environmentally responsible way (Shahbazi & Rezaei Nasab, 2016). CCS encompasses a variety of methods to capture and store CO_2 emissions from industrial processes and energy production. Key techniques include precombustion, postcombustion, and oxy-fuel

TABLE 4.3 Overview of carbon capture and storage.

Category	Method	Description	Key points	References
Carbon capture	Precombustion	CO_2 removed before combustion	Gasification, water-gas shift	Martens et al. (2017)
	Postcombustion	CO_2 captured from flue gases	Solvents/sorbents, regeneration, compression	Zhao et al. (2018)
	Oxy-fuel combustion	Burning in pure oxygen	Water vapor condensation, direct CO_2 capture	Nemitallah et al. (2017)
Transport	Pipelines	CO_2 transported via pipelines	Supercritical state, similar to natural gas	Onyebuchi et al. (2018)
	Shipping	CO_2 shipped in liquid form	Specially designed tankers	Al Baroudi et al. (2021)
Storage	Geological storage	CO_2 injected into underground formations	Depleted oil fields, saline aquifers	Ajayi et al. (2019)
	Mineral carbonation	CO_2 converted to stable minerals	Magnesium and calcium-bearing minerals	Yoo et al. (2022)
Advances in CCS technologies	Advanced solvents and sorbents	More efficient, less energy-intensive materials	Increased efficiency, reduced energy use	Burnham et al. (2017)
	Modular systems	Scalable capture systems	Retrofits, reduced capital costs	Hills et al. (2016)
	Direct air capture (DAC)	Capturing CO_2 from the atmosphere	Removes legacy emissions, currently expensive	Chowdhury et al. (2023)
	Bioenergy with CCS (BECCS)	Combines bioenergy with CCS	Negative emissions, energy generation	Almena et al. (2022)
Integration with renewables	Utilization of excess energy	Uses surplus renewable energy for CCS	Reduces CCS carbon footprint	Koytsoumpa et al. (2018)
Challenges and prospects	Economic viability	High costs, policy support needed	Cost reduction through R&D, carbon pricing	Burnham et al. (2017), Shen et al. (2022)
	Environmental concerns	Risk of CO_2 leakage	Long-term monitoring, management	Damen et al. (2006)
	Public acceptance	Importance of public support	Safety of transport and storage	Abdulla et al. (2020)

combustion capture methods. The captured CO_2 is then transported via pipelines or shipping and stored through geological storage or mineral carbonation. Advances in CCS technologies aim to improve efficiency and reduce costs, with innovations such as advanced solvents, modular systems, direct air capture, and bioenergy with CCS. Despite the potential, CCS faces economic, environmental, and public acceptance challenges (Table 4.3).

4.5.3 Impact of digitalization on carbon emissions (e.g., smart grids, IoT in energy management)

Digitalization, characterized by the integration of digital technologies into everyday life, is profoundly impacting carbon emissions. Technologies such as smart grids, the Internet of Things (IoT), and artificial intelligence (AI) are transforming energy management and consumption patterns, providing significant opportunities to reduce GHG emissions (Li et al., 2023). Smart grids utilize digital communication technology to detect and react to local changes in electricity usage. Unlike traditional grids, smart grids can manage the demands of connected homes, businesses, and other systems in real-time, enhancing the efficiency of electricity distribution and reducing wastage (Dileep, 2020). By optimizing power generation and distribution based on real-time data, smart grids can significantly reduce energy waste, leading to lower emissions from power plants, particularly those relying on fossil fuels (Panda & Das, 2021). Additionally, smart grids facilitate the integration of renewable energy sources such as wind and solar by managing their intermittent nature and smoothing the supply of electricity to the grid, thereby increasing the share of renewables in the energy mix (Worighi et al., 2019).

IoT devices can monitor and control the energy consumption of appliances and systems within homes and businesses. These devices collect data that can be used to optimize energy use and reduce unnecessary consumption (Li et al., 2023). IoT applications allow for precise control over heating, cooling, and lighting, adapting these functions to actual occupancy and use patterns, which reduces overconsumption and lowers carbon emissions (Naylor et al., 2018). Moreover, IoT sensors can predict when machinery or equipment requires maintenance before it fails, ensuring optimal performance and energy use, which helps in reducing emissions due to inefficiency (Li et al., 2023).

AI algorithms analyze large volumes of data from various sources, including weather stations, power plants, and consumers, to predict energy demand and optimize energy supply from different sources (Devaraj et al., 2021). AI can improve the predictability of energy demand and the efficiency of energy supply, particularly in integrating diverse energy sources with varying output levels, such as wind and solar power (Ahmad et al., 2021). AI-enhanced systems can significantly cut emissions by optimizing the operation of heating, ventilation, and air conditioning (HVAC) systems in buildings, route planning in logistics, and even in the operation of electric and hybrid vehicles (Haleem et al., 2023).

The increased connectivity and extensive use of data analytics raise concerns about data security and privacy. Protecting these data from breaches is critical, as the consequences could impact the reliability of energy systems (Moreno et al., 2016). Privacy concerns regarding the collection of detailed usage data from individuals and businesses must be managed through robust data protection regulations (Martin & Murphy, 2017). Implementing smart grids, IoT devices, and AI in energy systems involves significant initial investments, which can be a barrier for widespread adoption, particularly in developing countries (Serban & Lytras, 2020). The integration of these technologies into existing infrastructure requires overcoming significant technical challenges, including compatibility and interoperability issues (Janssen et al., 2020).

4.6 Socioeconomic impacts of carbon emissions

4.6.1 Health impacts related to air quality and climate change

The health impacts of poor air quality and climate change are significant and multifaceted, affecting communities globally (Khraishah et al., 2022). Climate change contributes to deteriorating air quality in many ways, such as increasing the frequency and intensity of wildfires, altering patterns of air pollution, and exacerbating heatwaves. These environmental changes have direct and indirect effects on human health (Otorkpa et al., 2024). The health impacts of poor air quality and climate change, detailing specific pollutants and sources, their health effects, and corresponding mitigation strategies. Key references for each aspect are included for further reading (Table 4.4).

4.6.2 Economic impacts of emission reductions and climate adaptation

The push to reduce emissions and adapt to climate change involves substantial economic considerations (Zilberman et al., 2012). These measures, while often costly in the short term, are generally aimed at avoiding the far more severe costs associated with unmitigated climate change impacts. Here, we explore both the direct and indirect economic effects of these initiatives (Riahi et al., 2015). Investment in clean technologies is a major area of expenditure. Transitioning to low-carbon energy sources, such as solar, wind, and nuclear power, requires significant upfront investments. For example, building renewable energy infrastructure like wind farms or solar panels involves high initial costs but tends to have lower operational costs compared to fossil fuel-based technologies (King & Van Den Bergh, 2018). Moreover, funding research and development for new technologies that reduce emissions in industries such as manufacturing, agriculture, and transportation also demands substantial investment (Ganda, 2019). Regulatory and policy measures also impose costs. Implementing carbon taxes or cap-and-trade systems can increase operating costs for businesses, particularly in high-emission industries such as power generation, manufacturing, and transportation. While this raises prices for consumers, it encourages investment in cleaner technologies and practices (Ji et al., 2023). Additionally, adhering to stricter emissions standards can lead to increased costs for businesses as they upgrade equipment, improve efficiency, or purchase emission credits (Ouyang et al., 2020).

Despite the initial costs, emission reduction efforts can yield significant long-term savings. Investments in energy efficiency and renewable energy can lower operational costs over time. Renewable energy sources, for instance, do not

TABLE 4.4 Health impacts and mitigation strategies for air quality deterioration and climate change.

Aspect	Pollutant/ factor	Source/ description	Health impact	Mitigation strategy	References
Air quality deterioration	Particulate matter (PM)	Industrial emissions, vehicles	Respiratory, cardiovascular diseases	Emission control, air filters	Chen et al. (2016)
	Ground-level ozone (O3)	Cars, power plants, sunlight reaction	Asthma, bronchitis, emphysema	Reduce vehicle emissions, regulate industrial output	Karthik et al. (2017)
	Nitrogen oxides (NOx)	Vehicles, industrial activities	Respiratory problems, smog formation	Stricter emission standards, cleaner technologies	Manisalidis et al. (2020)
	Sulfur dioxide (SO_2)	Fossil fuel combustion	Respiratory issues, acid rain	Switch to low-sulfur fuels, scrubbers in power plants	Manisalidis et al. (2020)
Climate change exacerbation	Heatwaves	Global warming	Increased ozone levels, heat stress	Urban cooling strategies, public health advisories	Wang et al. (2024)
	Wildfires	Droughts, high temperatures	Respiratory problems, particulate exposure	Forest management, emergency response plans	Knorr et al. (2017)
	Pollen and allergens	Higher CO_2, warmer temperatures	Aggravated asthma, allergies	Allergy forecasting, urban greenery management	Choi et al. (2021)
Health impacts	Respiratory diseases	Air pollutants, allergens	Asthma, infections, reduced lung function	Air quality monitoring, healthcare system strengthening	Deng et al. (2020)
	Cardiovascular diseases	Prolonged pollutant exposure	Heart attacks, stroke	Public health campaigns, emission reduction	Franklin et al. (2015)
	Mortality	Poor air quality	Premature deaths	Comprehensive air quality policies	Bălă et al. (2021)
Vulnerable populations	Children, elderly	Higher susceptibility	Increased health risks	Targeted health interventions, support services	Borchers et al. (2020)
	Low-income communities	Limited access to healthcare	Higher exposure, poorer outcomes	Equity-focused policies, improved healthcare access	Borchers et al. (2020)
Adaptation strategies	Healthcare system	Infrastructure, training	Better management of climate-related health	Enhanced healthcare infrastructure, training programs	Ofremu et al. (2024)
	Policy interventions	Regulations, urban planning	Improved air quality, reduced emissions	Stricter standards, green urban planning	Gunawardena et al. (2017), Jonidi Jafari et al. (2021)
	Public awareness	Education, campaigns	Reduced exposure, informed actions	Air quality index (AQI) education, heatwave advisories	Di Virgilio et al. (2024)
	Research and monitoring	Data collection, analysis	Timely health interventions	Advanced monitoring systems, ongoing research	Tong and Ebi (2019), Orsetti et al. (2022)

require fuel purchases, which can result in substantial savings compared to fossil fuel-based energy (Dranka et al., 2020). Furthermore, lower emissions lead to better air quality, which can significantly reduce healthcare costs related to air pollution-related diseases such as asthma, chronic bronchitis, and heart disease (Li et al., 2021; Yang, et al., 2021). Job creation is another benefit of emission reduction initiatives. The renewable energy sector has been a significant job creator in many regions, often offsetting job losses in declining fossil fuel industries. Jobs in solar and wind energy, energy efficiency retrofitting, and EV manufacturing contribute to a growing green economy (Meyer & Sommer, 2016). Additionally, by diversifying energy sources and reducing dependence on imported fuels, countries can enhance their energy security and economic stability, reducing their vulnerability to oil price fluctuations (Aslanturk & Kıprızlı, 2020).

Investments in climate-resilient infrastructure are crucial for mitigating the impacts of climate change. Upgrades and resilience building, such as flood defenses, hardened power grids, and water conservation systems, require significant expenditure but are essential for avoiding greater losses from climate impacts like extreme weather events (Shakou et al., 2019). In agriculture, adapting practices to changing climatic conditions can involve costs for new seeds, irrigation systems, and altered farming practices. However, these costs are critical for maintaining agricultural productivity and food security (Hertel & Lobell, 2014).

Failing to reduce emissions and adapt to climate changes can lead to severe economic consequences. Natural disasters, such as hurricanes, floods, and droughts, will become more frequent and severe, imposing heavy costs on economies from disaster response to rebuilding efforts (Banholzer et al., 2014). Additionally, heatwaves and rising temperatures can reduce worker productivity, particularly in outdoor and uncooled environments, impacting sectors such as construction, agriculture, and manufacturing (Oppermann et al., 2021). Changes in climate can also lead to water shortages, reduced agricultural output, and disruptions to supply chains, further impacting economic activities and food prices (Davis et al., 2021).

4.6.3 Social inequalities exacerbated by environmental degradation

Environmental degradation, including climate change, pollution, and the depletion of natural resources, disproportionately impacts marginalized and vulnerable populations, exacerbating existing social inequalities. This occurs through various mechanisms, including differential exposure to environmental hazards, unequal access to resources, and varied capacities to respond to environmental challenges (Holden et al., 2019). Marginalized communities often face geographic vulnerabilities due to their residential locations. They are frequently situated in areas more susceptible to environmental hazards, such as flood plains, regions prone to industrial pollution, or areas with poor water quality. For example, low-income neighborhoods may be located near polluting factories or in low-lying areas vulnerable to flooding (Tate et al., 2021). Additionally, individuals from disadvantaged backgrounds are more likely to work in hazardous conditions, such as agriculture with exposure to toxic pesticides or waste recycling involving harmful materials (Damalas & Eleftherohorinos, 2011). The cumulative effect of exposure to pollutants leads to higher rates of respiratory and cardiovascular diseases, cancer, and other health issues in these communities. Children in poorer areas are particularly at risk of developmental delays and health problems due to exposure to lead, air pollutants, and other toxic substances (Kim et al., 2020).

Economic disparities contribute significantly to the unequal impact of environmental degradation. Marginalized communities often face reduced access to critical natural resources such as clean water, arable land, and healthy food. Water scarcity, for instance, disproportionately affects poor regions with less infrastructure to handle shortages (Jhariya et al., 2022). The degradation of natural resources can also erode the economic base of communities relying on agriculture, fishing, or natural resources for their livelihoods, leading to increased poverty and reduced opportunities for upward mobility (Robinson, 2016). Wealthier nations and individuals have greater means to invest in infrastructure that can mitigate the impacts of environmental degradation, such as sea walls, air conditioning, or water purification systems. In contrast, poorer communities often lack the resources to protect themselves effectively (Alamgir et al., 2017). Access to insurance and the ability to recover financially from environmental disasters are also skewed toward those with greater economic resources (Chang et al., 2022).

There is often a disparity in environmental awareness and education between socioeconomic groups. Wealthier individuals tend to have better access to information on protecting themselves from environmental hazards and participating more effectively in environmental decision-making processes (Morello-Frosch et al., 2011). Marginalized groups frequently have less representation in political forums where decisions about environmental policy or resource allocation are made, leading to policies that do not consider their specific needs and vulnerabilities (Hinkel, 2011). Strong community networks can enhance resilience to environmental shocks, but these networks can be strained by continuous or severe environmental pressures. Social isolation is more likely in transient or highly stressed communities (Bouwer et al., 2021). Environmental degradation can force migration, which is a significant upheaval that poorer populations are less equipped to manage. Migration due to environmental reasons often leads to social fragmentation and further economic hardship (Warner et al., 2010).

Developing environmental policies that explicitly consider the needs and vulnerabilities of marginalized groups can help ensure that these communities are protected and that their conditions do not worsen (Shi et al., 2016). Empowering underrepresented communities to have a voice in environmental governance can help tailor responses to the specific needs and priorities of these populations (Butler & Adamowski, 2015). Investing in health services and infrastructure improvements specifically in vulnerable areas can reduce the impact of environmental degradation on these populations (Barbier & Hochard, 2018). Providing education about environmental risks and mitigation strategies in disadvantaged communities can improve their ability to cope with and adapt to environmental changes (Jabareen, 2013). Understanding and addressing the ways in which environmental degradation exacerbates social inequalities is crucial for creating sustainable and equitable societies. By recognizing these linkages, policymakers can design interventions that not only address environmental issues but also promote social justice (Van Niekerk, 2020).

4.7 Policy and regulatory frameworks

4.7.1 International agreements and their effectiveness (e.g., Kyoto Protocol, Paris Agreement)

International agreements on climate change are critical as climate issues transcend national borders and require cooperative global action. Two of the most significant agreements are the Kyoto Protocol and the Paris Agreement. Their effectiveness can be measured in terms of their design, implementation, and the real-world impacts on global carbon emissions (van Asselt & Zelli, 2014). Adopted in December 1997 and entered into force in February 2005, the Kyoto Protocol was the first international treaty to mandate country-specific emissions reductions. It set binding targets for 37 industrialized countries and the European community to reduce GHG emissions (Cirman et al., 2009). The Protocol introduced mechanisms such as emissions trading, known as the "carbon market," allowing countries with spare emission units to sell them to countries exceeding their targets (Raufer et al., 2022). The clean development mechanism (CDM) encouraged investment in sustainable development projects in developing countries, generating "certified emission reduction" (CER) credits (Lim & Lam, 2014). Joint implementation (JI) allowed countries with emission reduction commitments to earn emission reduction units from projects in other committed countries (Fichtner et al., 2001). While the Kyoto Protocol marked a historic step in global efforts to combat climate change, its impact was limited by several factors. The United States, a major emitter, never ratified the treaty, and Canada withdrew in 2011. Additionally, the Protocol only covered a small portion of global GHG emissions as it did not include developing countries like China and India, whose emissions have grown substantially (Cirman et al., 2009).

Adopted in December 2015, the Paris Agreement marked a significant shift in global climate policy by including 196 countries and setting the broader goal of limiting global warming to well below 2°C above preindustrial levels, aiming for 1.5°C (Höhne et al., 2021). Unlike the Kyoto Protocol, the Paris Agreement requires all parties, regardless of economic status, to submit nationally determined contributions (NDCs)—plans to reduce national emissions and adapt to climate impacts. These NDCs are to be updated every 5 years, becoming progressively more ambitious (Winning et al., 2019). Every 5 years, a global stocktake assesses the collective progress toward the Agreement's goals, informing further actions by Parties (Zahar, 2019). The Paris Agreement's flexibility has been praised for encouraging broader participation. However, the voluntary nature of NDCs has led to challenges in enforcement and variability in national ambition levels (Raiser et al., 2022). The Agreement relies on political and moral pressure to drive compliance. It has been instrumental in galvanizing global action and investment in renewable energy, establishing a framework for transparency and accountability in emissions reporting (Held & Roger, 2018).

Both the Kyoto Protocol and the Paris Agreement face challenges related to compliance and the sufficiency of country-level ambitions to meet global temperature goals (Hermwille et al., 2019). Developing countries often require financial and technical support to achieve ambitious mitigation and adaptation goals. The effectiveness of international agreements hinges on the ability of richer countries to fulfill their financial commitments to climate finance (Hofman & van der Gaast, 2019). The United States has had a fluctuating role in these agreements, significantly impacting their effectiveness. Its commitment and leadership—or lack thereof—can influence other countries' attitudes toward implementation (Vaccaro et al., 2012). As global economic and political dynamics evolve, international agreements must adapt to maintain relevance and effectiveness. Increasing participation from nonstate actors, such as cities, businesses, and civil society, is also crucial for advancing goals (Bozhikin et al., 2019).

4.7.2 National policies for emission reduction and their economic implications

National policies aimed at reducing GHG emissions are crucial in the fight against climate change. These policies vary widely depending on the country's economic structure, political landscape, and environmental priorities. Their economic implications are equally diverse, affecting industries, employment, public health, and overall economic growth (Eskander et al., 2020). Carbon pricing mechanisms, such as carbon taxes and cap-and-trade systems, are essential tools. A carbon tax imposes a direct tax on the carbon content of fossil fuels, effectively increasing the cost of using coal, oil, and natural gas, thereby incentivizing businesses and consumers to reduce consumption and invest in cleaner alternatives (Lin & Jia, 2020). Cap-and-trade systems establish a cap on total emissions and allow companies to buy and sell permits to emit CO_2, creating a market price for carbon emissions and a financial incentive to reduce emissions (Yang et al., 2018). Regulatory measures like emission standards and renewable portfolio standards (RPS) also play a crucial role. Emission standards set limits on the amount of pollutants that can be emitted from specific sources, such as vehicles or industrial plants, often requiring the use of certain pollution control technologies (Saxena & Sonwani, 2019). RPS require a certain percentage of electricity to be generated from renewable energy sources, driving investment in technologies like wind and solar power (Lyon, 2016). Subsidies and financial incentives support the development and adoption of renewable energy and energy efficiency technologies. These include tax credits, rebates, and grants (Abdmouleh et al., 2015). Government funding for research and development in low-carbon technologies can speed up innovation and lower the cost of clean technologies (Pigato et al., 2020).

The impact of these policies on industries is significant. High-emission industries such as coal mining, oil extraction, and traditional manufacturing may contract, leading to job losses and economic decline in regions dependent on these industries (Tian et al., 2022). Conversely, these policies can drive growth in renewable energy sectors, environmental consulting, sustainable agriculture, and green manufacturing, creating new jobs and business opportunities (Ikram et al., 2021). Macroeconomic effects include shifts in investment from fossil fuel-based assets to clean technologies, potentially leading to a "green boom" in certain economies (Le Billon & Kristoffersen, 2020). Nations that move early to adopt green technologies may gain a competitive edge in the rapidly growing global market for clean energy products and services (Gupta et al., 2008). Public health and social benefits are also significant. Reducing emissions can lower air pollution, leading to fewer health problems like asthma, heart disease, and lung cancer, thereby reducing healthcare costs and increasing workforce productivity (Hadley et al., 2018). Additionally, by decreasing reliance on imported fossil fuels and increasing the use of domestic renewable energy sources, countries can enhance their energy security and stability (Gökgöz & Güvercin, 2018).

Revenue from carbon taxes can be substantial, providing governments with funds that can be reinvested in clean energy projects, public transportation, or even returned to the public through rebates or reduced taxes (Carl & Fedor, 2016). Investment in renewable energy infrastructure and energy efficiency projects can act as an economic stimulus, creating jobs and stimulating economic activity during periods of economic downturn (Mundaca & Luth Richter, 2015). By reducing emissions, countries can mitigate long-term risks associated with climate change, such as extreme weather events and sea-level rise, which pose serious threats to economic stability (Abbass et al., 2022). Ensuring an equitable transition is crucial, as policies must consider socio-economic equity to ensure that vulnerable populations do not bear the brunt of the changes and that the benefits of green growth are widely distributed (Allam et al., 2020). Effective emission reduction requires coherent policies that integrate energy, environmental, and economic planning at all levels of government (Nguyen, 2023). National policies are more effective when aligned with international goals and coordinated with other countries, particularly in managing border tax adjustments and avoiding carbon leakage (Fourie, 2018).

4.7.3 The role of local governance in implementing sustainable practices

Local governments play a pivotal role in addressing climate change and implementing sustainable practices. As the level of government closest to the people, local authorities can tailor responses to the specific environmental, social, and economic conditions within their jurisdictions. Their efforts are crucial in achieving national and international sustainability goals, particularly in areas such as urban planning, waste management, energy use, and transportation (Table 4.5).

4.8 Future outlook and projections

4.8.1 Modeling future emission scenarios and their climatic impacts

Modeling future emission scenarios and predicting their climatic impacts are critical components of climate science, helping policymakers and the public understand the potential long-term consequences of different paths of GHG emissions. These models involve complex simulations that account for various factors, including economic growth, energy

TABLE 4.5 Local government strategies for climate change mitigation.

Area	Action	Example measures	Impact	Challenges	Support needed	Reference
Urban planning	Building codes	Green building practices, solar-ready roofs	Reduced energy consumption, sustainability	Enforcement, adoption	Technical guidance, funding	Dogru et al. (2019)
	Zoning laws	Mixed-use development, public transit zones	Reduced vehicle travel, preserved green spaces	Legal adjustments, community acceptance	Legal support, community engagement	Jacobsen et al. (2022)
Waste management	Recycling programs	Curbside recycling, composting programs	Reduced landfill use, lower emissions	Infrastructure, public participation	Grants, public awareness programs	Zheng et al. (2021)
	Sustainable disposal	Waste-to-energy plants, advanced composting	Lower emissions from waste	Technological investment, operational costs	Technology funding, subsidies	Zheng et al. (2021)
Renewable energy	Incentives for renewables	Tax incentives, rebates, subsidies	Increased renewable energy adoption	Financial resources, public uptake	Financial aid, policy support	Kostka (2014)
	Public projects	Solar panels on public buildings, geothermal	Renewable energy usage, reduced GHG emissions	Project funding, technical expertise	Federal grants, technical assistance	Kostka (2014)
Transportation	Public transit	Bus fleets, light rail, bike lanes	Reduced vehicle emissions, improved transit	Infrastructure costs, public use	Infrastructure funding, planning aid	Dogru et al. (2019)
	Traffic management	Car-free zones, carpool lanes	Decreased congestion, lower emissions	Implementation, public cooperation	Policy alignment, public awareness	Dogru et al. (2019)
Community	Awareness campaigns	Carbon footprint education, recycling programs	Increased public awareness, behavioral change	Outreach efforts, engagement strategies	Communication resources, training	Aronson et al. (2017)
	Stakeholder engagement	Involving businesses, resident groups	Effective policy implementation, local support	Coordination, maintaining interest	Facilitation support, incentives	Aronson et al. (2017)
Challenges	Resource constraints	Financial, technical, human resources	Limited adoption of advanced strategies	Funding, skilled personnel	Grants, technical training	Lowndes and Pratchett (2012)
	Policy coordination	Aligning with state/national regulations	Smooth implementation, policy coherence	Discrepancies in methods and objectives	Policy guidance, alignment tools	Jacobsen et al. (2022)
	Data availability	Environmental data access	Effective planning, impact measurement	Data scarcity, measurement tools	Data collection support, tools	Zheng et al. (2021)
Enhancements	Intermunicipal collaboration	Sharing resources, best practices	Enhanced program effectiveness, regional benefits	Coordination, resource sharing	Facilitation, cooperative frameworks	Aronson et al. (2017)
	Public–private partnerships	Collaborations with private sector	Additional resources, innovation	Partnership management, funding agreements	Partnership facilitation, funding	Kostka (2014)
	State and federal support	Grants, guidelines, technical assistance	Effective policy implementation, innovation	Access to support, regulatory compliance	Federal and state cooperation	Lowndes and Pratchett (2012)

TABLE 4.6 Modeling future emission scenarios and predicting climatic impacts.

Component	Description	Impact/outcome	References
Climate models			
General circulation models	Simulate Earth's climate systems (atmosphere, oceans, land, ice)	Understanding energy and matter interactions	Balaji et al. (2022)
Integrated assessment models	Integrate data from climatology, economics, technology	Evaluate impacts of climate change and mitigation	Pastor et al. (2020), Devaraj et al. (2021)
Emission scenarios			
Representative concentration pathways (RCPs)	Four pathways describing possible climate futures based on GHG emissions	Range from significant emission reductions (RCP 2.6) to continued increases (RCP 8.5)	Gütschow et al. (2021), Huard et al. (2022)
Shared socioeconomic pathways (SSPs)	Combine with RCPs to include demographic, economic, social changes	Help understand global societal evolution with climate change	O'Neill et al. (2017), Welch (2024)
Projecting climatic impacts			
Temperature increases	Predicts global temperature rise under different scenarios	Varies from <2°C to >4°C above preindustrial levels	Rogelj et al. (2018)
Sea level rise	Expansion of warmer sea waters, melting land ice	Greater sea-level rises linked to higher emissions	Cazenave and Cozannet (2014)
Extreme weather events	Increased frequency/intensity of hurricanes, droughts, heavy precipitation	Devastates ecosystems, agriculture, water resources, infrastructure	Aghakouchak et al. (2020)
Ecological and biodiversity impacts	Higher warming levels affecting ecosystems and biodiversity	Ocean acidification, forest diebacks, species extinctions	Mora et al. (2018)
Using models for policy and planning			
Adaptation strategies	Develop strategies to protect communities, infrastructure	Flood defenses, water conservation, heat-resistant designs	Jacobsen et al. (2022)
Mitigation efforts	Set GHG reduction targets, framework for international climate policies	Plan transition to renewable energy sources	Waisman et al. (2019)
Public awareness and engagement	Raise awareness, motivate sustainable practices	Highlight consequences of inaction	Gota et al. (2019)

production, technological advances, and policy changes (Han et al., 2024). Overview of climate models, emission scenarios, projected climatic impacts, and their applications in policy and planning is provided in Table 4.6.

4.8.2 The role of population growth, economic development, and consumption patterns

The dynamics of population growth, economic development, and consumption patterns are integral to understanding the drivers of climate change (Han et al., 2017). These factors not only influence the scale of GHG emissions but also affect the capacity of societies to manage and mitigate their impacts. Their interrelationship shapes climate policies, summarizing the impact of population growth, economic development, and consumption patterns on emissions. Table 4.7 highlights the associated challenges and opportunities across various categories.

4.8.3 Strategies for achieving net-zero emissions: global and regional perspectives

Achieving net-zero emissions—a balance between the greenhouse gases put into the atmosphere and those taken out—by mid-century is essential to mitigate the worst impacts of climate change. This goal requires an unprecedented

TABLE 4.7 Impact of population growth, economic development, and consumption patterns on emissions.

Category	Impact	Examples	Challenges	Opportunities	References
Population growth	Increases energy consumption and emissions	More housing, transportation, food	Higher CO_2 emissions	Advancements in efficiency, sustainable practices	Mohsin et al. (2019)
Urbanization trends	Expansion of cities, dual effects on environment	Efficient resource use, local pollution	Heat island effects without green planning	Green planning initiatives	Buhaug and Urdal (2013)
Industrialization	Increased carbon emissions	Growing industries	High energy consumption from fossil fuels	Transition to cleaner technologies	Rehman et al. (2021)
Income levels and energy demand	Higher energy demand and emissions	Larger homes, more goods, transportation	Sustainable policies to manage consumption	Renewable energy adoption	Sorrell (2015)
Transition to service economies	Lower direct emissions compared to manufacturing	Service-based economies	Managing overall global emissions (carbon leakage)	Sustainable service sector development	Haug and Ucal (2019), Wiedmann et al. (2021)
Consumerism	Continuous growth of emissions	Fast fashion, single-use products, meat	Severe climate impact	Promotion of sustainable consumption habits	Schandl et al. (2016), Le Quéré et al. (2019)
Technological advancements	Dual impact on consumption and emissions	Digital devices, energy-efficient tech	Increased electricity demand	Reduced per capita energy consumption	Subramanian et al. (2023)
Sustainable consumption	Efficient resource use, reduced waste	Eco-friendly products	Changing consumer behavior	Decoupling economic growth from environmental degradation	King and Van Den Bergh (2018)
Integrated approaches	Effective, equitable climate policies	Urban density, clean energy incentives	Tailoring solutions to different demographics	Comprehensive policy frameworks	Nores and Fernandez (2018)
International cooperation	Managing global impacts of local actions	Climate agreements, trade policies	Ensuring cooperation among nations	Global emission reduction strategies	Zaidi et al. (2019)
Educational and cultural shifts	Changing consumption patterns through awareness	Sustainable practices education	Overcoming cultural resistance	Broad public engagement campaigns	Blackburn et al. (2024)

transformation of how energy is produced, industries are run, cities are built, and land is used. Here, we explore both global and regional strategies that can facilitate this profound transition (Huang & Zhai, 2021). Decarbonizing energy production is a cornerstone of achieving net-zero emissions. Rapidly scaling up renewable energy technologies like solar, wind, hydro, and geothermal is vital, involving the phasing out of coal-fired power plants and reducing dependency on oil and natural gas (Aslanturk & Kıprızlı, 2020). Some advocate for the expansion of nuclear energy as a stable, low-carbon alternative to fossil fuels, particularly for countries with high energy demands (King & Van Den Bergh, 2018). Improving energy efficiency is another critical strategy. Implementing stricter energy efficiency standards for buildings, appliances, and industrial processes can significantly reduce energy consumption (Li & Shui, 2015). Developing smart grid technologies and expanding the capacity of energy storage solutions are crucial to managing the intermittency issues of renewable sources (Hepburn et al., 2021).

Transforming the transportation sector is essential. Promoting the adoption of EVs through incentives, building charging infrastructure, and setting timelines for phasing out internal combustion engine vehicles can reduce emissions significantly (Mai et al., 2018). Additionally, investing in robust public transit systems and infrastructure for walking and cycling can drastically cut urban transportation emissions (Martin & Murphy, 2017). Carbon removal strategies include both natural solutions and technological interventions. Protecting and restoring forests, wetlands, and other natural ecosystems that sequester carbon dioxide effectively is crucial (Aslanturk & Kıprızlı, 2020). Advancing technologies such as direct air capture and carbon storage can remove CO_2 from the atmosphere and store it underground (Subramanian et al., 2023). In agriculture, transitioning to sustainable practices such as no-till farming, organic production, and precision agriculture can reduce methane and nitrous oxide emissions. Encouraging a shift toward plant-based diets can also reduce the carbon footprint of food production (Reid et al., 2010).

Developed countries are expected to lead the way in cutting emissions, investing in clean technology, and funding climate change mitigation and adaptation efforts globally (Hepburn et al., 2021). They are also key players in developing new technologies and facilitating the transfer of these technologies to developing nations (Kruger & Steyn, 2020). For many developing countries, the energy strategy may include a mix of low-carbon technologies and efficient fossil fuel technologies in the short term, moving toward more renewable sources as they become economically feasible (Albino et al., 2014). Achieving net-zero in developing regions often requires substantial international financial and technical support, including investments in infrastructure and capacity building (Das & Ghosh, 2023). Regions especially vulnerable to climate change effects, such as small island developing states, focus heavily on adaptation strategies. Building resilience against rising sea levels, increasing hurricane intensity, and other impacts is crucial (Mycoo, 2018). These regions often need to transition their economies away from climate-vulnerable sectors like traditional agriculture and tourism to more sustainable alternatives (Mahat et al., 2019). Creating strong legal and regulatory frameworks at both national and international levels can guide and compel the necessary changes across all sectors (Nores & Fernandez, 2018). Climate financing is essential, particularly for poor and vulnerable countries, and includes grants, loans, and innovative funding solutions like green bonds and carbon credit trading (Bhandary et al., 2021). Both the public and private sectors play pivotal roles; governments need to set ambitious targets and policies, while businesses must innovate and adapt their operations toward sustainability goals (Mazzucato et al., 2020). Finally, achieving net-zero is not just a governmental or corporate task but involves communities and individuals actively participating through lifestyle changes and pressure on institutions to reform (Rosenbloom & Meadowcroft, 2022).

4.9 Conclusion

The comprehensive analysis of carbon emission patterns highlights several critical insights: first, the dominant role of energy production, industrial activities, and transportation in global emissions; second, the significant geographic and sectoral disparities in emissions; and third, the escalating impact of economic development and urbanization on global carbon output. These findings underscore the necessity of coordinated global action—no single nation can combat climate change alone. Effective mitigation requires international treaties like the Paris Agreement, coupled with national strategies and local initiatives that address specific regional needs and capabilities. Therefore there is a clear call to action for policymakers to enact and enforce robust climate policies, for businesses to innovate and invest in sustainable practices and technologies, and for individuals to adopt more sustainable lifestyles. Together, through a unified approach that spans governmental, corporate, and personal levels, significant progress can be made in reducing global carbon emissions and mitigating the effects of climate change.

References

Abbass, K., Qasim, M. Z., Song, H., Murshed, M., Mahmood, H., & Younis, I. (2022). A review of the global climate change impacts, adaptation, and sustainable mitigation measures. *Environmental Science and Pollution Research*, *29*(28), 42539–42559. Available from https://doi.org/10.1007/s11356-022-19718-6.

Abdmouleh, Z., Alammari, R. A. M., & Gastli, A. (2015). Review of policies encouraging renewable energy integration & best practices. *Renewable and Sustainable Energy Reviews*, *45*, 249–262. Available from https://doi.org/10.1016/j.rser.2015.01.035.

Abdulla, A., Hanna, R., Schell, K. R., Babacan, O., & Victor, D. G. (2020). Explaining successful and failed investments in U.S. carbon capture and storage using empirical and expert assessments. *Environmental Research Letters*, *16*(1), 014036. Available from https://doi.org/10.1088/1748-9326/abd19e.

Acker, A., Chatterjee, E., Becker, L., Shutzer, M., & Capellini, N. (2023). *Fossil fuels from extraction to emissions. The Routledge handbook of environmental history* (pp. 229–243). Switzerland: Taylor and Francis. Available from https://doi.org/10.4324/9781003189350-19.

Adegbeye, M. J., Ravi Kanth Reddy, P., Obaisi, A. I., Elghandour, M. M. M. Y., Oyebamiji, K. J., Salem, A. Z. M., Morakinyo-Fasipe, O. T., Cipriano-Salazar, M., & Camacho-Díaz, L. M. (2020). Sustainable agriculture options for production, greenhouse gasses and pollution alleviation, and nutrient recycling in emerging and transitional nations - an overview. *Journal of Cleaner Production*, *242*, 118319. Available from https://doi.org/10.1016/j.jclepro.2019.118319.

Adshead, D., Roman, O., Thacker, S., & Hall, J. W. (2021). Infrastructure strategies for achieving the global development agendas in small islands. *Earth's Future*, *9*(2). Available from https://doi.org/10.1029/2020ef001699.

Aghakouchak, A., Chiang, F., Huning, L. S., Love, C. A., Mallakpour, I., Mazdiyasni, O., Moftakhari, H., Papalexiou, S. M., Ragno, E., & Sadegh, M. (2020). Climate extremes and compound hazards in a warming world. *Annual Review of Earth and Planetary Sciences*, *48*(1), 519–548. Available from https://doi.org/10.1146/annurev-earth-071719-055228.

Ahmad, T., Zhang, D., Huang, C., Zhang, H., Dai, N., Song, Y., & Chen, H. (2021). Artificial intelligence in sustainable energy industry: Status quo, challenges and opportunities. *Journal of Cleaner Production.*, *289*. Available from https://doi.org/10.1016/j.jclepro.2021.125834.

Ajayi, T., Gomes, J. S., & Bera, A. (2019). A review of CO2 storage in geological formations emphasizing modeling, monitoring and capacity estimation approaches. *Petroleum Science*, *16*(5), 1028–1063. Available from https://doi.org/10.1007/s12182-019-0340-8.

Al Baroudi, H., Awoyomi, A., Patchigolla, K., Jonnalagadda, K., & Anthony, E. J. (2021). A review of large-scale CO2 shipping and marine emissions management for carbon capture, utilisation and storage. *Applied Energy*, *287*. Available from https://doi.org/10.1016/j.apenergy.2021.116510.

Alamgir, M., Campbell, M. J., Sloan, S., Goosem, M., Clements, G. R., Mahmoud, M. I., & Laurance, W. F. (2017). Economic, socio-political and environmental risks of road development in the tropics. *Current Biology*, *27*(20), R1130–R1140. Available from https://doi.org/10.1016/j.cub.2017.08.067.

Albayrak, T., Atodiresei, D., & Popa, C. (2021). Cross border cooperation for sustainable environment protection. *Scientific Bulletin "Mircea cel Batran" Naval Academy*, *24*(2), 1–13.

Albino, V., Ardito, L., Dangelico, R. M., & Messeni Petruzzelli, A. (2014). Understanding the development trends of low-carbon energy technologies: A patent analysis. *Applied Energy*, *135*, 836–854. Available from https://doi.org/10.1016/j.apenergy.2014.08.012.

Allam, Z., Jones, D., & Thondoo, M. (2020). *Cities and climate change: Climate policy, economic resilience and urban sustainability*. Springer Nature.

Almena, A., Thornley, P., Chong, K., & Röder, M. (2022). Carbon dioxide removal potential from decentralised bioenergy with carbon capture and storage (BECCS) and the relevance of operational choices. *Biomass and Bioenergy*, *159*. Available from https://doi.org/10.1016/j.biombioe.2022.106406.

Aminzadegan, S., Shahriari, M., Mehranfar, F., & Abramović, B. (2022). Factors affecting the emission of pollutants in different types of transportation: A literature review. *Energy Reports*, *8*, 2508–2529. Available from https://doi.org/10.1016/j.egyr.2022.01.161.

Arneth, A., Sitch, S., Pongratz, J., Stocker, B. D., Ciais, P., Poulter, B., Bayer, A. D., Bondeau, A., Calle, L., Chini, L. P., Gasser, T., Fader, M., Friedlingstein, P., Kato, E., Li, W., Lindeskog, M., Nabel, J. E. M. S., Pugh, T. A. M., Robertson, E., & Zaehle, S. (2017). Historical carbon dioxide emissions caused by land-use changes are possibly larger than assumed. *Nature Geoscience*, *10*(2), 79–84. Available from https://doi.org/10.1038/ngeo2882.

Aronson, M. F. J., Lepczyk, C. A., Evans, K. L., Goddard, M. A., Lerman, S. B., MacIvor, J. S., Nilon, C. H., & Vargo, T. (2017). Biodiversity in the city: Key challenges for urban green space management. *Frontiers in Ecology and the Environment*, *15*(4), 189–196. Available from https://doi.org/10.1002/fee.1480.

Aslanturk, O., & Kıprızlı, G. (2020). The role of renewable energy in ensuring energy security of supply and reducing energy-related import. *International Journal of Energy Economics and Policy*, *10*(2), 354–359. Available from https://doi.org/10.32479/ijeep.8414.

Astarita, C. (2016) The ambiguities in the China-India relationship, in The role of the European Union in Asia: China and India as strategic partners. Hong Kong: Taylor and Francis, pp. 189–209. Available from https://doi.org/10.4324/9781315553283-11.

Aste, N., Manfren, M., & Marenzi, G. (2017). Building automation and control systems and performance optimization: A framework for analysis. *Renewable and Sustainable Energy Reviews*, *75*, 313–330. Available from https://doi.org/10.1016/j.rser.2016.10.072.

Avila, N., Carvallo, J. P., Shaw, B., & Kammen, D. M. (2017). The energy challenge in sub-Saharan Africa: A guide for advocates and policy makers,. *Generating Energy for Sustainable and Equitable Development, Part 1*, 1–79.

Azam, A., Rafiq, M., Shafique, M., & Yuan, J. (2022). Mitigating carbon emissions in China: The role of clean energy, technological innovation, and political-institutional quality. *Frontiers in Environmental Science*, *10*. Available from https://doi.org/10.3389/fenvs.2022.814439.

Azarpour, A., Mohammadzadeh, O., Rezaei, N., & Zendehboudi, S. (2022). Current status and future prospects of renewable and sustainable energy in North America: Progress and challenges. *Energy Conversion and Management*, *269*, 115945. Available from https://doi.org/10.1016/j.enconman.2022.115945.

Balaji, V., Couvreux, F., Deshayes, J., Gautrais, J., Hourdin, F., & Rio, C. (2022). Are general circulation models obsolete? *Proceedings of the National Academy of Sciences*, *119*(47). Available from https://doi.org/10.1073/pnas.2202075119.

Banholzer, S., Kossin, J., & Donner, S. (2014). *The impact of climate change on natural disasters. Reducing disaster: Early warning systems for climate change* (pp. 21–49). Canada: Springer Netherlands. Available from 10.1007/978-94-017-8598-3_2.

Barbier, E. B., & Hochard, J. P. (2018). The impacts of climate change on the poor in disadvantaged regions. *Review of Environmental Economics and Policy*, *12*(1), 26–47. Available from https://doi.org/10.1093/reep/rex023.

Barma, M. C., Saidur, R., Rahman, S. M. A., Allouhi, A., Akash, B. A., & Sait, S. M. (2017). A review on boilers energy use, energy savings, and emissions reductions. *Renewable and Sustainable Energy Reviews*, *79*, 970–983. Available from https://doi.org/10.1016/j.rser.2017.05.187.

Bento, N., & Fontes, M. (2019). Emergence of floating offshore wind energy: Technology and industry. *Renewable and Sustainable Energy Reviews*, *99*, 66–82. Available from https://doi.org/10.1016/j.rser.2018.09.035.

Bersalli, G., Tröndle, T., & Lilliestam, J. (2023). Most industrialised countries have peaked carbon dioxide emissions during economic crises through strengthened structural change. *Communications Earth & Environment*, *4*(1). Available from https://doi.org/10.1038/s43247-023-00687-8.

Besen, M. R., Ribeiro, R. H., Bratti, F., Locatelli, J. L., Schmitt, D. E., & Piva, J. T. (2024). Cover cropping associated with no-tillage system promotes soil carbon sequestration and increases crop yield in Southern Brazil. *Soil and Tillage Research*, *242*, 106162. Available from https://doi.org/10.1016/j.still.2024.106162.

Bhandary, R. R., Gallagher, K. S., & Zhang, F. (2021). Climate finance policy in practice: A review of the evidence. *Climate Policy*, *21*(4), 529–545. Available from https://doi.org/10.1080/14693062.2020.1871313.

Blackburn, R., Leviston, Z., Walker, I., & Schram, A. (2024). Could a minimalist lifestyle reduce carbon emissions and improve wellbeing? A review of minimalism and other low consumption lifestyles. *Wiley Interdisciplinary Reviews: Climate Change*, *15*(2). Available from https://doi.org/10.1002/wcc.865.

Borchers, A., Bowman, N., Palmer, D. M., J, A., & Johnston, F. H. (2020). *Climate change, wildfires, heatwaves and health impacts in Australia. Extreme weather events and human health: International case studies* (pp. 99–116).

Bouwer, R., Pasquini, L., & Baudoin, M.-A. (2021). Breaking down the silos: Building resilience through cohesive and collaborative social networks. *Environmental Development*, *39*. Available from https://doi.org/10.1016/j.envdev.2021.100646.

Bozhikin, I., Macke, J., & da Costa, L. F. (2019). The role of government and key non-state actors in social entrepreneurship: A systematic literature review. *Journal of Cleaner Production*, *226*, 730–747. Available from https://doi.org/10.1016/j.jclepro.2019.04.076.

Brown, C., Alexander, P., Arneth, A., Holman, I., & Rounsevell, M. (2019). Achievement of Paris climate goals unlikely due to time lags in the land system. *Nature Climate Change*, *9*(3), 203–208. Available from https://doi.org/10.1038/s41558-019-0400-5.

Buhaug, H., & Urdal, H. (2013). An urbanization bomb? Population growth and social disorder in cities. *Global Environmental Change*, *23*(1), 1–10. Available from https://doi.org/10.1016/j.gloenvcha.2012.10.016.

Burke, P. J., Beck, F. J., Aisbett, E., Baldwin, K. G. H., Stocks, M., Pye, J., Venkataraman, M., Hunt, J., & Bai, X. (2022). Contributing to regional decarbonization: Australia's potential to supply zero-carbon commodities to the Asia-Pacific. *Energy*, *248*, 123563. Available from https://doi.org/10.1016/j.energy.2022.123563.

Burnham, A., Dufek, E. J., Stephens, T., Francfort, J., Michelbacher, C., Carlson, R. B., Zhang, J., Vijayagopal, R., Dias, F., Mohanpurkar, M., Scoffield, D., Hardy, K., Shirk, M., Hovsapian, R., Ahmed, S., Bloom, I., Jansen, A. N., Keyser, M., Kreuzer, C., & Tanim, T. R. (2017). Enabling fast charging – infrastructure and economic considerations. *Journal of Power Sources*, *367*, 237–249. Available from https://doi.org/10.1016/j.jpowsour.2017.06.079.

Butler, C., & Adamowski, J. (2015). Empowering marginalized communities in water resources management: Addressing inequitable practices in participatory model building. *Journal of Environmental Management*, *153*, 153–162. Available from https://doi.org/10.1016/j.jenvman.2015.02.010.

Bălă, G. P., Râjnoveanu, R-M., Tudorache, E., Motişan, R., & Oancea, C. (2021). Air pollution exposure—the (in)visible risk factor for respiratory diseases. *Environmental Science and Pollution Research*, *28*(16), 19615–19628. Available from https://doi.org/10.1007/s11356-021-13208-x.

Cadez, S., & Czerny, A. (2016). Climate change mitigation strategies in carbon-intensive firms. *Journal of Cleaner Production*, *112*, 4132–4143. Available from https://doi.org/10.1016/j.jclepro.2015.07.099.

Cantarero, M. M. V. (2020). Of renewable energy, energy democracy, and sustainable development: A roadmap to accelerate the energy transition in developing countries,. *Energy Research and Social Science*, *70*.

Cao, X., Dai, X., & Liu, J. (2016). Building energy-consumption status worldwide and the state-of-the-art technologies for zero-energy buildings during the past decade. *Energy and Buildings*, *128*, 198–213. Available from https://doi.org/10.1016/j.enbuild.2016.06.089.

Carl, J., & Fedor, D. (2016). Tracking global carbon revenues: A survey of carbon taxes versus cap-and-trade in the real world. *Energy Policy*, *96*, 50–77. Available from https://doi.org/10.1016/j.enpol.2016.05.023.

Cazenave, A., & Cozannet, G. L. (2014). Sea level rise and its coastal impacts, Earth's Future. *American Geophysical Union*, *2*(2), 15–34. Available from https://doi.org/10.1002/2013ef000188.

Chang, S. E., Brown, C., Handmer, J., Helgeson, J., Kajitani, Y., Keating, A., Noy, I., Watson, M., Derakhshan, S., Kim, J., & Roa-Henriquez, A. (2022). Business recovery from disasters: Lessons from natural hazards and the COVID-19 pandemic. *International Journal of Disaster Risk Reduction*, *80*, 103191. Available from https://doi.org/10.1016/j.ijdrr.2022.103191.

Chapman, M., Walker, W. S., Cook-Patton, S. C., Ellis, P. W., Farina, M., Griscom, B. W., & Baccini, A. (2020). Large climate mitigation potential from adding trees to agricultural lands. *Global Change Biology*, *26*(8), 4357–4365. Available from https://doi.org/10.1111/gcb.15121.

Chen, R., Hu, B., Liu, Y., Xu, J., Yang, G., Xu, D., & Chen, C. (2016). Beyond PM2.5: The role of ultrafine particles on adverse health effects of air pollution. *Biochimica et Biophysica Acta*, *1860*(12), 2844–2855. Available from https://doi.org/10.1016/j.bbagen.2016.03.019.

Cheshmehzangi, A., Butters, C., Xie, L., & Dawodu, A. (2021). Green infrastructures for urban sustainability: Issues, implications, and solutions for underdeveloped areas. *Urban Forestry and Urban Greening*, *59*, 127028. Available from https://doi.org/10.1016/j.ufug.2021.127028.

Chishti, M. Z., Ahmad, M., Rehman, A., & Khan, M. K. (2021). Mitigations pathways towards sustainable development: Assessing the influence of fiscal and monetary policies on carbon emissions in BRICS economies. *Journal of Cleaner Production*, *292*, 126035. Available from https://doi.org/10.1016/j.jclepro.2021.126035.

Choi, Y. J., Lee, K. S., & Oh, J. W. (2021). The impact of climate change on pollen season and allergic sensitization to pollens. *Immunology and Allergy Clinics of North America*, *41*(1), 97–109. Available from https://doi.org/10.1016/j.iac.2020.09.004.

Chowdhury, S., Kumar, Y., Shrivastava, S., Patel, S. K., & Sangwai, J. S. (2023). A review on the recent scientific and commercial progress on the direct air capture technology to manage atmospheric CO2 concentrations and future perspectives. *Energy and Fuels*, *37*(15), 10733–10757. Available from https://doi.org/10.1021/acs.energyfuels.2c03971.

Cirman, A., Domadenik, P., Koman, M., & Redek, T. (2009). The Kyoto protocol in a global perspective. *Economic and Business Review*, *11*(1). Available from https://doi.org/10.15458/2335-4216.1259.

Cooke, S., & Behrens, R. (2017). Correlation or cause? The limitations of population density as an indicator for public transport viability in the context of a rapidly growing developing city. *Transportation Research Procedia*, *25*, 3003–3016. Available from https://doi.org/10.1016/j.trpro.2017.05.229.

Cordovil, C. M., Bittman, S., Brito, L. M., Goss, M. J., Hunt, D., Serra, J., . . . Hutchings, N. (2020). *Climate-resilient and smart agricultural management tools to cope with climate change-induced soil quality decline. Climate change and soil interactions* (pp. 613–662). Elsevier.

Costa, F. N., & Ribeiro, D. V. (2020). Reduction in CO2 emissions during production of cement, with partial replacement of traditional raw materials by civil construction waste (CCW). *Journal of Cleaner Production*, *276*. Available from https://doi.org/10.1016/j.jclepro.2020.123302.

Crystal-Ornelas, R., Thapa, R., & Tully, K. L. (2021). Soil organic carbon is affected by organic amendments, conservation tillage, and cover cropping in organic farming systems: A meta-analysis. *Agriculture, Ecosystems and Environment*, *312*. Available from https://doi.org/10.1016/j.agee.2021.107356.

Damalas, C. A., & Eleftherohorinos, I. G. (2011). Pesticide exposure, safety issues, and risk assessment indicators. *International Journal of Environmental Research and Public Health*, *8*(5), 1402–1419. Available from https://doi.org/10.3390/ijerph8051402.

Damen, K., Faaij, A., & Turkenburg, W. (2006). Health, safety and environmental risks of underground Co2 storage – overview of mechanisms and current knowledge. *Climatic Change*, *74*(1–3), 289–318. Available from https://doi.org/10.1007/s10584-005-0425-9.

Das, A., & Ghosh, A. (2023). Vision net zero: A review of decarbonisation strategies to minimise climate risks of developing countries. *Environment, Development and Sustainability*. Available from https://doi.org/10.1007/s10668-023-03318-6.

Davis, K. F., Downs, S., & Gephart, J. A. (2021). Towards food supply chain resilience to environmental shocks. *Nature Food*, *2*(1), 54–65. Available from https://doi.org/10.1038/s43016-020-00196-3.

Deng, S. Z., Jalaludin, B. B., Antó, J. M., Hess, J. J., & Huang, C. R. (2020). Climate change, air pollution, and allergic respiratory diseases: A call to action for health professionals. *Chinese Medical Journal*, *133*(13), 1552–1560. Available from https://doi.org/10.1097/CM9.0000000000000861.

Devaraj, J., Madurai Elavarasan, R., Shafiullah, G. M., Jamal, T., & Khan, I. (2021). A holistic review on energy forecasting using big data and deep learning models. *International Journal of Energy Research*, *45*(9), 13489–13530. Available from https://doi.org/10.1002/er.6679.

Di, D., La Gennusa, S., Peri, M., Rizzo, G., & Vinci, I. (2018). Involving people in the building up of smart and sustainable cities: How to influence commuters' behaviors through a mobile app game. *Sustainable Cities and Society*, *42*, 325–336.

Di Virgilio, G., Hart, M., Maharaj, A., & Green, D. (2024). The impact of air quality on schoolchildren. The impact of extreme weather on school education (pp. 38-59). Routledge.

Dileep, G. (2020). A survey on smart grid technologies and applications. *Renewable Energy*, *146*, 2589–2625. Available from https://doi.org/10.1016/j.renene.2019.08.092.

Dodson, J., Li, T., & Sipe, N. (2018). *The socioeconomic equity dimensions of a transition in suburban motor vehicle fuel and technology. Urban sustainability transitions: Australian cases-international perspectives* (pp. 233–249).

Dogru, T., Marchio, E. A., Bulut, U., & Suess, C. (2019). Climate change: Vulnerability and resilience of tourism and the entire economy. *Tourism Management*, *72*, 292–305. Available from https://doi.org/10.1016/j.tourman.2018.12.010.

Dranka, G. G., Ferreira, P., & Vaz, A. I. F. (2020). Cost-effectiveness of energy efficiency investments for high renewable electricity systems. *Energy*, *198*. Available from https://doi.org/10.1016/j.energy.2020.117198.

Dulal, H. B., Brodnig, G., & Onoriose, C. G. (2011). Climate change mitigation in the transport sector through urban planning: A review. *Habitat International*, *35*(3), 494–500. Available from https://doi.org/10.1016/j.habitatint.2011.02.001.

Egan, P. J., & Mullin, M. (2017). Climate change: US public opinion. *Annual Review of Political Science*, *20*, 209–227. Available from https://doi.org/10.1146/annurev-polisci-051215-022857.

Eskander., Shaikh, M. S. U., & Fankhauser, S. (2020). Reduction in greenhouse gas emissions from national climate legislation. *Nature Climate Change*, *10*(8), 750–756. Available from https://doi.org/10.1038/s41558-020-0831-z.

Fais, B., Sabio, N., & Strachan, N. (2016). The critical role of the industrial sector in reaching long-term emission reduction, energy efficiency and renewable targets. *Applied Energy*, *162*, 699–712. Available from https://doi.org/10.1016/j.apenergy.2015.10.112.

Fekete, H., Kuramochi, T., Roelfsema, M., den Elzen, M., Forsell, N., Höhne, N., Luna, L., Hans, F., Sterl, S., Olivier, J., van Soest, H., Frank, S., & Gusti, M. (2021). A review of successful climate change mitigation policies in major emitting economies and the potential of global replication. *Renewable and Sustainable Energy Reviews*, *137*, 110602. Available from https://doi.org/10.1016/j.rser.2020.110602.

Fichtner, W., Goebelt, M., & Rentz, O. (2001). The efficiency of international cooperation in mitigating climate change: Analysis of joint implementation, the clean development mechanism and emission trading for the Federal Republic of Germany, the Russian Federation and Indonesia. *Energy Policy*, *29*(10), 817–830. Available from https://doi.org/10.1016/s0301-4215(00)00139-7.

Foster, V., Dim, J. U., Vollmer, S., & Zhang, F. (2023). Understanding the challenge of decoupling transport-related CO2 emissions from economic growth in developing countries. *World Development Sustainability*, *3*, 100111. Available from https://doi.org/10.1016/j.wds.2023.100111.

Fourie, W. (2018). Aligning South Africa's national development plan with the 2030 agenda's sustainable development goals: Guidelines from the policy coherence for development movement. *Sustainable Development*, *26*(6), 765–771. Available from https://doi.org/10.1002/sd.1745.

Franklin, B. A., Brook, R., & Arden Pope, C. (2015). Air pollution and cardiovascular disease. *Current Problems in Cardiology*, *40*(5), 207–238. Available from https://doi.org/10.1016/j.cpcardiol.2015.01.003.

Frölicher, T. L., Fischer, E. M., & Gruber, N. (2018). Marine heatwaves under global warming. *Nature*, *560*(7718), 360–364. Available from https://doi.org/10.1038/s41586-018-0383-9.

Ganda, F. (2019). The impact of innovation and technology investments on carbon emissions in selected organisation for economic co-operation and development countries. *Journal of Cleaner Production*, *217*, 469–483. Available from https://doi.org/10.1016/j.jclepro.2019.01.235.

Gielen, D., Boshell, F., Saygin, D., Bazilian, M. D., Wagner, N., & Gorini, R. (2019). The role of renewable energy in the global energy transformation. *Energy Strategy Reviews*, *24*, 38–50. Available from https://doi.org/10.1016/j.esr.2019.01.006.

Glaeser, E. L., & Kahn, M. E. (2004). *Sprawl and urban growth. Handbook of regional and urban economics* (pp. 2481–2527). United States: Elsevier. Available from https://doi.org/10.1016/S1574-0080(04)80013-0.

Gökgöz, F., & Güvercin, M. T. (2018). Energy security and renewable energy efficiency in EU. *Renewable and Sustainable Energy Reviews*, *96*, 226–239. Available from https://doi.org/10.1016/j.rser.2018.07.046.

Gomes da Silva, F. J., Gouveia, R. M., Gomes da Silva, F. J., & Gouveia, R. M. (2020). Global population growth and industrial impact on the environment. *Cleaner production: Toward a better future*, 33–75.

Gota, S., Huizenga, C., Peet, K., Medimorec, N., & Bakker, S. (2019). Decarbonising transport to achieve Paris Agreement targets. *Energy Efficiency*, *12*(2), 363–386. Available from https://doi.org/10.1007/s12053-018-9671-3.

Griffin, P. W., Hammond, G. P., & Norman, J. B. (2018). Industrial energy use and carbon emissions reduction in the chemicals sector: A UK perspective. *Applied Energy*, *227*, 587–602. Available from https://doi.org/10.1016/j.apenergy.2017.08.010.

Gulzar, A., Islam, T., Gulzar, R., & Hassan, T. (2021). Climate change and impacts of extreme events on human health: An overview. *Indonesian Journal of Social and Environmental Issues*, *2*(1), 68–77.

Gunawardena, K. R., Wells, M. J., & Kershaw, T. (2017). Utilising green and bluespace to mitigate urban heat island intensity. *Science of the Total Environment*, *584–585*, 1040–1055. Available from https://doi.org/10.1016/j.scitotenv.2017.01.158.

Gupta, A. K., Govindarajan, V., & Wang, H. (2008). *The quest for global dominance: Transforming global presence into global competitive advantage*. John Wiley & Sons.

Gür, T. M. (2022). Carbon Dioxide Emissions, Capture, Storage and Utilization: Review of Materials, Processes and Technologies. *Progress in Energy and Combustion Science*, *89*, 100965. Available from https://doi.org/10.1016/j.pecs.2021.100965.

Gütschow, J., Jeffery, M. L., Günther, A., & Meinshausen, M. (2021). Country-resolved combined emission and socio-economic pathways based on the representative concentration pathway (RCP) and shared socio-economic pathway (SSP) scenarios. *Earth System Science Data*, *13*(3), 1005–1040. Available from https://doi.org/10.5194/essd-13-1005-2021.

Guttikunda, S. K., Nishadh, K. A., Gota, S., Singh, P., Chanda, A., Jawahar, P., & Asundi, J. (2019). Air quality, emissions, and source contributions analysis for the Greater Bengaluru region of India. *Atmospheric Pollution Research*, *10*(3), 941–953. Available from https://doi.org/10.1016/j.apr.2019.01.002.

Hadley, M. B., Baumgartner, J., & Vedanthan, R. (2018). Developing a clinical approach to air pollution and cardiovascular health. *Circulation*, *137* (7), 725–742. Available from https://doi.org/10.1161/CIRCULATIONAHA.117.030377.

Haleem, A., Javaid, M., Khan, I. H., & Mohan, S. (2023). Significant applications of artificial intelligence towards attaining sustainability. *Journal of Industrial Integration and Management*, *8*(4), 489–520. Available from https://doi.org/10.1142/S2424862223500331.

Han, H., Zeeshan, Z., Talpur, B. A., Sadiq, T., Bhatti, U. A., Awwad, E. M., Al-Razgan, M., & Ghadi, Y. Y. (2024). Studying long term relationship between carbon emissions, soil, and climate change: Insights from a global Earth modeling Framework. *International Journal of Applied Earth Observation and Geoinformation*, *130*103902.

Han, J., Meng, X., Zhou, X., Yi, B., Liu, M., & Xiang, W. N. (2017). A long-term analysis of urbanization process, landscape change, and carbon sources and sinks: A case study in 'China's Yangtze River Delta region. *Journal of Cleaner Production*, *141*, 1040–1050.

Han, S., Yao, R., & Li, N. (2021). The development of energy conservation policy of buildings in China: A comprehensive review and analysis. *Journal of Building Engineering*, *38*102229.

Haug, A. A., & Ucal, M. (2019). The role of trade and FDI for CO2 emissions in Turkey: Nonlinear relationships. *Energy Economics*, *81*, 297–307. Available from https://doi.org/10.1016/j.eneco.2019.04.006.

Hegerl, G. C., Brönnimann, S., Schurer, A., & Cowan, T. (2018). The early 20th century warming: Anomalies, causes, and consequences. *Wiley Interdisciplinary Reviews: Climate Change*, *9*(4). Available from https://doi.org/10.1002/wcc.522.

Held, D., & Roger, C. (2018). Three models of global climate governance: From Kyoto to Paris and beyond. *Global Policy*, *9*(4), 527–537. Available from https://doi.org/10.1111/1758-5899.12617.

Hepburn, C., Qi, Y., Stern, N., Ward, B., Xie, C., & Zenghelis, D. (2021). Towards carbon neutrality and China's 14th Five-Year Plan: Clean energy transition, sustainable urban development, and investment priorities. *Environmental Science and Ecotechnology*, *8*. Available from https://doi.org/10.1016/j.ese.2021.100130.

Hermwille, L., Siemons, A., Förster, H., & Jeffery, L. (2019). Catalyzing mitigation ambition under the Paris Agreement: Elements for an effective Global Stocktake. *Climate Policy*, *19*(8), 988–1001. Available from https://doi.org/10.1080/14693062.2019.1624494.

Hertel, T. W., & Lobell, D. B. (2014). Agricultural adaptation to climate change in rich and poor countries: Current modeling practice and potential for empirical contributions. *Energy Economics*, *46*, 562–575.

Higham, J., Cohen, S. A., Cavaliere, C. T., Reis, A., & Finkler, W. (2016). Climate change, tourist air travel and radical emissions reduction. *Journal of Cleaner Production*, *111*, 336–347. Available from https://doi.org/10.1016/j.jclepro.2014.10.100.

Hills, T., Leeson, D., Florin, N., & Fennell, P. (2016). Carbon capture in the cement industry: Technologies, progress, and retrofitting. *Environmental Science and Technology*, *50*(1), 368–377. Available from https://doi.org/10.1021/acs.est.5b03508.

Hinkel, J. (2011). 'Indicators of vulnerability and adaptive capacity': Towards a clarification of the science–policy interface. *Global Environmental Change*, *21*(1), 198–208. Available from https://doi.org/10.1016/j.gloenvcha.2010.08.002.

Hoegh-Guldberg, O., Poloczanska, E. S., Skirving, W., & Dove, S. (2017). Coral reef ecosystems under climate change and ocean acidification. *Frontiers in Marine Science*, *4*, 158.

Hofman, E., & van der Gaast, W. (2019). Enhancing ambition levels in nationally determined contributions—learning from technology needs assessments. *Wiley Interdisciplinary Reviews: Energy and Environment*, *8*(1). Available from https://doi.org/10.1002/wene.311.

Höhne, N., Kuramochi, T., Warnecke, C., Röser, F., Fekete, H., Hagemann, M., Day, T., Tewari, R., Kurdziel, M., Sterl, S., & Gonzales, S. (2021). *The Paris Agreement: Resolving the inconsistency between global goals and national contributions. Climate policy after the 2015 Paris climate conference* (pp. 30–46). Routledge.

Holden, W. N. (2019). Endogenous exacerbation of an exogenous problem: Climate change, environmental degradation, and unsustainable development practices in the Philippines. *Asian Geographer*, *36*(1), 1–27. Available from https://doi.org/10.1080/10225706.2018.1483831.

Holechek, J. L., Geli, H. M., Sawalhah, M. N., & Valdez, R. (2022). A global assessment: Can renewable energy replace fossil fuels by 2050? *Sustainability*, *14*(8). Available from https://doi.org/10.3390/su14084792.

Hong, C., Burney, J. A., Pongratz, J., Nabel, J. E., Mueller, N. D., Jackson, R. B., & Davis, S. J. (2021). Global and regional drivers of land-use emissions in 1961–2017. *Nature*, *589*(7843), 554–561. Available from https://doi.org/10.1038/s41586-020-03138-y.

Huang, M. T., & Zhai, P. M. (2021). Achieving Paris Agreement temperature goals requires carbon neutrality by middle century with far-reaching transitions in the whole society. *Advances in Climate Change Research*, *12*(2), 281–286. Available from https://doi.org/10.1016/j.accre.2021.03.004.

Huang, Y., Shen, H., Chen, Y., Zhong, Q., Chen, H., Wang, R., Shen, G., Liu, J., Li, B., & Tao, S. (2015). Global organic carbon emissions from primary sources from 1960 to 2009. *Atmospheric Environment*, *122*, 505–512.

Huard, D., Fyke, J., Capellán-Pérez, I., Matthews, H. D., & Partanen, A. I. (2022). Estimating the likelihood of GHG concentration scenarios from probabilistic integrated assessment model simulations. *Earth's Future*, *10*(10). Available from https://doi.org/10.1029/2022EF002715.

Hui, C. X., Dan, G., Alamri, S., & Toghraie, D. (2023). Greening smart cities: An investigation of the integration of urban natural resources and smart city technologies for promoting environmental sustainability. *Sustainable Cities and Society*, *99*. Available from https://doi.org/10.1016/j.scs.2023.104985.

Hunt, J. D., Zakeri, B., Lopes, R., Barbosa, P. S. F., Nascimento, A., de Castro, N. J., Brandão, R., Schneider, P. S., & Wada, Y. (2020). Existing and new arrangements of pumped-hydro storage plants. *Renewable and Sustainable Energy Reviews*, *129*, 109914. Available from https://doi.org/10.1016/j.rser.2020.109914.

Igliński, H., & Babiak, M. (2017). Analysis of the potential of autonomous vehicles in reducing the emissions of greenhouse gases in road transport. *Procedia Engineering*, *192*, 353–358. Available from https://doi.org/10.1016/j.proeng.2017.06.061.

Ikram, M., Ferasso, M., Sroufe, R., & Zhang, Q. (2021). Assessing green technology indicators for cleaner production and sustainable investments in a developing country context. *Journal of Cleaner Production*, *322*, 129090. Available from https://doi.org/10.1016/j.jclepro.2021.129090.

Islam, M. T., Huda, N., Abdullah, A. B., & Saidur, R. (2018). A comprehensive review of state-of-the-art concentrating solar power (CSP) technologies: Current status and research trends. *Renewable and Sustainable Energy Reviews*, *91*, 987–1018. Available from https://doi.org/10.1016/j.rser.2018.04.097.

Jabareen, Y. (2013). Planning the resilient city: Concepts and strategies for coping with climate change and environmental risk. *Cities*, *31*, 220–229. Available from https://doi.org/10.1016/j.cities.2012.05.004.

Jacobs, M. (2023). The political ecology of the entropy principle: 19th century physics and the electrification of Germany. *Energy Research and Social Science*, *98*. Available from https://doi.org/10.1016/j.erss.2023.103008.

Jacobsen, A. P., Khiew, Y. C., Duffy, E., O'Connell, J., Brown, E., Auwaerter, P. G., Blumenthal, R. S., Schwartz, B. S., & McEvoy, J. W. (2022). Climate change and the prevention of cardiovascular disease. *American Journal of Preventive Cardiology*, *12*. Available from https://doi.org/10.1016/j.ajpc.2022.100391.

Janssen, M., Weerakkody, V., Ismagilova, E., Sivarajah, U., & Irani, Z. (2020). A framework for analysing blockchain technology adoption: Integrating institutional, market and technical factors. *International Journal of Information Management*, *50*, 302–309. Available from https://doi.org/10.1016/j.ijinfomgt.2019.08.012.

Jhariya, M. K., Banerjee, A., & Meena, R. S. (2022). *Importance of natural resources conservation: Moving toward the sustainable world. Natural resources conservation and advances for sustainability* (pp. 3–27). Elsevier.

Ji, J., Li, T., & Yang, L. (2023). Pricing and carbon reduction strategies for vertically differentiated firms under Cap-and-Trade regulation. *Transportation Research Part E: Logistics and Transportation Review*, *171*. Available from https://doi.org/10.1016/j.tre.2023.103064.

Jogdand, O. K. (2020). *Study on the effect of global warming and greenhouse gases on environmental system* (pp. 275–306). Informa UK Limited. Available from 10.1201/9780367808310-12.

Jones, C., & Kammen, D. M. (2014). Spatial distribution of U.S. household carbon footprints reveals suburbanization undermines greenhouse gas benefits of urban population density. *Environmental Science and Technology*, *48*(2), 895–902. Available from https://doi.org/10.1021/es4034364.

Jonidi Jafari, A., Charkhloo, E., & Pasalari, H. (2021). Urban air pollution control policies and strategies: A systematic review. *Journal of Environmental Health Science and Engineering*, *19*, 1911–1940.

Jung, S. P., & Pandit, S. (2018). *Important factors influencing microbial fuel cell performance. Biomass, biofuels, biochemicals: Microbial electrochemical technology: Sustainable Platform for fuels, chemicals and remediation* (pp. 377–406). South Korea: Elsevier. Available from 10.1016/B978-0-444-64052-9.00015-7.

Kabeyi, M. J. B., & Olanrewaju, O. A. (2022). Sustainable energy transition for renewable and low carbon grid electricity generation and supply. *Frontiers in Energy Research*, *9*. Available from https://doi.org/10.3389/fenrg.2021.743114.

Kabir, M., Habiba, U. E., Khan, W., Shah, A., Rahim, S., De los Rios-Escalante, P. R., Farooqi, Z-U-R., Ali, L., & Shafiq, M. (2023). Climate change due to increasing concentration of carbon dioxide and its impacts on environment in 21st century; a mini review. *Journal of King Saud University - Science*, *35*(5), 102693. Available from https://doi.org/10.1016/j.jksus.2023.102693.

Kalghatgi, G., Levinsky, H., & Colket, M. (2018). Future transportation fuels. *Progress in Energy and Combustion Science*, *69*, 103–105. Available from 10.1016/j.pecs.2018.06.003.

Kamyab, H., SaberiKamarposhti, M., Hashim, H., & Yusuf, M. (2024). Carbon dynamics in agricultural greenhouse gas emissions and removals: A comprehensive review. *Carbon Letters*, *34*(1), 265–289. Available from https://doi.org/10.1007/s42823-023-00647-4.

Karlen, D. L., & Cambardella, C. A. (2020). *Conservation strategies for improving soil quality and organic matter storage* (pp. 395–420). Informa UK Limited. Available from 10.1201/9781003075561-15.

Karner, K., Theissing, M., & Kienberger, T. (2016). Energy efficiency for industries through synergies with urban areas. *Journal of Cleaner Production*, *119*, 167–177. Available from https://doi.org/10.1016/j.jclepro.2016.02.010.

Karthik L, B., Sujith, B., Rizwan, A. S., & Sehgal, M. (2017). Characteristics of the ozone pollution and its health effects in India. *International Journal of Medicine and Public Health*, *7*(1), 56–60. Available from https://doi.org/10.5530/ijmedph.2017.1.10.

Kerimray, A., Rojas-Solórzano, L., Torkmahalleh, M. A., Hopke, P. K., & Gallachóir, B. P. Ó. (2017). Coal use for residential heating: Patterns, health implications and lessons learned. *Energy for Sustainable Development*, *40*, 19–30. Available from https://doi.org/10.1016/j.esd.2017.05.005.

Khalili, S., Rantanen, E., Bogdanov, D., & Breyer, C. (2019). Global transportation demand development with impacts on the energy demand and greenhouse gas emissions in a climate-constrained world. *Energies*, *12*(20). Available from https://doi.org/10.3390/en12203870.

Khan, I., Hou, F., Irfan, M., Zakari, A., & Le, H. P. (2021). Does energy trilemma a driver of economic growth? The roles of energy use, population growth, and financial development. *Renewable and Sustainable Energy Reviews*, *146*. Available from https://doi.org/10.1016/j.rser.2021.111157.

Khraishah, H., Alahmad, B., Ostergard, R. L., AlAshqar, A., Albaghdadi, M., Vellanki, N., Chowdhury, M. M., Al-Kindi, S. G., Zanobetti, A., Gasparrini, A., & Rajagopalan, S. (2022). Climate change and cardiovascular disease: Implications for global health. *Nature Reviews Cardiology*, *19*(12), 798–812.

Kim, J. B., Prunicki, M., Haddad, F., Dant, C., Sampath, V., Patel, R., Smith, E., Akdis, C., Balmes, J., Snyder, M. P., Wu, J. C., & Nadeau, K. C. (2020). Cumulative lifetime burden of cardiovascular disease from early exposure to air pollution. *Journal of the American Heart Association*, *9* (6). Available from https://doi.org/10.1161/JAHA.119.014944.

King, L. C., & Van Den Bergh, J. C. (2018). Implications of net energy-return-on-investment for a low-carbon energy transition. *Nature Energy*, *3*(4), 334–340.

Knorr, W., Dentener, F., Lamarque, J. F., Jiang, L., & Arneth, A. (2017). Wildfire air pollution hazard during the 21st century. *Atmospheric Chemistry and Physics*, *17*(14), 9223–9236. Available from https://doi.org/10.5194/acp-17-9223-2017.

Kostka, G. (2014). Barriers to the implementation of environmental policies at the local level in China.

Koytsoumpa, E. I., Bergins, C., & Kakaras, E. (2018). The CO2 economy: Review of CO2 capture and reuse technologies. *Journal of Supercritical Fluids*, *132*, 3–16. Available from https://doi.org/10.1016/j.supflu.2017.07.029.

Kruger, S., & Steyn, A. A. (2020). Enhancing technology transfer through entrepreneurial development: practices from innovation spaces. *Journal of Technology Transfer*, *45*(6), 1655–1689. Available from https://doi.org/10.1007/s10961-019-09769-2.

Kuzemko, C., Blondeel, M., Dupont, C., & Brisbois, M. C. (2022). Russia's war on Ukraine, European energy policy responses & implications for sustainable transformations. *Energy Research and Social Science*, *93*. Available from https://doi.org/10.1016/j.erss.2022.102842.

Lall, S. (2004). *The employment impact of globalization in developing countries* (pp. 73–101). Palgrave Macmillan.

Lamb, W. F., Wiedmann, T., Pongratz, J., Andrew, R., Crippa, M., Olivier, J. G. J., Wiedenhofer, D., Mattioli, G., Khourdajie, A. A., House, J., Pachauri, S., Figueroa, M., Saheb, Y., Slade, R., Hubacek, K., Sun, L., Ribeiro, S. K., Khennas, S., de la Rue du Can, S., & Minx, J. (2021). A review of trends and drivers of greenhouse gas emissions by sector from 1990 to 2018. *Environmental Research Letters*, *16*(7)073005.

Larch, M., & Wanner, J. (2024). The consequences of non-participation in the Paris Agreement. *European Economic Review*, *163*. Available from https://doi.org/10.1016/j.euroecorev.2024.104699.

Lau, H. C., Ramakrishna, S., Zhang, K., & Radhamani, A. V. (2021). The role of carbon capture and storage in the energy transition. *Energy and Fuels*, *35*(9), 7364–7386. Available from https://doi.org/10.1021/acs.energyfuels.1c00032.

Le Billon, P., & Kristoffersen, B. (2020). Just cuts for fossil fuels? Supply-side carbon constraints and energy transition. *Environment and Planning A: Economy and Space*, *52*(6), 1072–1092. Available from https://doi.org/10.1177/0308518x18816702.

Le Quéré, C., Korsbakken, J. I., Wilson, C., Tosun, J., Andrew, R., Andres, R. J., Canadell, J. G., Jordan, A., Peters, G. P., & van Vuuren, D. P. (2019). Drivers of declining CO2 emissions in 18 developed economies. *Nature Climate Change*, *9*(3), 213–217. Available from https://doi.org/10.1038/s41558-019-0419-7.

Le Treut, G., Lefevre, J., Lallana, F., & Bravo, G. (2021). The multi-level economic impacts of deep decarbonization strategies for the energy system. *Energy Policy*, *156*112423.

Lewis, N. S. (2016). Research opportunities to advance solar energy utilization. *Science*, *351*(6271). Available from https://doi.org/10.1126/science.aad1920.

Li, J., Herdem, M. S., Nathwani, J., & Wen, J. Z. (2023). Methods and applications for Artificial Intelligence, Big Data, Internet of Things, and Blockchain in smart energy management. *Energy and AI*, *11*. Available from https://doi.org/10.1016/j.egyai.2022.100208.

Li, J., Kagawa, S., & Lin, C. (2020). China's CO2 emission structure for 1957–2017 through transitions in economic and environmental policies. *Journal of Cleaner Production*, *255*.

Li, J., & Shui, B. (2015). A comprehensive analysis of building energy efficiency policies in China: Status quo and development perspective. *Journal of Cleaner Production*, *90*, 326–344. Available from https://doi.org/10.1016/j.jclepro.2014.11.061.

Li, W., Yang, M., Long, R., He, Z., Zhang, L., & Chen, F. (2021). Assessment of greenhouse gasses and air pollutant emissions embodied in cross-province electricity trade in China. *Resources, Conservation and Recycling*, *171*105623.

Li, X., Li, P., Wu, Z., Luo, D., Yu, H. Y., & Lu, Z. H. (2021). Review and perspective of materials for flexible solar cells. *Materials Reports: Energy*, *1*(1). Available from https://doi.org/10.1016/j.matre.2020.09.001.

Lim, X. L., & Lam, W. H. (2014). Review on clean development mechanism (CDM) implementation in Malaysia. *Renewable and Sustainable Energy Reviews*, *29*, 276–285. Available from https://doi.org/10.1016/j.rser.2013.08.106.

Lin, B., & Jia, Z. (2020). Supply control vs. demand control: Why is resource tax more effective than carbon tax in reducing emissions? *Humanities and Social Sciences Communications*, *7*(1). Available from https://doi.org/10.1057/s41599-020-00569-w.

Lin, Q., Zhang, X., Wang, T., Zheng, C., & Gao, X. (2022). Technical perspective of carbon capture, utilization, and storage. *Engineering*, *14*, 27–32.

Lowndes, V., & Pratchett, L. (2012). Local governance under the coalition government: Austerity, localism and the 'big society. *Local Government Studies*, *38*(1), 21–40. Available from https://doi.org/10.1080/03003930.2011.642949.

Lu, Y., Yuan, J., Lu, X., Su, C., Zhang, Y., Wang, C., Cao, X., Li, Q., Su, J., Ittekkot, V., Garbutt, R. A., Bush, S., Fletcher, S., Wagey, T., Kachur, A., & Sweijd, N. (2018). Major threats of pollution and climate change to global coastal ecosystems and enhanced management for sustainability. *Environmental Pollution*, *239*, 670–680. Available from https://doi.org/10.1016/j.envpol.2018.04.016.

Lyon, T. P. (2016). Drivers and impacts of renewable portfolio standards. *Annual Review of Resource Economics*, *8*(1), 141–155. Available from https://doi.org/10.1146/annurev-resource-100815-095432.

Løvendahl, P., Difford, G. F., Li, B., Chagunda, M. G. G., Huhtanen, P., Lidauer, M. H., Lassen, J., & Lund, P. (2018). Selecting for improved feed efficiency and reduced methane emissions in dairy cattle. *Animal*, *12*(s2), s336–s349.

Mahat, T. J., Bláha, L., Uprety, B., & Bittner, M. (2019). Climate finance and green growth: Reconsidering climate-related institutions, investments, and priorities in Nepal. *Environmental Sciences Europe*, *31*(1). Available from https://doi.org/10.1186/s12302-019-0222-0.

Mai, T. T., Jadun, P., Logan, J. S., McMillan, C. A., Muratori, M., Steinberg, D. C., Vimmerstedt, L. J., Haley, B., Jones, R., & Nelson, B. (2018). *Electrification futures study: Scenarios of electric technology adoption and power consumption for the United States*. Office of Scientific and Technical Information.

Makarova, I., Pashkevich, A., Shubenkova, K., & Mukhametdinov, E. (2017). Ways to increase population mobility through the transition to sustainable transport. *Procedia Engineering*, *187*, 756–762. Available from https://doi.org/10.1016/j.proeng.2017.04.434.

Malik, K., Rahman, S. M., Khondaker, A. N., Abubakar, I. R., Aina, Y. A., & Hasan, M. A. (2019). Renewable energy utilization to promote sustainability in GCC countries: Policies, drivers, and barriers. *Environmental Science and Pollution Research*, *26*(20), 20798–20814. Available from https://doi.org/10.1007/s11356-019-05337-1.

Manigandan, P., Alam, M. S., Alagirisamy, K., Pachiyappan, D., Murshed, M., & Mahmood, H. (2023). Realizing the sustainable development goals through technological innovation: Juxtaposing the economic and environmental effects of financial development and energy use. *Environmental Science and Pollution Research*, *30*(3), 8239–8256. Available from https://doi.org/10.1007/s11356-022-22692-8.

Manisalidis, I., Stavropoulou, E., Stavropoulos, A., & Bezirtzoglou, E. (2020). Environmental and health impacts of air pollution: A review. *Frontiers in Public Health*, *8*. Available from https://doi.org/10.3389/fpubh.2020.00014.

Martens, J. A., Bogaerts, A., De Kimpe, N., Jacobs, P. A., Marin, G. B., Rabaey, K., & Verhelst, S. (2017). The chemical route to a carbon dioxide neutral world. *ChemSusChem*, *10*(6), 1039–1055. Available from https://doi.org/10.1002/cssc.201601051.

Martin, K. D., & Murphy, P. E. (2017). The role of data privacy in marketing. *Journal of the Academy of Marketing Science*, *45*(2), 135–155. Available from https://doi.org/10.1007/s11747-016-0495-4.

Mazzucato, M., Kattel, R., & Ryan-Collins, J. (2020). Challenge-driven innovation policy: Towards a new policy toolkit. *Journal of Industry, Competition and Trade*, *20*(2), 421–437. Available from https://doi.org/10.1007/s10842-019-00329-w.

Meng, J., Yang, H., Yi, K., Liu, J., Guan, D., Liu, Z., Mi, Z., Coffman, D. M., Wang, X., Zhong, Q., Huang, T., Meng, W., & Tao, S. (2019). The slowdown in global air-pollutant emission growth and driving factors. *One Earth*, *1*(1), 138–148. Available from https://doi.org/10.1016/j.oneear.2019.08.013.

Metz, B. (2013). The legacy of the kyoto Protocol: A view from the policy world. *Wiley Interdisciplinary Reviews: Climate Change*, *4*(3), 151–158. Available from https://doi.org/10.1002/wcc.216.

Meyer, I., & Sommer, M. W. (2016). Employment effects of renewable energy deployment - a review. *International Journal of Sustainable Development*, *19*(3), 217–245. Available from https://doi.org/10.1504/IJSD.2016.078274.

Meyer, K., Newman, P., Meyer, K., & Newman, P. (2020). *The science of anthropogenic climate change. Planetary accounting: Quantifying how to live within planetary limits at different scales of human activity* (pp. 11–33).

Milojevic-Dupont, N., & Creutzig, F. (2021). Machine learning for geographically differentiated climate change mitigation in urban areas. *Sustainable Cities and Society*, *64*. Available from https://doi.org/10.1016/j.scs.2020.102526.

Ming, T., De_richter, R., Shen, S., & Caillol, S. (2016). Fighting global warming by greenhouse gas removal: Destroying atmospheric nitrous oxide thanks to synergies between two breakthrough technologies. *Environmental Science and Pollution Research*, *23*(7), 6119–6138. Available from https://doi.org/10.1007/s11356-016-6103-9.

Mohsin, M., Abbas, Q., Zhang, J., Ikram, M., & Iqbal, N. (2019). Integrated effect of energy consumption, economic development, and population growth on CO2 based environmental degradation: A case of transport sector. *Environmental Science and Pollution Research*, *26*(32), 32824–32835. Available from https://doi.org/10.1007/s11356-019-06372-8.

Mora, C., Spirandelli, D., Franklin, E. C., Lynham, J., Kantar, M. B., Miles, W., Smith, C. Z., Freel, K., Moy, J., Louis, L. V., Barba, E. W., Bettinger, K., Frazier, A. G., Colburn, J. F., IX, Hanasaki, N., Hawkins, E., Hirabayashi, Y., Knorr, W., Little, C. M., . . . Hunter, C. L. (2018). Broad threat to humanity from cumulative climate hazards intensified by greenhouse gas emissions. *Nature Climate Change*, *8*(12), 1062–1071. Available from https://doi.org/10.1038/s41558-018-0315-6.

Morello-Frosch, R., Zuk, M., Jerrett, M., Shamasunder, B., & Kyle, A. D. (2011). Understanding the cumulative impacts of inequalities in environmental health: Implications for policy. *Health Affairs*, *30*(5), 879–887. Available from https://doi.org/10.1377/hlthaff.2011.0153.

Moreno, J., Serrano, M. A., & Fernández-Medina, E. (2016). Main issues in Big Data security. *Future Internet*, *8*(3). Available from https://doi.org/10.3390/fi8030044.

Mundaca, L., & Luth Richter, J. (2015). Assessing 'green energy economy' stimulus packages: Evidence from the U.S. programs targeting renewable energy. *Renewable and Sustainable Energy Reviews*, *42*, 1174–1186. Available from https://doi.org/10.1016/j.rser.2014.10.060.

Mycoo, M. A. (2018). Beyond 1.5 °C: Vulnerabilities and adaptation strategies for Caribbean Small Island Developing States. *Regional Environmental Change*, *18*(8), 2341–2353. Available from https://doi.org/10.1007/s10113-017-1248-8.

Nacke, L., Cherp, A., & Jewell, J. (2022). Phases of fossil fuel decline: Diagnostic framework for policy sequencing and feasible transition pathways in resource dependent regions. *Oxford Open Energy*, *1*. Available from https://doi.org/10.1093/ooenergy/oiac002.

Nalule, V. R. (2019). *Transitioning to a low carbon economy: Is africa ready to bid farewell to fossil fuels? The Palgrave handbook of managing fossil fuels and energy transitions* (pp. 261–286). Available from http://doi.org/10.1007/978-3-030-28076-5_10.

Nawaz, Z., Imran, M., Nawaz, S., Ali, A., & Sangi, A. R. (2022). Strategic prioritization of action plan towards de-carbonization and sustainable energy transition for developing nations. *International Journal of Sustainable Development and Planning*, *17*(2), 549–558. Available from https://doi.org/10.18280/ijsdp.170220.

Naylor, S., Gillott, M., & Lau, T. (2018). A review of occupant-centric building control strategies to reduce building energy use. *Renewable and Sustainable Energy Reviews*, *96*, 1–10. Available from https://doi.org/10.1016/j.rser.2018.07.019.

Nemitallah, M. A., Habib, M. A., Badr, H. M., Said, S. A., Jamal, A., Ben-Mansour, R., Mokheimer, E. M. A., & Mezghani, K. (2017). Oxy-fuel combustion technology: Current status, applications, and trends. *International Journal of Energy Research*, *41*(12), 1670–1708. Available from https://doi.org/10.1002/er.3722.

Nerini, F. F., Broad, O., Mentis, D., Welsch, M., Bazilian, M., Howells, M., et al. (2016). A cost comparison of technology approaches for improving access to electricity services. *Energy*, *95*, 255–265. Available from https://doi.org/10.1016/j.energy.2015.11.068.

Nguyen, H. K. (2023). Climate justice in Vietnam's green transformation: A three-dimensional analysis of the green growth and climate change mitigation sectors. *Ministry of Science and Technology, Vietnam*, *65*(1), 90–99. Available from https://doi.org/10.31276/vmostjossh.65(1).90-99.

Van Niekerk, A. J. (2020). Inclusive economic sustainability: SDGs and global inequality. *Sustainability*, *12*(13).

Nilsson, L. J., Bauer, F., Åhman, M., Andersson, F. N. G., Bataille, C., de la Rue du Can, S., Ericsson, K., Hansen, T., Johansson, B., Lechtenböhmer, S., van Sluisveld, M., & Vogl, V. (2021). An industrial policy framework for transforming energy and emissions intensive industries towards zero emissions. *Climate Policy*, *21*(8), 1053–1065. Available from https://doi.org/10.1080/14693062.2021.1957665.

Nores, M., & Fernandez, C. (2018). Building capacity in health and education systems to deliver interventions that strengthen early child development. *Annals of the New York Academy of Sciences*, *1419*(1), 57–73. Available from https://doi.org/10.1111/nyas.13682.

Nwaka, I. D., Nwogu, M. U., Uma, K. E., & Ike, G. N. (2020). Agricultural production and CO2 emissions from two sources in the ECOWAS region: New insights from quantile regression and decomposition analysis. *Science of the Total Environment*, *748*. Available from https://doi.org/10.1016/j.scitotenv.2020.141329.

Oberschelp, C., Pfister, S., Raptis, C. E., & Hellweg, S. (2019). Global emission hotspots of coal power generation. *Nature Sustainability*, *2*(2), 113–121. Available from https://doi.org/10.1038/s41893-019-0221-6.

Obschonka, M., Stuetzer, M., Rentfrow, P. J., Shaw-Taylor, L., Satchell, M., Silbereisen, R. K., Potter, J., & Gosling, S. D. (2018). In the shadow of coal: How large-scale industries contributed to present-day regional differences in personality and well-being. *Journal of Personality and Social Psychology*, *115*(5), 903–927. Available from https://doi.org/10.1037/pspp0000175.

Ofremu, G. O., Raimi, B. Y., Yusuf, S. O., Dziwornu, B. A., Nnabuife, S. G., Eze, A. M., & Nnajiofor, C. A. (2024). Exploring the relationship between climate change, air pollutants and human health: impacts, adaptation, and mitigation strategies. *Green Energy and Resources*. Available from https://doi.org/10.1016/j.gerr.2024.100074.

Oke, D., Dunn, J. B., & Hawkins, T. R. (2024). Reducing economy-wide greenhouse gas emissions with electrofuels and biofuels as the grid decarbonizes. *Energy and Fuels*, *38*(7), 6048–6061. Available from https://doi.org/10.1021/acs.energyfuels.3c04833.

Olivier, J. G., Schure, K. M., & Peters, J. A. H. W. (2017). Trends in global CO2 and total greenhouse gas emissions. *PBL Netherlands Environmental Assessment Agency*, *5*, 1–11.

O'Neill, B. C., Kriegler, E., Ebi, K. L., Kemp-Benedict, E., Riahi, K., Rothman, D. S., van Ruijven, B. J., van Vuuren, D. P., Birkmann, J., Kok, K., Levy, M., & Solecki, W. (2017). The roads ahead: Narratives for shared socioeconomic pathways describing world futures in the 21st century. *Global Environmental Change*, *42*, 169–180. Available from https://doi.org/10.1016/j.gloenvcha.2015.01.004.

Onyebuchi, V. E., Kolios, A., Hanak, D. P., Biliyok, C., & Manovic, V. (2018). A systematic review of key challenges of CO2 transport via pipelines. *Renewable and Sustainable Energy Reviews*, *81*, 2563–2583. Available from https://doi.org/10.1016/j.rser.2017.06.064.

Oppermann, E., Kjellstrom, T., Lemke, B., Otto, M., & Lee, J. K. W. (2021). Establishing intensifying chronic exposure to extreme heat as a slow onset event with implications for health, wellbeing, productivity, society and economy. *Current Opinion in Environmental Sustainability*, *50*, 225–235. Available from https://doi.org/10.1016/j.cosust.2021.04.006.

Orsetti, E., Tollin, N., Lehmann, M., Valderrama, V. A., & Morató, J. (2022). Building resilient cities: Climate change and health interlinkages in the planning of public spaces. *International Journal of Environmental Research and Public Health*, *19*(3). Available from https://doi.org/10.3390/ijerph19031355.

Otorkpa, O. J., Emmanuel, S., Obuye, F., & Otorkpa, C. O. (2024). Impact of climate change on global health: A comprehensive review. *International Journal of Environment and Climate Change*, *14*(5), 342–351.

Ouerghi, F. H., Omri, M., Nisar, K. S., Abd El-Aziz, R. M., & Taloba, A. I. (2024). Investigating the potential of geothermal energy as a sustainable replacement for fossil fuels in commercial buildings. *Alexandria Engineering Journal*, *97*, 215–229. Available from https://doi.org/10.1016/j.aej.2024.03.094.

Ouyang, X., Li, Q., & Du, K. (2020). How does environmental regulation promote technological innovations in the industrial sector? Evidence from Chinese provincial panel data. *Energy Policy*, *139*. Available from https://doi.org/10.1016/j.enpol.2020.111310.

Panda, D. K., & Das, S. (2021). Smart grid architecture model for control, optimization and data analytics of future power networks with more renewable energy. *Journal of Cleaner Production*, *301*. Available from https://doi.org/10.1016/j.jclepro.2021.126877.

Parihar, S. S., Saini, K. P. S., Lakhani, G. P., Jain, A., Roy, B., Ghosh, S., & Aharwal, B. (2019). Livestock waste management: A review. *Journal of Entomology and Zoology Studies*, *7*(3), 384–393.

Pastor, A. V., et al. (2020). How uncertainties are tackled in multi-disciplinary science? A review of integrated assessments under global change. *CATENA*, *186*. Available from https://doi.org/10.1016/j.catena.2019.104305.

Peng, H., Kan, S., Meng, J., Li, S., Cui, C., Tan, C., Wang, Z., Wen, Q., & Guan, D. (2024). Emission accounting and drivers in South American countries. *Applied Energy*, *358*. Available from https://doi.org/10.1016/j.apenergy.2023.122528.

Pigato, M., Black, S., Dussaux, D., Mao, Z., Rafaty, R., & Touboul, S. (2020). *Technology transfer and innovation for low-carbon development*. World Bank Publications.

Piñeiro, V., Arias, J., Dürr, J., Elverdin, P., Ibáñez, A. M., Kinengyere, A., Opazo, C. M., Owoo, N., Page, J. R., Prager, S. D., & Torero, M. (2020). A scoping review on incentives for adoption of sustainable agricultural practices and their outcomes. *Nature Sustainability*, *3*(10), 809–820.

Plaza-Bonilla, D., Álvaro-Fuentes, J., Lampurlanés, J., Arrúe, J. L., & Cantero-Martínez, C. (2020). *No-till farming systems to reduce nitrous oxide emissions and increase methane uptake. No-till farming systems for sustainable agriculture: Challenges and opportunities* (pp. 319–335). Spain: Springer International Publishing. Available from 10.1007/978-3-030-46409-7_19.

Pujara, Y., Pathak, P., Sharma, A., & Govani, J. (2019). Review on Indian municipal solid waste management practices for reduction of environmental impacts to achieve sustainable development goals. *Journal of Environmental Management*, *248*. Available from https://doi.org/10.1016/j.jenvman.2019.07.009.

Quader, M. A., Ahmed, S., Ghazilla, R. A. R., Ahmed, S., & Dahari, M. (2015). A comprehensive review on energy efficient CO2 breakthrough technologies for sustainable green iron and steel manufacturing. *Renewable and Sustainable Energy Reviews*, *50*, 594–614. Available from https://doi.org/10.1016/j.rser.2015.05.026.

Rahman, A., Farrok, O., & Haque, M. M. (2022). Environmental impact of renewable energy source based electrical power plants: Solar, wind, hydroelectric, biomass, geothermal, tidal, ocean, and osmotic. *Renewable and Sustainable Energy Reviews*, *161*112279.

Rahman, M. M., Khan, I., Field, D. L., Techato, K., & Alameh, K. (2022). Powering agriculture: Present status, future potential, and challenges of renewable energy applications. *Renewable Energy*, *188*, 731–749.

Raihan, A., Begum, R. A., Nizam, M., Said, M., & Pereira, J. J. (2022). Dynamic impacts of energy use, agricultural land expansion, and deforestation on CO2 emissions in Malaysia. *Environmental and Ecological Statistics*, *29*(3), 477–507.

Raiser, K., Çalı, B., & Flachsland, C. (2022). Understanding pledge and review: Learning from analogies to the Paris Agreement review mechanisms. *Climate Policy*, *22*(6), 711–727. Available from https://doi.org/10.1080/14693062.2022.2059436.

Ramachandra, T. V., Bharath, H. A., Kulkarni, G., & Han, S. S. (2018). Municipal solid waste: Generation, composition and GHG emissions in Bangalore, India. *Renewable and Sustainable Energy Reviews*, *82*, 1122–1136. Available from https://doi.org/10.1016/j.rser.2017.09.085.

Raufer, R., Coussy, P., & Freeman, C. (2022). *Emissions trading. Handbook of climate change mitigation and adaptation* (3rd ed., pp. 3237–3294). China: Springer International Publishing. Available from 10.1007/978-3-030-72579-2_8.

Rehman, A., Ma, H., & Ozturk, I. (2021). Do industrialization, energy importations, and economic progress influence carbon emission in Pakistan. *Environmental Science and Pollution Research*, *28*(33), 45840–45852. Available from https://doi.org/10.1007/s11356-021-13916-4.

Reid, W. V., Chen, D., Goldfarb, L., Hackmann, H., Lee, Y. T., Mokhele, K., Ostrom, E., Raivio, K., Rockström, J., Schellnhuber, H. J., & Whyte, A. (2010). Earth system science for global sustainability: Grand challenges. *Science*, *330*(6006), 916–917. Available from https://doi.org/10.1126/science.1196263.

Riahi, K., Kriegler, E., Johnson, N., Bertram, C., den Elzen, M., Eom, J., Schaeffer, M., Edmonds, J., Isaac, M., Krey, V., Longden, T., Luderer, G., Méjean, A., McCollum, D. L., Mima, S., Turton, H., van Vuuren, D. P., Wada, K., Bosetti, V., ... Edenhofer, O. (2015). Locked into Copenhagen pledges—implications of short-term emission targets for the cost and feasibility of long-term climate goals. *Technological Forecasting and Social Change*, *90*, 8–23. Available from https://doi.org/10.1016/j.techfore.2013.09.016.

Robinson, E. J. Z. (2016). Resource-dependent livelihoods and the natural resource base. *Annual Review of Resource Economics*, *8*(1), 281–301. Available from https://doi.org/10.1146/annurev-resource-100815-095521.

Rode, P., Floater, G., Thomopoulos, N., Docherty, J., Schwinger, P., Mahendra, A., & Fang, W. (2017). *Accessibility in cities: Transport and urban form* (pp. 239–273). Springer Science and Business Media LLC. Available from https://doi.org/10.1007/978-3-319-51602-8_15.

Rogelj, J., Popp, A., Calvin, K. V., Luderer, G., Emmerling, J., Gernaat, D., Fujimori, S., Strefler, J., Hasegawa, T., Marangoni, G., Krey, V., Kriegler, E., Riahi, K., van Vuuren, D. P., Doelman, J., Drouet, L., Edmonds, J., Fricko, O., Harmsen, M., ... Tavoni, M. (2018). Scenarios towards limiting global mean temperature increase below 1.5 °C. *Nature Climate Change*, *8*(4), 325–332. Available from https://doi.org/10.1038/s41558-018-0091-3.

Rosenbloom, D., & Meadowcroft, J. (2022). Accelerating pathways to net zero: Governance strategies from transition studies and the transition accelerator. *Current Climate Change Reports*, *8*(4), 104–114. Available from https://doi.org/10.1007/s40641-022-00185-7.

Roy Choudhury, A. K. (2014). Environmental Impacts of the Textile Industry and Its Assessment through Life Cycle Assessment. *Roadmap to Sustainable Textiles and Clothing*, 1–39. Available from https://doi.org/10.1007/978-981-287-110-7_1.

Rummukainen, M. (2013). Climate change: Changing means and changing extremes. *Climatic Change*, *121*(1), 3–13. Available from https://doi.org/10.1007/s10584-013-0888-z.

Sahu, S., Kunj, P., Kaur, A., Khatri, M., Singh, G., & Arya, S. K. (2024). Catalytic strategies for algal-based carbon capture and renewable energy: A review on a sustainable approach. *Energy Conversion and Management*, *310*. Available from https://doi.org/10.1016/j.enconman.2024.118467.

Sakadevan, K., & Nguyen, M. L. (2017). Livestock production and its impact on nutrient pollution and greenhouse gas emissions. *Advances in Agronomy*, 147–184. Available from https://doi.org/10.1016/bs.agron.2016.10.002.

Salahodjaev, R., & Isaeva, A. (2022). Post-Soviet states and CO2 emissions: The role of foreign direct investment. *Post-Communist Economies*, *34*(7), 944–965. Available from https://doi.org/10.1080/14631377.2021.1965360.

Sandaka, B. P., & Kumar, J. (2023). Alternative vehicular fuels for environmental decarbonization: A critical review of challenges in using electricity, hydrogen, and biofuels as a sustainable vehicular fuel. *Chemical Engineering Journal Advances*, *14*. Available from https://doi.org/10.1016/j.ceja.2022.100442.

Santarromana, R., Mendonça, J., & Dias, A. M. (2020). The effectiveness of decarbonizing the passenger transport sector through monetary incentives. *Transportation Research Part A: Policy and Practice*, *138*, 442–462. Available from https://doi.org/10.1016/j.tra.2020.06.020.

Saros, J. E., Arp, C. D., Bouchard, F., Comte, J., Couture, R.-M., Dean, J. F., Lafrenière, M., MacIntyre, S., McGowan, S., Rautio, M., Prater, C., Tank, S. E., Walvoord, M., Wickland, K. P., Antoniades, D., Ayala-Borda, P., Canario, J., Drake, T. W., Folhas, D., ... Vincent, W. F. (2022). Sentinel responses of Arctic freshwater systems to climate: Linkages, evidence, and a roadmap for future research. *Arctic Science*, *9*(2), 356–392. Available from https://doi.org/10.1139/as-2022-0021.

Saxena, P., & Sonwani, S. (2019). *Policy regulations and future recommendations. Criteria air pollutants and their impact on environmental health* (pp. 127–157).

Subramanian, A., Nagarajan, A. M., Vinod, S., Chakraborty, S., Sivagami, K., Theodore, T., Sathyanarayanan, S. S., Tamizhdurai, P., & Mangesh, V. L. (2016). Decoupling global environmental pressure and economic growth: Scenarios for energy use, materials use and carbon emissions. *Journal of Cleaner Production*, *132*, 45–56. Available from https://doi.org/10.1016/j.jclepro.2015.06.100.

Schwarzman, M., Schildroth, S., Bhetraratana, M., Alvarado, Á., & Balmes, J. (2021). Raising standards to lower diesel emissions. *Science*, *371* (6536), 1314–1316. Available from https://doi.org/10.1126/science.abf8159.

Sen Gupta, G. and Tiwari, S. (2021) Past and present events of climate change: Natural versus anthropogenic causes, in Global climate change. India: Elsevier, pp. 267–295. Available from https://doi.org/10.1016/B978-0-12-822928-6.00014-9.

Serban, A. C., & Lytras, M. D. (2020). Artificial intelligence for smart renewable energy sector in europe - Smart energy infrastructures for next generation smart cities. *IEEE Access*, *8*, 77364–77377. Available from https://doi.org/10.1109/ACCESS.2020.2990123.

Shahbazi, A., & Rezaei Nasab, B. (2016). Carbon capture and storage (CCS) and its impacts on climate change and global warming. *Journal of Petroleum and Environmental Biotechnology*, *7*(4). Available from https://doi.org/10.4172/2157-7463.1000291.

Shahsavari, A., & Akbari, M. (2018). Potential of solar energy in developing countries for reducing energy-related emissions. *Renewable and Sustainable Energy Reviews*, *90*, 275–291. Available from https://doi.org/10.1016/j.rser.2018.03.065.

Shakou, L. M., Wybo, J. L., Reniers, G., & Boustras, G. (2019). Developing an innovative framework for enhancing the resilience of critical infrastructure to climate change. *Safety Science*, *118*, 364–378. Available from https://doi.org/10.1016/j.ssci.2019.05.019.

Sharma, D., Saini, A., Sharma, D., & Saini, A. (2020). *Cellulosic ethanol feedstock: diversity and potential. Lignocellulosic ethanol production from a biorefinery perspective: Sustainable valorization of waste* (pp. 23–63).

Sharmina, M., Edelenbosch, O. Y., Wilson, C., Freeman, R., Gernaat, D. E. H. J., Gilbert, P., Larkin, A., Littleton, E. W., Traut, M., van Vuuren, D. P., Vaughan, N. E., Wood, F. R., & Le Quéré, C. (2021). Decarbonising the critical sectors of aviation, shipping, road freight and industry to limit warming to 1.5–2°C. *Climate Policy*, *21*(4), 455–474. Available from https://doi.org/10.1080/14693062.2020.1831430.

Shen, M., Kong, F., Tong, L., Luo, Y., Yin, S., Liu, C., Zhang, P., Wang, L., Chu, P. K., & Ding, Y. (2022). Carbon capture and storage (CCS): Development path based on carbon neutrality and economic policy. *Carbon Neutrality*, *1*(1). Available from https://doi.org/10.1007/s43979-022-00039-z.

Shi, L., Chu, E., Anguelovski, I., Aylett, A., Debats, J., Goh, K., Schenk, T., Seto, K. C., Dodman, D., Roberts, D., Roberts, J. T., & VanDeveer, S. D. (2016). Roadmap towards justice in urban climate adaptation research. *Nature Climate Change*, *6*(2), 131–137. Available from https://doi.org/10.1038/nclimate2841.

Simoni, M., Wilkes, M. D., Brown, S., Provis, J. L., Kinoshita, H., & Hanein, T. (2022). Decarbonising the lime industry: State-of-the-art. *Renewable and Sustainable Energy Reviews*, *168*. Available from https://doi.org/10.1016/j.rser.2022.112765.

Singh, R. M., Sani, A. K., & Amis, T. (2019). An overview of ground-source heat pump technology. *Managing Global Warming*, 455–485.

Skendžic, S., Zovko, M., Živkovic., Lešic, I. P., & Lemic, V. (2021). The impact of climate change on agricultural insect pests. *Insects*, *12*(5). Available from https://doi.org/10.3390/insects12050440.

Snyder, C. S., Davidson, E. A., Smith, P., & Venterea, R. T. (2014). Agriculture: Sustainable crop and animal production to help mitigate nitrous oxide emissions. *Current Opinion in Environmental Sustainability*, *9–10*, 46–54. Available from https://doi.org/10.1016/j.cosust.2014.07.005.

Song, X., Li, G., Huang, Z., Shi, Y., Wang, G., Song, G., & Xu, F. (2023). Review of high-temperature geothermal drilling and exploitation technologies. *Gondwana Research*, *122*, 315–330. Available from https://doi.org/10.1016/j.gr.2022.10.013.

Sorrell, S. (2015). Reducing energy demand: A review of issues, challenges and approaches. *Renewable and Sustainable Energy Reviews*, *47*, 74–82. Available from https://doi.org/10.1016/j.rser.2015.03.002.

Subramanian, A., Nagarajan, A. M., Vinod, S., Chakraborty, S., Sivagami, K., Theodore, T., Sathyanarayanan, S. S., Tamizhdurai, P., & Mangesh, V. L. (2023). Long-term impacts of climate change on coastal and transitional eco-systems in India: an overview of its current status, future projections, solutions, and policies. *RSC Advances*, *13*(18), 12204–12228. Available from https://doi.org/10.1039/d2ra07448f.

Suman, A. (2021). Role of renewable energy technologies in climate change adaptation and mitigation: A brief review from Nepal. *Renewable and Sustainable Energy Reviews*, *151*. Available from https://doi.org/10.1016/j.rser.2021.111524.

Suzuki, M. (2015). Identifying roles of international institutions in clean energy technology innovation and diffusion in the developing countries: Matching barriers with roles of the institutions. *Journal of Cleaner Production*, *98*, 229–240. Available from https://doi.org/10.1016/j.jclepro.2014.08.070.

Tan, K. M., Babu, T. S., Ramachandaramurthy, V. K., Kasinathan, P., Solanki, S. G., & Raveendran, S. K. (2021). Empowering smart grid: A comprehensive review of energy storage technology and application with renewable energy integration. *Journal of Energy Storage*, *39*. Available from https://doi.org/10.1016/j.est.2021.102591.

Tate, E., Rahman, M. A., Emrich, C. T., & Sampson, C. C. (2021). Flood exposure and social vulnerability in the United States. *Natural Hazards*, *106* (1), 435–457. Available from https://doi.org/10.1007/s11069-020-04470-2.

Tian, J., Yu, L., Xue, R., Zhuang, S., & Shan, Y. (2022). Global low-carbon energy transition in the post-COVID-19 era. *Applied Energy*, *307*118205.

Tian, X., Bai, F., Jia, J., Liu, Y., & Shi, F. (2019). Realizing low-carbon development in a developing and industrializing region: Impacts of industrial structure change on CO2 emissions in southwest China. *Journal of Environmental Management*, *233*, 728–738.

Tong, C. (2018). *Advanced materials enable renewable wind energy capture and generation* (pp. 379–444). Springer Science and Business Media LLC. Available from 10.1007/978-3-319-98002-7_6.

Tong, S., & Ebi, K. (2019). Preventing and mitigating health risks of climate change. *Environmental Research*, *174*, 9–13. Available from https://doi.org/10.1016/j.envres.2019.04.012.

Turner, M. G., Calder, W. J., Cumming, G. S., Hughes, T. P., Jentsch, A., LaDeau, S. L., Lenton, T. M., Shuman, B. N., Turetsky, M. R., Ratajczak, Z., Williams, J. W., Williams, A. P., & Carpenter, S. R. (2020). Climate change, ecosystems and abrupt change: science priorities. *Philosophical Transactions of the Royal Society B*, *375*(1794)20190105.

Vaccaro, I. G., Jansen, J. J., Van Den Bosch, F. A., & Volberda, H. W. (2012). Management innovation and leadership: The moderating role of organizational size. *Journal of Management Studies*, *49*(1), 28–51. Available from https://doi.org/10.1111/j.1467-6486.2010.00976.x.

Valenciano-Salazar, J. A., André, F. J., & Martín-de Castro, G. (2022). Sustainability and firms' mission in a developing country: the case of voluntary certifications and programs in Costa Rica. *Journal of Environmental Planning and Management*, *65*(11), 2029–2053. Available from https://doi.org/10.1080/09640568.2021.1950658.

van Asselt, H., & Zelli, F. (2014). Connect the dots: Managing the fragmentation of global climate governance. *Environmental Economics and Policy Studies*, *16*(2), 137–155. Available from https://doi.org/10.1007/s10018-013-0060-z.

Verma, B. C., Pramanik, P., & Bhaduri, D. (2020). *Organic fertilizers for sustainable soil and environmental management. Nutrient dynamics for sustainable crop production* (pp. 289–313).

Vijayan, D. S., Sivasuriyan, A., Patchamuthu, P., & Jayaseelan, R. (2022). Thermal performance of energy-efficient buildings for sustainable development. *Environmental Science and Pollution Research*, *29*(34), 51130–51142. Available from https://doi.org/10.1007/s11356-021-17602-3.

Vimmerstedt, L., Brown, A., Heath, G., Mai, T., Melaina, M., Newes, E., Ruth, M., Simpkins, T., Warner, E., Bertram, K., Plotkin, S., Patel, D., Stephens, T., & Vyas, A. (2013). Potential reductions in emissions and petroleum use in transportation. *Transportation Research Record*, *2375*, 37–44. Available from https://doi.org/10.3141/2375-05.

von Borgstede, C., Andersson, M., & Johnsson, F. (2013). Public attitudes to climate change and carbon mitigation—implications for energy-associated behaviours. *Energy Policy*, *57*, 182–193. Available from https://doi.org/10.1016/j.enpol.2013.01.051.

Vos, R. (2019). *Agriculture, the rural sector, and development* (pp. 160–185). Oxford University Press (OUP). Available from 10.1093/oso/9780198844938.003.0007.

Waisman, H., Bataille, C., Winkler, H., Jotzo, F., Shukla, P., Colombier, M., Buira, D., Criqui, P., Fischedick, M., Kainuma, M., La Rovere, E., Pye, S., Safonov, G., Siagian, U., Teng, F., Virdis, M.-R., Williams, J., Young, S., Anandarajah, G., & Trollip, H. (2019). A pathway design framework for national low greenhouse gas emission development strategies. *Nature Climate Change*, *9*(4), 261–268. Available from https://doi.org/10.1038/s41558-019-0442-8.

Walker, I. S., Less, B. D., & Casquero-Modrego, N. (2022). Carbon and energy cost impacts of electrification of space heating with heat pumps in the US. *Energy and Buildings*, *259*. Available from https://doi.org/10.1016/j.enbuild.2022.111910.

Wang, N., Du, Y., Chen, D., Meng, H., Chen, X., Zhou, L., Shi, G., Zhan, Y., Feng, M., Li, W., Chen, M., Li, Z., & Yang, F. (2024). Spatial disparities of ozone pollution in the Sichuan Basin spurred by extreme, hot weather. *Atmospheric Chemistry and Physics*, *24*(5), 3029–3042.

Wang, Q., Su, M., & Li, R. (2018). Toward to economic growth without emission growth: The role of urbanization and industrialization in China and India. *Journal of Cleaner Production*, *205*, 499–511.

Wang, Z., Li, F., Xie, Z., Li, Q., Zhang, Y., & Dai, M. (2023). Decoupling CO2 emissions from economic growth in China's cities from 2000 to 2020: A case study of the pearl river delta agglomeration. *Land*, *12*(9), 1804.

Wang, Y., Yang, L., Han, S., Li, C., & Ramachandra, T. V. (2017). Urban CO2 emissions in Xi'an and Bangalore by commuters: implications for controlling urban transportation carbon dioxide emissions in developing countries. *Mitigation and Adaptation Strategies for Global Change*, *22*(7), 993–1019. Available from https://doi.org/10.1007/s11027-016-9704-1.

Warner, K., Hamza, M., Oliver-Smith, A., Renaud, F., & Julca, A. (2010). Climate change, environmental degradation and migration. *Natural Hazards*, *55*(3), 689–715. Available from https://doi.org/10.1007/s11069-009-9419-7.

Welch, I. (2024). The Ipcc shared socioeconomic pathways (Ssps): Explained, evaluated, replaced? United States: SSRN. Available from https://doi.org/10.2139/ssrn.4724099.

Wiedmann, T., Chen, G., Owen, A., Lenzen, M., Doust, M., Barrett, J., & Steele, K. (2021). Three-scope carbon emission inventories of global cities. *Journal of Industrial Ecology*, *25*(3), 735–750. Available from https://doi.org/10.1111/jiec.13063.

Winning, M., Price, J., Ekins, P., Pye, S., Glynn, J., Watson, J., & McGlade, C. (2019). Nationally determined contributions under the Paris Agreement and the costs of delayed action. *Climate Policy*, *19*(8), 947–958. Available from https://doi.org/10.1080/14693062.2019.1615858.

Worighi, I., Maach, A., Hafid, A., Hegazy, O., & Van Mierlo, J. (2019). Integrating renewable energy in smart grid system: Architecture, virtualization and analysis. *Sustainable Energy, Grids and Networks*, *18*. Available from https://doi.org/10.1016/j.segan.2019.100226.

Wu, Z., Zhou, T., Zhang, N., Choi, Y., & Kong, F. (2023). A hidden risk in climate change: The effect of daily rainfall shocks on industrial activities. *Economic Analysis and Policy*, *80*, 161–180. Available from https://doi.org/10.1016/j.eap.2023.08.002.

Xin, L., Ahmad, M., & Khattak, S. I. (2023). Impact of innovation in hybrid electric vehicles-related technologies on carbon dioxide emissions in the 15 most innovative countries. *Technological Forecasting and Social Change*, *196*122859.

Yaashikaa, P. R., Kumar, P. S., Varjani, S. J., & Saravanan, A. (2019). A review on photochemical, biochemical and electrochemical transformation of CO2 into value-added products. *Journal of CO2 Utilization*, *33*, 131–147. Available from https://doi.org/10.1016/j.jcou.2019.05.017.

Yang, W., Pan, Y., Ma, J., Zhou, M., Chen, Z., & Zhu, W. (2018). Optimization on emission permit trading and green technology implementation under cap-and-trade scheme. *Journal of Cleaner Production*, *194*, 288–299. Available from https://doi.org/10.1016/j.jclepro.2018.05.123.

Yoo, Y., Kim, I., Lee, D., Yong., Choi, W., Choi, J., Jang, K., Park, J., & Kang, D. (2022). Review of contemporary research on inorganic CO2 utilization via CO2 conversion into metal carbonate-based materials. *Journal of Industrial and Engineering Chemistry*, *116*, 60–74. Available from https://doi.org/10.1016/j.jiec.2022.09.007.

Yoro, K. O., Amosa, M. K., Sekoai, P. T., Mulopo, J., & Daramola, M. O. (2020). Diffusion mechanism and effect of mass transfer limitation during the adsorption of CO2by polyaspartamide in a packed-bed unit. *International Journal of Sustainable Engineering*, *13*(1), 54–67. Available from https://doi.org/10.1080/19397038.2020.1592261.

Yoro, K. O., & Daramola, M. O. (2020). *CO2 emission sources, greenhouse gases, and the global warming effect. Advances in carbon capture: Methods, technologies and applications* (pp. 3–28). South Africa: Elsevier. Available from https://doi.org/10.1016/B978-0-12-819657-1.00001-3.

Yu, Y., Li, S., Sun, H., & Taghizadeh-Hesary, F. (2021). Energy carbon emission reduction of China's transportation sector: An input–output approach. *Economic Analysis and Policy*, *69*, 378–393. Available from https://doi.org/10.1016/j.eap.2020.12.014.

Zahar, A. (2019). Collective progress in the light of equity under the global stocktake. *Climate Law*, *9*(1–2), 101–121. Available from https://doi.org/10.1163/18786561-00901006.

Zaidi, S. A. H., Zafar, M. W., Shahbaz, M., & Hou, F. (2019). Dynamic linkages between globalization, financial development and carbon emissions: Evidence from Asia Pacific Economic Cooperation countries. *Journal of Cleaner Production*, *228*, 533–543. Available from https://doi.org/10.1016/j.jclepro.2019.04.210.

Zhang, W., Li, G., & Guo, F. (2022). Does carbon emissions trading promote green technology innovation in China? *Applied Energy*, *315*. Available from https://doi.org/10.1016/j.apenergy.2022.119012.

Zhao, X., Cui, Q., Wang, B., Yan, X., Singh, S., Zhang, F., Gao, X., & Li, Y. (2018). Recent progress of amine modified sorbents for capturing CO2 from flue gas. *Chinese Journal of Chemical Engineering*, *26*(11), 2292–2302. Available from https://doi.org/10.1016/j.cjche.2018.04.009.

Zheng, B., Zhang, Q., Geng, G., Chen, C., Shi, Q., Cui, M., Lei, Y., & He, K. (2021). Changes in China's anthropogenic emissions and air quality during the COVID-19 pandemic in 2020. *Earth System Science Data*, *13*(6), 2895–2907.

Zheng, X., Lu, Y., Yuan, J., Baninla, Y., Zhang, S., Stenseth, N. C., Hessen, D. O., Tian, H., Obersteiner, M., & Chen, D. (2020). Drivers of change in China's energy-related CO2 emissions. *Proceedings of the National Academy of Sciences*, *117*(1), 29–36.

Zheng, X., Streimikiene, D., Balezentis, T., Mardani, A., Cavallaro, F., & Liao, H. (2019). A review of greenhouse gas emission profiles, dynamics, and climate change mitigation efforts across the key climate change players. *Journal of Cleaner Production*, *234*, 1113–1133.

Zhou, D., & Deng, Z. (2017). Ultra-low-head hydroelectric technology: A review. *Renewable and Sustainable Energy Reviews*, *78*, 23–30. Available from https://doi.org/10.1016/j.rser.2017.04.086.

Zhou, L., Li, X., Zhang, P., Liu, J., Zhou, M., Dang, R., Vo, D.-V. N., & Kumar, P. S. (2024). Dimension does matter: Carbon-based substances with different dimensions exhibit opposite effects on soil CO2 and CH4 production. *International Journal of Environmental Science and Technology*, *21*(4), 3819–3826. Available from https://doi.org/10.1007/s13762-023-05200-4.

Zhou, X., Zhang, M., Zhou, M., & Zhou, M. (2017). A comparative study on decoupling relationship and influence factors between China's regional economic development and industrial energy–related carbon emissions. *Journal of Cleaner Production*, *142*, 783–800.

Zilberman, D., Zhao, J., & Heiman, A. (2012). Adoption versus adaptation, with emphasis on climate change, annual review of resource economics. *Annual Reviews*, *4*(1), 27–53. Available from https://doi.org/10.1146/annurev-resource-083110-115954.

Section II

Deep learning for ecological patterns

Chapter 5

Application of geographic information system and remote sensing technology in ecosystem services and biodiversity conservation

Maqsood Ahmed Khaskheli[1,*], Mir Muhammad Nizamani[1,*], Umed Ali Laghari[2], Abdul Hafeez Laghari[3], Anab Khanzada[4], Faria Sundas[1], Rashid Ali Khaskheli[5], Raheel Ahmed Khaskheli[6], Fahad Ali Kalhoro[7], Muhammad Rashid Nizamani[8] and Yong Wang[*,1]

[1]*Department of Plant Pathology, College of Agriculture, Guizhou University, Guiyang, Guizhou, P.R. China,* [2]*Department of Agriculture, Mir Chakar Khan Rind University, Sibi, Balochistan, Pakistan,* [3]*Department of Agronomy, Sindh Agriculture University Sub Campus Umerkot, Umerkot, Sindh, Pakistan,* [4]*State Key Laboratory of Crop Biology, Agronomy College, Shandong Agricultural University, Taian, Shandong, P.R. China,* [5]*Department of Crop Physiology, Faculty of Crop Production, Sindh Agriculture University, Tando Jam, Sindh, Pakistan,* [6]*Department of Plant Pathology, Faculty of Crop Protection, Sindh Agriculture University, Tando Jam, Sindh, Pakistan,* [7]*Department of Plant Breeding and Genetics, Sindh Agriculture University, Tando Jam, Sindh, Pakistan,* [8]*Crop Science Research Institute, Agriculture Research Centre, Tando Jam, Sindh, Pakistan*

5.1 Introduction

Conserving biodiversity is not just an ecological necessity but also a critical economic strategy. The rich variety of life on Earth underpins the provisioning services that are crucial to human survival, such as food and water, and supports regulating services that mitigate environmental hazards (Smith et al., 2017). When biodiversity is robust, ecosystems are more resilient and adaptable, better able to withstand and recover from stresses like climate change or invasive species. This ecological stability, in turn, supports economic stability by ensuring the continuity of resources that industries depend on, from agriculture to pharmaceuticals, thereby securing livelihoods and creating economic opportunities (Sekaran et al., 2021; Weiskopf et al., 2020). Moreover, by preserving biodiversity, we also maintain the esthetic and cultural values that contribute significantly to human well-being and societal health (Sandifer et al., 2015). Thus biodiversity conservation is fundamentally intertwined with sustaining the natural capital that underpins economic growth and human prosperity, highlighting the importance of integrated conservation strategies that align ecological health with economic and social goals (Rees et al., 2018).

Machine learning (ML) is revolutionizing environmental science by providing tools that enhance our ability to analyze complex data, predict ecological outcomes, and manage natural resources more effectively (Rao et al., 2023). ML's role in this field is multifaceted, spanning predictive analytics, image recognition, and data modeling, each contributing uniquely to better environmental stewardship. ML algorithms excel in identifying patterns and predicting future events from historical and real-time data (Pichler & Hartig, 2023). In environmental science, predictive analytics can forecast environmental impacts such as pollution levels, the spread of invasive species, or the effects of climate change on biodiversity. For example, ML models can predict which areas are most at risk of deforestation or how changing sea temperatures might affect marine life. This predictive power is crucial for preemptive conservation efforts, allowing policymakers and scientists to focus resources and protection measures more strategically (Demertzis & Iliadis, 2018; Vishwakarma et al., 2022).

*Maqsood Ahmed Khaskheli and Mir Muhammad Nizamani contributed equally to the work.

Deep Learning for Earth Observation and Climate Monitoring. DOI: https://doi.org/10.1016/B978-0-443-24712-5.00006-3

ML also plays a critical role in monitoring biodiversity and ecosystem health through advanced image recognition technologies. Cameras and sensors deployed in various ecosystems capture vast amounts of visual data, which ML algorithms can analyze to track wildlife populations, plant health, and changes in landscape features (Bhatti et al., 2019; Buchelt et al., 2024). For instance, automated image recognition has been used to monitor endangered species, count animal populations in aerial or satellite images, and detect illegal logging activities in remote forests. This capability not only enhances the efficiency and coverage of monitoring efforts but also reduces human error and the need for labor-intensive field work (Han et al., 2023; Mukherjee, 2022). ML's ability to model complex data relationships is invaluable in understanding ecosystem dynamics and interactions. These models can simulate various scenarios, such as the impact of different conservation strategies or climate change mitigation efforts. By integrating diverse data sources, such as weather patterns, land use changes, and species migration data, ML models provide a holistic view of ecosystem health and help scientists and decision-makers devise more effective environmental policies and management practices (Rudd et al., 2018; Wiebe et al., 2015).

Through these applications, ML is proving to be an indispensable tool in the field of environmental science, offering innovative solutions to some of the most pressing ecological challenges. Its ability to process and analyze large datasets quickly and accurately is essential for advancing our understanding of natural systems and improving the effectiveness of conservation efforts (Dash et al., 2019; Liu et al., 2022). The primary objective of this chapter is to examine the application of ML in the field of ecosystem services and biodiversity conservation. As environmental challenges grow in complexity and scale, traditional methods of monitoring, predicting, and managing ecological systems are increasingly insufficient (Fernandes et al., 2020; Manley et al., 2022). This chapter aims to highlight how ML, with its advanced computational capabilities, offers innovative solutions that can significantly enhance conservation efforts.

By leveraging ML's ability to analyze large datasets, predict ecological outcomes, and automate monitoring processes, we can achieve a deeper understanding of ecosystem dynamics and respond more effectively to environmental threats (Sun & Scanlon, 2019). This chapter will systematically explore various ML applications—from predictive analytics and image recognition to complex data modeling—in different conservation contexts. The goal is to elucidate how these technologies not only improve the efficiency and accuracy of conservation strategies but also enable proactive measures to safeguard biodiversity and ecosystem health (Amiri et al., 2023; Edeigba et al., 2024).

Ultimately, this chapter will underscore the critical need for integrating advanced ML solutions into conservation practices, offering insights into the potential benefits and challenges of such integration. By doing so, it aims to inform researchers, practitioners, and policymakers about the transformative potential of ML in making conservation efforts more adaptive, resilient, and effective in the face of global environmental changes.

5.2 Methodology

5.2.1 Data sources

The integration of ML into biodiversity conservation and ecosystem service management represents a significant advancement in environmental sciences, addressing complex ecological challenges with unprecedented precision and efficiency (Shiva Prakash et al., 2022). By leveraging databases like Web of Science, Google Scholar, and Scopus for gathering research, ML techniques such as predictive analytics and image recognition can be extensively reviewed and evaluated for their effectiveness in various conservation tasks (Pandey et al., 2023). For instance, predictive analytics allows for real-time data processing and forecasting, crucial for anticipating environmental changes and mitigating potential impacts. Similarly, image recognition facilitates the monitoring of species and habitats, enhancing the accuracy and scope of conservation efforts (Bhatti et al., 2023; Zhang et al., 2022). Consequently, a detailed chapter of these applications, drawn from diverse and authoritative sources, will not only demonstrate ML's potential to transform conservation practices but also highlight areas where further technological and methodological innovations are needed. This approach ensures that ML contributions are not only acknowledged but also critically examined, fostering more targeted and sustainable conservation strategies (Barbosa Júnior et al., 2024; Srivastava & Maity, 2023).

5.2.2 Search strategy

The meticulous selection of keywords and phrases such as "machine learning," "ecosystem services," and "biodiversity conservation" forms the backbone of an effective literature review strategy. By combining these core terms with related concepts like "artificial intelligence" and "habitat monitoring," the search can be broadened to encompass a wider array of relevant studies, thereby ensuring a comprehensive analysis (Lahoz-Monfort & Magrath, 2021). Furthermore, the use of compound search queries enables the targeting of specific intersections between ML and environmental science, such as the application of predictive analytics in ecosystem management (Bhatti et al., 2020). This strategic approach not only maximizes the retrieval of pertinent information but also refines the focus to current and impactful research, providing a robust foundation upon which to explore the transformative potential of ML in advancing biodiversity conservation and the sustainability of ecosystem services (Liao et al., 2020; Wu et al., 2022).

5.2.3 Selection criteria

Establishing precise inclusion and exclusion criteria is essential for a systematic and focused literature review on the application of ML in ecosystem services and biodiversity conservation (Almeida et al., 2024). By limiting included studies to those published within the last decade, the chapter ensures that the findings reflect the latest technological advancements and contemporary challenges in the field. Focusing on peer-reviewed articles that specifically employ ML techniques like predictive analytics, image recognition, and data modeling guarantees that the chapter remains relevant to its core subject matter (Fernandes et al., 2022; Peters et al., 2020). Moreover, restricting the scope to articles published in English and excluding nonpeer-reviewed gray literature ensures the reliability and accessibility of the data analyzed. This methodical approach not only streamlines the research process but also enhances the validity and applicability of the review's conclusions, making it a valuable resource for informing future conservation strategies and technological innovations (Barbhuiya & Das, 2023; Murphy et al., 2021).

5.2.4 Data extraction

The data extraction process for a chapter on the application of ML in ecosystem services and biodiversity conservation is meticulously designed to ensure comprehensive and relevant insights (Huettmann, 2018). By employing a standardized data extraction form, consistency is maintained across the chapter, capturing essential details such as publication details, ML techniques, and conservation outcomes. This approach facilitates a focused analysis on specific ML applications, such as predictive analytics or image recognition, and their direct impacts on conservation efforts (Dinis et al., 2021; Sun & Scanlon, 2019). Each study's conservation outcomes are closely examined, noting the success metrics and any comparative data that highlights the efficacy of ML over traditional methods. Furthermore, qualitative insights from the studies are synthesized to address practical challenges and future research directions, enriching the chapter's depth (Caskurlu et al., 2021; González et al., 2020). This systematic extraction and analysis not only underscore the transformative potential of ML in enhancing conservation practices but also identify gaps and limitations, paving the way for targeted future advancements. This methodical approach ensures that the chapter provides a robust foundation for understanding the current landscape and future potential of ML in environmental conservation (Alqahtani & Kumar, 2024; Li et al., 2023).

5.3 Literature review

5.3.1 Species distribution modeling

ML models are instrumental in enhancing species distribution modeling (SDM), providing sophisticated tools to predict how species habitats might shift in response to climate change and human activities. Models such as Random Forests and Boosted Regression Trees excel in managing complex datasets and capturing intricate interactions within ecosystems (Barnard et al., 2019; Mechenich & Žliobaitė, 2023). These models can accurately predict changes in species distributions, offering vital insights for conservation planning. For instance, Random Forests are valued for their robustness against overfitting and their capacity to handle numerous predictors, making them ideal for analyzing the multifaceted effects of environmental changes on species distributions (Angelieri et al., 2016; Simon et al., 2023).

Further, artificial neural networks (ANNs) and support vector machines (SVMs) play crucial roles in understanding nonlinear ecological dynamics and defining habitat suitability boundaries under varying climatic scenarios (Su et al., 2018). ANNs' ability to adapt to new, dynamic data sets them apart in environments where species responses to changing climates are unpredictable, while SVMs are particularly adept at delineating presence–absence boundaries, crucial for assessing habitat fragmentation and suitability under future conditions (Claerhout et al., 2023; Gupta et al., 2022). Together, these ML models contribute not only to theoretical advancements in ecology but also to practical conservation efforts, such as designing effective strategies to manage invasive species and creating adaptive environmental policies that anticipate future ecological scenarios (Huettmann et al., 2018). Through the predictive power of ML, SDM becomes a cornerstone in proactive biodiversity conservation, enabling stakeholders to make informed decisions to mitigate the adverse effects of environmental changes on vital ecosystems (Alibabaei et al., 2022; Attri et al., 2023).

5.3.2 Habitat monitoring

The integration of ML with remote sensing technologies has revolutionized habitat monitoring by enabling the analysis of complex ecological data on an unprecedented scale (Bhatti et al., 2021; Stupariu et al., 2022). Remote sensing provides detailed, continuous data on various environmental parameters, such as vegetation, land use, and water conditions, from satellite and aerial imagery. When coupled with the analytical power of ML, these data become a potent tool for tracking and understanding changes in ecosystems over time (MohanRajan et al., 2020). ML models like convolutional neural networks (CNNs) and supervised algorithms such as Random Forests are adept at processing this imagery to classify land cover types and detect changes, such as deforestation, urban expansion, or shifts in wetland areas (Aziz et al., 2024; Praveen et al., 2023). For example, CNNs, with their deep learning capabilities, excel in identifying patterns in spatial data, making them ideal for detailed and nuanced tasks like monitoring forest degradation or urban sprawl. These models are trained on vast datasets of labeled images, allowing them to learn and predict land cover changes with high accuracy (Roy et al., 2024; Zhang et al., 2020). Furthermore, unsupervised learning algorithms play a crucial role in identifying unanticipated changes and emerging trends in ecosystem health without prior labeling (Bhatti et al., 2022). Techniques like K-means clustering and principal component analysis (PCA) analyze data to reveal hidden patterns, such as emergent stresses on habitats, which are critical for initiating proactive conservation measures (Boroujeni et al., 2024; Zhong et al., 2021). The practical applications of these ML-enhanced remote sensing tools are vast. They support critical conservation efforts by providing actionable insights into habitat degradation, enabling stakeholders to respond quickly to environmental threats. Studies leveraging these technologies have made significant contributions to fields such as deforestation monitoring, wetland restoration, and biodiversity conservation, highlighting areas in urgent need of intervention (Cooke et al., 2021; Das et al., 2023; Goetz et al., 2015). Thus the confluence of ML and remote sensing is not just enhancing our ability to monitor habitats but is also transforming conservation strategies. It allows for a more dynamic, informed, and timely response to environmental challenges, ensuring that efforts to protect and manage ecosystems are based on robust, data-driven insights (Chisom et al., 2024; Tomsett & Leyland, 2019).

5.3.3 Threat detection

ML has emerged as a pivotal technology in enhancing threat detection within conservation areas, particularly for combating poaching and illegal logging. Utilizing ML algorithms to analyze data from cameras, drones, and acoustic sensors transforms traditional surveillance methods into dynamic, real-time monitoring systems (Han et al., 2023; Panduman et al., 2024). CNNs are particularly effective in processing visual data from remote or difficult-to-access locations, allowing for the identification of illegal activities such as unauthorized human entry or tree cutting. These images and videos can be analyzed almost instantaneously, enabling quick response actions to mitigate these threats (Pather et al., 2020; Shinde et al., 2024). Furthermore, anomaly detection algorithms like One-Class SVM and isolation forests excel in recognizing patterns that deviate from the norm, which is crucial in vast wilderness areas where direct human surveillance is challenging (Bałdyga et al., 2024). These ML techniques can detect subtle signs of illegal activities, such as the sound of chainsaws or gunshots, captured by strategically placed acoustic sensors (Brickson et al., 2023). The ability of ML to analyze and classify different sounds adds an additional layer of surveillance that operates continuously, covering areas that are not visually monitored (Domingos et al., 2022). Predictive analytics also play a vital role by using historical data to forecast potential future incidents of poaching or logging. This predictive insight allows conservation managers to preemptively allocate resources more efficiently, focusing efforts on high-risk areas and times. This not only optimizes the use of limited conservation resources but also enhances the overall effectiveness of protection efforts (Maestas et al., 2023; Pan et al., 2021). The integration of ML into wildlife and forest conservation strategies represents a significant technological leap forward. It not only increases the efficiency and coverage of monitoring efforts but also supports a more informed and proactive approach to conservation (Shiva Prakash et al., 2022). As ML technology continues to evolve, its incorporation into environmental protection efforts promises to significantly bolster the ability to safeguard vulnerable ecosystems and species from human-induced threats (Aslam et al., 2024).

5.3.4 Climate impact prediction

ML serves as a pivotal tool in forecasting the impacts of climate change on biodiversity and ecosystem services, leveraging its capacity to analyze large and complex datasets to predict future environmental scenarios (Almeida et al., 2024; Chang, 2023). Supervised learning models, such as Random Forests and Neural Networks, are particularly effective for predicting how species distributions and ecosystem functions might change in response to varying climate variables like temperature and precipitation. These models are trained on historical climate and ecological data, enabling them to forecast future trends with significant accuracy (Bonannella et al., 2022; Deser et al., 2020; Morera et al., 2021). Unsupervised learning methods, including clustering and PCA, further complement these predictions by identifying hidden patterns in environmental data, which may indicate emerging changes in biodiversity and ecosystem characteristics (Borowiec et al., 2022; Nieto-Mora et al., 2023). This capability is crucial for detecting subtle, yet critical, shifts that might not be evident through traditional observational methods (Zhao et al., 2019). Deep learning techniques, such as CNNs and recurrent neural networks (RNNs), also play a crucial role, particularly in processing spatial and temporal data (Goel et al., 2023; Sun et al., 2019). CNNs are adept at analyzing sequences of satellite imagery to monitor changes in land cover or vegetation health over time, which are key indicators of ecosystem responses to climate change. RNNs, on the other hand, handle data that involve sequences, such as weather patterns, providing forecasts that help in understanding long-term ecological impacts and aiding in disaster risk reduction strategies (Alijoyo et al., 2024; Huang et al., 2021; Peng et al., 2024). Moreover, the application of ML in climate impact studies extends to practical conservation efforts (Untaru et al., 2016). For example, ML-driven species distribution models inform conservation planning by predicting shifts in habitat ranges due to climate change, guiding the establishment of new protected areas and the management of species migration (D'Aloia et al., 2019). Additionally, ML models that predict changes in ecosystem services under different climate scenarios help in valuing these services accurately, emphasizing the economic impact of biodiversity loss and ecosystem degradation (Meraj et al., 2022). Thus ML not only enhances our understanding of the potential impacts of climate change on natural systems but also equips policymakers and conservationists with the tools to plan more effectively (Kumar et al., 2021). By forecasting changes and their implications, ML enables proactive measures that aim to mitigate the adverse effects of climate change, ensuring the preservation and resilience of biodiversity and ecosystem services (Bibri et al., 2024; Brown et al., 2022). This strategic use of ML underscores its importance in adapting to and managing the complexities of environmental conservation in the face of global climate dynamics (Pan et al., 2022; Shankar & Gupta, 2024).

5.4 Discussion

5.4.1 Integrative insights

The effectiveness of ML in conservation efforts is profound, encompassing a range of applications from species distribution modeling to habitat monitoring and threat detection. By synthesizing the findings from various studies, we can critically assess the role and impact of ML in enhancing conservation strategies across multiple dimensions (Huettmann et al., 2018; Pereira et al., 2023).

5.4.1.1 Synthesis of machine learning applications in conservation

1. Species distribution and habitat monitoring: ML models have significantly enhanced the accuracy and efficiency of predicting changes in species distributions and habitats (Bhatti et al., 2023). Traditional methods often struggled with the complexity and volume of ecological and geographical data, but modern ML approaches have overcome these limitations. Tools like Random Forests and CNNs have been particularly effective (Lan et al., 2020). Random Forests, with their ensemble learning approach, can handle a large number of predictor variables and their interactions, making them ideal for ecological data that often involves numerous environmental factors (Gobeyn et al., 2019). CNNs, originally developed for image recognition tasks, have also found applications in ecology. CNNs can process spatial data by learning patterns and features directly from raw input data, such as satellite images or sensor data, without extensive preprocessing (Kattenborn et al., 2021). This capability is crucial for habitat monitoring, where spatial heterogeneity and temporal dynamics are significant. These ML models aid in identifying potential shifts in species habitats due to environmental changes, such as climate change, deforestation, and urbanization. For instance, they can predict how a species' habitat might move in response to changing temperature and precipitation patterns (Matthews et al., 2011). This predictive power is vital for conservation efforts, as it enables proactive planning rather than reactive responses. Conservationists can use these predictions to establish protected areas tailored to future conditions, ensuring that critical habitats are preserved even as environmental conditions evolve (Rilov et al., 2019). Furthermore, these models can integrate various data sources, including climate models, land use maps, and biodiversity records, to provide a comprehensive understanding of habitat dynamics (Randin et al., 2020). This integrative approach ensures that conservation strategies are based on robust, data-driven insights, ultimately enhancing the effectiveness of biodiversity preservation efforts. In summary, ML models, particularly Random Forests and CNNs, have revolutionized species distribution and habitat monitoring by improving prediction accuracy and

efficiency. These advancements facilitate proactive conservation planning, helping to safeguard biodiversity in the face of rapid environmental changes (Pichler & Hartig, 2023; Zhang et al., 2024).

2. Threat detection: ML has revolutionized the detection of illegal activities such as poaching and logging, providing new tools and techniques to enhance conservation efforts (Singh et al., 2022). Traditionally, the monitoring of protected areas relied heavily on manual patrols and basic surveillance systems, which were often limited in scope and efficiency. ML has transformed this landscape by enabling the analysis of vast amounts of data from various sources to detect threats more effectively (Muneer et al., 2024). By leveraging imagery from drones and satellites, ML algorithms can process high-resolution images to identify suspicious activities or changes in land use indicative of illegal logging. These algorithms can analyze patterns and detect anomalies that might be missed by human observers (Howitt et al., 2022). For instance, changes in vegetation cover or the presence of vehicles in restricted areas can be flagged for further investigation (Papyan et al., 2024). Acoustic sensors provide another valuable data source, especially in dense forest environments where visual surveillance is challenging. ML models can analyze sound data to detect noises associated with poaching, such as gunshots or chainsaws. These models can distinguish between normal forest sounds and those indicating illegal activities, allowing for real-time threat detection and response (Nuttall et al., 2022). The integration of these technologies significantly enhances the effectiveness of law enforcement in protected areas. One of the key advantages of ML-based threat detection is the reduction in time between threat identification and response. Real-time analysis means that alerts can be sent immediately to rangers or enforcement agencies, who can then take swift action to prevent irreversible damage to wildlife and habitats (Geldmann et al., 2019; Imran et al., 2023). Moreover, ML algorithms can be trained to predict potential threats by analyzing historical data and identifying patterns associated with illegal activities. This predictive capability allows for the deployment of resources more strategically, focusing efforts on areas with a higher likelihood of illegal activities occurring (Bharadiya, 2023; Kennedy et al., 2011). The application of ML in threat detection has significantly improved the monitoring and protection of wildlife and habitats. By analyzing imagery from drones and satellites, and sound data from acoustic sensors, ML algorithms can detect anomalies and predict potential threats in real-time, enhancing law enforcement effectiveness and reducing response times. This proactive approach is critical in preventing irreversible damage to ecosystems and ensuring the preservation of biodiversity (Nagendra et al., 2013; Papyan et al., 2024).
3. Climate impact prediction: ML is a powerful tool for predicting the impacts of climate change on biodiversity, offering critical insights that support effective conservation strategies (Shameer & Sanil, 2023). The predictive capabilities of ML models allow researchers and policymakers to anticipate how ecosystems will respond to different climate scenarios, thereby informing long-term planning efforts (Chang, 2023; Rilov et al., 2019). ML models can integrate vast datasets, including climate variables, species distribution data, and ecological interactions, to simulate future conditions. These models can forecast shifts in species ranges, changes in habitat suitability, and potential disruptions in ecosystem services (Pecchi et al., 2019). By understanding these potential changes, conservationists can develop strategies to mitigate negative impacts and enhance ecosystem resilience. One of the primary applications of ML in climate impact prediction is the identification of climate refugia (Seidl et al., 2016). These are areas that are expected to remain relatively stable despite changing climatic conditions and thus serve as vital havens for biodiversity. Protecting these refugia ensures that species have a better chance of surviving and adapting to new environmental realities (Yousefzadeh et al., 2024). ML models also help prioritize conservation efforts by identifying critical areas for intervention. For instance, enhancing connectivity between habitats can be a crucial strategy to facilitate species migration and gene flow, which are essential for adaptation to changing climates (Aitken & Whitlock, 2013; Jones et al., 2016). By predicting which regions will become more or less suitable for particular species, ML can guide the creation of wildlife corridors and other connectivity-enhancing measures (Shaver et al., 2022). Furthermore, ML-driven predictions support adaptive management practices. Conservation plans can be regularly updated based on the latest model outputs, ensuring that strategies remain effective as new data and better predictions become available (Jawad & Balázs, 2024). This dynamic approach is essential for dealing with the inherent uncertainties of climate change and its impacts on biodiversity. ML's predictive power is invaluable for understanding and mitigating the effects of climate change on ecosystems (Rangwala et al., 2021). By forecasting responses to various climate scenarios, ML models provide essential insights for long-term conservation planning. These insights help prioritize efforts to bolster ecosystem resilience, such as protecting climate refugia and enhancing habitat connectivity, ensuring the survival of species under changing climatic conditions (Garden et al., 2015; Morán-Ordóñez et al., 2018).
4. Integrative insights and decision making: ML significantly enhances conservation efforts by integrating diverse data streams to provide a comprehensive view of ecological health and environmental threats (Sun & Scanlon, 2019). This integrative approach is vital for strategic planning and resource allocation, ensuring that conservation initiatives are both effective and efficient (Chams & García-Blandón, 2019; Nishant et al., 2020). ML models can process and

analyze data from various sources, including satellite imagery, climate data, species distribution records, and sensor data. By synthesizing this information, ML provides a holistic understanding of ecological dynamics, enabling conservationists to identify patterns and trends that might not be apparent when considering data in isolation. For example, ML can combine data on land use changes, climate projections, and biodiversity surveys to predict future hotspots of biodiversity loss (Huettmann et al., 2018; Salcedo-Sanz et al., 2020). This allows conservationists to prioritize areas where interventions will have the greatest impact, optimizing the use of limited resources. Furthermore, integrating socio-economic data with ecological data can help in designing conservation strategies that are not only environmentally sound but also socially and economically viable (Bowgen et al., 2022; Koundouri et al., 2016). The ability of ML to handle large and complex datasets also facilitates the identification of indirect and cumulative impacts on ecosystems. For instance, changes in land use or climate can have cascading effects on food webs and ecosystem services. ML models can trace these impacts across different ecological levels, providing insights that are essential for developing comprehensive conservation strategies (Schuwirth et al., 2019; Wood, 2021). Moreover, ML supports adaptive management by continuously updating predictions and recommendations based on new data. This dynamic approach ensures that conservation plans remain relevant and effective in the face of changing conditions and new information (Benos et al., 2021; Prober et al., 2019). Decision-makers can use these integrative insights to adjust strategies in real time, enhancing the resilience and sustainability of conservation efforts (Cook et al., 2014). ML's ability to integrate various data streams provides a holistic view of ecological health and environmental threats. This comprehensive perspective is crucial for strategic planning and resource allocation, ensuring that conservation efforts are both effective and efficient. By leveraging integrative insights, ML enhances decision-making processes, ultimately contributing to more robust and adaptive conservation strategies (Bibri et al., 2024; Ramyar et al., 2021).

5.4.1.2 *Effectiveness and limitations*

While the effectiveness of ML in conservation is evident, several limitations and challenges must be acknowledged to fully leverage its potential. The performance of ML models is heavily dependent on the quality and availability of ecological data (Bennett et al., 2013; Tuia et al., 2022). Often, ecological datasets are incomplete, outdated, or biased due to limited sampling efforts. Inaccuracies in data can lead to significant errors in ML predictions, which in turn can adversely affect conservation outcomes (Bubac et al., 2019; Ploton et al., 2020). For instance, if the data used to train a model does not adequately represent the variability within an ecosystem, the model's predictions will be less reliable. Ecological systems are inherently complex, involving numerous interactions across multiple scales. This complexity can exceed the current capabilities of ML models, necessitating ongoing refinement and adaptation of these tools (Bhatti et al., 2022; Pichler & Hartig, 2023). For example, models may struggle to accurately capture rare events or subtle interactions that are critical to ecosystem dynamics. As a result, continuous improvement and validation of ML models are essential to enhance their applicability and reliability in conservation contexts. Another significant challenge is the need for increased transparency and interpretability of ML models (Balch et al., 2019; Musuamba et al., 2021). Conservation decision-makers must understand how ML models generate predictions and the potential uncertainties involved. This understanding is crucial for building trust in ML applications and ensuring that they complement traditional ecological knowledge and conservation practices (Cha et al., 2021; Williams & Johnson, 2013). Black-box models, which provide little insight into their internal workings, can be particularly problematic. To address this, there is a growing emphasis on developing interpretable ML models and communicating their findings clearly to stakeholders. Effective conservation requires the integration of ML insights with traditional ecological knowledge and conservation practices (Bodria et al., 2023; Chisom et al., 2024). This integration can be challenging, as it requires collaboration between data scientists, ecologists, and local communities. Ensuring that ML tools are used to complement rather than replace traditional methods is essential for their successful implementation (Bibri et al., 2024). ML offers powerful tools for conservation, its application is constrained by challenges related to data quality, the complexity of ecological systems, and the need for transparency and interpretability. Addressing these challenges requires ongoing efforts to improve data collection, refine ML models, and foster interdisciplinary collaboration to integrate ML insights with traditional conservation practices (Pichler & Hartig, 2023; Udegbe et al., 2024).

5.4.2 Challenges and limitations

The integration of ML in conservation has showcased significant benefits, yet it also faces notable challenges and limitations. Addressing these hurdles is crucial for enhancing the efficacy and reliability of ML applications in conservation efforts (Allal et al., 2024; Waleed et al., 2023).

5.4.2.1 Data availability and quality

One of the foremost challenges in applying ML to conservation is the availability and quality of data. Environmental and ecological data are often sparse, incomplete, or unevenly distributed across geographical regions and time periods (Kumar et al., 2023; Martínez-Minaya et al., 2018). This limitation can severely impact the training and performance of ML models, leading to biased or inaccurate predictions. For instance, in species distribution modeling, a lack of historical occurrence data for certain species or regions can result in their underrepresentation in predictive analyses. This underrepresentation can skew conservation priorities and efforts, potentially overlooking critical areas that require protection (Christodoulou et al., 2019; Lentini & Wintle, 2015). If key habitats or species are not adequately represented in the data, the resulting ML models may fail to identify them as conservation priorities, leading to gaps in protection measures. Data sparsity can also affect the temporal dimension, where changes in ecosystems over time are not captured due to insufficient longitudinal data. This can hinder the ability to model trends accurately and predict future changes, thus impacting long-term conservation planning (Kodikara et al., 2022; Nagendra et al., 2013). Moreover, data that are unevenly distributed across different geographical areas can lead to models that are overly specific to well-studied regions, with poor generalizability to other areas. The issues of data quality and availability are further complicated by the challenges of data collection in remote or inaccessible regions (Zhao et al., 2020). Collecting high-quality ecological data often requires significant resources, including time, funding, and expertise, which may not always be available. To mitigate these challenges, several strategies can be employed. Techniques such as data augmentation can be used to synthetically increase the size of the dataset, improving the robustness of ML models (Khosla & Saini, 2020; Kosmala et al., 2016). Engaging the public in data collection through citizen science initiatives can help gather more extensive and diverse datasets. Combining datasets from multiple sources, including remote sensing, field surveys, and historical records, can provide a more comprehensive dataset. Developing and using advanced ML algorithms that can handle missing data and bias correction techniques can improve model performance despite data limitations (Cho et al., 2020; de Sherbinin et al., 2021). The availability and quality of data are significant challenges in applying ML to conservation. Sparse, incomplete, and unevenly distributed data can lead to biased or inaccurate predictions, affecting conservation efforts. Addressing these challenges requires a combination of advanced techniques and collaborative efforts to enhance data collection and integration, ensuring that ML models can provide reliable insights for effective conservation planning (Lotfian et al., 2021; Tuia et al., 2022).

5.4.2.2 Model accuracy and generalization

The accuracy and generalization of ML models are crucial in conservation, especially in complex ecological contexts where multiple variables and interactions need to be considered (Aamir et al., 2022; Pichler & Hartig, 2023). ML models often face challenges such as overfitting, where they perform well on training data but fail to generalize to new, unseen conditions (Kernbach & Staartjes, 2022). This issue is particularly problematic in conservation, where conditions are continuously changing due to factors like climate change, habitat destruction, and invasive species. Overfitting occurs when an ML model learns the noise and specifics of the training data rather than the underlying patterns (Oliver & Morecroft, 2014). This makes the model highly accurate on the training dataset but less effective on new data. In conservation, where predictive accuracy across varied and changing conditions is essential, overfitting can lead to misleading conclusions and ineffective strategies. For example, a model trained on data from a specific period or region might not accurately predict species distribution in different climatic or geographic contexts (Fois et al., 2018; Khan et al., 2022). To ensure that ML models are robust and adaptable to changing conditions, several strategies can be employed. Applying regularization methods such as L1 and L2 regularization can help prevent overfitting by penalizing overly complex models. Using cross-validation techniques, such as k-fold cross-validation, helps in assessing the model's performance on different subsets of the data, ensuring that it generalizes well to unseen data (Hanelt et al., 2020; Mahesh et al., 2022). Combining predictions from multiple models through ensemble methods like bagging, boosting, and stacking can improve generalization and robustness. Enhancing the training dataset through data augmentation and resampling techniques can help the model learn more generalized patterns. This includes generating synthetic data or using techniques like bootstrapping (Cao et al., 2020; Yin et al., 2021). Developing and employing adaptive learning algorithms that can update and improve their predictions as new data becomes available is crucial. This is especially important in dynamic ecological contexts where conditions change rapidly. Utilizing transfer learning, where a model trained on one problem is adapted for a different but related problem, can help leverage existing knowledge and improve model performance in new conditions (Colchester et al., 2016; Seyfollahi et al., 2023). Implementing systems for continuous monitoring and updating of ML models ensures they remain relevant and accurate as new data and conditions emerge. The dynamic nature of ecosystems adds another layer of complexity (Diez-Olivan et al., 2019).

Factors such as climate change, habitat destruction, and invasive species continuously alter the conditions under which ML models operate. Ensuring that ML models are not only accurate but also adaptable to these changes is essential. Models must be designed to incorporate new data and adapt to shifting patterns quickly and effectively (Butt et al., 2016; Parisi et al., 2019). The accuracy and generalization of ML models are pivotal in conservation, ensuring these models remain robust and adaptable in the face of changing ecological conditions is a challenging task. Employing strategies to prevent overfitting, leveraging ensemble and adaptive methods, and continuously updating models are essential steps in addressing these challenges. By doing so, ML models can provide reliable and actionable insights for effective conservation planning and decision-making (Bharadiya, 2023; Yin et al., 2024).

5.4.2.3 *Interdisciplinary approaches*

Conservation is inherently multidisciplinary, involving fields such as ecology, geography, sociology, and economics, which requires not only technical expertise in data science but also a deep understanding of ecological dynamics and conservation principles (Lapeyrolerie et al., 2022; Leenhardt et al., 2015). Effective ML applications in conservation depend on interdisciplinary collaboration to integrate diverse perspectives and knowledge bases. However, this poses challenges due to communication gaps between data scientists and field ecologists, potentially hindering the translation of ML findings into actionable strategies (Bammer et al., 2020; Bibri et al., 2024). To overcome these challenges, forming interdisciplinary teams, developing cross-training programs, utilizing collaborative platforms, and fostering clear communication are essential. Ensuring that ML findings are translated into practical conservation efforts requires accessible ML tools, transparent communication of model uncertainties, and close collaboration with policymakers and practitioners to integrate these insights into real-world conservation policies and practices (Brown et al., 2023; Huettmann et al., 2018).

5.4.2.4 *Ethical and practical considerations*

The deployment of ML in conservation must consider ethical implications, particularly regarding surveillance and privacy in monitored areas (Franchini et al., 2022). Additionally, while ML and advanced technologies offer powerful tools for conservation, they should not overshadow traditional ecological methods and indigenous knowledge, which are invaluable for a comprehensive approach to conservation (Neaves et al., 2024). Balancing technological interventions with respect and integration of conventional practices is crucial for effective and ethical conservation efforts (Bhatti et al., 2023). By incorporating traditional knowledge alongside ML, conservation strategies can be more holistic, culturally sensitive, and sustainable (Neethirajan, 2024).

5.4.2.5 *Scalability and implementation*

Scaling ML solutions from pilot studies or controlled environments to wide-scale applications poses significant challenges (Shaikh & Birajdar, 2024). Issues such as the need for extensive computational resources, maintaining technology in remote or harsh environments, and the high costs associated with data collection and processing can limit the feasibility and sustainability of ML-driven conservation projects (Villa-Henriksen et al., 2020). Addressing these challenges requires robust infrastructure, investment in durable and cost-effective technologies, and innovative approaches to data management and resource allocation to ensure that ML applications can effectively support large-scale conservation efforts (Sun & Scanlon, 2019).

5.4.2.6 *Addressing the challenges*

To overcome these challenges, ongoing efforts are necessary to improve data collection methods, such as deploying more extensive and consistent monitoring systems and encouraging the open sharing of ecological data (Fraisl et al., 2022). Enhancing model robustness through advanced ML techniques that focus on generalization and adaptability is also crucial. Furthermore, fostering interdisciplinary collaborations and training among conservationists, ecologists, and data scientists can bridge gaps in understanding and the application of ML technologies (Alqahtani & Kumar, 2024; De Koning et al., 2023). By addressing these challenges, the potential of ML to contribute to effective and sustainable conservation strategies can be more fully realized, ensuring that technological advancements align with the overarching goals of biodiversity preservation and ecosystem resilience (Breed et al., 2019; Obaideen et al., 2024).

5.4.3 Future directions

Exploring future directions for ML in conservation offers exciting possibilities for advancing our understanding and management of ecosystems and biodiversity (Sharma et al., 2022). As technology evolves, several areas stand out as particularly promising for further research and development.

5.4.3.1 *Integration of deep learning for complex environmental modeling*

Deep learning, a subset of ML characterized by sophisticated neural networks, holds significant potential for modeling complex environmental systems with higher accuracy (Yuan et al., 2020). Future research can focus on developing and refining deep learning models that can process and interpret the vast amounts of data generated by remote sensing technologies, such as satellite imagery and aerial photos (Li et al., 2024). These advancements could lead to the creation of more robust predictive models capable of analyzing intricate patterns in environmental data. For instance, deep learning algorithms could be employed to monitor ecosystem dynamics in real-time, providing critical insights into how various factors influence biodiversity and ecosystem health (Cui et al., 2023; Zhong et al., 2021). By analyzing changes in land use and vegetation cover, these models could identify early signs of habitat degradation, allowing for timely interventions to preserve endangered ecosystems. Additionally, deep learning models can enhance our understanding of climate change impacts by integrating data from multiple sources, including weather stations, ocean buoys, and climate models (Salcedo-Sanz et al., 2020; Waleed et al., 2023). This comprehensive approach can improve the accuracy of climate predictions, helping policymakers develop more effective strategies to combat climate change. Moreover, the application of deep learning in environmental sciences can extend to pollution monitoring. By analyzing air and water quality data, these models can identify pollution hotspots and predict the spread of contaminants, aiding in the formulation of targeted mitigation measures (Gardner-Frolick et al., 2022; Weaver et al., 2013; Yuan et al., 2020). Overall, the integration of deep learning with remote sensing technologies presents a promising avenue for advancing environmental research. By providing more nuanced insights into ecosystem dynamics, land use changes, and habitat degradation, these models can significantly enhance our ability to predict and mitigate environmental impacts, ultimately contributing to the sustainability of our planet (Li et al., 2022; Liu et al., 2024).

5.4.3.2 *Improved predictive models for climate change effects*

As the global climate continues to change, there is a critical need for improved predictive models that can accurately forecast the long-term effects on biodiversity and ecosystem services (Tulloch et al., 2020). Future research should aim to develop ML models that can integrate various data types, including meteorological, geological, and biological data, to predict changes in species distributions, migration patterns, and ecosystem resilience. These advanced ML models could synthesize large datasets from diverse sources, creating comprehensive simulations of environmental scenarios (Amindin et al., 2024; Zhong et al., 2021). For instance, integrating meteorological data with biological observations can help predict how climate change will alter habitats and influence species' geographic ranges. This holistic approach can reveal the cascading effects of climate shifts on ecosystems, providing critical insights for biodiversity conservation (Iverson et al., 2011; Prober et al., 2019). Additionally, ML models can be used to analyze geological data alongside climate projections, offering a deeper understanding of how changes in physical landscapes, such as erosion or land degradation, affect ecosystem services. By incorporating biological data, these models can also track migration patterns and population dynamics, which are essential for identifying species at risk and designing effective conservation measures. Such models would be invaluable in planning conservation strategies that are adaptive and robust in the face of climate uncertainty (Leigh et al., 2021; Rilov et al., 2019). For example, they can inform the creation of wildlife corridors that facilitate species migration or the implementation of habitat restoration projects that enhance ecosystem resilience. Moreover, these predictive tools can help prioritize conservation efforts in regions most vulnerable to climate impacts, ensuring resources are allocated efficiently (Summers et al., 2012; Timpane-Padgham et al., 2017). The integration of various data types into sophisticated ML models represents a crucial advancement in our ability to predict and mitigate the impacts of climate change on biodiversity and ecosystem services. By providing accurate and actionable forecasts, these models can support the development of adaptive, science-based conservation strategies that safeguard our natural world for future generations (Fernandes et al., 2020; van Kerkhoff et al., 2019).

5.4.3.3 *Real-time monitoring and automated response systems*

Leveraging ML for real-time environmental monitoring and automated response systems represents a significant advancement in conservation technology (Borah et al., 2024). Future research could explore the development of ML algorithms capable of processing real-time data streams from sensors and cameras to detect immediate threats, such as illegal logging or poaching activities (Zeller Zigaitis & Robinson, 2023). These advanced ML systems can analyze vast amounts of data in real-time, identifying patterns and anomalies that indicate potential threats. For instance, by continuously monitoring sensor data from protected areas, ML algorithms can detect unusual activities that may signal illegal logging (Gupta et al., 2020). Similarly, integrating camera feeds with ML models can help identify poachers or unauthorized vehicles in wildlife reserves. The potential benefits of these systems extend beyond detection. They could trigger automatic alerts to relevant authorities, enabling rapid response to threats. For example, upon detecting signs of poaching, the system could immediately notify rangers and deploy drones to the area, increasing the chances of intercepting illegal activities (Doull et al., 2021; Nandutu et al., 2022). This automated response capability can significantly enhance the effectiveness of conservation efforts by reducing response times and improving coordination. Moreover, these ML-driven systems can be integrated with existing conservation infrastructure to optimize resource allocation (Hossain et al., 2024). By prioritizing areas with high threat levels, conservation organizations can deploy their resources more efficiently, focusing on regions that need immediate attention. This strategic approach can maximize the impact of conservation initiatives, ensuring that efforts are concentrated where they are most needed (Maxwell et al., 2020). In addition to terrestrial applications, similar ML technologies can be applied to marine environments. For instance, real-time monitoring of underwater sensors and cameras can help detect illegal fishing activities, monitor coral reef health, and track the movements of marine species (Glaviano et al., 2022). These insights can inform conservation strategies and policies, contributing to the preservation of marine ecosystems. Overall, the development of ML algorithms for real-time environmental monitoring and automated response systems holds great promise for enhancing conservation efforts (Long et al., 2015). By enabling proactive threat detection and rapid response, these technologies can play a crucial role in protecting biodiversity and preserving natural habitats (Yuan et al., 2017). Future research should focus on refining these algorithms and integrating them with a wide range of data sources to build comprehensive and effective conservation tools.

5.4.3.4 Cross-disciplinary approaches and integrative data analysis

There is a growing recognition of the value of interdisciplinary approaches in environmental science (Reed et al., 2016). Future research should focus on integrating ML with other scientific disciplines such as ecology, climatology, and social sciences to develop comprehensive models that consider ecological, atmospheric, and human factors (Konya & Nematzadeh, 2024). This integrative approach could lead to more holistic environmental management practices and better conservation outcomes (Reed et al., 2018). Integrating ML with ecology can enhance our understanding of complex biological interactions and ecosystem dynamics. By analyzing data from ecological studies, ML algorithms can identify patterns and relationships that might be missed using traditional methods. For example, ML can be used to model species interactions, predict population dynamics, and assess the impacts of environmental changes on biodiversity (Anderson et al., 2013; Pickett et al., 2017). These insights can inform conservation strategies, such as habitat restoration and species protection plan. In climatology, ML can process vast amounts of atmospheric data to improve climate models and predictions (Jose et al., 2022). By integrating data from weather stations, satellite observations, and climate simulations, ML models can provide more accurate forecasts of climate change impacts. This information is crucial for developing mitigation and adaptation strategies to address the challenges posed by global warming, such as sea-level rise, extreme weather events, and shifts in climate zones (Schwalbert et al., 2020; Tang, 2019). The integration of social sciences is equally important, as human activities and behaviors play a significant role in environmental changes. ML can analyze socio-economic data to understand the drivers of environmental degradation, such as deforestation, pollution, and resource overexploitation (Bennett et al., 2017). By incorporating human factors into environmental models, researchers can develop policies that address both ecological and social dimensions of sustainability. For instance, ML can help identify the most effective interventions to promote sustainable land use practices, reduce carbon emissions, and enhance community resilience to environmental hazards (Chams & García-Blandón, 2019). Moreover, interdisciplinary approaches can facilitate stakeholder engagement and collaboration across different sectors. By involving ecologists, climatologists, social scientists, policymakers, and local communities in the development and application of ML models, researchers can ensure that environmental management practices are inclusive, equitable, and context-specific (Salim et al., 2024). This collaborative approach can lead to more effective and sustainable conservation outcomes, as it takes into account the diverse perspectives and needs of various stakeholders. The integration of ML with ecology, climatology, and social sciences represents a promising direction for future environmental research (Gavin et al., 2018). By developing comprehensive models that consider ecological, atmospheric, and human factors, researchers can achieve a more holistic understanding of environmental systems and devise better strategies for conservation and sustainability. This interdisciplinary approach has the potential to significantly enhance environmental management practices and contribute to the preservation of our planet's natural resources and biodiversity (Bibri et al., 2024; Reed et al., 2018).

5.4.3.5 Ethical artificial intelligence use in conservation

As artificial intelligence (AI) and ML play increasingly prominent roles in conservation, ethical considerations must be addressed, particularly regarding data privacy, surveillance, and the impact of automated decisions on local communities and indigenous populations (Ullah et al., 2023). Future research should include the development of guidelines and frameworks that ensure ethical practices in the application of ML in conservation (Griffin et al., 2020). One critical ethical concern is data privacy. Conservation projects often involve collecting and analyzing large amounts of data, including images and geolocation information, which can inadvertently capture sensitive information about individuals and communities (Anhalt-Depies et al., 2019). Ensuring that data collection and use respect privacy rights is paramount. This can be achieved by implementing strict data anonymization techniques, obtaining informed consent from affected communities, and establishing clear data governance policies. Surveillance is another significant issue (Adekugbe & Ibeh, 2024). While real-time monitoring technologies can enhance conservation efforts by detecting illegal activities like poaching, they can also be used for unwarranted surveillance of local populations. To mitigate this, it is essential to balance the need for environmental monitoring with the protection of individual privacy (Du et al., 2023). Developing transparent protocols for data collection and usage, as well as involving local communities in decision-making processes, can help maintain this balance. The impact of automated decisions on local communities and indigenous populations is a further ethical consideration (Treweek et al., 2013). AI and ML systems can make decisions that affect land use, resource allocation, and conservation priorities, which may have significant consequences for people living in or near conservation areas. Ensuring that these systems are designed and implemented in a way that respects the rights and knowledge of indigenous and local communities is crucial (Dudley et al., 2018; Tengö et al., 2017). This includes engaging these communities in the development and deployment of ML technologies, recognizing their traditional knowledge, and ensuring that conservation initiatives do not undermine their livelihoods or cultural practices (Cárdenas & Monterroso, 2022).

To address these ethical challenges, future research should focus on developing comprehensive guidelines and frameworks that promote ethical practices in the application of ML in conservation. These frameworks should include:

1. Ethical data management: Establishing protocols for data collection, storage, and sharing that prioritize privacy and consent. This includes anonymizing data, securing informed consent, and providing transparency about how data will be used (Beaugard et al., 2024).
2. Community engagement: Involving local and indigenous communities in the development and implementation of ML projects. This can be achieved through participatory approaches that recognize and incorporate traditional knowledge and ensure that conservation efforts benefit these communities (Kakekagumick et al., 2013).
3. Transparent decision-making: Creating transparent algorithms and decision-making processes that allow for accountability and scrutiny. This involves making the workings of AI and ML systems understandable to nonexperts and ensuring that there are mechanisms for addressing grievances and concerns (Wilson & van der Velden, 2022).
4. Equity and fairness: Ensuring that the benefits and burdens of conservation technologies are distributed fairly. This means designing ML systems that do not disproportionately impact vulnerable communities and actively working to mitigate any negative effects (Khreis et al., 2016).
5. Regulatory compliance: Adhering to relevant legal and ethical standards, including international human rights frameworks, in the deployment of AI and ML in conservation (Boldt & Orrù, 2022).

AI and ML technologies become integral to conservation efforts, it is essential to develop and adhere to ethical guidelines and frameworks. By addressing issues of data privacy, surveillance, and the impact on local communities, we can ensure that these technologies are used responsibly and equitably, ultimately contributing to more effective and just conservation outcomes (Griffin et al., 2020).

5.4.3.6 Enhanced models for ecosystem service valuation

Understanding and valuing ecosystem services are crucial for effective conservation planning. ML models that can accurately assess and predict the economic impacts of biodiversity loss and ecosystem degradation are needed (Meraj et al., 2021). Future research could focus on creating sophisticated models that quantify ecosystem services in monetary terms, helping policymakers and conservationists prioritize and justify investments in biodiversity protection

(Tinch et al., 2019). These advanced ML models can integrate various types of environmental, economic, and social data to provide a comprehensive valuation of ecosystem services. For instance, models could analyze data on water purification, carbon sequestration, pollination, and recreational benefits to estimate their economic value (Turner et al., 2016). By translating these services into monetary terms, ML models can highlight the financial benefits of conserving natural habitats, making a compelling case for biodiversity protection (Karolyi & Tobin-de la Puente, 2023). Moreover, these models can help predict the economic costs associated with ecosystem degradation and biodiversity loss. For example, the loss of pollinators due to habitat destruction can significantly impact agricultural productivity, leading to increased food prices and economic instability (Hristov et al., 2020). By forecasting such outcomes, ML models can inform policies aimed at mitigating these impacts through conservation efforts. In addition to economic assessments, ML models can enhance our understanding of the complex interactions within ecosystems (Mishra et al., 2024). By analyzing large datasets from ecological studies, remote sensing, and climate models, these tools can identify critical areas for conservation, optimize resource allocation, and monitor the effectiveness of conservation strategies (Nagendra et al., 2013). This holistic approach can lead to more informed decision-making and better conservation outcomes. Future research should also focus on developing user-friendly tools and platforms that make these sophisticated ML models accessible to policymakers, conservationists, and local communities (Shiva Prakash et al., 2022). By providing practical tools and strategies for sustainable ecosystem management, these models can support the implementation of effective conservation plans and policies (DeAngelis et al., 2021).

The integration of ML in valuing ecosystem services offers several key benefits:

1. Informed decision-making: Providing policymakers with quantitative data on the economic value of ecosystem services can guide investment and policy decisions, ensuring resources are allocated efficiently and effectively (Laurans et al., 2013).
2. Enhanced public awareness: Quantifying the economic benefits of ecosystem services can raise public awareness about the importance of biodiversity and conservation, fostering greater support for environmental protection efforts (Patterson & Coelho, 2009).
3. Strategic conservation planning: Identifying high-value ecosystems and predicting the impacts of biodiversity loss can help prioritize conservation actions, ensuring the most critical areas are protected (Li et al., 2022).
4. Adaptive management: Continuous monitoring and analysis of ecosystem services can enable adaptive management, allowing conservation strategies to be adjusted based on real-time data and changing conditions (Naseer et al., 2024).

By focusing on these future directions, research in ML can significantly enhance our ability to understand, protect, and manage the natural environment (Nishant et al., 2020). This will not only advance scientific knowledge but also provide practical tools and strategies for sustainable ecosystem management, ultimately contributing to the long-term preservation of biodiversity and ecosystem services (Duru et al., 2015).

5.5 Conclusion

The chapter of ML applications in ecosystem services and biodiversity conservation reveals its transformative potential to significantly enhance environmental monitoring, management, and protection efforts. ML's ability to process vast and complex datasets has led to more accurate predictions of species distribution, better monitoring of habitat changes, and effective detection of environmental threats such as illegal logging and poaching. Moreover, the predictive prowess of ML in assessing the impacts of climate change on biodiversity provides crucial insights that can guide adaptive conservation strategies. The insights gained from ML applications in conservation have substantial implications for policymaking. They provide a strong evidence base for crafting more informed and effective environmental policies and conservation strategies. For instance, predictive models can help prioritize conservation areas and actions that maximize biodiversity preservation and ecosystem resilience. Additionally, real-time monitoring capabilities of ML can enhance enforcement of environmental laws and regulations by providing authorities with timely and accurate information on illegal activities and ecological changes. Policymakers can leverage these capabilities to allocate resources more efficiently, targeting areas at greatest risk and where intervention can yield the most significant impact. Furthermore, ML-driven data can support the creation of dynamic environmental policies that adapt to new threats and changing conditions, ensuring long-term sustainability and conservation success. As we move forward, the importance of continuing innovation and research in the application of ML to conservation cannot be overstated. Ongoing advancements in ML technologies promise even greater capabilities and opportunities for tackling some of the most pressing environmental challenges. However, it is crucial to approach these innovations with a commitment to ethical practices and inclusivity, ensuring that the benefits of ML in conservation are widely distributed and contribute positively to all communities, including indigenous and local populations. The journey of integrating ML into conservation is just beginning, and it is paved with opportunities for significant impacts. By continuing to invest in research and development and fostering collaborative, interdisciplinary efforts that combine technology with traditional ecological knowledge, we can enhance our ability to safeguard our planet's invaluable biodiversity and ecosystem services for future generations.

References

Aamir, M., Bhatti, M. A., Bazai, S. U., Marjan, S., Mirza, A. M., Wahid, A., Hasnain, A., & Bhatti, U. A. (2022). Predicting the environmental change of carbon emission patterns in South Asia: A deep learning approach using BiLSTM. *Atmosphere*, *13*(12). Available from https://doi.org/10.3390/atmos13122011, http://www.mdpi.com/journal/atmosphere.

Adekugbe, A. P., & Ibeh, C. V. (2024). Navigating ethical challenges in data management for U.S. program development: Best practices and recommendations. *International Journal of Management and Entrepreneurship Research*, *6*(4), 1023–1033. Available from https://doi.org/10.51594/ijmer.v6i4.982.

Aitken, S. N., & Whitlock, M. C. (2013). Assisted gene flow to facilitate local adaptation to climate change. *Annual Review of Ecology, Evolution, and Systematics*, *44*, 367–388. Available from https://doi.org/10.1146/annurev-ecolsys-110512-135747, http://arjournals.annualreviews.org/loi/ecolsys.

Alibabaei, K., Gaspar, P. D., Lima, T. M., Campos, R. M., Girão, I., Monteiro, J., & Lopes, C. M. (2022). A review of the challenges of using deep learning algorithms to support decision-making in agricultural activities. *Remote Sensing*, *14*(3). Available from https://doi.org/10.3390/rs14030638, https://www.mdpi.com/2072-4292/14/3/638/pdf.

Alijoyo, F. A., Gongada, T. N., Kaur, C., Mageswari, N., Sekhar, J. C., Ramesh, J. V. N., El-Ebiary, Y. A. B., & Ulmas, Z. (2024). Advanced hybrid CNN-Bi-LSTM model augmented with GA and FFO for enhanced cyclone intensity forecasting. *Alexandria Engineering Journal*, *92*, 346–357. Available from https://doi.org/10.1016/j.aej.2024.02.062, https://www.sciencedirect.com/science/journal/11100168.

Allal, Z., Noura, H. N., Salman, O., & Chahine, K. (2024). Machine learning solutions for renewable energy systems: Applications, challenges, limitations, and future directions. *Journal of Environmental Management*, *354*. Available from https://doi.org/10.1016/j.jenvman.2024.120392, https://www.sciencedirect.com/science/journal/03014797.

Almeida, B., David, J., Campos, F. S., & Cabral, P. (2024). Satellite-based machine learning modelling of ecosystem services indicators: A review and meta-analysis. *Applied Geography*, *165*, 103249. Available from https://doi.org/10.1016/j.apgeog.2024.103249.

Alqahtani, H., & Kumar, G. (2024). Machine learning for enhancing transportation security: A comprehensive analysis of electric and flying vehicle systems. *Engineering Applications of Artificial Intelligence*, *129*, 107667. Available from https://doi.org/10.1016/j.engappai.2023.107667.

Amindin, A., Siamian, N., Kariminejad, N., Clague, J. J., & Pourghasemi, H. R. (2024). An integrated GEE and machine learning framework for detecting ecological stability under land use/land cover changes. *Global Ecology and Conservation*, *53*. Available from https://doi.org/10.1016/j.gecco.2024.e03010, https://www.sciencedirect.com/science/journal/23519894.

Amiri, Z., Heidari, A., Navimipour, N. J., Unal, M., & Mousavi, A. (2023). Adventures in data analysis: a systematic review of Deep Learning techniques for pattern recognition in cyber-physical-social systems. *Multimedia Tools and Applications*, *83*(8), 22909–22973. Available from https://doi.org/10.1007/s11042-023-16382-x.

Anderson, J. J., Gurarie, E., Bracis, C., Burke, B. J., & Laidre, K. L. (2013). Modeling climate change impacts on phenology and population dynamics of migratory marine species. *Ecological Modelling*, *264*, 83–97. Available from https://doi.org/10.1016/j.ecolmodel.2013.03.009.

Angelieri, C. C. S., Adams-Hosking, C., Paschoaletto, K. M., Ferraz, M. D. B., De Souza, M. P., & McAlpine, C. A. (2016). Using species distribution models to predict potential landscape restoration effects on puma conservation. *PLoS ONE*, *11*(1). Available from https://doi.org/10.1371/journal.pone.0145232, http://www.plosone.org/article/fetchObject.action?uri = info:doi/10.1371/journal.pone.0145232&representation = PDF.

Anhalt-Depies, C., Stenglein, J. L., Zuckerberg, B., Townsend, P. M., & Rissman, A. R. (2019). Tradeoffs and tools for data quality, privacy, transparency, and trust in citizen science. *Biological Conservation*, *238*. Available from http://www.elsevier.com/inca/publications/store/4/0/5/8/5/3, https://doi.org/10.1016/j.biocon.2019.108195.

Aslam, R. W., Shu, H., Naz, I., Quddoos, A., Yaseen, A., Gulshad, K., & Alarifi, S. S. (2024). Machine learning-based wetland vulnerability assessment in the Sindh Province Ramsar site using remote sensing data. *Remote Sensing*, *16*(5). Available from https://doi.org/10.3390/rs16050928, http://www.mdpi.com/journal/remotesensing.

Attri, I., Awasthi, L. K., & Sharma, T. P. (2023). Machine learning in agriculture: a review of crop management applications. *Multimedia Tools and Applications*, *83*(5), 12875–12915. Available from https://doi.org/10.1007/s11042-023-16105-2.

Aziz, G., Minallah, N., Saeed, A., Frnda, J., & Khan, W. (2024). Remote sensing based forest cover classification using machine learning. *Scientific Reports*, *14*(1). Available from https://doi.org/10.1038/s41598-023-50863-1, https://www.nature.com/srep/.

Balch, J., Iglesias, V., Braswell, A., Rossi, M., Joseph, M. B., Mahood, A. L., Travis, W., Shrum, T., White, C. T., Scholl, V., McGuire, B., Karban, C., & Buckland, M. (2019). Socio-environmental extremes: Rethinking extraordinary events as outcomes of interacting biophysical and social systems. *Earth's Future*. Available from https://doi.org/10.7287/peerj.preprints.27877v12019.

Bammer, G., O'Rourke, M., O'Connell, D., Neuhauser, L., Midgley, G., Klein, J. T., Grigg, N. J., Gadlin, H., Elsum, I. R., Bursztyn, M., Fulton, E. A., Pohl, C., Smithson, M., Vilsmaier, U., Bergmann, M., Jaeger, J., Merkx, F., Vienni Baptista, B., Burgman, M. A., … Richardson, G. P. (2020). Expertise in research integration and implementation for tackling complex problems: When is it needed, where can it be found and how can it be strengthened. *Palgrave Communications*, *6*(1). Available from https://doi.org/10.1057/s41599-019-0380-0, http://www.palgrave-journals.com/palcomms/.

Barbhuiya, S., & Das, B. B. (2023). Life cycle assessment of construction materials: Methodologies, applications and future directions for sustainable decision-making. *Case Studies in Construction Materials*, *19*. Available from https://doi.org/10.1016/j.cscm.2023.e02326, http://www.journals.elsevier.com/case-studies-in-construction-materials/.

Barbosa Júnior, M. R., Moreira, B. Rd. A., Carreira, Vd. S., Brito Filho, A. Ld, Trentin, C., Souza, F. L. Pd, Tedesco, D., Setiyono, T., Flores, J. P., Ampatzidis, Y., Silva, R. Pd, & Shiratsuchi, L. S. (2024). Precision agriculture in the United States: A comprehensive meta-review inspiring further research, innovation, and adoption. *Computers and Electronics in Agriculture*, *221*. Available from https://doi.org/10.1016/j.compag.2024.108993, https://www.sciencedirect.com/science/journal/01681699.

Barnard, D. M., Germino, M. J., Pilliod, D. S., Arkle, R. S., Applestein, C., Davidson, B. E., & Fisk, M. R. (2019). Cannot see the random forest for the decision trees: Selecting predictive models for restoration ecology. *Restoration Ecology*, *27*(5), 1053–1063. Available from https://doi.org/10.1111/rec.12938, http://onlinelibrary.wiley.com/journal/10.1111/(ISSN)1526-100X.

Bałdyga, M., Barański, K., Belter, J., Kalinowski, M., & Weichbroth, P. (2024). Anomaly detection in railway sensor data environments: State-of-the-art methods and empirical performance evaluation. *Sensors*, *24*(8), 2633. Available from https://doi.org/10.3390/s24082633.

Beaugard, C. A., Formica, S. W., Cummins, E. R., Bagley, S. M., Beletskey, L., Green, T. C., Murray, S. P., Yan, S., Xuan, Z., Walley, A. Y., & Carroll, J. J. (2024). Privacy and confidentiality in Massachusetts' post-overdose outreach programs: Mixed methods analysis of outreach staff surveys and interviews. *International Journal of Drug Policy*, *124*. Available from https://doi.org/10.1016/j.drugpo.2023.104310, https://www.sciencedirect.com/science/journal/09553959.

Bennett, N. D., Croke, B. F. W., Guariso, G., Guillaume, J. H. A., Hamilton, S. H., Jakeman, A. J., Marsili-Libelli, S., Newham, L. T. H., Norton, J. P., Perrin, C., Pierce, S. A., Robson, B., Seppelt, R., Voinov, A. A., Fath, B. D., & Andreassian, V. (2013). Characterising performance of environmental models. *Environmental Modelling and Software*, *40*, 1–20. Available from https://doi.org/10.1016/j.envsoft.2012.09.011.

Bennett, N. J., Roth, R., Klain, S. C., Chan, K., Christie, P., Clark, D. A., Cullman, G., Curran, D., Durbin, T. J., Epstein, G., Greenberg, A., Nelson, M. P., Sandlos, J., Stedman, R., Teel, T. L., Thomas, R., Veríssimo, D., & Wyborn, C. (2017). Conservation social science: Understanding and integrating human dimensions to improve conservation. *Biological Conservation*, *205*, 93–108. Available from https://doi.org/10.1016/j.biocon.2016.10.006, http://www.elsevier.com/inca/publications/store/4/0/5/8/5/3.

Benos, L., Tagarakis, A. C., Dolias, G., Berruto, R., Kateris, D., & Bochtis, D. (2021). Machine learning in agriculture: A comprehensive updated review. *Sensors*, *21*(11). Available from https://doi.org/10.3390/s21113758, https://www.mdpi.com/1424-8220/21/11/3758/pdf.

Bharadiya, J. P. (2023). Machine learning and AI in business intelligence: Trends and opportunities. *International Journal of Computer*, *48*(1), 123–134, 2023.

Bhatti, U. A., Huang, M., Wu, D., Zhang, Y., Mehmood, A., & Han, H. (2019). Recommendation system using feature extraction and pattern recognition in clinical care systems. *Enterprise Information Systems*, *13*(3), 329–351. Available from https://doi.org/10.1080/17517575.2018.1557256, http://www.tandf.co.uk/journals/titles/17517575.asp.

Bhatti, U. A., Marjan, S., Wahid, A., Syam, M. S., Huang, M., Tang, H., & Hasnain, A. (2023). The effects of socioeconomic factors on particulate matter concentration in China's: New evidence from spatial econometric model. *Journal of Cleaner Production*, *417*, 137969. Available from https://doi.org/10.1016/j.jclepro.2023.137969.

Bhatti, U. A., Yan, Y., Zhou, M., Ali, S., Hussain, A., Qingsong, H., Yu, Z., & Yuan, L. (2021). Time series analysis and forecasting of air pollution particulate matter (PM2.5): An SARIMA and factor analysis approach. *IEEE Access.*, *9*, 41019–41031. Available from https://doi.org/10.1109/ACCESS.2021.3060744, http://ieeexplore.ieee.org/xpl/RecentIssue.jsp?punumber = 6287639.

Bhatti, U. A., Yu, Z., Hasnain, A., Nawaz, S. A., Yuan, L., Wen, L., & Bhatti, M. A. (2022). Evaluating the impact of roads on the diversity pattern and density of trees to improve the conservation of species. *Environmental Science and Pollution Research*, *29*(10), 14780–14790. Available from https://doi.org/10.1007/s11356-021-16627-y, https://link.springer.com/journal/11356.

Bhatti, U. A., Yu, Z., Li, J., Nawaz, S. A., Mehmood, A., Zhang, K., & Yuan, L. (2020). Hybrid watermarking algorithm using clifford algebra with arnold scrambling and chaotic encryption. *IEEE Access*, *8*, 76386–76398. Available from https://doi.org/10.1109/ACCESS.2020.2988298, http://ieeexplore.ieee.org/xpl/RecentIssue.jsp?punumber = 6287639.

Bibri, S. E., Krogstie, J., Kaboli, A., & Alahi, A. (2024). Smarter eco-cities and their leading-edge artificial intelligence of things solutions for environmental sustainability: A comprehensive systematic review. *Environmental Science and Ecotechnology*, *19*. Available from http://www.journals.elsevier.com/environmental-science-and-ecotechnology, https://doi.org/10.1016/j.ese.2023.100330.

Bodria, F., Giannotti, F., Guidotti, R., Naretto, F., Pedreschi, D., & Rinzivillo, S. (2023). Benchmarking and survey of explanation methods for black box models. *Data Mining and Knowledge Discovery*, *37*(5), 1719–1778. Available from https://doi.org/10.1007/s10618-023-00933-9, https://www.springer.com/journal/10618.

Boldt, J., & Orrù, E. (2022). Towards a unified list of ethical principles for emerging technologies. An analysis of four European reports on molecular biotechnology and artificial intelligence. *Sustainable Futures*, *4*, 100086. Available from https://doi.org/10.1016/j.sftr.2022.100086.

Bonannella, C., Hengl, T., Heisig, J., Parente, L., Wright, M. N., Herold, M., & de Bruin, S. (2022). Forest tree species distribution for Europe 2000–2020: Mapping potential and realized distributions using spatiotemporal machine learning. *PeerJ*, *10*. Available from https://doi.org/10.7717/peerj.13728, https://peerj.com/articles/13728/.

Borah, S. S., Khanal, A., & Sundaravadivel, P. (2024). Emerging Technologies for Automation in Environmental Sensing: Review. *Applied Sciences*, *14*(8), 3531. Available from https://doi.org/10.3390/app14083531.

Boroujeni, S. P. H., Razi, A., Khoshdel, S., Afghah, F., Coen, J. L., O'Neill, L., Fule, P., Watts, A., Kokolakis, N. M. T., & Vamvoudakis, K. G. (2024). A comprehensive survey of research towards AI-enabled unmanned aerial systems in pre-, active-, and post-wildfire management. *Information Fusion*, *108*. Available from https://doi.org/10.1016/j.inffus.2024.102369, https://www.sciencedirect.com/science/journal/15662535.

Borowiec, M. L., Dikow, R. B., Frandsen, P. B., McKeeken, A., Valentini, G., & White, A. E. (2022). Deep learning as a tool for ecology and evolution. *Methods in Ecology and Evolution*, *13*(8), 1640–1660. Available from https://doi.org/10.1111/2041-210X.13901, http://onlinelibrary.wiley.com/journal/10.1111/(ISSN)2041-210X.

Bowgen, K. M., Kettel, E. F., Butchart, S. H. M., Carr, J. A., Foden, W. B., Magin, G., Morecroft, M. D., Smith, R. K., Stein, B. A., Sutherland, W. J., Thaxter, C. B., & Pearce-Higgins, J. W. (2022). Conservation interventions can benefit species impacted by climate change. *Biological Conservation*, *269*, 109524. Available from https://doi.org/10.1016/j.biocon.2022.109524.

Breed, M. F., Harrison, P. A., Blyth, C., Byrne, M., Gaget, V., Gellie, N. J. C., Groom, S. V. C., Hodgson, R., Mills, J. G., Prowse, T. A. A., Steane, D. A., & Mohr, J. J. (2019). The potential of genomics for restoring ecosystems and biodiversity. *Nature Reviews Genetics*, *20*(10), 615–628. Available from https://doi.org/10.1038/s41576-019-0152-0, http://www.nature.com/reviews/genetics.

Brickson, L., Zhang, L., Vollrath, F., Douglas-Hamilton, I., & Titus, A. J. (2023). Elephants and algorithms: A review of the current and future role of AI in elephant monitoring. *Journal of the Royal Society Interface*, *20*(208). Available from https://doi.org/10.1098/rsif.2023.0367, http://rsif.royalsocietypublishing.org/.

Brown, M. B., Morrison, J. C., Schulz, T. T., Cross, M. S., Püschel-Hoeneisen, N., Suresh, V., & Eguren, A. (2022). Using the conservation standards framework to address the effects of climate change on biodiversity and ecosystem services. *Climate*, *10*(2). Available from https://doi.org/10.3390/cli10020013, http://www.mdpi.com/journal/climate.

Brown, S. A., Sparapani, R., Osinski, K., Zhang, J., Blessing, J., Cheng, F., Hamid, A., MohamadiPour, M. B., Lal, J. C., Kothari, A. N., Caraballo, P., Noseworthy, P., Johnson, R. H., Hansen, K., Sun, L. Y., Crotty, B., Cheng, Y. C., Echefu, G., Doshi, K., & Olson, J. (2023). Team principles for successful interdisciplinary research teams. *American Heart Journal Plus: Cardiology Research and Practice*, *32*. Available from https://doi.org/10.1016/j.ahjo.2023.100306, https://www.journals.elsevier.com/american-heart-journal-plus-cardiology-research-and-practice.

Bubac, C. M., Johnson, A. C., Fox, J. A., & Cullingham, C. I. (2019). Conservation translocations and post-release monitoring: Identifying trends in failures, biases, and challenges from around the world. *Biological Conservation*, *238*. Available from https://doi.org/10.1016/j.biocon.2019.108239, http://www.elsevier.com/inca/publications/store/4/0/5/8/5/3.

Buchelt, A., Adrowitzer, A., Kieseberg, P., Gollob, C., Nothdurft, A., Eresheim, S., Tschiatschek, S., Stampfer, K., & Holzinger, A. (2024). Exploring artificial intelligence for applications of drones in forest ecology and management. *Forest Ecology and Management*, *551*, 121530. Available from https://doi.org/10.1016/j.foreco.2023.121530.

Butt, N., Possingham, H. P., De Los Rios, C., Maggini, R., Fuller, R. A., Maxwell, S. L., & Watson, J. E. M. (2016). Challenges in assessing the vulnerability of species to climate change to inform conservation actions. *Biological Conservation*, *199*, 10–15. Available from https://doi.org/10.1016/j.biocon.2016.04.020.

Cao, P., Li, X., Mao, K., Lu, F., Ning, G., Fang, L., & Pan, Q. (2020). A novel data augmentation method to enhance deep neural networks for detection of atrial fibrillation. *Biomedical Signal Processing and Control.*, *56*, 101675. Available from https://doi.org/10.1016/j.bspc.2019.101675.

Caskurlu, S., Richardson, J. C., Maeda, Y., & Kozan, K. (2021). The qualitative evidence behind the factors impacting online learning experiences as informed by the community of inquiry framework: A thematic synthesis. *Computers and Education*, *165*. Available from https://doi.org/10.1016/j.compedu.2020.104111, http://www.journals.elsevier.com/computers-and-education/.

Cha, Y. K., Shin, J., Go, B. G., Lee, D. S., Kim, Y. W., Kim, T. H., & Park, Y. S. (2021). An interpretable machine learning method for supporting ecosystem management: Application to species distribution models of freshwater macroinvertebrates. *Journal of Environmental Management*, *291*. Available from https://doi.org/10.1016/j.jenvman.2021.112719, https://www.sciencedirect.com/journal/journal-of-environmental-management.

Chams, N., & García-Blandón, J. (2019). On the importance of sustainable human resource management for the adoption of sustainable development goals. *Resources, Conservation and Recycling*, *141*, 109–122. Available from https://doi.org/10.1016/j.resconrec.2018.10.006, http://www.elsevier.com/locate/resconrec.

Chang, G. J. (2023). Biodiversity estimation by environment drivers using machine/deep learning for ecological management. *Ecological Informatics*, *78*, 102319. Available from https://doi.org/10.1016/j.ecoinf.2023.102319.

Chisom, O. N., Biu, P. W., Umoh, A. A., Obaedo, B. O., Adegbite, A. O., & Abatan, A. (2024). Reviewing the role of AI in environmental monitoring and conservation: A data-driven revolution for our planet. *World Journal of Advanced Research and Reviews*, *21*(1), 161–171. Available from https://doi.org/10.30574/wjarr.2024.21.1.2720.

Cho, D., Yoo, C., Im, J., & Cha, D. H. (2020). Comparative assessment of various machine learning-based bias correction methods for numerical weather prediction model forecasts of extreme air temperatures in urban areas. *Earth and Space Science*, *7*(4). Available from https://doi.org/10.1029/2019EA000740, http://agupubs.onlinelibrary.wiley.com/hub/journal/10.1002/(ISSN)2333-5084/.

Christodoulou, E., Ma, J., Collins, G. S., Steyerberg, E. W., Verbakel, J. Y., & Van Calster, B. (2019). A systematic review shows no performance benefit of machine learning over logistic regression for clinical prediction models. *Journal of Clinical Epidemiology*, *110*, 12–22. Available from https://doi.org/10.1016/j.jclinepi.2019.02.004, http://www.elsevier.com/locate/jclinepi.

Claerhout, T., Strubbe, D., Castillo Torres, R., Samain, M. S., Chatrou, L. W., & Veltjen, E. (2023). An integrative study of species distribution modelling and conservation genetics: Magnolia in Hispaniola. *Biodiversity and Conservation*, *32*(4), 1205–1231. Available from https://doi.org/10.1007/s10531-023-02546-6, https://www.springer.com/journal/10531.

Colchester, K., Hagras, H., Alghazzawi, D., & Aldabbagh, G. (2016). A Survey of Artificial Intelligence Techniques Employed for Adaptive Educational Systems within E-Learning Platforms. *Journal of Artificial Intelligence and Soft Computing Research*, *7*(1), 47–64. Available from https://doi.org/10.1515/jaiscr-2017-0004.

Cook, C. N., Inayatullah, S., Burgman, M. A., Sutherland, W. J., & Wintle, B. A. (2014). Strategic foresight: How planning for the unpredictable can improve environmental decision-making. *Trends in Ecology and Evolution*, *29*(9), 531−541. Available from http://www.elsevier.com/inca/publications/store/3/0/3/3/9, https://doi.org/10.1016/j.tree.2014.07.005.

Cooke, S. J., Bergman, J. N., Madliger, C. L., Cramp, R. L., Beardall, J., Burness, G., Clark, T. D., Dantzer, B., De La Barrera, E., Fangue, N. A., Franklin, C. E., Fuller, A., Hawkes, L. A., Hultine, K. R., Hunt, K. E., Love, O. P., MacMillan, H. A., Mandelman, J. W., Mark, F. C., … Chown, S. L. (2021). One hundred research questions in conservation physiology for generating actionable evidence to inform conservation policy and practice. *Conservation Physiology*, *9*(1). Available from https://doi.org/10.1093/conphys/coab009, https://academic.oup.com/conphys.

Cui, S., Gao, Y., Huang, Y., Shen, L., Zhao, Q., Pan, Y., & Zhuang, S. (2023). Advances and applications of machine learning and deep learning in environmental ecology and health. *Environmental Pollution*, *335*, 122358. Available from https://doi.org/10.1016/j.envpol.2023.122358.

Cárdenas, O. B., & Monterroso, I. (2022). *Achieving political rights, enhancing forest livelihoods* (pp. 338−354). Informa UK Limited. Available from 10.4324/9780367488710-27.

D'Aloia, C. C., Naujokaitis-Lewis, I., Blackford, C., Chu, C., Curtis, J. M. R., Darling, E., Guichard, F., Leroux, S. J., Martensen, A. C., Rayfield, B., Sunday, J. M., Xuereb, A., & Fortin, M. J. (2019). Coupled networks of permanent protected areas and dynamic conservation areas for biodiversity conservation under climate change. *Frontiers in Ecology and Evolution*, *7*. Available from https://doi.org/10.3389/fevo.2019.00027, https://doi.org/10.3389/fevo.2019.00027.

Das, S. K., Benkhelifa, F., Sun, Y., Abumarshoud, H., Abbasi, Q. H., Imran, M. A., & Mohjazi, L. (2023). Comprehensive review on ML-based RIS-enhanced IoT systems: Basics, research progress and future challenges. *Computer Networks*, *224*. Available from https://doi.org/10.1016/j.comnet.2023.109581, http://www.journals.elsevier.com/computer-networks/.

Dash, S., Shakyawar, S. K., Sharma, M., & Kaushik, S. (2019). Big data in healthcare: Management, analysis and future prospects. *Journal of Big Data*, *6*(1). Available from https://doi.org/10.1186/s40537-019-0217-0, https://journalofbigdata.springeropen.com.

DeAngelis, D. L., Franco, D., Hastings, A., Hilker, F. M., Lenhart, S., Lutscher, F., Petrovskaya, N., Petrovskii, S., & Tyson, R. C. (2021). Towards building a sustainable future: Positioning ecological modelling for impact in ecosystems management. *Bulletin of Mathematical Biology*, *83*(10). Available from https://doi.org/10.1007/s11538-021-00927-y, http://www.springer.com.

De Koning, K., Broekhuijsen, J., Kühn, I., Ovaskainen, O., Taubert, F., Endresen, D., Schigel, D., & Grimm, V. (2023). Digital twins: dynamic model-data fusion for ecology. *Trends in ecology and evolution*, *38*(10), 916−926. Available from https://doi.org/10.1016/j.tree.2023.04.010.

de Sherbinin, A., Bowser, A., Chuang, T. R., Cooper, C., Danielsen, F., Edmunds, R., Elias, P., Faustman, E., Hultquist, C., Mondardini, R., Popescu, I., Shonowo, A., & Sivakumar, K. (2021). The critical importance of citizen science data. *Frontiers in Climate*, *3*. Available from https://doi.org/10.3389/fclim.2021.650760, http://www.frontiersin.org/journals/climate.

Demertzis, K., & Iliadis, L. (2018). *The impact of climate change on biodiversity: The ecological consequences of invasive species in Greece Climate Change Management* (pp. 15−38). Greece: Springer. Available from springer.com/series/8740, https://doi.org/10.1007/978-3-319-69838-0_2.

Deser, C., Lehner, F., Rodgers, K. B., Ault, T., Delworth, T. L., DiNezio, P. N., Fiore, A., Frankignoul, C., Fyfe, J. C., Horton, D. E., Kay, J. E., Knutti, R., Lovenduski, N. S., Marotzke, J., McKinnon, K. A., Minobe, S., Randerson, J., Screen, J. A., Simpson, I. R., & Ting, M. (2020). Insights from Earth system model initial-condition large ensembles and future prospects. *Nature Climate Change*, *10*(4), 277−286. Available from https://doi.org/10.1038/s41558-020-0731-2.

Diez-Olivan, A., Del Ser, J., Galar, D., & Sierra, B. (2019). Data fusion and machine learning for industrial prognosis: Trends and perspectives towards Industry 4.0. *Information Fusion*, *50*, 92−111. Available from https://doi.org/10.1016/j.inffus.2018.10.005, http://www.elsevier.com/inca/publications/store/6/2/0/8/6/2/index.htt.

Dinis, F. M., Poças Martins, J., Guimarães, A. S., & Rangel, B. (2021). BIM and Semantic Enrichment Methods and Applications: A Review of Recent Developments. *Archives of Computational Methods in Engineering*, *29*(2), 879−895. Available from https://doi.org/10.1007/s11831-021-09595-6.

Domingos, L. C. F., Santos, P. E., Skelton, P. S. M., Brinkworth, R. S. A., & Sammut, K. (2022). A survey of underwater acoustic data classification methods using deep learning for shoreline surveillance. *Sensors*, *22*(6). Available from https://doi.org/10.3390/s22062181, https://www.mdpi.com/1424-8220/22/6/2181/pdf.

Doull, K. E., Chalmers, C., Fergus, P., Longmore, S., Piel, A. K., & Wich, S. A. (2021). An evaluation of the factors affecting 'poacher' detection with drones and the efficacy of machine-learning for detection. *Sensors*, *21*(12). Available from https://doi.org/10.3390/s21124074, https://www.mdpi.com/1424-8220/21/12/4074/pdf.

Du, N., Fathollahi-Fard, A. M., & Wong, K. Y. (2023). Wildlife resource conservation and utilization for achieving sustainable development in China: main barriers and problem identification. *Environmental Science and Pollution Research*, 1−20. Available from https://doi.org/10.1007/s11356-023-26982-7.

Dudley, N., Jonas, H., Nelson, F., Parrish, J., Pyhälä, A., Stolton, S., & Watson, J. E. M. (2018). The essential role of other effective area-based conservation measures in achieving big bold conservation targets. *Global Ecology and Conservation*, *15*. Available from https://doi.org/10.1016/j.gecco.2018.e00424.

Duru, M., Therond, O., Martin, G., Martin-Clouaire, R., Magne, M. A., Justes, E., Journet, E. P., Aubertot, J. N., Savary, S., Bergez, J. E., & Sarthou, J. P. (2015). How to implement biodiversity-based agriculture to enhance ecosystem services: A review. *Agronomy for Sustainable Development*, *35*(4), 1259−1281. Available from https://doi.org/10.1007/s13593-015-0306-1, http://www.springerlink.com/content/1773-0155.

Edeigba, B. A., Ashinze, U. K., Umoh, A. A., Biu, P. W., & Daraojimba, A. I. (2024). Urban green spaces and their impact on environmental health: A global review. *World Journal of Advanced Research and Reviews*, *21*(2), 917−927. Available from https://doi.org/10.30574/wjarr.2024.21.2.0518.

Fernandes, M., Corchado, J. M., & Marreiros, G. (2022). Machine learning techniques applied to mechanical fault diagnosis and fault prognosis in the context of real industrial manufacturing use-cases: a systematic literature review. *Applied Intelligence*, *52*(12), 14246–14280. Available from https://doi.org/10.1007/s10489-022-03344-3.

Fernandes, A. C. M., Gonzalez, R. Q., Lenihan-Clarke, M. A., Trotter, E. F. L., & Arsanjani, J. J. (2020). Machine learning for conservation planning in a changing climate. *Sustainability*, *12*(18), 7657. Available from https://doi.org/10.3390/su12187657.

Fois, M., Cuena-Lombraña, A., Fenu, G., & Bacchetta, G. (2018). Using species distribution models at local scale to guide the search of poorly known species: Review, methodological issues and future directions. *Ecological Modelling*, *385*, 124–132. Available from https://doi.org/10.1016/j.ecolmodel.2018.07.018, http://www.elsevier.com/inca/publications/store/5/0/3/3/0/6.

Fraisl, D., Hager, G., Bedessem, B., Gold, M., Hsing, P. Y., Danielsen, F., Hitchcock, C. B., Hulbert, J. M., Piera, J., Spiers, H., Thiel, M., & Haklay, M. (2022). Citizen science in environmental and ecological sciences. *Nature Reviews Methods Primers*, *2*(1). Available from https://doi.org/10.1038/s43586-022-00144-4, https://www.nature.com/nrmp/journal-information.

Franchini, M., Rullman, S., & Claramunt-López, B. (2022). A questionnaire-based investigation to explore the social and legal implications derived from the use of camera traps for wildlife monitoring and conservation. *European Journal of Wildlife Research*, *68*(4). Available from https://doi.org/10.1007/s10344-022-01593-8.

Garden, J. G., O'Donnell, T., & Catterall, C. P. (2015). Changing habitat areas and static reserves: challenges to species protection under climate change. *Landscape Ecology*, *30*(10), 1959–1973. Available from https://doi.org/10.1007/s10980-015-0223-3.

Gardner-Frolick, R., Boyd, D., & Giang, A. (2022). Selecting Data Analytic and Modeling Methods to Support Air Pollution and Environmental Justice Investigations: A Critical Review and Guidance Framework. *Environmental Science and Technology*, *56*(5), 2843–2860. Available from https://doi.org/10.1021/acs.est.1c01739.

Gavin, M. C., McCarter, J., Berkes, F., Mead, A. T. P., Sterling, E. J., Tang, R., & Turner, N. J. (2018). Effective biodiversity conservation requires dynamic, pluralistic, partnership-based approaches. *Sustainability*, *10*(6), 1846. Available from https://doi.org/10.3390/su10061846.

Geldmann, J., Manica, A., Burgess, N. D., Coad, L., & Balmford, A. (2019). A global-level assessment of the effectiveness of protected areas at resisting anthropogenic pressures. *Proceedings of the National Academy of Sciences*, *116*(46), 23209–23215. Available from https://doi.org/10.1073/pnas.1908221116.

Glaviano, F., Esposito, R., Cosmo, A. D., Esposito, F., Gerevini, L., Ria, A., Molinara, M., Bruschi, P., Costantini, M., & Zupo, V. (2022). Management and sustainable exploitation of marine environments through smart monitoring and automation. *Journal of Marine Science and Engineering*, *10*(2), 297. Available from https://doi.org/10.3390/jmse10020297.

Gobeyn, S., Mouton, A. M., Cord, A. F., Kaim, A., Volk, M., & Goethals, P. L. M. (2019). Evolutionary algorithms for species distribution modelling: A review in the context of machine learning. *Ecological Modelling*, *392*, 179–195. Available from https://doi.org/10.1016/j.ecolmodel.2018.11.013, http://www.elsevier.com/inca/publications/store/5/0/3/3/0/6.

Goel, A., Goel, A. K., & Kumar, A. (2023). The role of artificial neural network and machine learning in utilizing spatial information. *Spatial Information Research*, *31*(3), 275–285. Available from https://doi.org/10.1007/s41324-022-00494-x.

Goetz, S. J., Hansen, M., Houghton, R. A., Walker, W., Laporte, N., & Busch, J. (2015). Measurement and monitoring needs, capabilities and potential for addressing reduced emissions from deforestation and forest degradation under REDD + . *Environmental Research Letters*, *10*(12). Available from https://doi.org/10.1088/1748-9326/10/12/123001, http://iopscience.iop.org/article/10.1088/1748-9326/10/12/123001/pdf.

González, S., García, S., Del Ser, J., Rokach, L., & Herrera, F. (2020). A practical tutorial on bagging and boosting based ensembles for machine learning: Algorithms, software tools, performance study, practical perspectives and opportunities. *Information Fusion*, *64*, 205–237. Available from https://doi.org/10.1016/j.inffus.2020.07.007, http://www.elsevier.com/inca/publications/store/6/2/0/8/6/2/index.htt.

Griffin, A. S., Callen, A., Klop-Toker, K., Scanlon, R. J., & Hayward, M. W. (2020). Compassionate conservation clashes with conservation biology: Should empathy, compassion, and deontological moral principles drive conservation practice? *Frontiers in Psychology*, *11*. Available from https://doi.org/10.3389/fpsyg.2020.01139, http://www.frontiersin.org/Psychology.

Gupta, D., Gujre, N., Singha, S., & Mitra, S. (2022). Role of existing and emerging technologies in advancing climate-smart agriculture through modeling: A review. *Ecological Informatics*, *71*, 101805. Available from https://doi.org/10.1016/j.ecoinf.2022.101805.

Gupta, R., Tanwar, S., Tyagi, S., & Kumar, N. (2020). Machine learning models for secure data analytics: A taxonomy and threat model. *Computer Communications*, *153*, 406–440. Available from https://doi.org/10.1016/j.comcom.2020.02.008, http://www.journals.elsevier.com/computer-communications/.

Han, W., Zhang, X., Wang, Y., Wang, L., Huang, X., Li, J., Wang, S., Chen, W., Li, X., Feng, R., Fan, R., Zhang, X., & Wang, Y. (2023). A survey of machine learning and deep learning in remote sensing of geological environment: Challenges, advances, and opportunities. *ISPRS Journal of Photogrammetry and Remote Sensing*, *202*, 87–113. Available from https://doi.org/10.1016/j.isprsjprs.2023.05.032, http://www.elsevier.com/inca/publications/store/5/0/3/3/4/0.

Hanelt, A., Bohnsack, R., Marz, D., & Antunes Marante, C. (2020). A Systematic Review of the Literature on Digital Transformation: Insights and Implications for Strategy and Organizational Change. *Journal of Management Studies*, *58*(5), 1159–1197. Available from https://doi.org/10.1111/joms.12639.

Hossain, M. A., Mazumder, M. S. A., Bari, M. H., & Mahi, R. (2024). Impact assessment of machine learning algorithms on resource efficiency and management in urban developments. *Global Mainstream Journal*, *1*(2), 1–9. Available from https://doi.org/10.62304/ijbm.v1i2.129.

Howitt, R., Karp, L., & Rausser, G. (2022). Remote sensing technologies: Implications for agricultural and resource economics. In Modern Agricultural and Resource Economics and Policy. *Essays in Honor of Gordon Rausser. Cham: Springer International Publishing*, 183–217. Available from https://doi.org/10.1007/978-3-030-77760-9_9.

Hristov, P., Neov, B., Shumkova, R., & Palova, N. (2020). Significance of apoidea as main pollinators. Ecological and economic impact and implications for human nutrition. *Diversity*, *12*(7), 280. Available from https://doi.org/10.3390/d12070280.

Huang, D., Wang, S., & Liu, Z. (2021). A systematic review of prediction methods for emergency management. *International Journal of Disaster Risk Reduction*, *62*, 102412. Available from https://doi.org/10.1016/j.ijdrr.2021.102412.

Huettmann, F. (2018). *Machine learning for 'strategic conservation and planning': Patterns, applications. Thoughts and urgently needed global progress for sustainability* (pp. 315–333). Springer Science and Business Media LLC. Available from http://doi.org/10.1007/978-3-319-96978-7_16.

Huettmann, F., Craig, E. H., Herrick, K. A., Baltensperger, A. P., Humphries, G. R. W., Lieske, D. J., Miller, K., Mullet, T. C., Oppel, S., Resendiz, C., Rutzen, I., Schmid, M. S., Suwal, M. K., & Young, B. D. (2018). *Use of machine learning (ML) for predicting and analyzing ecological and 'presence only' data: An overview of applications and a good outlook. Machine learning for ecology and sustainable natural resource management* (pp. 27–61). Springer Science and Business Media LLC. Available from http://doi.org/10.1007/978-3-319-96978-7_2.

Imran, M., Siddiqui, H. U. R., Raza, A., Raza, M. A., Rustam, F., & Ashraf, I. (2023). A performance overview of machine learning-based defense strategies for advanced persistent threats in industrial control systems. *Computers and Security*, *134*, 103445. Available from https://doi.org/10.1016/j.cose.2023.103445.

Iverson, L. R., Prasad, A. M., Matthews, S. N., & Peters, M. P. (2011). Lessons learned while integrating habitat, dispersal, disturbance, and life-history traits into species habitat models under climate change. *Ecosystems*, *14*(6), 1005–1020. Available from https://doi.org/10.1007/s10021-011-9456-4.

Jawad, Z. N., & Balázs, V. (2024). Machine learning-driven optimization of enterprise resource planning (ERP) systems: A comprehensive review. *Beni-Suef University Journal of Basic and Applied Sciences*, *13*(1). Available from https://doi.org/10.1186/s43088-023-00460-y, https://www.springer.com/journal/43088.

Jones, K. R., Watson, J. E. M., Possingham, H. P., & Klein, C. J. (2016). Incorporating climate change into spatial conservation prioritisation: A review. *Biological Conservation*, *194*, 121–130. Available from https://doi.org/10.1016/j.biocon.2015.12.008, http://www.elsevier.com/inca/publications/store/4/0/5/8/5/3.

Jose, D. M., Vincent, A. M., & Dwarakish, G. S. (2022). Improving multiple model ensemble predictions of daily precipitation and temperature through machine learning techniques. *Scientific Reports*, *12*(1). Available from https://doi.org/10.1038/s41598-022-08786-w.

Kakekagumick, K. E., Hayward, M. N., Harris, S. B., Saksvig, B., Gittelsohn, J., Manokeesic, G., Goodman, S., & Hanley, A. J. (2013). Sandy lake health and diabetes project: A community-based intervention targeting type 2 diabetes and its risk factors in a first nations community. *Frontiers in Endocrinology*, *4*. Available from https://doi.org/10.3389/fendo.2013.00170Canada, http://www.frontiersin.org/Journal/FullText.aspx?ART_DOI = 10.3389/fendo.2013.00170&x = y#.

Karolyi, G. A., & Tobin-de la Puente, J. (2023). Biodiversity finance: A call for research into financing nature. *Financial Management*, *52*(2), 231–251. Available from https://doi.org/10.1111/fima.12417, http://onlinelibrary.wiley.com/journal/10.1111/(ISSN)1755-053X.

Kattenborn, T., Leitloff, J., Schiefer, F., & Hinz, S. (2021). Review on convolutional neural networks (CNN) in vegetation remote sensing. *ISPRS Journal of Photogrammetry and Remote Sensing*, *173*, 24–49. Available from https://doi.org/10.1016/j.isprsjprs.2020.12.010, http://www.elsevier.com/inca/publications/store/5/0/3/3/4/0.

Kennedy, L. W., Caplan, J. M., & Piza, E. (2011). Risk clusters, hotspots, and spatial intelligence: risk terrain modeling as an algorithm for police resource allocation strategies. *Journal of quantitative criminology*, *27*, 339–362. Available from https://doi.org/10.1007/s10940-010-9126-2.

Kernbach, J. M., & Staartjes, V. E. (2022). Foundations of machine learning-based clinical prediction modeling: Part II—generalization and overfitting. *Germany Acta Neurochirurgica, Supplementum*, *134*, 15–21. Available from https://doi.org/10.1007/978-3-030-85292-4_3, http://www.springerlink.com/content/p7rw02/.

Khan, N., Kamaruddin, M. A., Ullah Sheikh, U., Zawawi, M. H., Yusup, Y., Bakht, M. P., & Noor, N. M. (2022). Prediction of oil palm yield using machine learning in the perspective of fluctuating weather and soil moisture conditions: Evaluation of a generic workflow. *Plants*, *11*(13). Available from https://doi.org/10.3390/plants11131697, https://www.mdpi.com/2223-7747/11/13/1697/pdf?version = 1656326514.

Khosla, C., & Saini, B.S. (2020). Enhancing performance of deep learning models with different data augmentation techniques: A survey. *Proceedings of international conference on intelligent engineering and management*, ICIEM. http://ieeexplore.ieee.org/xpl/mostRecentIssue.jsp?punumber = 9151448.

Khreis, H., Warsow, K. M., Verlinghieri, E., Guzman, A., Pellecuer, L., Ferreira, A., Jones, I., Heinen, E., Rojas-Rueda, D., Mueller, N., Schepers, P., Lucas, K., & Nieuwenhuijsen, M. (2016). The health impacts of traffic-related exposures in urban areas: Understanding real effects, underlying driving forces and co-producing future directions. *Journal of Transport and Health*, *3*(3), 249–267. Available from https://doi.org/10.1016/j.jth.2016.07.002, http://www.journals.elsevier.com/journal-of-transport-and-health/.

Kodikara, S., Ellul, S., & Lê Cao, K. A. (2022). Statistical challenges in longitudinal microbiome data analysis. *Briefings in Bioinformatics*, *23*(4). Available from https://doi.org/10.1093/bib/bbac273.

Konya, A., & Nematzadeh, P. (2024). Recent applications of AI to environmental disciplines: A review. *Science of the Total Environment*, *906*, 167705. Available from https://doi.org/10.1016/j.scitotenv.2023.167705.

Kosmala, M., Wiggins, A., Swanson, A., & Simmons, B. (2016). Assessing data quality in citizen science. *Frontiers in Ecology and the Environment*, *14*(10), 551–560. Available from https://doi.org/10.1002/fee.1436, http://onlinelibrary.wiley.com/journal/10.1002/(ISSN)1540-9309.

Koundouri, P., Ker Rault, P., Pergamalis, V., Skianis, V., & Souliotis, I. (2016). Development of an integrated methodology for the sustainable environmental and socio-economic management of river ecosystems. *Science of the Total Environment*, *540*, 90–100. Available from https://doi.org/10.1016/j.scitotenv.2015.07.082.

Kumar, P., Debele, S. E., Sahani, J., Rawat, N., Marti-Cardona, B., Alfieri, S. M., Basu, B., Basu, A. S., Bowyer, P., Charizopoulos, N., Jaakko, J., Loupis, M., Menenti, M., Mickovski, S. B., Pfeiffer, J., Pilla, F., Pröll, J., Pulvirenti, B., Rutzinger, M., ... Zieher, T. (2021). An overview of monitoring methods for assessing the performance of nature-based solutions against natural hazards. *Earth-Science Reviews*, *217*. Available from https://doi.org/10.1016/j.earscirev.2021.103603, http://www.sciencedirect.com/science/journal/00128252.

Kumar, S., Gopi, T., Harikeerthana, N., Gupta, M. K., Gaur, V., Krolczyk, G. M., & Wu, C. S. (2023). Machine learning techniques in additive manufacturing: A state of the art review on design, processes and production control. *Journal of Intelligent Manufacturing*, *34*(1), 21–55. Available from https://doi.org/10.1007/s10845-022-02029-5, https://www.springer.com/journal/10845.

Lahoz-Monfort, J. J., & Magrath, M. J. L. (2021). A comprehensive overview of technologies for species and habitat monitoring and conservation. *BioScience*, *71*(10), 1038–1062. Available from https://doi.org/10.1093/biosci/biab073, http://bioscience.oxfordjournals.org/.

Lan, T., Hu, H., Jiang, C., Yang, G., & Zhao, Z. (2020). A comparative study of decision tree, random forest, and convolutional neural network for spread-F identification. *Advances in Space Research*, *65*(8), 2052–2061. Available from https://doi.org/10.1016/j.asr.2020.01.036, http://www.journals.elsevier.com/advances-in-space-research/.

Lapeyrolerie, M., Chapman, M. S., Norman, K. E. A., & Boettiger, C. (2022). Deep reinforcement learning for conservation decisions. *Methods in Ecology and Evolution*, *13*(11), 2649–2662. Available from https://doi.org/10.1111/2041-210X.13954, http://onlinelibrary.wiley.com/journal/10.1111/(ISSN)2041-210X.

Laurans, Y., Rankovic, A., Billé, R., Pirard, R., & Mermet, L. (2013). Use of ecosystem services economic valuation for decision making: Questioning a literature blindspot. *Journal of Environmental Management*, *119*, 208–219. Available from https://doi.org/10.1016/j.jenvman.2013.01.008, https://www.sciencedirect.com/journal/journal-of-environmental-management.

Leenhardt, P., Teneva, L., Kininmonth, S., Darling, E., Cooley, S., & Claudet, J. (2015). Challenges, insights and perspectives associated with using social-ecological science for marine conservation. *Ocean and Coastal Management*, *115*, 49–60. Available from https://doi.org/10.1016/j.ocecoaman.2015.04.018, http://www.elsevier.com/inca/publications/store/4/0/5/8/8/9.

Leigh, D. M., van Rees, C. B., Millette, K. L., Breed, M. F., Schmidt, C., Bertola, L. D., Hand, B. K., Hunter, M. E., Jensen, E. L., Kershaw, F., Liggins, L., Luikart, G., Manel, S., Mergeay, J., Miller, J. M., Segelbacher, G., Hoban, S., & Paz-Vinas, I. (2021). Opportunities and challenges of macrogenetic studies. *Nature Reviews Genetics*, *22*(12), 791–807. Available from https://doi.org/10.1038/s41576-021-00394-0, http://www.nature.com/reviews/genetics.

Lentini, P. E., & Wintle, B. A. (2015). Spatial conservation priorities are highly sensitive to choice of biodiversity surrogates and species distribution model type. *Ecography*, *38*(11), 1101–1111. Available from https://doi.org/10.1111/ecog.01252Australia, http://onlinelibrary.wiley.com/journal/10.1111/(ISSN)1600-0587.

Li, Z., Chen, B., Wu, S., Su, M., Chen, J. M., & Xu, B. (2024). Deep learning for urban land use category classification: A review and experimental assessment. *Remote Sensing of Environment*, *311*, 114290. Available from https://doi.org/10.1016/j.rse.2024.114290.

Li, J., Hong, D., Gao, L., Yao, J., Zheng, K., Zhang, B., & Chanussot, J. (2022). Deep learning in multimodal remote sensing data fusion: A comprehensive review. *International Journal of Applied Earth Observation and Geoinformation*, *112*, 102926. Available from https://doi.org/10.1016/j.jag.2022.102926.

Li, F., Yigitcanlar, T., Nepal, M., Nguyen, K., & Dur, F. (2023). Machine learning and remote sensing integration for leveraging urban sustainability: A review and framework. *Sustainable Cities and Society*, *96*, 104653. Available from https://doi.org/10.1016/j.scs.2023.104653.

Liao, C., Qiu, J., Chen, B., Chen, D., Fu, B., Georgescu, M., He, C., Jenerette, G. D., Li, X., Li, X., Li, X., Qiuying, B., Shi, P., & Wu, J. (2020). Advancing landscape sustainability science: Theoretical foundation and synergies with innovations in methodology, design, and application. *Landscape Ecology*, *35*(1). Available from https://doi.org/10.1007/s10980-020-00967-0, http://www.springerlink.com/content/103025/.

Liu, Y., Huang, X., & Liu, Y. (2024). Detection of long-term land use and ecosystem services dynamics in the Loess Hilly-Gully region based on artificial intelligence and multiple models. *Journal of Cleaner Production*, *447*, 141560. Available from https://doi.org/10.1016/j.jclepro.2024.141560.

Liu, X., Lu, D., Zhang, A., Liu, Q., & Jiang, G. (2022). Data-driven machine learning in environmental pollution: Gains and problems. *Environmental Science and Technology*, *56*(4), 2124–2133. Available from https://doi.org/10.1021/acs.est.1c06157, http://pubs.acs.org/journal/esthag.

Long, R. D., Charles, A., & Stephenson, R. L. (2015). Key principles of marine ecosystem-based management. *Marine Policy*, *57*, 53–60. Available from https://doi.org/10.1016/j.marpol.2015.01.013.

Lotfian, M., Ingensand, J., & Brovelli, M. A. (2021). The Partnership of Citizen Science and Machine Learning: Benefits, Risks, and Future Challenges for Engagement, Data Collection, and Data Quality. *Sustainability*, *13*(14), 8087. Available from https://doi.org/10.3390/su13148087.

Maestas, J. D., Smith, J. T., Allred, B. W., Naugle, D. E., Jones, M. O., O'Connor, C., Boyd, C. S., Davies, K. W., Crist, M. R., & Olsen, A. C. (2023). Using dynamic, fuels-based fire probability maps to reduce large wildfires in the great basin. *Rangeland Ecology and Management*, *89*, 33–41. Available from https://doi.org/10.1016/j.rama.2022.08.002, http://www.bioone.org/loi/rama.

Mahesh, T. R., Dhilip Kumar, V., Kumar, V. V., Asghar, J., Geman, O., Arulkumaran, G., Arun, N., & Ahmad, M. (2022). AdaBoost ensemble methods using K-fold cross validation for survivability with the early detection of heart disease. *Computational Intelligence and Neuroscience*, *2022*, 1–11. Available from https://doi.org/10.1155/2022/9005278.

Manley, K., Nyelele, C., & Egoh, B. N. (2022). A review of machine learning and big data applications in addressing ecosystem service research gaps. *Ecosystem Services*, *57*, 101478. Available from https://doi.org/10.1016/j.ecoser.2022.101478.

Martínez-Minaya, J., Cameletti, M., Conesa, D., & Pennino, M. G. (2018). Species distribution modeling: A statistical review with focus in spatio-temporal issues. *Stochastic Environmental Research and Risk Assessment*, *32*(11), 3227–3244. Available from https://doi.org/10.1007/s00477-018-1548-7, http://link.springer-ny.com/link/service/journals/00477/index.htm.

Matthews, S. N., Iverson, L. R., Prasad, A. M., & Peters, M. P. (2011). Changes in potential habitat of 147 North American breeding bird species in response to redistribution of trees and climate following predicted climate change. *Ecography*, *34*(6), 933–945. Available from https://doi.org/10.1111/j.1600-0587.2011.06803.x.

Maxwell, S. L., Cazalis, V., Dudley, N., Hoffmann, M., Rodrigues, A. S. L., Stolton, S., Visconti, P., Woodley, S., Kingston, N., Lewis, E., Maron, M., Strassburg, B. B. N., Wenger, A., Jonas, H. D., Venter, O., & Watson, J. E. M. (2020). Area-based conservation in the twenty-first century. *Nature Research, Australia Nature*, *586*(7828), 217–227. Available from https://doi.org/10.1038/s41586-020-2773-z, http://www.nature.com/nature/index.html.

Mechenich, M. F., & Žliobaitė, I. (2023). Eco-ISEA3H, a machine learning ready spatial database for ecometric and species distribution modeling. *Scientific Data*, *10*(1). Available from https://doi.org/10.1038/s41597-023-01966-x, https://www.nature.com/sdata/.

Meraj, G., Kanga, S., Kranjčić, N., Đurin, B., & Singh, S. K. (2021). Role of natural capital economics for sustainable management of earth resources. *Earth*, *2*(3), 622–634. Available from https://doi.org/10.3390/earth2030036.

Meraj, G., Singh, S. K., Kanga, S., & Islam, M. N. (2022). Modeling on comparison of ecosystem services concepts, tools, methods and their ecological-economic implications: A review. *Modeling Earth Systems and Environment*, *8*(1), 15–34. Available from https://doi.org/10.1007/s40808-021-01131-6, springer.com/journal/40808.

Mishra, M., Guria, R., Baraj, B., Nanda, A. P., Santos, C. A. G., Silva, R. Md, & Laksono, F. A. T. (2024). Spatial analysis and machine learning prediction of forest fire susceptibility: A comprehensive approach for effective management and mitigation. *Science of the Total Environment*, *926*. Available from https://doi.org/10.1016/j.scitotenv.2024.171713, https://www.sciencedirect.com/science/journal/00489697.

MohanRajan, S. N., Loganathan, A., & Manoharan, P. (2020). Survey on Land Use/Land Cover (LU/LC) change analysis in remote sensing and GIS environment: Techniques and Challenges. *Environmental Science and Pollution Research*, *27*(24), 29900–29926. Available from https://doi.org/10.1007/s11356-020-09091-7.

Morán-Ordóñez, A., Briscoe, N. J., & Wintle, B. A. (2018). Modelling species responses to extreme weather provides new insights into constraints on range and likely climate change impacts for Australian mammals. *Ecography*, *41*(2), 308–320. Available from https://doi.org/10.1111/ecog.02850.

Morera, A., Martínez de Aragón, J., Bonet, J. A., Liang, J., & de-Miguel, S. (2021). Performance of statistical and machine learning-based methods for predicting biogeographical patterns of fungal productivity in forest ecosystems. *Forest Ecosystems*, *8*(1). Available from https://doi.org/10.1186/s40663-021-00297-w, https://forestecosyst.springeropen.com/.

Mukherjee, A. N. (2022). Application of artificial intelligence: Benefits and limitations for human potential and labor-intensive economy – An empirical investigation into pandemic ridden Indian industry. *Management Matters*, *19*(2), 149–166. Available from https://doi.org/10.1108/manm-02-2022-0034.

Muneer, S., Farooq, U., Athar, A., Raza, M. A., Ghazal, T. M., Sakib, S., & Madhukumar, A. S. (2024). A critical review of artificial intelligence based approaches in intrusion detection: A comprehensive analysis. *Journal of Engineering*, *2024*, 1–16. Available from https://doi.org/10.1155/2024/3909173.

Murphy, K., Di Ruggiero, E., Upshur, R., Willison, D. J., Malhotra, N., Cai, J. C., Malhotra, N., Lui, V., & Gibson, J. (2021). Artificial intelligence for good health: A scoping review of the ethics literature. *BMC Medical Ethics*, *22*(1). Available from https://doi.org/10.1186/s12910-021-00577-8, http://www.biomedcentral.com/bmcmedethics/.

Musuamba, F. T., Skottheim Rusten, I., Lesage, R., Russo, G., Bursi, R., Emili, L., Wangorsch, G., Manolis, E., Karlsson, K. E., Kulesza, A., Courcelles, E., Boissel, J. P., Rousseau, C. F., Voisin, E. M., Alessandrello, R., Curado, N., Dall'ara, E., Rodriguez, B., Pappalardo, F., & Geris, L. (2021). Scientific and regulatory evaluation of mechanistic in silico drug and disease models in drug development: Building model credibility. *Netherlands CPT: Pharmacometrics and Systems Pharmacology*, *10*(8), 804–825. Available from https://doi.org/10.1002/psp4.12669, http://onlinelibrary.wiley.com/journal/10.1002/(ISSN)2163-8306.

Nagendra, H., Lucas, R., Honrado, J. P., Jongman, R. H. G., Tarantino, C., Adamo, M., & Mairota, P. (2013). Remote sensing for conservation monitoring: Assessing protected areas, habitat extent, habitat condition, species diversity, and threats. *Ecological Indicators*, *33*, 45–59. Available from https://doi.org/10.1016/j.ecolind.2012.09.014, http://www.elsevier.com/locate/ecolind.

Nandutu, I., Atemkeng, M., & Okouma, P. (2022). Intelligent Systems Using Sensors and/or Machine Learning to Mitigate Wildlife–Vehicle Collisions: A Review, Challenges, and New Perspectives. *Sensors*, *22*(7), 2478. Available from https://doi.org/10.3390/s22072478.

Naseer, H., Desouza, K., Maynard, S. B., & Ahmad, A. (2024). Enabling cybersecurity incident response agility through dynamic capabilities: The role of real-time analytics. *European Journal of Information Systems*, *33*(2), 200–220. Available from https://doi.org/10.1080/0960085X.2023.2257168, https://www.tandfonline.com/loi/tjis20.

Neaves, L. E., Brockett, B. M., Evans, M. J., Pierson, J. C., & Sarre, S. D. (2024). Technological advances have enhanced and expanded conservation genomics research but are yet to be integrated fully into biodiversity management. *Biological Conservation*, *299*, 110788. Available from https://doi.org/10.1016/j.biocon.2024.110788.

Neethirajan, S. (2024). Artificial intelligence and sensor innovations: Enhancing livestock welfare with a human-centric approach. *Human-Centric Intelligent Systems*, *4*(1), 77–92. Available from https://doi.org/10.1007/s44230-023-00050-2.

Nieto-Mora, D. A., Rodríguez-Buritica, S., Rodríguez-Marín, P., Martínez-Vargaz, J. D., & Isaza-Narváez, C. (2023). Systematic review of machine learning methods applied to ecoacoustics and soundscape monitoring. *Heliyon*, *9*(10), e20275. Available from https://doi.org/10.1016/j.heliyon.2023.e20275.

Nishant, R., Kennedy, M., & Corbett, J. (2020). Artificial intelligence for sustainability: Challenges, opportunities, and a research agenda. *International Journal of Information Management*, *53*, 102104. Available from https://doi.org/10.1016/j.ijinfomgt.2020.102104.

Nuttall, M. N., Griffin, O., Fewster, R. M., McGowan, P. J. K., Abernethy, K., O'Kelly, H., Nut, M., Sot, V., & Bunnefeld, N. (2022). Long-term monitoring of wildlife populations for protected area management in Southeast Asia. *Conservation Science and Practice*, *4*(2). Available from https://doi.org/10.1111/csp2.614, conbio.onlinelibrary.wiley.com/journal/25784854.

Obaideen, K., Albasha, L., Iqbal, U., & Mir, H. (2024). Wireless power transfer: Applications, challenges, barriers, and the role of AI in achieving sustainable development goals - a bibliometric analysis. *Energy Strategy Reviews*, *53*, 101376. Available from https://doi.org/10.1016/j.esr.2024.101376.

Oliver, T. H., & Morecroft, M. D. (2014). Interactions between climate change and land use change on biodiversity: Attribution problems, risks, and opportunities. *WIREs Climate Change*, *5*(3), 317–335. Available from https://doi.org/10.1002/wcc.271.

Pan, S. L., Carter, L., Tim, Y., & Sandeep, M. S. (2022). Digital sustainability, climate change, and information systems solutions: Opportunities for future research. *International Journal of Information Management*, *63*. Available from https://doi.org/10.1016/j.ijinfomgt.2021.102444, https://www.journals.elsevier.com/international-journal-of-information-management.

Pan, S. L., Li, M., Pee, L. G., & Sandeep, M. S. (2021). Sustainability design principles for a wildlife management analytics system: An action design research. *European Journal of Information Systems*, *30*(4), 452–473. Available from https://doi.org/10.1080/0960085X.2020.1811786, https://www.tandfonline.com/loi/tjis20.

Pandey, D. K., Hunjra, A. I., Bhaskar, R., & Al-Faryan, M. A. S. (2023). Artificial intelligence, machine learning and big data in natural resources management: A comprehensive bibliometric review of literature spanning 1975–2022. *Resources Policy*, *86*. Available from https://doi.org/10.1016/j.resourpol.2023.104250, http://www.elsevier.com/inca/publications/store/3/0/4/6/7/.

Panduman, Y. Y. F., Funabiki, N., Fajrianti, E. D., Fang, S., & Sukaridhoto, S. (2024). A survey of AI techniques in IoT applications with use case investigations in the smart environmental monitoring and analytics in real-time IoT platform. *Information*, *15*(3), 153. Available from https://doi.org/10.3390/info15030153.

Papyan, N., Kulhandjian, M., Kulhandjian, H., & Aslanyan, L. (2024). AI-based drone assisted human rescue in disaster environments: Challenges and opportunities. *Pattern Recognition and Image Analysis*, *34*(1), 169–186. Available from https://doi.org/10.1134/S1054661824010152, https://www.springer.com/journal/11493.

Parisi, G. I., Kemker, R., Part, J. L., Kanan, C., & Wermter, S. (2019). Continual lifelong learning with neural networks: A review. *Neural Networks*, *113*, 54–71. Available from https://doi.org/10.1016/j.neunet.2019.01.012, http://www.elsevier.com/locate/neunet.

Pather, N., Blyth, P., Chapman, J. A., Dayal, M. R., Flack, N. A. M. S., Fogg, Q. A., Green, R. A., Hulme, A. K., Johnson, I. P., Meyer, A. J., Morley, J. W., Shortland, P. J., Štrkalj, G., Štrkalj, M., Valter, K., Webb, A. L., Woodley, S. J., & Lazarus, M. D. (2020). Forced disruption of anatomy education in Australia and New Zealand: An acute response to the Covid-19 pandemic. *Anatomical Sciences Education*, *13*(3), 284–300. Available from https://doi.org/10.1002/ase.1968, http://onlinelibrary.wiley.com/journal/10.1002/(ISSN)1935-9780.

Patterson, T. M., & Coelho, D. L. (2009). Ecosystem services: Foundations, opportunities, and challenges for the forest products sector. *Forest Ecology and Management*, *257*(8), 1637–1646. Available from https://doi.org/10.1016/j.foreco.2008.11.010.

Pecchi, M., Marchi, M., Burton, V., Giannetti, F., Moriondo, M., Bernetti, I., Bindi, M., & Chirici, G. (2019). Species distribution modelling to support forest management. A literature review. *Ecological Modelling*, *411*, 108817. Available from https://doi.org/10.1016/j.ecolmodel.2019.108817.

Peng, M., Liu, Y., Khan, A., Ahmed, B., Sarker, S. K., Ghadi, Y. Y., Bhatti, U. A., Al-Razgan, M., & Ali, Y. A. (2024). Crop monitoring using remote sensing land use and land change data: Comparative analysis of deep learning methods using pre-trained CNN models. *Big Data Research*, *36*. Available from https://doi.org/10.1016/j.bdr.2024.100448, https://www.sciencedirect.com/science/journal/22145796.

Pereira, V., Hadjielias, E., Christofi, M., & Vrontis, D. (2023). A systematic literature review on the impact of artificial intelligence on workplace outcomes: A multi-process perspective. *Human Resource Management Review*, *33*(1), 100857. Available from https://doi.org/10.1016/j.hrmr.2021.100857.

Peters, M. D. J., Marnie, C., Tricco, A. C., Pollock, D., Munn, Z., Alexander, L., McInerney, P., Godfrey, C. M., & Khalil, H. (2020). Updated methodological guidance for the conduct of scoping reviews. *JBI Evidence Synthesis*, *18*(10), 2119–2126. Available from https://doi.org/10.11124/JBIES-20-00167, https://journals.lww.com/jbisrir/Pages/issuelist.aspx.

Pichler, M., & Hartig, F. (2023). Machine learning and deep learning—A review for ecologists. *Methods in Ecology and Evolution*, *14*(4), 994–1016. Available from https://doi.org/10.1111/2041-210X.14061, http://onlinelibrary.wiley.com/journal/10.1111/(ISSN)2041-210X.

Pickett, S. T. A., Cadenasso, M. L., Rosi-Marshall, E. J., Belt, K. T., Groffman, P. M., Grove, J. M., Irwin, E. G., Kaushal, S. S., LaDeau, S. L., Nilon, C. H., Swan, C. M., & Warren, P. S. (2017). Dynamic heterogeneity: A framework to promote ecological integration and hypothesis generation in urban systems. *Urban Ecosystems*, *20*(1), 1–14. Available from https://doi.org/10.1007/s11252-016-0574-9.

Ploton, P., Mortier, F., Réjou-Méchain, M., Barbier, N., Picard, N., Rossi, V., Dormann, C., Cornu, G., Viennois, G., Bayol, N., Lyapustin, A., Gourlet-Fleury, S., & Pélissier, R. (2020). Spatial validation reveals poor predictive performance of large-scale ecological mapping models. *Nature Communications*, *11*(1). Available from https://doi.org/10.1038/s41467-020-18321-y, http://www.nature.com/ncomms/index.html.

Praveen, B., Parveen, S., & Akram, V. (2023). Urban Change Detection Analysis Using Big Data and Machine Learning: A Review. *GIScience and Geo-Environmental Modelling*, 125–133. Available from https://doi.org/10.1007/978-3-031-21587-2_9.

Prober, S. M., Doerr, V. A. J., Broadhurst, L. M., Williams, K. J., & Dickson, F. (2019). Shifting the conservation paradigm: A synthesis of options for renovating nature under climate change. *Ecological Monographs*, *89*(1). Available from https://doi.org/10.1002/ecm.1333, http://onlinelibrary.wiley.com/journal/10.1002/(ISSN)1557-7015.

Ramyar, R., Ackerman, A., & Johnston, D. M. (2021). Adapting cities for climate change through urban green infrastructure planning. *Cities*, *117*, 103316. Available from https://doi.org/10.1016/j.cities.2021.103316.

Randin, C. F., Ashcroft, M. B., Bolliger, J., Cavender-Bares, J., Coops, N. C., Dullinger, S., Dirnböck, T., Eckert, S., Ellis, E., Fernández, N., Giuliani, G., Guisan, A., Jetz, W., Joost, S., Karger, D., Lembrechts, J., Lenoir, J., Luoto, M., Morin, X., ... Payne, D. (2020). Monitoring biodiversity in the Anthropocene using remote sensing in species distribution models. *Remote Sensing of Environment*, *239*. Available from https://doi.org/10.1016/j.rse.2019.111626, http://www.elsevier.com/inca/publications/store/5/0/5/7/3/3.

Rangwala, I., Moss, W., Wolken, J., Rondeau, R., Newlon, K., Guinotte, J., & Travis, W. R. (2021). Uncertainty, complexity and constraints: How do we robustly assess biological responses under a rapidly changing climate. *Climate*, *9*(12). Available from https://doi.org/10.3390/cli9120177, https://www.mdpi.com/2225-1154/9/12/177/pdf.

Rao, A., Talan, A., Abbas, S., Dev, D., & Taghizadeh-Hesary, F. (2023). The role of natural resources in the management of environmental sustainability: Machine learning approach. *Resources Policy*, *82*, 103548. Available from https://doi.org/10.1016/j.resourpol.2023.103548.

Reed, J., Van Vianen, J., Deakin, E. L., Barlow, J., & Sunderland, T. (2016). Integrated landscape approaches to managing social and environmental issues in the tropics: Learning from the past to guide the future. *Global Change Biology*, *22*(7), 2540–2554. Available from https://doi.org/10.1111/gcb.13284, http://onlinelibrary.wiley.com/journal/10.1111/(ISSN)1365-2486.

Reed, M. S., Vella, S., Challies, E., de Vente, J., Frewer, L., Hohenwallner-Ries, D., Huber, T., Neumann, R. K., Oughton, E. A., Sidoli del Ceno, J., & van Delden, H. (2018). A theory of participation: What makes stakeholder and public engagement in environmental management work? *Restoration Ecology*, *26*, S7. Available from https://doi.org/10.1111/rec.12541, http://www.blacksci.co.uk/~cgilib/bsinc.bin?Journal = ecology.

Rees, S. E., Foster, N. L., Langmead, O., Pittman, S., & Johnson, D. E. (2018). Defining the qualitative elements of Aichi Biodiversity Target 11 with regard to the marine and coastal environment in order to strengthen global efforts for marine biodiversity conservation outlined in the United Nations Sustainable Development Goal 14. *Marine Policy*, *93*, 241–250. Available from https://doi.org/10.1016/j.marpol.2017.05.016, http://www.elsevier.com/inca/publications/store/3/0/4/5/3/.

Rilov, G., Mazaris, A. D., Stelzenmüller, V., Helmuth, B., Wahl, M., Guy-Haim, T., Mieszkowska, N., Ledoux, J. B., & Katsanevakis, S. (2019). Adaptive marine conservation planning in the face of climate change: What can we learn from physiological, ecological and genetic studies? *Global Ecology and Conservation*, *17*. Available from https://doi.org/10.1016/j.gecco.2019.e00566.

Roy, P. P., Abdullah, M. S., & Siddique, I. M. (2024). Machine learning empowered geographic information systems: Advancing Spatial analysis and decision making. *World Journal of Advanced Research and Reviews*, *22*(1), 1387–1397. Available from https://doi.org/10.30574/wjarr.2024.22.1.1200.

Rudd, M. A., Moore, A. F. P., Rochberg, D., Bianchi-Fossati, L., Brown, M. A., D'Onofrio, D., Furman, C. A., Garcia, J., Jordan, B., Kline, J., Risse, L. M., Yager, P. L., Abbinett, J., Alber, M., Bell, J. E., Bhedwar, C., Cobb, K. M., Cohen, J., Cox, M., . . . Worley, A. N. (2018). Climate research priorities for policy-makers, practitioners, and scientists in Georgia, USA. *Environmental Management*, *62*(2), 190–209. Available from https://doi.org/10.1007/s00267-018-1051-4, https://www.springer.com/journal/267.

Salcedo-Sanz, S., Ghamisi, P., Piles, M., Werner, M., Cuadra, L., Moreno-Martínez, A., Izquierdo-Verdiguier, E., Muñoz-Marí, J., Mosavi, A., & Camps-Valls, G. (2020). Machine learning information fusion in Earth observation: A comprehensive review of methods, applications and data sources. *Information Fusion*, *63*, 256–272. Available from https://doi.org/10.1016/j.inffus.2020.07.004.

Salim, M. Z., Choudhari, N., Kafy, A. A., Nath, H., Alsulamy, S., Rahaman, Z. A., Aldosary, A. S., Rahmand, M. T., & Al-Ramadan, B. (2024). A comprehensive review of navigating urbanization induced climate change complexities for sustainable groundwater resources management in the Indian subcontinent. *Groundwater for Sustainable Development*, *25*. Available from https://doi.org/10.1016/j.gsd.2024.101115, https://www.sciencedirect.com/science/journal/2352801X.

Sandifer, P. A., Sutton-Grier, A. E., & Ward, B. P. (2015). Exploring connections among nature, biodiversity, ecosystem services, and human health and well-being: Opportunities to enhance health and biodiversity conservation. *Ecosystem Services*, *12*, 1–15. Available from https://doi.org/10.1016/j.ecoser.2014.12.007.

Schuwirth, N., Borgwardt, F., Domisch, S., Friedrichs, M., Kattwinkel, M., Kneis, D., Kuemmerlen, M., Langhans, S. D., Martínez-López, J., & Vermeiren, P. (2019). How to make ecological models useful for environmental management. *Ecological Modelling*, *411*. Available from https://doi.org/10.1016/j.ecolmodel.2019.108784, http://www.elsevier.com/inca/publications/store/5/0/3/3/0/6.

Schwalbert, R. A., Amado, T., Corassa, G., Pott, L. P., Prasad, P. V. V., & Ciampitti, I. A. (2020). Satellite-based soybean yield forecast: Integrating machine learning and weather data for improving crop yield prediction in southern Brazil. *Agricultural and Forest Meteorology*, *284*. Available from https://doi.org/10.1016/j.agrformet.2019.107886, http://www.elsevier.com/inca/publications/store/5/0/3/2/9/5.

Seidl, R., Spies, T. A., Peterson, D. L., Stephens, S. L., & Hicke, J. A. (2016). Searching for resilience: Addressing the impacts of changing disturbance regimes on forest ecosystem services. *Journal of Applied Ecology*, *53*(1), 120–129. Available from https://doi.org/10.1111/1365-2664.12511, http://www.blackwellpublishing.com/journals/JPE.

Sekaran, U., Lai, L., Ussiri, D. A. N., Kumar, S., & Clay, S. (2021). Role of integrated crop-livestock systems in improving agriculture production and addressing food security – A review. *Journal of Agriculture and Food Research*, *5*. Available from http://www.journals.elsevier.com/journal-of-agriculture-and-food-research, https://doi.org/10.1016/j.jafr.2021.100190.

Seyfollahi, A., Taami, T., & Ghaffari, A. (2023). Towards developing a machine learning-metaheuristic-enhanced energy-sensitive routing framework for the internet of things. *Microprocessors and Microsystems*, *96*, 104747. Available from https://doi.org/10.1016/j.micpro.2022.104747.

Shaikh, M., & Birajdar, F. (2024). Artificial intelligence in groundwater management: Innovations, challenges, and future prospects. *International Journal of Science and Research Archive*, *11*(1), 502–512. Available from https://doi.org/10.30574/ijsra.2024.11.1.0105.

Shameer, T. T., & Sanil, R. (2023). *Machine learning-based predictive modelling approaches for effective understanding of evolutionary history, distribution, and niche occupancy: Western Ghats as a model. Ecosystem and species habitat modeling for conservation and restoration* (pp. 41–57). India: Springer Nature. Available from https://link.springer.com/book/10.1007/978-981-99-0131-9, https://doi.org/10.1007/978-981-99-0131-9_3.

Shankar, R., & Gupta, L. (2024). An integrated AI framework for managing organizational risk and climate change concerns in B2B market. *Industrial Marketing Management*, *117*, 173–187. Available from https://doi.org/10.1016/j.indmarman.2023.12.019, https://www.sciencedirect.com/science/journal/00198501.

Sharma, V., Tripathi, A. K., & Mittal, H. (2022). Technological revolutions in smart farming: Current trends, challenges & future directions. *Computers and Electronics in Agriculture*, *201*, 107217. Available from https://doi.org/10.1016/j.compag.2022.107217.

Shaver, E. C., McLeod, E., Hein, M. Y., Palumbi, S. R., Quigley, K., Vardi, T., Mumby, P. J., Smith, D., Montoya-Maya, P., Muller, E. M., Banaszak, A. T., McLeod, I. M., & Wachenfeld, D. (2022). A roadmap to integrating resilience into the practice of coral reef restoration. *Global Change Biology*, *28*(16), 4751–4764. Available from https://doi.org/10.1111/gcb.16212, http://onlinelibrary.wiley.com/journal/10.1111/(ISSN)1365-2486.

Shinde, R., Patil, S., Kotecha, K., Potdar, V., Selvachandran, G., & Abraham, A. (2024). Securing AI-based healthcare systems using blockchain technology: A state-of-the-art systematic literature review and future research directions. *Transactions on Emerging Telecommunications Technologies*, *35*(1). Available from https://doi.org/10.1002/ett.4884, http://onlinelibrary.wiley.com/journal/10.1002/(ISSN)2161-3915.

Shiva Prakash, K. N., Swami, N., Mysorekar, S., Arora, R., Gangadharan, A., Vohra, K., Jadeyegowda, M., & Kiesecker, J. M. (2022). Potential for Artificial Intelligence (AI) and Machine Learning (ML) Applications in Biodiversity Conservation, Managing Forests, and Related Services in India. *Sustainability*, *14*(12), 7154. Available from https://doi.org/10.3390/su14127154.

Simon, S. M., Glaum, P., & Valdovinos, F. S. (2023). Interpreting random forest analysis of ecological models to move from prediction to explanation. *Scientific Reports*, *13*(1). Available from https://doi.org/10.1038/s41598-023-30313-8.

Singh, R., Gehlot, A., Vaseem Akram, S., Kumar Thakur, A., Buddhi, D., & Kumar Das, P. (2022). Forest 4.0: Digitalization of forest using the Internet of Things (IoT). *Journal of King Saud University - Computer and Information Sciences*, *34*(8), 5587–5601. Available from https://doi.org/10.1016/j.jksuci.2021.02.009, http://www.journals.elsevier.com/journal-of-king-saud-university-computer-and-information-sciences/.

Smith, A. C., Harrison, P. A., Pérez Soba, M., Archaux, F., Blicharska, M., Egoh, B. N., Erős, T., Fabrega Domenech, N., György, Á. I., Haines-Young, R., Li, S., Lommelen, E., Meiresonne, L., Miguel Ayala, L., Mononen, L., Simpson, G., Stange, E., Turkelboom, F., Uiterwijk, M., ... Wyllie de Echeverria, V. (2017). How natural capital delivers ecosystem services: A typology derived from a systematic review. *Ecosystem Services*, *26*, 111–126. Available from https://doi.org/10.1016/j.ecoser.2017.06.006.

Srivastava, A., & Maity, R. (2023). Assessing the potential of AI–ML in urban climate change adaptation and sustainable development. *Sustainability*, *15*(23), 16461. Available from https://doi.org/10.3390/su152316461.

Stupariu, M. S., Cushman, S. A., Pleşoianu, A. I., Pătru-Stupariu, I., & Fürst, C. (2022). Machine learning in landscape ecological analysis: A review of recent approaches. *Landscape Ecology*, *37*(5), 1227–1250. Available from https://doi.org/10.1007/s10980-021-01366-9, http://www.springerlink.com/content/103025/.

Su, J. H., Piao, Y. C., Luo, Z., & Yan, B. P. (2018). Modeling habitat suitability of migratory birds from remote sensing images using convolutional neural networks. *Animals*, *8*(5). Available from https://doi.org/10.3390/ani8050066, http://www.mdpi.com/2076-2615/8/5/66/pdf.

Summers, D. M., Bryan, B. A., Crossman, N. D., & Meyer, W. S. (2012). Species vulnerability to climate change: Impacts on spatial conservation priorities and species representation. *Global Change Biology*, *18*(7), 2335–2348. Available from https://doi.org/10.1111/j.1365-2486.2012.02700.x, http://onlinelibrary.wiley.com/journal/10.1111/(ISSN)1365-2486.

Sun, Z., Di, L., & Fang, H. (2019). Using long short-term memory recurrent neural network in land cover classification on Landsat and Cropland data layer time series. *International Journal of Remote Sensing*, *40*(2), 593–614. Available from https://doi.org/10.1080/01431161.2018.1516313.

Sun, A. Y., & Scanlon, B. R. (2019). How can Big Data and machine learning benefit environment and water management: A survey of methods, applications, and future directions. *Environmental Research Letters*, *14*(7). Available from https://doi.org/10.1088/1748-9326/ab1b7d, https://iopscience.iop.org/article/10.1088/1748-9326/ab1b7d/pdf.

Tang, K. H. D. (2019). Climate change in Malaysia: Trends, contributors, impacts, mitigation and adaptations. *Science of the Total Environment*, *650*, 1858–1871. Available from https://doi.org/10.1016/j.scitotenv.2018.09.316, http://www.elsevier.com/locate/scitotenv.

Tengö, M., Hill, R., Malmer, P., Raymond, C. M., Spierenburg, M., Danielsen, F., Elmqvist, T., & Folke, C. (2017). Weaving knowledge systems in IPBES, CBD and beyond—lessons learned for sustainability. *Current Opinion in Environmental Sustainability*, *26-27*, 17–25. Available from https://doi.org/10.1016/j.cosust.2016.12.005, http://www.elsevier.com/wps/find/journaldescription.cws_home/718675/description#description.

Timpane-Padgham, B. L., Beechie, T., Klinger, T., & (Gee) Geraldine Chapman, M. (2017). A systematic review of ecological attributes that confer resilience to climate change in environmental restoration. *PLoS ONE*, *12*(3), e0173812. Available from https://doi.org/10.1371/journal.pone.0173812.

Tinch, R., Beaumont, N., Sunderland, T., Ozdemiroglu, E., Barton, D., Bowe, C., Börger, T., Burgess, P., Cooper, C. N., Faccioli, M., Failler, P., Gkolemi, I., Kumar, R., Longo, A., McVittie, A., Morris, J., Park, J., Ravenscroft, N., Schaafsma, M., ... Ziv, G. (2019). Economic valuation of ecosystem goods and services: A review for decision makers. *Journal of Environmental Economics and Policy*, *8*(4), 359–378. Available from https://doi.org/10.1080/21606544.2019.1623083, https://www.tandfonline.com/action/journalInformation?journalCode = teep20.

Tomsett, C., & Leyland, J. (2019). Remote sensing of river corridors: A review of current trends and future directions. *River Research and Applications*, *35*(7), 779–803. Available from https://doi.org/10.1002/rra.3479, http://onlinelibrary.wiley.com/journal/10.1002/(ISSN)1535-1467.

Treweek, S., Oxman, A. D., Alderson, P., Bossuyt, P. M., Brandt, L., Brozek, J., Davoli, M., Flottorp, S., Harbour, R., Hill, S., Liberati, A., Liira, H., Schünemann, H. J., Rosenbaum, S., Thornton, J., Vandvik, P. O., & Alonso-Coello, P. (2013). Developing and evaluating communication strategies to support informed decisions and practice based on evidence (DECIDE): Protocol and preliminary results. *Implementation Science.*, *8*(1). Available from https://doi.org/10.1186/1748-5908-8-6United, http://www.implementationscience.com/content/8/1/6.

Tuia, D., Kellenberger, B., Beery, S., Costelloe, B. R., Zuffi, S., Risse, B., Mathis, A., Mathis, M. W., van Langevelde, F., Burghardt, T., Kays, R., Klinck, H., Wikelski, M., Couzin, I. D., van Horn, G., Crofoot, M. C., Stewart, C. V., & Berger-Wolf, T. (2022). Perspectives in machine learning for wildlife conservation. *Nature Communications*, *13*(1). Available from https://doi.org/10.1038/s41467-022-27980-y, http://www.nature.com/ncomms/index.html.

Tulloch, A. I. T., Hagger, V., & Greenville, A. C. (2020). Ecological forecasts to inform near-term management of threats to biodiversity. *Global Change Biology*, *26*(10), 5816–5828. Available from https://doi.org/10.1111/gcb.15272.

Turner, K. G., Anderson, S., Gonzales-Chang, M., Costanza, R., Courville, S., Dalgaard, T., Dominati, E., Kubiszewski, I., Ogilvy, S., Porfirio, L., Ratna, N., Sandhu, H., Sutton, P. C., Svenning, J.-C., Turner, G. M., Varennes, Y.-D., Voinov, A., & Wratten, S. (2016). A review of methods, data, and models to assess changes in the value of ecosystem services from land degradation and restoration. *Ecological Modelling*, *319*, 190–207. Available from https://doi.org/10.1016/j.ecolmodel.2015.07.017.

Udegbe, F. C., Ebulue, O. R., Ebulue, C. C., & Ekesiobi, C. S. (2024). Machine learning in drug discovery: A critical review of applications and challenges. *Computer Science and IT Research Journal*, *5*(4), 892–902. Available from https://doi.org/10.51594/csitrj.v5i4.1048.

Ullah, A., Anwar, S. M., Li, J., Nadeem, L., Mahmood, T., Rehman, A., & Saba, T. (2023). Smart cities: the role of Internet of Things and machine learning in realizing a data-centric smart environment. *Complex & Intelligent Systems*, *10*(1), 1607–1637. Available from https://doi.org/10.1007/s40747-023-01175-4.

Untaru, E. N., Ispas, A., Candrea, A. N., Luca, M., & Epuran, G. (2016). Predictors of individuals' intention to conserve water in a lodging context: The application of an extended Theory of Reasoned Action. *International Journal of Hospitality Management*, *59*, 50–59. Available from https://doi.org/10.1016/j.ijhm.2016.09.001, http://www.elsevier.com/locate/ijhosman.

van Kerkhoff, L., Munera, C., Dudley, N., Guevara, O., Wyborn, C., Figueroa, C., Dunlop, M., Hoyos, M. A., Castiblanco, J., & Becerra, L. (2019). Towards future-oriented conservation: Managing protected areas in an era of climate change. *Ambio*, *48*(7), 699–713. Available from https://doi.org/10.1007/s13280-018-1121-0, http://www.bioone.org/loi/ambi.

Villa-Henriksen, A., Edwards, G. T. C., Pesonen, L. A., Green, O., & Sørensen, C. A. G. (2020). Internet of Things in arable farming: Implementation, applications, challenges and potential. *Biosystems Engineering*, *191*, 60–84. Available from https://doi.org/10.1016/j.biosystemseng.2019.12.013, http://www.elsevier.com/inca/publications/store/6/2/2/7/9/5/index.htt.

Vishwakarma, D. K., Ali, R., Bhat, S. A., Elbeltagi, A., Kushwaha, N. L., Kumar, R., Rajput, J., Heddam, S., & Kuriqi, A. (2022). Pre- and post-dam river water temperature alteration prediction using advanced machine learning models. *Environmental Science and Pollution Research*, *29*(55), 83321–83346. Available from https://doi.org/10.1007/s11356-022-21596-x, https://www.springer.com/journal/11356.

Waleed, M., Sajjad, M., Shazil, M. S., Tariq, M., & Alam, M. T. (2023). Machine learning-based spatial-temporal assessment and change transition analysis of wetlands: An application of Google Earth Engine in Sylhet, Bangladesh (1985–2022. *Ecological Informatics*, *75*. Available from https://doi.org/10.1016/j.ecoinf.2023.102075, http://www.elsevier.com/wps/find/journaldescription.cws_home/705192/description#description.

Weaver, C. P., Lempert, R. J., Brown, C., Hall, J. A., Revell, D., & Sarewitz, D. (2013). Improving the contribution of climate model information to decision making: The value and demands of robust decision frameworks. *Wiley Interdisciplinary Reviews: Climate Change*, *4*(1), 39–60. Available from https://doi.org/10.1002/wcc.202, http://onlinelibrary.wiley.com/journal/10.1002/(ISSN)1757-7799.

Weiskopf, S. R., Rubenstein, M. A., Crozier, L. G., Gaichas, S., Griffis, R., Halofsky, J. E., Hyde, K. J. W., Morelli, T. L., Morisette, J. T., Muñoz, R. C., Pershing, A. J., Peterson, D. L., Poudel, R., Staudinger, M. D., Sutton-Grier, A. E., Thompson, L., Vose, J., Weltzin, J. F., & Whyte, K. P. (2020). Climate change effects on biodiversity, ecosystems, ecosystem services, and natural resource management in the United States. *Science of the Total Environment*, *733*. Available from https://doi.org/10.1016/j.scitotenv.2020.137782, http://www.elsevier.com/locate/scitotenv.

Wiebe, K., Lotze-Campen, H., Sands, R., Tabeau, A., van der Mensbrugghe, D., Biewald, A., Bodirsky, B., Islam, S., Kavallari, A., Mason-D'Croz, D., Müller, C., Popp, A., Robertson, R., Robinson, S., van Meijl, H., & Willenbockel, D. (2015). Climate change impacts on agriculture in 2050 under a range of plausible socioeconomic and emissions scenarios. *Environmental Research Letters*, *10*(8), 085010. Available from https://doi.org/10.1088/1748-9326/10/8/085010.

Williams, B. K., & Johnson, F. A. (2013). Confronting dynamics and uncertainty in optimal decision making for conservation. *Environmental Research Letters*, *8*(2). Available from https://doi.org/10.1088/1748-9326/8/2/025004, http://iopscience.iop.org/1748-9326/8/2/025004/pdf/1748-9326_8_2_025004.pdf.

Wilson, C., & van der Velden, M. (2022). Sustainable AI: An integrated model to guide public sector decision-making. *Technology in Society*, *68*, 101926. Available from https://doi.org/10.1016/j.techsoc.2022.101926.

Wood, D. A. (2021). Net ecosystem carbon exchange prediction and insightful data mining with an optimized data-matching algorithm. *Ecological Indicators*, *124*, 107426. Available from https://doi.org/10.1016/j.ecolind.2021.107426.

Wu, X., Zhou, Q., Mu, L., & Hu, X. (2022). Machine learning in the identification, prediction and exploration of environmental toxicology: Challenges and perspectives. *Journal of Hazardous Materials*, *438*, 129487. Available from https://doi.org/10.1016/j.jhazmat.2022.129487.

Yin, X., Liu, Q., Pan, Y., Huang, X., Wu, J., & Wang, X. (2021). Strength of stacking technique of ensemble learning in rockburst prediction with imbalanced data: Comparison of eight single and ensemble models. *Natural Resources Research*, *30*(2), 1795–1815. Available from https://doi.org/10.1007/s11053-020-09787-0, https://link.springer.com/journal/11053.

Yin, Y., Ahmadianfar, I., Karim, F. K., & Elmannai, H. (2024). Advanced forecasting of COVID-19 epidemic: Leveraging ensemble models, advanced optimization, and decomposition techniques. *Computers in Biology and Medicine*, *175*, 108442. Available from https://doi.org/10.1016/j.compbiomed.2024.108442.

Yousefzadeh, H., Walas, Ł., Amirchakhmaghi, N., Alipour, S., Pouramin, M., Song, Y.-G., & Kozlowski, G. (2024). Potential effects of climate change on future distribution of an endangered tree species, Acer mazandaranicum, in the Hyrcanian forest. *Forest Ecology and Management*, *555*, 121654. Available from https://doi.org/10.1016/j.foreco.2023.121654.

Yuan, C., Liu, Z., & Zhang, Y. (2017). Aerial Images-Based Forest Fire Detection for Firefighting Using Optical Remote Sensing Techniques and Unmanned Aerial Vehicles. *Journal of Intelligent and Robotic Systems*, *88*(2–4), 635–654. Available from https://doi.org/10.1007/s10846-016-0464-7.

Yuan, Q., Shen, H., Li, T., Li, Z., Li, S., Jiang, Y., Xu, H., Tan, W., Yang, Q., Wang, J., Gao, J., & Zhang, L. (2020). Deep learning in environmental remote sensing: Achievements and challenges. *Remote Sensing of Environment*, *241*, 111716. Available from https://doi.org/10.1016/j.rse.2020.111716.

Zhang, F., Zhou, L., Wang, Z., Lv, C., Zhang, Q., Wang, J., Zhang, J., & Zhang, Y. (2024). Empowering urban energy transition through data-driven decision-making: A statistical examination of technological innovations in transportation and mobility. *Sustainable Cities and Society*, *106*, 105374. Available from https://doi.org/10.1016/j.scs.2024.105374.

Zeller Zigaitis, W. L., & Robinson, A. C. (2023). Sensemaking in the Wild: A Review of Practitioner Collected Geospatial Data and its Synthesis within Protected Areas for Poaching Mitigation. *Annals of GIS*, *29*(3), 319–335. Available from https://doi.org/10.1080/19475683.2023.2192761.

Zhang, W., Gu, X., Tang, L., Yin, Y., Liu, D., & Zhang, Y. (2022). Application of machine learning, deep learning and optimization algorithms in geoengineering and geoscience: Comprehensive review and future challenge. *Gondwana Research*, *109*, 1–17. Available from https://doi.org/10.1016/j.gr.2022.03.015, http://www.sciencedirect.com/science/journal/1342937X.

Zhang, X., Han, L., Han, L., & Zhu, L. (2020). How well do deep learning-based methods for land cover classification and object detection perform on high resolution remote sensing imagery? *Remote Sensing*, *12*(3), 417. Available from https://doi.org/10.3390/rs12030417.

Zhao, K., Wulder, M. A., Hu, T., Bright, R., Wu, Q., Qin, H., Li, Y., Toman, E., Mallick, B., Zhang, X., & Brown, M. (2019). Detecting change-point, trend, and seasonality in satellite time series data to track abrupt changes and nonlinear dynamics: A Bayesian ensemble algorithm. *Remote Sensing of Environment*, *232*. Available from https://doi.org/10.1016/j.rse.2019.04.034, http://www.elsevier.com/inca/publications/store/5/0/5/7/3/3.

Zhao, X., Song, M., Liu, A., Wang, Y., Wang, T., & Cao, J. (2020). Data-driven temporal-spatial model for the prediction of AQI in Nanjing. *Journal of Artificial Intelligence and Soft Computing Research*, *10*(4), 255–270. Available from https://doi.org/10.2478/jaiscr-2020-0017, http://www.degruyter.com/view/j/jaiscr.

Zhong, S., Zhang, K., Bagheri, M., Burken, J. G., Gu, A., Li, B., Ma, X., Marrone, B. L., Ren, Z. J., Schrier, J., Shi, W., Tan, H., Wang, T., Wang, X., Wong, B. M., Xiao, X., Yu, X., Zhu, J. J., & Zhang, H. (2021). Machine learning: New ideas and tools in environmental science and engineering. *Environmental Science and Technology*, *55*(19), 12741–12754. Available from https://doi.org/10.1021/acs.est.1c01339, http://pubs.acs.org/journal/esthag.

Chapter 6

Deep learning for environmental insights in India's earth observation and climate monitoring

Vinit Kumar[1,2]

[1]*National Key Laboratory of Green Pesticide, Key Laboratory of Green Pesticide and Agricultural Bioengineering, Ministry of Education, Center for Research and Development of Fine Chemicals, Guizhou University, Guizhou, Guiyang, P.R. China,* [2]*Department of Medical Lab Technology, Markham College of Commerce, Vinoba Bhave University, Hazaribagh, Jharkhand, India*

6.1 Introduction

Climate change stands as one of the most pressing challenges of our time, threatening ecosystems, economies, and human well-being on a global scale. As the world grapples with rising temperatures, extreme weather events, and shifting ecological patterns, the need for accurate environmental monitoring and disaster management (DM) has become increasingly critical. In this context, land cover (LC) and land use (LU) mapping emerge as fundamental tools for understanding the earth's surface and the human activities that impact it. LC refers to the natural elements such as plants, water bodies, soil, snow, and ice that constitute the earth's surface, while LU pertains to human activities conducted on the land, including forestry, agriculture, urbanization, and industrialization. Mapping LC and LU provides essential information on the spatial distribution and temporal changes of natural and human systems, which is invaluable for various applications in earth observation (EO) and climate monitoring (CM). These applications are crucial for addressing climate change, as they inform urban planning, water resource management, biodiversity conservation, and climate change mitigation strategies.

Deep learning (DL) has become a potent technique in LC and LU mapping from satellite data due to the acceleration of climate change. By improving the precision and dependability of environmental monitoring, its capacity to handle high-resolution, multispectral, multitemporal, and heterogeneous data has completely changed the industry. The accuracy of LC and LU mapping may be greatly increased by using DL to include other data sources such digital elevation models, census data, or information from social media. This cutting-edge expertise is especially useful in areas like India where the effects of climate change are becoming more noticeable. The absence of reliable and current ground truth data, which is required for DL model training and testing, poses a major obstacle to LC and LU mapping in these areas. To address this issue in the context of climate change monitoring, researchers have proposed using synthetic or semisynthetic labels derived from existing datasets or expert knowledge. Additionally, the diversity and complexity of landscapes in climate-vulnerable areas necessitate the use of context-sensitive and fine-grained classification methods to accurately track environmental changes over time (Table 6.1).

Effective DM is more important than ever as a result of the increasing frequency and intensity of natural catastrophes brought on by climate change. DM describes the actions done in advance, during, and after a disaster to mitigate its consequences and enhance recovery. Climate change is causing natural catastrophes like heat waves, wildfires, droughts, and floods to occur more frequently and intensely, endangering the lives and livelihoods of millions of people. DM largely depends on EO and CM data to monitor and assess the risk, severity, and damage caused by these climate-related catastrophes, as well as to support decision-making and response operations. With its ability to perform a wide range of tasks, including as detection, segmentation, classification, creation, and prediction, DL has demonstrated great promise in this field. This makes it an invaluable tool for DM in the face of increasing climate-related challenges. To address the growing frequency and intensity of disasters, some researchers have proposed methods to

Deep Learning for Earth Observation and Climate Monitoring. DOI: https://doi.org/10.1016/B978-0-443-24712-5.00007-5

TABLE 6.1 Application of deep learning in disaster management.

Disaster management phase	Task	Deep learning application	References
Preparedness	Hazard risk assessment	Using convolutional neural networks (CNNs) to estimate the probability of landslide occurrence from satellite images and terrain data.	Sun et al. (2020)
Response	Disaster detection	Using deep convolutional encoder-decoder (DCED) networks to detect and segment flood areas from synthetic aperture radar (SAR) images.	Shahabi et al. (2020)
Recovery	Damage assessment	Using deep residual networks (ResNets) to classify building damage levels from postdisaster satellite images.	Gupta et al. (2021)
Mitigation	Vulnerability assessment	Using deep belief networks (DBNs) to identify the socio-economic factors that influence the vulnerability of urban households to floods.	Basu et al. (2023)

TABLE 6.2 Application of deep learning in air quality assessment.

Task	Objective	Deep learning application	References
Air quality index (AQI) prediction	To provide timely and accurate information on the current and future air quality levels in different locations and regions of India.	Using recurrent neural networks (RNNs) or long short-term memory (LSTM) networks to model the temporal and spatial patterns of air pollutant concentrations and meteorological factors, and to forecast the AQI values for short- and long-term horizons.	Zaini et al. (2022)
Air pollution source identification	To identify the major sources and contributors of air pollution in different areas and seasons of India, and to quantify their relative impacts on the air quality.	Using convolutional neural networks (CNNs) or deep autoencoders to extract features from satellite images, ground-based measurements, emission inventories, and other data sources, and to perform source apportionment and attribution analysis.	Liao et al. (2020)
Air quality management and policy evaluation	To evaluate the effectiveness and efficiency of various air quality management strategies and policies implemented in India, and to provide recommendations for improvement.	Using deep reinforcement learning (DRL) or deep Q-learning to simulate the dynamic interactions between the air quality system, the human activities, and the policy interventions, and to optimize the policy outcomes under different scenarios and constraints.	Yan et al. (2021)

employ DL for near-real-time or online DM applications, which can provide the public and authorities with quick and reliable information crucial for climate change adaptation and mitigation efforts (Table 6.2).

Another problem is the unpredictable and variable nature of climate-related natural disasters, which calls for flexible and broadly applicable DL models. This issue has been addressed by researchers through the use of transfer learning, domain adaptation, and uncertainty estimation for DM applications in the context of climate change. For instance, in developing nations like India, where climate change is making water-related problems worse, DL approaches have been used to forecast, track, and manage water-related crises. DL models can anticipate climate-related disasters more accurately and quickly by utilizing large volumes of data. This allows authorities to take preventive action and lessen the effect of the disasters. These developments in DL for DM are critical for recovery operations, resource allocation optimization, and evaluating the long-term effects of climate-related disasters on infrastructure and communities. They go beyond prediction and reaction. Building resilience and adjusting to our changing globe will depend heavily on the incorporation of DL in environmental monitoring and catastrophe management as climate change continues to impact our world (Table 6.3).

TABLE 6.3 Application of deep learning in crop yield estimation.

Task	Objective	Deep learning application	References
Rice yield estimation	To provide timely and accurate information on the current and future rice yield levels in different locations and regions of India.	Using convolutional neural networks (CNNs) to estimate rice yield from analyzing preharvest photographs of rice canopy.	Mishra et al. (2016)
Crop yield prediction	To speed up the breeding process and make it resource-efficient by predicting the phenotype of the crops based on their genotype and environment.	Using a genetic algorithm-assisted deep learning approach to model the complex biological and physiologic relationships between genotype, environment, and crop yield.	Bi & Hu (2021)
Crop yield forecasting	To provide reliable and robust forecasts of crop yield for various crops and regions in India using historical and current data.	Using recurrent neural networks (RNNs) or long short-term memory (LSTM) networks to model the temporal and spatial patterns of crop yield and related factors, such as weather, soil, irrigation, etc.	Prashant et al. (2021)
Wheat yield estimation	To monitor and assess the wheat crop condition and productivity in different agro-climatic zones of India.	Using a deep convolutional encoder-decoder (DCED) network to estimate wheat yield from multitemporal Sentinel-2 satellite images.	Zhao et al. (2020)
Maize yield estimation	To evaluate the impact of climate change and variability on maize production and food security in India.	Using a deep neural network (DNN) with attention mechanism to estimate maize yield from multisource data, such as meteorological data, soil data, crop management data, etc.	Muruganantham et al. (2022)
Cotton yield estimation	To improve the cotton quality and profitability by optimizing the crop management practices and inputs in India.	Using a deep belief network (DBN) to estimate cotton yield from hyperspectral images acquired by unmanned aerial vehicles (UAVs).	Rodriguez-Sanchez et al. (2022)
Sugarcane yield estimation	To support the sugarcane industry and farmers by providing timely and accurate information on sugarcane production and supply in India.	Using a deep residual network (ResNet) to estimate sugarcane yield from synthetic aperture radar (SAR) images.	Xu et al. (2020)
Millet yield estimation	To enhance the resilience and sustainability of millet cultivation in India under changing climatic conditions.	Using a deep convolutional generative adversarial network (DCGAN) to estimate millet yield from multispectral images acquired by satellites or UAVs.	Bandyopadhyay et al. (2017)
Oilseed yield estimation	To increase the oilseed production and reduce the dependency on imports in India.	Using a deep reinforcement learning (DRL) or deep Q-learning approach to estimate oilseed yield from remote sensing data and optimize the crop management decisions.	Rathod et al. (2018)
Pulse yield estimation	To improve the pulse production and consumption in India and address the protein malnutrition problem.	Using a deep attention-based recurrent neural network (DARNN) to estimate pulse yield from time series data, such as crop phenology, weather, soil moisture, etc.	Mishra et al. (2023)

6.2 Land cover and land use mapping

The ideas of LC and LU are essential to comprehending the earth's surface and how human activity affects it. The term "land cover" describes the organic materials that make up the surface of the earth, including plants, water, ice, and soil. Conversely, LU refers to the activities carried out by humans on the land, such as industry, forestry, urbanization, and agriculture. The geographical distribution and temporal variations of natural and human systems may be understood via mapping LC and LU. For a variety of EO and CM applications, this information is priceless. For example, precise LC and LU maps are essential for managing water resources, conserving biodiversity, urban planning, and developing mitigation plans for climate change. Because DL can handle high-resolution, multispectral, multitemporal, and heterogeneous data, it has become an effective method for LC and LU mapping from satellite imagery. To improve the precision and dependability of LC and LU mapping, DL can also combine data from other sources, such as digital elevation models, census information, or information from social media. The absence of reliable and current ground truth data, which is required for DL model training and testing, is a major obstacle to LC and LU mapping in places like India. To solve this issue, researchers have suggested utilizing synthetic or semisynthetic labels obtained from preexisting datasets or expert knowledge. Furthermore, context-sensitive and fine-grained categorization techniques are necessary due to the richness and complexity of the Indian terrain.

Some examples of DL applications for LC and LU mapping in India are:

- The accuracy and effectiveness of LU mapping have been improved with the use of DL in remote sensing. In particular, researchers have created a strong framework that makes use of DL techniques to precisely identify and categorize agricultural land in India using Sentinel-2 satellite data. The European Space Agency deployed the Sentinel-2 satellites, which offer high-resolution optical images that is especially useful for monitoring agricultural fields. By collecting data in several spectral bands, these satellites allow for a thorough examination of the characteristics of the soil and plants. DL algorithms may be taught to identify distinctive patterns and traits connected to agricultural land by utilizing this vast dataset. The DL framework integrates various algorithms, such as convolutional neural networks (CNNs), which are particularly effective in image recognition tasks. These models process the satellite images, learning to distinguish agricultural fields from other LC types based on their spectral signatures and spatial patterns.

This method not only enhances the accuracy of LU maps but also expedites the mapping process compared to traditional techniques. In the context of India, where agriculture is a cornerstone of the economy and sustains a large portion of the population, accurate LU mapping is crucial for effective resource management and policy-making. The ability to monitor agricultural areas with high precision aids in crop management, irrigation planning, and assessment of agricultural productivity. The integration of Sentinel-2 data with DL frameworks represents a significant advancement in agricultural monitoring. It provides a scalable and efficient tool for stakeholders to make informed decisions, ultimately contributing to the sustainability and growth of the agricultural sector in India (Singh et al., 2022).

- An inventive method for tracking environmental changes is presented in the paper "Land Use and Cover Change Detection Using the Tunicate Swarm Algorithm and Deep Learning in the Nallamalla Forest in India (TSADL-LULCCD)." The technique, known as TSADL-LULCCD, makes use of the Tunicate Swarm Algorithm (TSA) and DL to precisely identify and assess changes in LU and LC inside India's important ecological area, the Nallamalla Forest. With the help of this technology, environmental monitoring will be more accurate and efficient, which is important for conservation efforts and the long-term sustainable management of forest resources. This work presents a state-of-the-art tool for monitoring and comprehending ecological processes through the integration of DL with sophisticated computational techniques (Lavanya et al., 2023).
- With an emphasis on deep recurrent networks in the context of India, this paragraph explores the use of DL techniques for LU categorization using Sentinel-2 time series data. Deep recurrent networks are able to capture the dynamic changes in LC and LU by utilizing the extensive temporal information supplied by Sentinel-2 frequent observations. This strategy is especially useful in India, where standard categorization techniques are severely challenged by the country's complex LU patterns and swift environmental changes. The study's goal is to show how these cutting-edge DL models may increase LU classification's precision and effectiveness while providing fresh perspectives and resources for sustainable land management and planning (Campos-Taberner et al., 2020).
- Effective environmental management and urban planning now depend more and more on the classification of land use and land cover (LULC). In order to improve the accuracy of LULC classification, we use sophisticated artificial intelligence (AI) approaches in this study, namely random forests (RFs) and artificial neural networks (ANNs). With the use of high-resolution satellite images from Sentinel-2A and Landsat-8, we want to enhance these machine learning (ML) approaches' accuracy and dependability in relation to India.

Sentinel-2A and Landsat-8 are well known for their ability to collect large amounts of data, including multispectral pictures that are essential for in-depth examination of land characteristics. Greater spatial resolution is provided by Sentinel-2A, which can take pictures at a 10-meter resolution, greatly assisting in the differentiation of various LC types. However, Landsat-8 offers a wider temporal coverage, guaranteeing that variations in LU over time are efficiently tracked. Given their reputation for modeling intricate patterns with layered structures, ANNs are well-suited to tackle the complexities of LULC classification. Because of its ensemble learning methodology, RFs reduce overfitting and increase generalization, which further improves classification accuracy. This study demonstrates how integrating data from both Sentinel-2A and Landsat-8 with sophisticated AI techniques can lead to more precise and reliable LULC classifications. Such advancements are crucial for various applications, including agricultural monitoring, urban development planning, and environmental conservation in India (Alshari et al., 2023).

- A major development in the mapping of LULC in India is the ML-based categorization of LANDSAT satellite pictures. In order to map the varied LULC around the nation, researchers recently assessed how well a number of ML algorithms interpreted and classified LANDSAT photos. Since the 1970s, NASA and the USGS have been using LANDSAT satellites to provide vital EO data. The LANDSAT photos are essential for agricultural management, urban planning, and environmental monitoring because of their precise multispectral data that can distinguish between different types of LC.

To categorize LANDSAT photos, the researchers used a variety of ML methods, such as CNNs, decision trees, and support vector machines (SVMs). These algorithms are renowned for their capacity to predict intricate patterns in the data and for being resilient when working with big datasets. Because SVMs may provide optimum decision boundaries, they work well for classification problems. Decision trees work well for preliminary classifications because they are simple to use and effective in managing categorical data. CNNs are particularly good for image processing jobs because of its DL design, which makes it possible to identify land characteristics more precisely. By combining various ML approaches, the LULC classification process should become more accurate and efficient. This improved mapping capability is crucial for India, given its vast and varied landscapes, from urban areas to agricultural fields and natural forests. Accurate LULC maps support better resource management, environmental protection, and policy-making (Singh et al., 2021).

- Understanding and controlling urbanization requires mapping the usage of urban land in Mexico and India via the use of remote sensing and ML. This study shows how to create and apply cutting-edge ML techniques to describe LULC in India over wide geographic and temporal ranges. Satellite photography is one type of remote sensing technology that can give precise and reliable data on LU trends. Researchers can learn more about how metropolitan areas grow and alter over time by examining this data. This work uses ML methods to analyze and interpret remote sensing data, enabling very accurate identification and classification of various LC categories.

The study makes use of a range of ML methods, such as RFs, SVMs, and CNNs. Every one of these techniques has a special advantage: SVMs are great at binary classification, CNNs are very good at image recognition, and RFs can handle big datasets with ease and are resistant to overfitting. The researchers mapped urban land usage in Mexico and India by using similar approaches on satellite photos from sources like Landsat and Sentinel. This dual emphasis makes comparative study possible, showcasing the parallels and discrepancies in the two nations' urbanization trends. Such insights are invaluable for urban planners and policymakers, who can use this information to make informed decisions about sustainable development and resource management. This research underscores the power of combining remote sensing technology with ML to produce detailed, accurate, and timely maps of urban LU. The findings contribute to a better understanding of urban growth dynamics, which is essential for addressing challenges related to urbanization, such as infrastructure development, environmental impact, and social equity (Kerins et al., 2021).

6.3 Disaster management

DM describes the actions taken in the lead-up to, during, and after a catastrophe to minimize its impacts and enhance recovery. Natural or man-made disasters include floods, droughts, earthquakes, landslides, cyclones, wildfires, and industrial catastrophes. DM depends on EO and CM data to monitor and assess the risk, severity, and damage brought about by disasters as well as to support decision-making and response activities. DL has been widely used for DM from EO and CM data because to its versatility in performing tasks such as detection, segmentation, classification, generation, and prediction. In order to improve DM applications' efficiency and dependability, DL may also incorporate data from other sources, such as text, audio, video, optical, radar, thermal, or LiDAR images.

One challenge encountered by DM in India is the increasing frequency and intensity of disasters that endanger the lives and livelihoods of millions of people. To solve this issue, several scholars have proposed using DL for online or near-real-time DM applications, which can provide the public and the authorities with accurate and timely information. Lists some reading recommendations, explains, and provides some instances of how DL may be used to various stages and activities of DM. Another challenge and the requirement for flexible and broadly applicable DL models is the unpredictable nature of natural disasters. This issue has been addressed by researchers through the use of transfer learning, domain adaptation, and uncertainty estimation for DM applications.

Some examples of DL applications for DM in India are:

DL is redefining many industries, and catastrophe management involving water in impoverished nations is no exception. With an emphasis on India specifically, this chapter explores the state of DL technology now being applied to address water-related disasters in developing countries. The chapter looks at how these sophisticated algorithms are used to forecast, track, and control natural catastrophes like cyclones, floods, and droughts that frequently occur in these areas. DL models can produce more precise and timely forecasts by using enormous volumes of data, assisting authorities in taking preventive action and lessening the effects of such calamities. The study sheds light on the prospects and difficulties of deploying these technologies in environments with limited resources, offering suggestions for future research topics and possible advantages (Kareem et al., 2022).

Current developments in ML techniques and applications for DM. The fields of ML and DL applications for DM have advanced significantly in recent years. This paper offers a thorough summary of studies on ML and DL methods to DM, with an emphasis on advancements made after 2017. In the end, these developments have helped to lessen the effects of natural and man-made catastrophes on infrastructure and people by creating new opportunities for better disaster prediction, response, and recovery.

Extensive use has been made of ML techniques, such as supervised, unsupervised, and reinforcement learning, to evaluate massive volumes of data produced during catastrophes. For example, ML algorithms have the ability to analyze sensor data, social media posts, and satellite images to forecast the likelihood and severity of natural catastrophes like hurricanes, floods, and earthquakes. ML's deep learning subset has shown especially adept in managing intricate data patterns, allowing for more precise and timely predictions. CNNs are one prominent application that makes use of satellite imagery to identify and evaluate damage following a disaster. Damage assessment is made possible by this technology in a timely and accurate manner, which is essential for efficient response and resource allocation. Furthermore, by examining sequential data like weather patterns and seismic activity, recurrent neural networks (RNNs) have been used to forecast the occurrence of disasters. Combining ML and DL with Geographic Information Systems (GIS) is another important development. By combining the two, the spatial analytic skills are improved, allowing for more accurate mapping and monitoring of regions that are vulnerable to disasters. For instance, ML systems may evaluate past catastrophe data to pinpoint high-risk areas and more accurately forecast future occurrences.

Emergency response has also been transformed by the creation of real-time catastrophe management systems driven by ML and DL. These systems are able to collect and analyze data in real time from a variety of sources, giving emergency responders the necessary knowledge to act swiftly and decisively. For example, ML algorithms may use geography, vegetation kinds, and meteorological information to forecast the progress of a wildfire and recommend the best routes for evacuation. The applications of ML and DL for DM have advanced beyond reaction and prediction. They are essential to the endeavors of rehabilitation as well. Aid may be distributed more efficiently using ML models, guaranteeing that it reaches the most impacted areas first. They may also evaluate the long-term effects of catastrophes on infrastructure and communities, which helps create more resilient systems. All things considered, the latest developments in DL and ML have greatly improved catastrophe management capacities. In the end, these technologies serve to lessen the impact of catastrophes on infrastructure and human lives by offering more precise forecasts, better reaction plans, and efficient recovery procedures. The application of ML and DL to DM is expected to advance in sophistication and spread as research progresses (Linardos et al., 2022).

An overview of the research on artificial neural network methods for disaster relief: we examine five main DM application areas in India in this chapter, with a particular emphasis on managing landslides, earthquakes, and floods. To offer readers a thorough grasp of the ways in which different technologies and tactics are used to mitigate and manage these natural disasters, these application areas have been methodically categorized into five main categories.

To begin with, DM in India includes a broad range of techniques meant to lessen the effects of natural disasters. This entails creating programs for community awareness, early warning system development, and the deployment of resilient infrastructure to survive such occurrences.

The second kind, flood management, is taking substantial steps to avoid and control floods, including building river embankments, using flood forecasting models, and implementing effective drainage systems. These initiatives are vital in a nation where the annual monsoons frequently cause catastrophic floods, especially in low-lying areas.

In the third category, earthquake management, the emphasis is on creating thorough emergency response protocols, modifying existing structures, and developing buildings that are resistant to earthquakes. These steps are essential for minimizing deaths and financial losses because of India's vulnerability to seismic activity, particularly in the Himalayan area.

The fourth category, "landslide management," consists of soil erosion prevention measures including reforestation, slope stabilization techniques, and real-time monitoring systems. In India's hilly regions, landslides are a common danger that need for concerted measures to anticipate and lessen their occurrence. All together, these five categories demonstrate the complex strategy needed for India's efficient catastrophe management. In order to increase resilience against natural catastrophes, each category represents a crucial area of concentration and emphasizes the necessity of continuing study, technological innovation, and community engagement (Guha et al., 2010).

Using DL algorithms to extract social media data for DM assistance. In this work, we suggest using DL techniques to extract useful data from social media platforms for DM aid, focusing on India. Social media has emerged as an important source of real-time information during catastrophes, offering rapid updates and on-the-ground insights that can be critical for successful response and relief operations. DL, a form of ML, uses ANNs to analyze and comprehend complicated data patterns. DL algorithms can automatically filter and categorize massive quantities of data, discovering pertinent postings that might help DM authorities. This includes detecting distress signals, mapping affected areas, and monitoring the spread of misinformation.

The use of DL in DM has various benefits. For starters, it facilitates the rapid analysis of enormous datasets, allowing authorities to react more quickly to growing situations. Second, it improves the accuracy of information extraction by eliminating the need for manual analysis, which may be time-consuming and error-prone. Third, DL models can constantly learn and adapt to new types of data, hence increasing their efficacy over time. In India, where natural catastrophes such as floods, earthquakes, and landslides are widespread, using social media data with DL techniques might dramatically enhance disaster response and recovery. For example, during the 2013 Uttarakhand floods, social media helped disseminate information and coordinate rescue efforts. By integrating DL models, the efficiency and accuracy of such efforts can be further enhanced, leading to better outcomes for affected communities. Overall, the integration of DL techniques in extracting and analyzing social media data represents a promising advancement in DM. It underscores the potential of cutting-edge technology to augment traditional methods, providing a more robust and agile response to natural calamities (Vishwanath et al., 2023).

A recent paper by Linardos et al. (2022) titled "Machine Learning in Disaster Management: A Review of Recent Methods and Applications" provides a thorough analysis of the applications of DL and ML approaches in the field of DM. With an emphasis on case studies from India, the study addresses a number of topics, including catastrophe prediction, response, preparedness, mitigation, and postdisaster analysis. The writers explore the most current developments in these techniques and their real-world uses, emphasizing how these technologies can revolutionize catastrophe management procedures.

According to Linardos et al. (2022), ML is crucial for anticipating natural catastrophes including earthquakes, floods, and cyclones. This may greatly improve community and government preparedness and response plans. These technologies can lessen the possible effect of such occurrences by identifying trends and issuing early warnings based on the analysis of enormous volumes of data. The paper delves into the application of DL models in real-time disaster response, which facilitates the expeditious and enhanced allocation of resources to impacted regions.

The use of ML in disaster preparedness and mitigation is covered in more detail in the article. ML models may aid in the development of efficient mitigation methods and the enhancement of infrastructure resilience by simulating diverse catastrophe scenarios. The authors also emphasize postdisaster analysis, where these tools help in planning for future emergencies, coordinating recovery operations, and determining the amount of damage. The paper demonstrates the real-world uses of these techniques using case studies from India, offering insightful information on both their advantages and disadvantages. The writers stress that in order to improve ML approaches' accuracy and dependability for DM, they must be continuously developed and improved. Overall, Linardos et al. (2022) provide a thorough review of the state-of-the-art in ML applications for DM, emphasizing the potential of these technologies to revolutionize how we predict, prepare for, respond to, and recover from disasters (Linardos et al., 2022).

ML has become a potent tool for pandemic and catastrophe management in recent years. In order to solve these pressing issues, this chapter looks at the most recent ML algorithms that have been used in several places, including India. This study's main conclusions include the significance of creating vision-based catastrophe detection systems. These methods greatly improve the speed and accuracy of disaster response operations by using DL algorithms to quickly identify impacted regions from aerial pictures. The use of ML algorithms to forecast, control, and lessen the effects of pandemics and disasters is growing. Large volumes of data have been analyzed, patterns have been found, and well-informed judgments have been made thanks to techniques like supervised learning, unsupervised learning, and reinforcement learning. For example, in India, ML models have been employed to predict the spread of COVID-19, optimize resource allocation, and streamline contact tracing efforts.

The potential of ML to improve the efficacy and efficiency of epidemic control has been shown by these applications. One cannot stress the importance of vision-based disaster detection systems. Conventional approaches to catastrophe assessment frequently rely on labor-intensive, slow, and error-prone human examination and reporting. On the other hand, impacted regions can be promptly and precisely identified by DL algorithms through the processing of aerial photographs taken by drones or satellites. These cutting-edge technologies scan visual data using CNNs to find abnormalities that point to potential calamities like building collapses, flooded areas, or wildfire breakouts. These technologies provide for improved coordination of disaster relief operations and speedier deployment of emergency services by automating the detection process. Overall, the integration of ML algorithms into disaster and epidemic management holds significant promise. By harnessing the power of vision-based detection systems and other advanced ML techniques, we can improve our ability to respond to emergencies and mitigate their impacts. Continued research and development in this field will be crucial for enhancing disaster preparedness and resilience (Chamola et al., 2021).

6.4 Air quality assessment

The term air quality assessment (AQA) refers to the measurement and assessment of the amount and composition of air pollutants, such as particulate matter (PM), nitrogen dioxide (NO_2), ozone (O_3), carbon monoxide (CO), sulfur dioxide (SO_2), etc. (Mkhize, 2020). As air quality analysis provides information on the consequences of air pollution on human health and the environment, it is an essential task for EO and CM applications. Numerous objectives, such as increasing public awareness, influencing legislation, lowering emissions, etc., might be assisted by AQA. DL has been widely used for AQA employing EO and CM data because of its versatility in performing tasks such as estimating, interpolation, extrapolation, and forecasting. To increase the precision and coverage of AQA, DL can also incorporate several data sources, such as satellite photos, ground measurements, meteorological data, or traffic data (Shen, 2018).

Millions of people's health and well-being are impacted by air pollution, which makes assessing the quality of the country's air (AQA) difficult in India due to its high level and fluctuation. In order to tackle this difficulty, scholars have suggested the subsequent approaches:

For fine-grained or high-resolution AQA, use DL to deliver precise and localized air quality data. The lack of reliable ground measurements, which are necessary for DL model calibration and validation, is another problem (Kumar et al., 2020). In order to tackle this difficulty, scholars have suggested the subsequent approaches:

Use DL for data fusion or data augmentation for AQA. While there are some resources available for AQA in India, such as the Air Quality Asia organization and the Directorate General of Aeronautical Quality Assurance, more research and development are needed to improve the accuracy and reliability of AQA methods (Sekar et al., 2021).

Some examples of DL applications for AQA concerning India are:

Predicting Delhi's air quality using LSTM deep learning models—with an emphasis on the COVID-19 lockdown Delhi, India's air quality has been a major concern in recent years because of the city's fast industrialization and urbanization. The novel method of air quality prediction presented in this chapter makes use of Long Short-Term Memory (LSTM) network models, a kind of recurrent neural network renowned for its efficiency in processing time-series data. The research not only looks at how well LSTM models can predict air pollution levels, but it also looks at how much the COVID-19 shutdown affected Delhi's air quality. Tight limitations were placed on industrial operations, vehicle movements, and other sources of pollution during the COVID-19 shutdown. This exceptional time offered a natural experiment to evaluate the impact of decreased human activity on air quality. The study's conclusions demonstrate how well the LSTM models predicted variations in air quality metrics like PM2.5 and PM10 concentrations, providing important information on the temporal dynamics of air pollution. The investigation highlighted the possible advantages of less human activity on environmental health by demonstrating a discernible increase in air quality during the lockdown time. These findings emphasize how crucial it is to have sustainable procedures and policies in place to fight air pollution, not just in extreme cases like pandemics but also on a constant basis to guarantee better air quality for Delhi's citizens (Tiwari et al., 2021).

Air pollution poses a serious threat to the environment and human health, especially in areas with high population densities and industrial activity. In "Air Pollution Prediction with Machine Learning: A Case Study of Indian Cities," scientists analyzed 6 years' worth of air pollution data from 23 major Indian towns in great detail. The main goal was to use cutting-edge ML techniques to create a forecast model for air quality. The study used the following five ML models: Support Vector Machine, Random Forest, Decision Tree, Linear Regression, and Extreme Gradient Boosting (XGBoost) in order to do this. The dataset was used to train and test each model in order to assess each one's performance and forecast accuracy. The results showed that the XGBoost model performed better than the other models and showed greater air pollution level prediction accuracy. The efficiency and scalability of XGBoost were especially evident in how well it handled the intricate and nonlinear correlations found in the air quality data. This model is a priceless tool for environmental monitoring and public health planning because of its capacity to handle big datasets and generate reliable forecasts. The study's findings highlight the ability of ML models—particularly XGBoost—to anticipate air quality and support the creation of practical pollution control plans. Policymakers and environmental agencies may better predict pollution trends and carry out appropriate measures to preserve public health by utilizing these cutting-edge analytical techniques (Kumar & Pande, 2023).

The application of ML and geospatial AI is revolutionizing the monitoring of air quality in Indian cities. The Indian Institute of Technology (IIT) Delhi, which is leading the way in creating state-of-the-art algorithms to track and predict air quality, is leading this creative approach. This project intends to deliver more precise, real-time data on air pollution levels by utilizing geospatial AI and ML, ultimately leading to more informed decision-making and more successful environmental policies. The project builds its models from a range of data sources. These consist of meteorological information, satellite imaging, and air quality sensors located on the ground. The models can detect pollution hotspots, evaluate temporal and geographical patterns of air pollution, and forecast future changes in air quality by combining these various information. In order to lessen the effects of air pollution on the environment and public health, a more sophisticated knowledge of the dynamics of air pollution is made possible by this all-encompassing approach. Processing large volumes of data fast and reliably is one of the main benefits of employing geospatial AI and ML in air quality monitoring. Conventional techniques for monitoring air quality frequently depend on a small number of ground-based sensors, which can only offer a piecemeal picture of air pollution. In contrast, geospatial AI and ML models can analyze data from numerous sources simultaneously, offering a more holistic and detailed picture of air quality across entire urban areas.

Moreover, these models' predictive powers allow authorities to foresee and address air quality problems before they worsen. Preventive measures, for example, can be put into place ahead of time if the models predict a surge in pollution levels owing to impending weather conditions or greater vehicle emissions. By taking preventative measures, the health hazards brought on by bad air quality may be decreased, and Indian cities' general quality of life can be raised. All things considered, the use of geospatial AI and ML to air quality monitoring marks a substantial breakthrough in environmental management. This IIT Delhi initiative is going to change the game in terms of air pollution monitoring and response by using technology to make cities healthier and more sustainable (Singh et al., 2021).

In the context of smart cities, air quality prediction is critical to public health and environmental sustainability. This paper presents a novel DL-based model that combines the advantages of CNNs and RNNs to improve air quality prediction. The model analyzes and predicts air quality using satellite imagery along with historical air quality data, providing a more advanced method of environmental monitoring. CNNs and RNNs are combined to leverage both spatial and temporal features of air quality data. CNNs are capable of processing and interpreting spatial patterns within images as supplied by satellite observations. These networks can effectively analyze satellite images to identify pollution sources, track changes in air quality over time, and recognize patterns that might indicate worsening conditions.

In contrast, RNNs—more especially, LSTM networks—are excellent at identifying temporal connections in data. They are quite good at analyzing sets of past air quality data to forecast patterns and variations in the future. Four Indian cities, Bengaluru, Chennai, Delhi, and Mumbai, were used to thoroughly evaluate the suggested approach. The various sources of pollution, climates, and densities of metropolitan areas that these cities have each provide different issues with regard to air quality. The researchers wanted to show how resilient and adaptable the model was, so they applied it to these various contexts. During the testing phase, the model was trained using satellite pictures and historical data on air quality from these cities. After that, its predictive accuracy for air quality levels was verified.

The study's findings show that the CNN-RNN model outperforms conventional techniques for predicting air quality in a major way. Through the integration of CNNs' spatial analysis and RNNs' temporal forecasting, the model provides a more thorough knowledge of the dynamics of air pollution. With the use of this method, more accurate forecasts can be made, which helps politicians and municipal planners to create plans to reduce air pollution and act quickly. In summary, this article highlights a significant advancement in air quality prediction technology. The integration of CNNs and RNNs into a unified DL model represents a promising step forward in smart city infrastructure. By accurately predicting air quality based on both satellite imagery and historical data, this model offers a valuable tool for improving urban environmental management and safeguarding public health (Bekkar et al., 2021).

Predicting air quality with accuracy is becoming increasingly important in many places, especially fast expanding cities like Delhi, India. This article examines a cutting-edge DL method for predicting PM2.5 concentrations, a kind of tiny PM that has a major negative influence on respiratory health. The research focuses on a DL model that uses a Multi-Layer Perceptron (MLP) and a variety of input factors, such as calendar information, traffic circumstances, and meteorological data, to forecast PM2.5 levels. This study's MLP model is a kind of artificial neural network made up of several layers of linked nodes, or "neurons." Each layer processes the input data and passes it through nonlinear activation functions, allowing the model to capture complex relationships between the inputs and the target variable—PM2.5 concentration. By integrating data from various sources, the MLP model aims to provide a comprehensive forecast of air pollution levels.

The inputs to the model include:

Meteorological Data: This includes variables such as temperature, humidity, wind speed, and atmospheric pressure, all of which can influence air quality. For instance, wind speed and direction can affect the dispersion of pollutants, while temperature and humidity can impact their formation and accumulation.

Traffic Data: Traffic volume and congestion are significant contributors to air pollution, particularly in a densely populated city like Delhi. Data on traffic patterns helps the model account for emissions from vehicles, which are a major source of PM2.5.

Calendar Data: Information related to time, such as day of the week and seasonal variations, can provide insights into periodic changes in air quality. For example, pollution levels may vary between weekdays and weekends or across different seasons due to changes in human activities and meteorological conditions.

The chapter also evaluates the performance of the MLP model by comparing it with other ML techniques. These comparisons are crucial for assessing the effectiveness and accuracy of the MLP approach in predicting PM2.5 concentrations. The study explores various metrics to measure model performance, such as mean absolute error (MAE), correlation coefficients, and root mean squared error (RMSE) offer a thorough examination of the MLP model's capabilities. The findings show that when it comes to PM2.5 level prediction, the MLP model performs better than a number of conventional ML methods. Better air quality management methods may be informed by the model, which delivers higher forecast accuracy through effective integration and analysis of meteorological, traffic, and calendar data. Consequently, this section illustrates how DL methods—more especially, MLP—can enhance air quality predictions. The study's conclusions highlight the value of combining several data sources to provide forecasts that are more precise and useful, which will eventually improve Delhi's public health and air quality management (Srivastava et al., 2019).

For efficient environmental management and public health protection, air quality prediction is crucial, particularly in varied and heavily populated areas like India. This chapter introduces a state-of-the-art DL algorithm that uses satellite images from different parts of the nation to forecast the air quality index (AQI). To improve prediction accuracy and model durability, the novel method combines the capabilities of a dense neural network with the power of a state-of-the-art Convolutional Neural Network (CNN), ResNet-50. ResNet-50, a deep residual network with exceptional performance on image classification tasks, is the basis of this model. ResNet-50 is well known for its deep architecture and residual learning framework, which enable it to extract complicated features from pictures with ease. This framework allows the network to train more efficiently by addressing the problem of vanishing gradients, which often hampers the training of very deep networks. By leveraging ResNet-50, the model can extract detailed and relevant features from satellite images, such as pollutant sources, LU patterns, and atmospheric conditions, which are crucial for accurate AQI prediction.

To handle the features that the CNN retrieved, the model incorporates a dense neural network in addition to ResNet-50. This dense network, which consists of layers with complete connections, learns the intricate correlations between the extracted features and the AQI values and refines the feature representations. The dense neural network and ResNet-50 work together to provide the program the ability to estimate AQI accurately using satellite photos. Through thorough testing, the model's excellent accuracy and resilience are demonstrated in this chapter. The model was tested in several Indian locations, taking into consideration diverse climatic and environmental factors. This thorough test procedure demonstrates the model's capacity to continue operating at a high level of performance and dependability despite localized fluctuations in air quality. The findings demonstrate how, because of its sophisticated feature extraction and processing capabilities, the DL model performs noticeably better than conventional AQI prediction techniques. The model provides useful information for environmental agencies, legislators, and urban planners by precisely forecasting AQI from satellite photos. This methodology enables prompt responses and enhanced methods for managing air quality, therefore promoting public health and environmental sustainability. In conclusion, this chapter presents a state-of-the-art DL model that predicts AQI from satellite data by utilizing a dense neural network and ResNet-50. The model has proven to be accurate and resilient, which highlights its potential as an effective tool for managing and monitoring air quality in various parts of India (Devi et al., 2022).

6.5 Crop yield estimation

Crop productivity and food security depend on crop yield estimate, or CYE. DL has been heavily utilized by CYE using data from EO and agricultural monitoring. Crop type classification, crop growth tracking, yield forecasting, and crop quality evaluation are tasks that DL can complete (Apolo-Apolo et al., 2020). DL may also make use of a variety of data sources, including soil, meteorological, agricultural, optical, radar, thermal, or hyperspectral imaging, to increase the precision and reliability of CYE (Raj et al., 2021). Nevertheless, CYE has challenges due to the scarcity and unpredictability of ground data. Researchers have suggested approaches to employ data fusion or data augmentation for AQA to overcome this problem. At several levels, EO and CM applications can support the development of food policy (Jiang et al., 2021). In the end, remote sensing applications can support several objectives, such as crop management, market analysis, and crop insurance.

CYE is essential in order to guarantee our access to sufficient food and ethical agricultural methods (Beske et al., 2014). Determining CYE can be difficult, though, because farming is so varied and crop yield varies widely in India. As remedies to these issues, some astute individuals have suggested the following: They suggest using DL to manage the vast and complicated data from CM and EO. This can help with general CYE calculations (Dasgupta et al., 2020). They also say that DL might be used to make CYE estimations that are explicable or unclear. This way, we may understand the causes for the unexpected findings and the accuracy of the estimations. To increase CYE, they are thinking of utilizing a range of data, such as information on the soil, the weather, and the crops themselves, as well as optical, radar, and thermal pictures from satellites (Roy, 2018). The accuracy and trustworthiness of CYE can be enhanced by all of this information. Using cutting-edge technology like DL and ML to gather data remotely is another idea. This gives us an idea of how well some crops will grow in the future. Furthermore, they assert that we may integrate data from remote regions with maps to ascertain the effects of climate change on food security. We can forecast what could happen to our food as a result of climate change by fusing technology and mapping (Savary et al., 2017). Thus combining all of these concepts might significantly improve predictions of agricultural yields in India. This helps us understand the market, it helps farmers know how to produce things, and it even helps with crop insurance.

In India, some instances of DL applications for CYE include:

- To predict crop yield for more than 30 crops across all Indian districts from 1997 to 2015, this paper proposes a novel DL model using ensemble neural networks and one-dimensional CNNs and LSTMs (Prashant et al., 2021).
- Crop yield prediction utilizing DL: In this research, a neural network model is presented to forecast crop yield for the state of Karnataka utilizing variables such as the agricultural area, crop, taluka, year, season, rainfall, and temperature (Mamatha et al., 2023).
- Crop yield estimation in India using ML: This work uses four ML algorithms to predict crop yield using area, yield, production, and area under irrigation: Decision Tree, Linear Regression, Lasso Regression, and Ridge Regression (Kavita & Mathur, 2020).
- This research proposes a DL-based crop yield prediction model for Indian agriculture utilizing CNNs and RNNs to capture the spatial and temporal characteristics of crop data (Gupta et al., 2021).
- A deep neural network (DNN) is used in this study to estimate the agricultural yields of wheat and rice in India using meteorological data and satellite photos (van Klompenburg et al., 2020).
- DL strategy for crop yield prediction utilizing Sentinel-2 satellite data: This research suggests a DL strategy for crop yield prediction utilizing Sentinel-2 satellite data for remote sensing. Data preparation, feature extraction, and regression make up the method's three phases (Muruganantham et al., 2022).

6.6 Conclusion

The chapter focuses on the promise and potential of DL for applications in India related to EO and crop monitoring. The chapter examines the most current developments and uses of DL concerning four key themes: LC and LU mapping, DM, AQA, and CYE. The difficulties and possibilities for DL research and development in this field are also covered. According to the observations, DL can manage the amount, diversity, and complexity of EO and CM data and get valuable insights from them. To increase the precision and dependability of EO and CM applications, DL may also make use of auxiliary data sources such as digital elevation models, census data, social media data, meteorological data, traffic data, soil data, etc. (Salcedo-Sanz et al., 2020). This shows that DL has the potential to revolutionize EO and CM in India by supplying more precise and trustworthy information for agricultural, defense, and sustainable rural development goals (Mondejar et al., 2021; Rasp et al., 2018). India can improve its EO and CM capabilities and help to accomplish sustainable development goals by utilizing DL methods and combining multiple data sources. However, the lack of consistent and up-to-date ground truth data, the diversity and complexity of the Indian landscape, the frequent and severe natural disasters, the high level and variability of air pollution, the size and diversity of the agricultural sector, the uncertainty and variability of crop yield, etc. are some of the limitations and difficulties that DL must overcome for EO and CM applications in India. To overcome these obstacles, further research and development activities are needed.

Some possible directions for future work are:

- Developing brand-new DL architectures and methods that can manage the very detailed, multispectral, multitemporal, heterogeneous, and noisy characteristics of EO and CM data.
- Developing brand-new DL techniques that can be used to EO and CM applications to take into account spatial dependencies, semantic hierarchies, attention processes, transfer learning, domain adaption, uncertainty estimates, etc.
- Developing new DL frameworks that can integrate a variety of data sources, including EO and CM applications, text, audio, video, digital elevation models, census data, social media data, meteorological data, traffic data, soil data, etc.
- Developing new DL applications that can deal with pressing or emergent EO and CM challenges, such as biodiversity loss, climate change adaptation, urban sprawl, water shortages, food security, etc.
- AI-based alarms to predict solar storms: In response to the rise in solar storms, a group of researchers from around the world, including those from India, have teamed up with NASA to create a new computer model that combines AI and satellite data to sound alarms for perilous space weather. Because the Sun is preparing for a peak in activity, which occurs roughly every 11 years and is anticipated to occur around 2025, scientists have projected an increase in solar storms.
- DL for air pollution prediction: DL algorithms have been used to forecast air pollution levels in Indian cities, where air pollution is a significant issue. The study combined DL and GoogleMaps satellite photos to estimate air pollution levels with the highest level of accuracy possible.

Acknowledgments

I thank Prof. Gefei Hao for his support of my research studies. I thank Dr. Mir Muhammad Nizamani for his ideas in helping me finish the chapter.

AI disclosure

During the preparation of this work the author(s) used ChatGPT in order to To improve English. After using this tool/service, the author(s) reviewed and edited the content as needed and take(s) full responsibility for the content of the publication.

References

Alshari, E. A., Abdulkareem, M. B., & Gawali, B. W. (2023). Classification of land use/land cover using artificial intelligence (ANN-RF). *Frontiers in Artificial Intelligence*, *5*, 964279. Available from https://doi.org/10.3389/frai.2022.964279, Retrieved from, http://www.frontiersin.org/journals/artificial-intelligence#.

Apolo-Apolo, O. E., Pérez-Ruiz, M., Martínez-Guanter, J., & Valente, J. (2020). A cloud-based environment for generating yield estimation maps from apple orchards using UAV imagery and a deep learning technique. *Frontiers in Plant Science*, *11*. Available from https://doi.org/10.3389/fpls.2020.01086.

Bandyopadhyay, T., Muthamilarasan, M., & Prasad, M. (2017). Millets for next generation climate-smart agriculture. *Frontiers in plant science*, *8*, 1266.

Basu, T., Mondal, B. K., Abdelrahman, K., Fnais, M. S., & Praharaj, S. (2023). Assessing Urban Flood Hazard Vulnerability Using Multi-Criteria Decision Making and Geospatial Techniques in Nabadwip Municipality, West Bengal in India. *Atmosphere, 14*(4), 669.

Bekkar, A., Hssina, B., Douzi, S., & Douzi, K. (2021). Air-pollution prediction in smart city, deep learning approach. *Journal of Big Data*, *8*(1), 161. Available from https://doi.org/10.1186/s40537-021-00548-1.

Beske, P., Land, A., & Seuring, S. (2014). Sustainable supply chain management practices and dynamic capabilities in the food industry: A critical analysis of the literature. *International Journal of Production Economics*, *152*, 131–143. Available from https://doi.org/10.1016/j.ijpe.2013.12.026.

Bi, L., & Hu, G. (2021). A genetic algorithm-assisted deep learning approach for crop yield prediction. *Soft Computing*, *25*, 10617–10628.

Campos-Taberner, M., García-Haro, F. J., Martínez, B., Izquierdo-Verdiguier, E., Atzberger, C., Camps-Valls, G., & Gilabert, M. A. (2020). Understanding deep learning in land use classification based on Sentinel-2 time series. *Scientific Reports*, *10*(1). Available from https://doi.org/10.1038/s41598-020-74215-5, Retrieved from, http://www.nature.com/srep/index.html.

Chamola, V., Hassija, V., Gupta, S., Goyal, A., Guizani, M., & Sikdar, B. (2021). Disaster and pandemic management using machine learning: A survey. *IEEE Internet of Things Journal*, *8*(21), 16047–16071. Available from https://doi.org/10.1109/JIOT.2020.3044966, Retrieved from, http://ieeexplore.ieee.org/servlet/opac?punumber = 6488907.

Dasgupta, I., Saha, J., Venkatasubbu, P., & Ramasubramanian, P. (2020). AI crop predictor and weed detector using wireless technologies: A smart application for farmers. *Arabian Journal for Science and Engineering*, *45*(12), 11115–11127. Available from https://doi.org/10.1007/s13369-020-04928-2.

Devi, K.M., Retnakumar, G.J., & Kavya, G. (2022). 1 2022/01 4th International Conference on Inventive Research in Computing Applications, ICIRCA 2022 – Proceedings. Institute of Electrical and Electronics Engineers Inc. India An Image based Air Quality Analysis using Transfer Learning Algorithms. 746–753. https://doi.org/10.1109/ICIRCA54612.2022.9985588 9781665497077. http://ieeexplore.ieee.org/xpl/mostRecentIssue.jsp?punumber = 9985450.

Guha, S., Jana, R. K., & Sanyal, M. K. (2010). Artificial neural network approaches for disaster management: A literature review. *International Journal of Disaster Risk Reduction*.

Gupta, P.K., Gandotra, E., Tyagi, V., & Sehgal, V.K. (2021). International Conference on Image Information Processing (ICIIP). IEEE, USA.

Gupta, et al. (2019). Building damage assessment after 2015 Nepal earthquake using VHR satellite images. *International Journal of Disaster Risk Reduction*, *41*, 101264.

Jiang, S., Ni, C., Chen, G., & Liu, Y. (2021). A novel data fusion strategy based on multiple intelligent sensory technologies and its application in the quality evaluation of Jinhua dry-cured hams. *Sensors and Actuators B: Chemical*, *344*, 130324. Available from https://doi.org/10.1016/j.snb.2021.130324.

Kareem, K. Y., Seong, Y., Bastola, S., & Jung, Y. (2022). Current state of deep learning application to water-related disaster management in developing countries. *Natural Hazards and Earth System Sciences Discussions*, 1–33.

Kavita, M., & Mathur, P. (2020). 10 30 2020/10/30 2020 IEEE 5th International Conference on Computing Communication and Automation, ICCCA 2020. Institute of Electrical and Electronics Engineers Inc. India Crop Yield Estimation in India Using Machine Learning, 220–224. 10.1109/ICCCA49541.2020.9250915 9781728163246. http://ieeexplore.ieee.org/xpl/mostRecentIssue.jsp?punumber = 9250743.

Kerins, P., Guzder-Williams, B., Mackres, E., Rashid, T., & Pietraszkiewicz, E. (2021). Mapping urban land use in India and Mexico using remote sensing and machine learning. *World Resources Institute*. Available from https://doi.org/10.46830/writn.20.00048.

van Klompenburg, T., Kassahun, A., & Catal, C. (2020). Crop yield prediction using machine learning: A systematic literature review. *Computers and Electronics in Agriculture*, *177*, 105709. Available from https://doi.org/10.1016/j.compag.2020.105709.

Kumar, D., Roshni, T., Singh, A., Kumar Jha, M., & Samui, P. (2020). Predicting groundwater depth fluctuations using deep learning, extreme learning machine and Gaussian process: A comparative study. *Earth Science Informatics*, *13*(4), 1237–1250. Available from https://doi.org/10.1007/s12145-020-00508-y.

Kumar, K., & Pande, B. P. (2023). Air pollution prediction with machine learning: A case study of Indian cities. *International Journal of Environmental Science and Technology*, *20*(5), 5333–5348. Available from https://doi.org/10.1007/s13762-022-04241-5.

Lavanya, K., Mahendran, A., Selvanambi, R., Mazzara, M., & Hemanth, J. D. (2023). Tunicate swarm algorithm with deep learning based land use and cover change detection in Nallamalla Forest India. *Applied Sciences*, *13*(2), 1173. Available from https://doi.org/10.3390/app13021173.

Liao, Q., Zhu, M., Wu, L., Pan, X., Tang, X., & Wang, Z. (2020). Deep learning for air quality forecasts: a review. *Current Pollution Reports*, *6*, 399–409.

Linardos, V., Drakaki, M., Tzionas, P., & Karnavas, Y. L. (2022). Machine learning in disaster management: Recent developments in methods and applications. *Machine Learning and Knowledge Extraction*, *4*(2), 446–473. Available from https://doi.org/10.3390/make4020020, Retrieved from, http://www.mdpi.com/journal/make.

Mamatha, K., Samantha, S., & Prasad, K. K. (2023). *Crop yield prediction using deep learning* (947, pp. 93–102). Springer Science and Business Media LLC. Available from http://doi.org/10.1007/978-981-19-5936-3_9.

Mkhize, P.L. (2020). Use of air dispersion modelling to determine the impact of gas emissions from coal-fired boilers in South African Durban basin.

Mishra, P., Al Khatib., Lal, P., Anwar, A., Nganvongpanit, K., Abotaleb, M., ... Punyapornwithaya, V. (2023). An Overview of Pulses Production in India: Retrospect and Prospects of the Future Food with an Application of Hybrid Models. *National Academy Science Letters*, 1–8.

Mishra, S., Mishra, D., & Santra, G. H. (2016). Applications of machine learning techniques in agricultural crop production: a review paper. *Indian J. Sci. Technol*, *9*(38), 1–14.

Mondejar, M. E., Avtar, R., Diaz, H. L. B., Dubey, R. K., Esteban, J., Gómez-Morales, A., Hallam, B., Mbungu, N. T., Okolo, C. C., Prasad, K. A., She, Q., & Garcia-Segura, S. (2021). Digitalization to achieve sustainable development goals: Steps towards a Smart Green Planet. *Science of the Total Environment*, *794*, 148539. Available from https://doi.org/10.1016/j.scitotenv.2021.148539, Retrieved from, http://www.elsevier.com/locate/scitotenv.

Muruganantham, P., Wibowo, S., Grandhi, S., Samrat, N. H., & Islam, N. (2022). A systematic literature review on crop yield prediction with deep learning and remote sensing. *Remote Sensing*, *14*(9), 1990.

Prashant, P., Ponkshe, K., Garg, C., Pendse, I., & Muley, P. (2021). 1 2021/01 Proceedings of the IEEE International Conference Image Information Processing. Institute of Electrical and Electronics Engineers Inc. India Crop Yield Prediction of Indian Districts Using Deep Learning 2021, 250–255. https://doi.org/10.1109/ICIIP53038.2021.9702573 9781665433617.

Raj, M., Gupta, S., Chamola, V., Elhence, A., Garg, T., Atiquzzaman, M., & Niyato, D. (2021). A survey on the role of Internet of Things for adopting and promoting Agriculture 4.0. *Journal of Network and Computer Applications*, *187*, 103107. Available from https://doi.org/10.1016/j.jnca.2021.103107.

Rasp, S., Pritchard, M. S., & Gentine, P. (2018). Deep learning to represent subgrid processes in climate models. *Proceedings of the National Academy of Sciences of the United States of America*, *115*(39), 9684–9689. Available from https://doi.org/10.1073/pnas.1810286115, Retrieved from, http://www.pnas.org/content/pnas/115/39/9684.full.pdf.

Rathod, S., Singh, K. N., Patil, S. G., Naik, R. H., Ray, M., & Meena, V. S. (2018). Modeling and forecasting of oilseed production of India through artificial intelligence techniques. *Indian J. Agric. Sci*, *88*(1), 22–27.

Rodriguez-Sanchez, J., Li, C., & Paterson, A. H. (2022). Cotton yield estimation from aerial imagery using machine learning approaches. *Frontiers in plant science*, *13*, 870181.

Roy, S. (2018). Distributions of phytoplankton carbohydrate, protein and lipid in the world oceans from satellite ocean colour. *The ISME Journal*, *12* (6), 1457–1472. Available from https://doi.org/10.1038/s41396-018-0054-8.

Salcedo-Sanz, S., Ghamisi, P., Piles, M., Werner, M., Cuadra, L., Moreno-Martínez, A., Izquierdo-Verdiguier, E., Muñoz-Marí, J., Mosavi, A., & Camps-Valls, G. (2020). Machine learning information fusion in Earth observation: A comprehensive review of methods, applications and data sources. *Information Fusion*, *63*, 256–272. Available from https://doi.org/10.1016/j.inffus.2020.07.004.

Savary, S., Bregaglio, S., Willocquet, L., Gustafson, D., Mason D'Croz, D., Sparks, A., Castilla, N., Djurle, A., Allinne, C., Sharma, M., Rossi, V., Amorim, L., Bergamin, A., Yuen, J., Esker, P., McRoberts, N., Avelino, J., Duveiller, E., Koo, J., & Garrett, K. (2017). Crop health and its global impacts on the components of food security. *Food Security*, *9*(2), 311–327. Available from https://doi.org/10.1007/s12571-017-0659-1.

Sekar, M., Praveen Kumar, T. R., Selva Ganesh Kumar, M., Vaníčková, R., & Maroušek, J. (2021). Techno-economic review on short-term anthropogenic emissions of air pollutants and particulate matter. *Fuel*, *305*, 121544. Available from https://doi.org/10.1016/j.fuel.2021.121544.

Shahabi, H., Shirzadi, A., Ghaderi, K., Omidvar, E., Al-Ansari, N., Clague, J. J., Geertsema, M., Khosravi, K., Amini, A., Bahrami, S., & Rahmati, O. (2020). Flood detection and susceptibility mapping using sentinel-1 remote sensing data and a machine learning approach: Hybrid intelligence of bagging ensemble based on k-nearest neighbor classifier. *Remote Sensing, 12*(2), 266.

Shen, C. (2018). A transdisciplinary review of deep learning research and its relevance for water resources scientists. *Water Resources Research*, *54* (11), 8558–8593. Available from https://doi.org/10.1029/2018WR022643, Retrieved from, http://agupubs.onlinelibrary.wiley.com/hub/journal/10.1002/(ISSN)1944-7973/.

Singh, D., Dahiya, M., Kumar, R., & Nanda, C. (2021). Sensors and systems for air quality assessment monitoring and management: A review. *Journal of Environmental Management*, *289*, 112510. Available from https://doi.org/10.1016/j.jenvman.2021.112510.

Singh, G., Singh, S., Sethi, G., & Sood, V. (2022). Deep learning in the mapping of agricultural land use using Sentinel-2 satellite data. *Geographies*, *2*(4), 691–700. Available from https://doi.org/10.3390/geographies2040042.

Singh, R. K., Singh, P., Drews, M., Kumar, P., Singh, H., Gupta, A. K., Govil, H., Kaur, A., & Kumar, M. (2021). A machine learning-based classification of LANDSAT images to map land use and land cover of India. *Remote Sensing Applications: Society and Environment*, *24*, 100624. Available from https://doi.org/10.1016/j.rsase.2021.100624, Retrieved from, http://www.journals.elsevier.com/remote-sensing-applications-society-and-environment/.

Srivastava, C., Singh, S., & Singh, A.P. (2019). 3 26 2019/03/26 2018 International Conference on Computing, Power and Communication Technologies, GUCON 2018. Institute of Electrical and Electronics Engineers Inc. India Estimation of air pollution in Delhi using machine learning techniques, 304–309. 10.1109/GUCON.2018.8675022 9781538644911. http://ieeexplore.ieee.org/xpl/mostRecentIssue.jsp?punumber = 8671767.

Sun, W., Bocchi, P., & Davison, BD. (2020). Applications of artificial intelligence for disaster management. *Natural Hazards*, *103*, 2631–2689.

Tiwari, A., Gupta, R., & Chandra, R. (2021). Delhi air quality prediction using LSTM deep learning models with a focus on COVID-19 lockdown. arXiv, India arXiv. Retrieved from https://arxiv.org.

Vishwanath, T., Shirwaikar, R. D., Jaiswal, W. M., & Yashaswini, M. (2023). Social media data extraction for disaster management aid using deep learning techniques. *Remote Sensing Applications: Society and Environment*, *30*, 100961. Available from https://doi.org/10.1016/j.rsase.2023.100961, Retrieved from, http://www.journals.elsevier.com/remote-sensing-applications-society-and-environment/.

Xu, J. X., Ma, J., Tang, Y. N., Wu, W. X., Shao, J. H., Wu, W. B., … Guo, H. Q. (2020). Estimation of sugarcane yield using a machine learning approach based on uav-lidar data. *Remote Sensing*, *12*(17), 2823.

Yan, X., Zang, Z., Jiang, Y., Shi, W., Guo, Y., Li, D., … Husi, L. (2021). A Spatial-Temporal Interpretable Deep Learning Model for improving interpretability and predictive accuracy of satellite-based PM2. 5. *Environmental Pollution*, *273*, 116459.

Zaini, N. A., Ean, L. W., Ahmed, A. N., & Malek, M. A. (2022). A systematic literature review of deep learning neural network for time series air quality forecasting. *Environmental Science and Pollution Research*, 1–33.

Zhao, Y., Potgieter, A. B., Zhang, M., Wu, B., & Hammer, G. L. (2020). Predicting wheat yield at the field scale by combining high-resolution Sentinel-2 satellite imagery and crop modelling. *Remote Sensing*, *12*(6), 1024.

Chapter 7

The application of geographic information systems and remote sensing technologies in urban ecology

Mir Muhammad Nizamani[1], Muhammad Awais[2], Muhammad Qayyum[3], Maqsood Ahmed Khaskheli[1], Entaj Tarafder[1] and Yong Wang[1]

[1]*Department of Plant Pathology, College of Agriculture, Guizhou University, Guiyang, Guizhou, P.R. China,* [2]*Institute of Plant Sciences, Faculty of Natural Sciences, University of Sindh, Jamshoro, Sindh, Pakistan,* [3]*School of Economics and Statistics, Guangzhou University, Guangzhou, P.R. China*

7.1 Introduction

Urban ecology is the scientific study of the interactions between living organisms and their urban environment (Alberti et al., 2017). It focuses on understanding how urban landscapes, which are predominantly shaped by human activity, influence and are influenced by the natural ecosystems they encompass and transform (Tan et al., 2020). This field of study has become increasingly important in contemporary environmental management as more of the world's population shifts to urban areas, making it imperative to integrate ecological principles into urban planning and development (Pandit et al., 2017; Sodiq et al., 2019). This chapter highlights critical gaps in the literature, particularly the need for longitudinal studies, integration of socio-economic data, evaluation of urban policy interventions, scalability of findings, real-time monitoring applications, and effective community engagement using geographic information systems (GIS) and remote sensing in urban ecology (Alaimo et al., 2022; Reed et al., 2021).

The significance of urban ecology lies in its ability to provide insights that lead to more sustainable urban environments (Angelidou et al., 2018; Zhou et al., 2017). By studying patterns and processes such as species distribution, green infrastructure, and the flow of energy and materials through cities, urban ecologists can help address issues like air and water quality, heat islands, and biodiversity loss (Farrell et al., 2022; Perrelet et al., 2024). The knowledge gained from urban ecological research guides policymakers, urban planners, and community groups in creating cities that are not only habitable but are also resilient and adaptive to environmental changes (De Luca et al., 2021; Díaz et al., 2024). As cities continue to grow, the principles of urban ecology will play a crucial role in ensuring that urban development is balanced with environmental conservation, contributing to the overall health and well-being of urban residents (Barot et al., 2019; Li et al., 2017; Pineo, 2022).

GIS and remote sensing are pivotal tools in urban ecological studies, offering sophisticated techniques to capture, store, analyze, and visualize environmental data (Ouchra et al., 2022; Pei et al., 2021; Sunny, 2024). GIS provides the framework for mapping and analyzing spatial relationships in urban environments, enabling researchers to synthesize large amounts of diverse data sets—from land use patterns and water systems to vegetation cover and wildlife habitats (Badach et al., 2020; Nedd et al., 2021). This capacity for detailed spatial analysis helps in understanding the complex dynamics of urban ecosystems and assists in planning and management decisions that account for ecological variables (Swan et al., 2021; Zhu et al., 2019).

Remote sensing, on the other hand, serves as a critical source of data for GIS applications, particularly through the collection of imagery and other data from satellites or aerial sensors (Guimarães et al., 2020; Wang & Xie, 2018). This technology is indispensable for monitoring large-scale environmental changes over time, such as urban expansion, vegetation changes, or surface temperature variations (Li, Pei, et al., 2020; Ravanelli et al., 2018). By providing up-to-date and high-resolution images, remote sensing allows for the continuous observation of ecological phenomena that are difficult to measure on the ground, making it a valuable tool for long-term ecological studies and immediate assessment needs following disturbances or disasters (Huylenbroeck et al., 2020; Mahabir et al., 2018; Ouma, 2016).

Deep Learning for Earth Observation and Climate Monitoring. DOI: https://doi.org/10.1016/B978-0-443-24712-5.00008-7

Together, GIS and remote sensing create a powerful toolkit for urban ecologists, allowing for an integrative approach to studying urban environments (Lacerda et al., 2021; Neyns & Canters, 2022). They enable a multiscalar analysis, from local to global levels, essential for effective urban ecological planning and conservation strategies (Hui et al., 2023). These technologies not only enhance our understanding of urban ecosystems but also improve our ability to predict outcomes of urban planning decisions and to implement more sustainable solutions tailored to specific environmental contexts (Pfadenhauer et al., 2017; Sepasgozar et al., 2019).

The primary aim of this chapter is to elucidate the integration and applications of GIS and remote sensing technologies in urban ecology. It endeavors to illustrate how these technologies have become indispensable tools in understanding and managing urban environments with a focus on sustainable development. The scope of the chapter extends to detailed examinations of the roles that GIS and remote sensing play in various urban ecological assessments such as land use and cover change detection, biodiversity and habitat analysis, urban climate monitoring, and water management. It also delves into the latest advancements in technology such as light detection and ranging (LiDAR), hyperspectral imaging, and real-time GIS capabilities that enhance these applications. Furthermore, the chapter addresses the challenges associated with the adoption and implementation of these technologies, including data quality, resolution, and the requisite specialized skills. It discusses the potential for public participation GIS (PPGIS) to enhance public engagement and underscores the importance of integrating socio-economic data to enrich urban ecological evaluations. Through this discourse, the chapter aims to chart a forward path for policymakers, urban planners, and researchers, emphasizing the need for continuous innovation and education in deploying GIS and remote sensing for resilient and sustainable urban development.

7.2 Fundamentals of geographic information systems and remote sensing technologies

7.2.1 Geographic information systems explained

GIS serve as essential tools for efficiently managing, analyzing, and visualizing spatial data. These systems synergize hardware, software, and data to capture, manage, analyze, and display geographically referenced information (Ashkezari et al., 2018; Carrasco et al., 2022). GIS enables the examination of spatial relationships, patterns, and trends across a variety of data types and sources, offering invaluable insights to both researchers and practitioners (Sharma et al., 2018; Song et al., 2017). The initial phase in GIS involves data input, where both spatial (location-based) and nonspatial data are gathered or imported (Loidl et al., 2016). Spatial data, which pinpoint specific Earth locations, may come from sources such as GPS, satellite imagery, maps, and surveys (Burke et al., 2021), while nonspatial data, describing the characteristics of these spatial points, typically originate from censuses, field observations, and databases. These datasets provide essential context and detail to spatial layers (Lloyd et al., 2019; Wang et al., 2020). Following data collection, the information must be systematically organized, stored, and retrieved within the GIS, using various database management tools that handle both spatial and nonspatial data (Bokefode et al., 2016; Vaitis et al., 2019). Effective data management is crucial for ensuring easy data access and optimizing performance during spatial analysis operations (Gomes et al., 2020; Guha & Kumar, 2018; Sun & Scanlon, 2019). The robust spatial analysis capabilities of GIS allow for detailed examination of the relationships among different data layers (Shreenithi et al., 2024). For instance, overlaying a city's water network with population density maps can identify areas with potential water shortages (Noori et al., 2022; Saha & Paul, 2021). These analyses can range from simple queries to complex simulations and future spatial scenario modeling (Fisher & Koven, 2020). Another significant feature of GIS is its powerful visualization capabilities, which transform complex geographic data sets into easily comprehensible visual formats through maps and graphic displays, aiding analysis, interpretation, and communication (Roth, 2021; Singleton & Arribas-Bel, 2021; Wu & Biljecki, 2022). Moreover, GIS supports interactive visualizations that allow users to explore and manipulate data, offering various perspectives (Balla et al., 2020; Staniszewska et al., 2018). By integrating these functions, GIS provides a dynamic and versatile approach that is crucial for understanding and managing urban environments and beyond, proving itself an indispensable tool in urban ecological studies (Pickett et al., 2020; Zhu & Wu, 2021).

7.2.2 Remote sensing basics

Remote sensing technologies are essential in urban ecology for monitoring and analyzing the dynamic changes within urban environments. These technologies, which include satellite imagery, aerial photography, and unmanned aerial vehicles (UAVs), provide crucial data without physical contact with the subject area (Kabisch et al., 2019; Zhang et al., 2022).

Satellite imagery: This involves sensors onboard satellites that capture data across various spectral bands, offering broad coverage and the ability to consistently monitor large urban areas over time (Zhang et al., 2022). Satellite imagery is particularly valuable for tracking urban sprawl, land-use change, and environmental compliance, with high-resolution satellites providing detailed images for mapping urban infrastructure, green spaces, and water bodies (Neyns & Canters, 2022; Stoian et al., 2019; Surya et al., 2021; Tomar et al., 2023).

Aerial photography: Captured from aircraft, aerial photography provides higher resolution images than most satellite sensors. It is especially useful for detailed urban planning and infrastructure projects, offering a clear, bird's-eye view of the land surface (Alganci et al., 2020; Eilola et al., 2019; Olson & Anderson, 2021). These photographs are instrumental in creating base maps, assessing property boundaries, and planning land use by providing a straightforward and easily interpretable visual perspective (Foody et al., 2024).

UAVs: UAVs, or drones, are increasingly popular due to their versatility and cost-effectiveness in capturing high-resolution images. They can be deployed quickly to provide real-time data, essential for monitoring short-term events like urban flooding or wildlife movements (Bowes et al., 2022; Hao & Wang, 2021; Kamarulzaman et al., 2023). UAVs are particularly relevant for localized studies that require detailed spatial data, such as inspecting vegetative health in urban parks or the conditions of rooftops and other inaccessible areas (Bayomi & Fernandez, 2023; Chrysoulakis et al., 2021).

The application of these technologies in urban ecological settings is profound. They enable continuous monitoring and analysis of urban ecosystems, crucial for sustainable urban planning and management (Bibri, 2021; Simon, 2024; Yu et al., 2024). By providing detailed, repetitive, and up-to-date geographical data, remote sensing aids urban ecologists in assessing the impact of urbanization on natural resources, tracking changes in urban biodiversity, and implementing effective green infrastructure strategies (Li, Liu, et al., 2020; Liu et al., 2019; Trinder and Liu, 2020). This suite of technologies supports the overarching goals of enhancing urban resilience and preparing cities to face environmental challenges (Gallotti et al., 2021; Mehryar et al., 2022).

7.2.3 Integration of geographic information systems and remote sensing

The synergistic integration of GIS and remote sensing technologies has markedly enhanced the efficacy of urban ecological assessments, enabling more thorough and accurate investigations into the complexities of urban environments (Kamran & Yamamoto, 2023; Tang et al., 2021). This combination harnesses the spatial analysis strengths of GIS with the broad data acquisition capabilities of remote sensing, offering urban ecologists a powerful toolkit for exploring urban dynamics (Deng et al., 2019; Pei et al., 2021).

Remote sensing contributes a diverse range of data—including spectral, spatial, and temporal information about urban surfaces—which can be precisely georeferenced and integrated into GIS platforms (Ghamisi et al., 2019; Zhu, Qiu, et al., 2022). This integration enriches GIS with layers of current data sourced from satellite imagery, aerial photographs, and UAV footage, making it a dynamic tool for urban analysis (Kamarulzaman et al., 2023; Karagiannopoulou et al., 2022). For example, remote sensing is instrumental in updating land cover data, which GIS then uses to analyze changes over time, evaluate urban expansion, and model potential impacts on local ecosystems (Aboelnour & Engel, 2018; Shrestha et al., 2022).

GIS excels at manipulating and analyzing multiple layers of data. By incorporating remote sensing data, urban ecologists can overlay diverse information such as vegetation indices, surface temperatures, and hydrological data to craft comprehensive models of urban ecological conditions (Bai et al., 2022; Hall & Hossain, 1980; Safaei et al., 2023). This multilayered analysis is critical for identifying urban heat islands, mapping areas prone to flooding, or understanding the connectivity between urban green spaces (Ghalehteimouri et al., 2024; Razzaghi Asl & Pearsall, 2023).

The combination of these technologies significantly aids in informed decision-making within urban planning and environmental management. Integrating time-series satellite data into GIS, for instance, allows planners to observe historical transformations and anticipate future trends, which is pivotal for strategic resource allocation, including green infrastructure development and mitigation of urbanization impacts (Abd-Elhamid et al., 2023; Koehler & Kuenzer, 2020; Lourdes et al., 2022; Tran et al., 2020).

Moreover, the synergy between GIS and remote sensing supports dynamic monitoring of urban ecosystems. Urban managers can utilize these integrated tools not only to monitor the health of urban forests and track environmental restoration efforts but also to conduct real-time surveillance of environmental compliance (Bibri, 2023; Uysal et al., 2021). The ability to rapidly analyze and visually represent environmental data facilitates effective responses to ecological crises or opportunities as they emerge, enhancing both proactive and reactive urban environmental management (Bayulken et al., 2021; Lopez et al., 2019).

7.3 Applications in urban ecology

7.3.1 Land use and land cover change detection

Land use and land cover (LULC) change detection through GIS and remote sensing is pivotal for understanding the dynamics of urban environments, providing crucial insights into how land areas evolve over time in their usage and physical cover (Abebe et al., 2022; Arif et al., 2023). This process is particularly vital in urban areas, where rapid development drives significant transformations, making LULC change detection essential for informed urban planning, environmental management, and sustainability assessments (Karimi & Sultana, 2024).

Satellite imagery plays a key role in tracking urban expansion. Time-series analyses of Landsat images, for example, have been extensively used to monitor the growth of cities, revealing shifts from vegetative states to built-up zones and helping assess the rate of urban growth and its ecological impacts (Fu et al., 2019; Liu et al., 2021). Such studies have been crucial in documenting the urban sprawl in cities like Beijing, where decades of remote sensing data show substantial encroachment on agricultural and forest lands (Guo et al., 2023; Yang et al., 2024).

The conversion of permeable surfaces to impermeable ones in urban contexts significantly affects hydrological systems. GIS enables the overlay of historical and current remote sensing data to trace changes in surface runoff patterns and identify potential flood zones (Naeini et al., 2024; Tang et al., 2024). This capability was instrumental in places like Atlanta, USA, where increased storm water runoff due to rapid urbanization has necessitated revisions in water management strategies (Rezvani et al., 2023).

Additionally, remote sensing is crucial for monitoring urban vegetation cover, which plays a significant role in mitigating urban heat islands and supporting biodiversity (Cai et al., 2019; Silveira et al., 2024). High-resolution imagery from UAVs allows for detailed mapping of urban green spaces, providing data on their health and extent over time. For example, in Melbourne, Australia, UAV-based imagery and GIS were used to monitor tree canopy cover, supporting urban greening initiatives to improve livability and ecological balance (Timilsina et al., 2020).

LULC change detection also supports the enforcement of environmental regulations. In cities like San Francisco, GIS and remote sensing are utilized to ensure development compliance with environmental impact assessments and protect against encroachment on protected areas (Contreras-Alonso et al., 2020). These applications are crucial for maintaining sustainable urban growth boundaries and safeguarding sensitive habitats from unauthorized development (Caro-Borrero et al., 2021; Xia et al., 2020).

The integration of machine learning and artificial intelligence (AI) has significantly enhanced the precision and efficiency of LULC change detection. Automated classification algorithms process large volumes of remote sensing data swiftly, identifying changes with increased accuracy (Gu & Zeng, 2024; Kaur & Afaq, 2024). These advancements not only streamline the detection process but also enable near real-time monitoring of land-use changes, facilitating more proactive urban environment management (Bibri et al., 2024; Mbuh et al., 2019; Shaikh & Birajdar, 2024).

7.3.2 Biodiversity and habitat analysis

GIS and remote sensing technologies are crucial in enhancing our understanding and preservation of urban biodiversity, offering vital data on habitat distribution, species diversity, and ecological threats in urban landscapes (Hawken et al., 2021; Quamar et al., 2023). These tools serve as indispensable assets for biologists, ecologists, and urban planners committed to bolstering biodiversity within densely populated areas (Dobbs et al., 2018; Nitoslawski et al., 2019). Remote sensing provides expansive and detailed views of extensive and often inaccessible urban areas, facilitating the mapping of habitats and green spaces including parks, gardens, riparian zones, and green roofs (Chen et al., 2021; Herath et al., 2023). Utilizing different spectral bands, ecologists can distinguish between various vegetation types and surface materials, evaluate habitat quality, and monitor ecological changes over time (Wachendorf et al., 2018).

Innovations such as multispectral and hyperspectral imaging are crucial; they enable the identification of plant species and health, reflecting habitat richness and pinpointing areas needing conservation or restoration (Lassalle, 2021; Tang et al., 2022). GIS is instrumental in developing species distribution models (SDMs) that predict species locations based on environmental conditions and observed data, identifying potential wildlife corridors and areas under threat from urban development (Ma & You, 2022; Rodríguez et al., 2019). Such models are essential for urban planning, guiding the development of strategies to mitigate urban sprawl's impacts on wildlife habitats and promoting the creation of interconnected green spaces (Chang et al., 2023; Penteado, 2023).

Integrated into GIS platforms, remote sensing data facilitates detailed impact assessments for urban developments. Urban ecologists leverage these tools to simulate land-use changes' effects on biodiversity and devise mitigation strategies like establishing protected areas or designing infrastructure to minimize ecological disruptions (Coleman et al., 2020; Simkin et al., 2022). For example, LiDAR data used to assess canopy cover assists cities in managing and planning their urban forests, crucial for species dependent on tree cover for movement and survival (Pretzsch et al., 2023).

The advancement of UAV technology enhances real-time urban biodiversity monitoring. Drones, equipped with cameras and sensors, provide immediate habitat and species data, especially in hard-to-reach areas (Dronova et al., 2021; Kamarulzaman et al., 2023). This capability is vital for evaluating the success of biodiversity conservation initiatives, such as species reintroduction or the effectiveness of green infrastructure improvements like bio-swales and green roofs (Bo et al., 2018; Ventura et al., 2023).

Moreover, GIS and remote sensing facilitate public engagement in biodiversity conservation. Platforms like iNaturalist, which employ GIS to record and display wildlife sightings, not only involve the community in biodiversity monitoring but also enhance data collection, raising ecological awareness and encouraging community-driven conservation efforts (Marchante et al., 2024; McHale et al., 2018; Muhamad Khair et al., 2020). This democratization of data collection fosters a broader understanding of urban ecology and strengthens community-based approaches to sustaining biodiversity.

7.3.3 Urban climate and environmental monitoring

The deployment of GIS and remote sensing technologies has profoundly transformed urban climate and environmental monitoring, enhancing our capacity to understand and address issues such as urban heat islands, deteriorating air quality, and climatic impacts. These tools deliver essential data that bolster sustainable urban development and public health initiatives (Bikis & Wang, 2023; Diener & Mudu, 2021; Halder et al., 2022). A prime example is their application in studying urban heat islands—urban areas that register higher temperatures than their rural counterparts due to human activities and dense construction (Halder et al., 2021; Halder et al., 2022). Thermal infrared sensing, particularly, is instrumental in capturing urban surface temperatures, which, when analyzed via GIS, pinpoint heat hotspots, guiding urban planning efforts aimed at mitigating extreme temperatures (Mavrakou et al., 2018; Rezvani et al., 2023).

Cities like Los Angeles and New York have leveraged these data to strategically place green roofs and reflective pavements to counter urban heat island effects (AzariJafari et al., 2021; Gonzalez-Trevizo et al., 2021). Additionally, satellite and aerial imagery monitor air pollutants—including nitrogen dioxide, sulfur dioxide, and particulate matter—providing a temporal and spatial map that identifies pollution sources like busy roads and industrial zones (Chen et al., 2022; Song et al., 2019). This capability is vital for urban health studies, which depend on precise, geographically detailed data to link pollution levels with health outcomes (Brook et al., 2018; Southerland et al., 2021).

During the COVID-19 pandemic, remote sensing data revealed significant reductions in air pollution during lockdowns, offering insights into the link between urban activities and air quality (Elshorbany et al., 2021; Galán-Madruga, 2023; Zhang et al., 2020). Furthermore, remote sensing is crucial for monitoring other urban climatic factors like precipitation, wind speed, and humidity (Almeida et al., 2021; Zhou et al., 2018). These data are key in assessing risks and impacts of climatic events such as storms and floods (Clarke et al., 2021; Shen et al., 2019). GIS integrates these data with urban infrastructure layouts to create predictive models for disaster preparedness and response (Abid et al., 2021; Rezvani et al., 2023). Cities prone to flooding, for instance, utilize GIS and remote sensing to generate essential flood hazard maps that inform emergency planning and community alerts (Bilaşco et al., 2022; Tomar et al., 2024).

Beyond monitoring, GIS and remote sensing are pivotal in designing and managing sustainable urban environments. These technologies help craft urban layouts that incorporate green spaces, water features, and tree canopies to alleviate harsh climatic conditions (Štrbac et al., 2023; Zhu, Yang, et al., 2022). Urban planners use these data to simulate scenarios that balance development with climate resilience, ensuring that new or renovated developments improve environmental conditions (Mauree et al., 2019; Yang, Javanroodi, et al., 2021; Yang, Sanesi, et al., 2021).

The insights derived from remote sensing and GIS increasingly inform public policy and regulations (Shaikh & Birajdar, 2024; Sunny, 2024). Urban administrators employ these tools to ensure compliance with environmental standards and to refine policies based on solid empirical evidence (Bibri, 2022; Francesco et al., 2020). Remote sensing, for example, has been used to monitor the efficacy of policies aimed at mitigating urban heat island effects or enhancing air quality, creating a feedback loop that fine-tunes urban environmental policies (Mathew et al., 2024; Odongo, 2023; Shi et al., 2021).

7.3.4 Water management and quality assessment

The utilization of GIS and remote sensing technologies significantly advances the management of urban water resources, encompassing both surface and groundwater (Shaikh & Birajdar, 2024; Shinde et al., 2023). These tools enable dynamic, comprehensive assessment, monitoring, and management of water systems in densely populated regions, effectively addressing issues such as pollution, allocation, and conservation (Makanda et al., 2022; Nova, 2023). Particularly, remote sensing excels at mapping and monitoring urban water bodies like rivers, lakes, and reservoirs (Rahaman et al., 2023). Through satellite imagery, it is possible to detect changes in water levels, identify sedimentation patterns, and monitor the health of aquatic ecosystems (Duan et al., 2019; Mukonza & Chiang, 2022). Notably, satellite-detected changes in water color can signal algae blooms, often linked to nutrient-rich pollution (Mortula et al., 2020; Rodríguez-López et al., 2023). GIS integrates these data points with additional environmental and urban datasets to support effective management decisions. This integration facilitates planning green infrastructure to improve water quality or redesigning urban landscapes to promote natural water drainage and mitigate flood risks (Apostolaki, 2024; Jayasooriya et al., 2020; Le et al., 2022).

Groundwater in urban settings demands meticulous management to prevent contamination from surface pollutants and excessive extraction (Ahmad & Al-Ghouti, 2020; Ishtiaq et al., 2024). Remote sensing and GIS track land subsidence, which can indicate groundwater overextraction (Chatterjee et al., 2023; Hussain et al., 2022). Moreover, GIS is combined with groundwater simulation models to visualize groundwater flow, recharge areas, and potential contamination sources (Rossetto et al., 2018; Shawaqfah et al., 2021). In cities like Jakarta, where significant subsidence and groundwater depletion are issues, these tools are essential for strategic water resource management and regulation of groundwater extraction (Batubara et al., 2023; Josset et al., 2024; Sumadi, 2024).

Remote sensing also monitors water quality, assessing indicators such as turbidity, thermal pollution, and chlorophyll levels (Adjovu et al., 2023; Kumar et al., 2024). These indicators pinpoint pollution sources and trace their impacts over time. Coupled with GIS, the collected data support detailed spatial analysis, evaluating the efficacy of water treatment facilities and the impacts of urban runoff (Olds et al., 2018; Shariat et al., 2019). This capability has proven effective in urban centers like Singapore, where GIS and remote sensing are integral to ongoing water quality assessments to ensure the safety and sustainability of the water supply (Chan et al., 2018; Dandge & Patil, 2022; Feizizadeh et al., 2021).

In urban planning, managing flood risk is critical, and GIS and remote sensing are invaluable in providing insights (Rezvani et al., 2023). These technologies map flood-prone areas and model water flow during heavy rains, essential for developing early warning systems and urban planning that minimizes flood risks (Fernández-Nóvoa et al., 2024; Stavropoulos et al., 2020). For example, in hurricane-prone cities like New Orleans, GIS extensively simulates flood scenarios to enhance water diversion and containment strategies (Bukvic et al., 2023).

GIS and remote sensing underpin integrated water resource management by offering a unified view of water resources and associated infrastructure (Alamgir et al., 2024). This holistic approach aids in the sustainable management of urban water resources, ensuring that development and conservation efforts are well aligned (Santos et al., 2023; Tang & Adesina, 2022). Decision-makers utilize GIS to merge hydrological data with socio-economic datasets, improving policy-making and fostering community engagement in water conservation (Castro & Rifai, 2021; Saravani et al., 2024).

7.3.5 Disaster management and mitigation

Remote sensing and GIS play crucial roles in the effective management and mitigation of disasters in urban areas (Alipouri et al., 2024; Cao et al., 2023; Damaševičius et al., 2023). These technologies are pivotal for disaster preparedness, response, recovery, and long-term resilience building. Remote sensing facilitates the monitoring of environmental conditions, potentially averting natural and man-made disasters by providing real-time data for tracking phenomena such as wildfires and hurricanes (McKenzie et al., 2023; Munawar et al., 2022; Schmelz & Psuty, 2022; Tsatsaris et al., 2021). For example, satellite imagery is instrumental in detecting conditions conducive to wildfires and tracking hurricane paths. In conjunction, GIS enhances these capabilities by amalgamating diverse data layers—such as population density, land use, and historical disaster records—to create models that predict potential impacts and simulate response scenarios (Batista e Silva et al., 2020; Cheng et al., 2021). This integrated approach is invaluable for urban planners and emergency managers in devising critical strategies like evacuation routes and resource distribution plans, thereby reducing disaster impacts on urban populations (Wachtel et al., 2021; Yu et al., 2020).

During disasters, the rapid response facilitated by GIS and remote sensing is indispensable. Real-time remote sensing data are critical for ongoing event monitoring, such as floods or storm developments, enabling responders to adjust their strategies effectively (Lopez et al., 2020; Pérez-González et al., 2019; Santos et al., 2020). GIS supports these efforts by mapping impacted zones and integrating this with real-time resource and responder data (Fang et al., 2023; Hao & Wang, 2021). For instance, GIS's role in coordinating rescue operations and managing logistics was notably demonstrated during the response to Hurricane Katrina (Bingqing et al., 2020; Nagendra et al., 2022).

Postdisaster, remote sensing is key in swiftly assessing damages. High-resolution aerial and satellite imagery, analyzed pre- and postdisaster, offers precise visualizations of the affected areas, aiding in prioritizing and directing recovery efforts (Mazzanti et al., 2022; Rahman et al., 2021). GIS further supports recovery and reconstruction by providing a platform for planning and managing rebuilding efforts, simulating reconstruction scenarios, and conducting risk reduction analyses to enhance urban resilience against future disasters (Dube, 2020; Heinzlef et al., 2020; Rezvani et al., 2023; Sarkissian et al., 2022). Moreover, GIS facilitates community engagement in disaster preparedness through public access to risk information and active participation in planning processes, thereby enhancing local resilience (Chan et al., 2020; Haworth et al., 2018; Pascapurnama et al., 2018; Rezvani et al., 2023; Samarakkody et al., 2023; Shafapourtehrany et al., 2023).

7.4 Case studies

7.4.1 Successful urban ecological project using GIS and remote sensing

The integration of GIS and remote sensing has revolutionized urban ecological projects by providing a comprehensive framework for environmental restoration, ecological security evaluation, and sustainable management of urban ecosystems. For instance, a practical application where GIS and remote sensing were utilized to facilitate environmental restoration through 3D virtual reconstruction, allowing for preimplementation visualization of landscape impacts (Martínez-Graña & Valdés Rodríguez, 2016). Similarly, these technologies are used to evaluate urban ecological security dynamically, providing a spatial-temporal analysis that underscores changes in ecological conditions due to urban development (Du et al., 2013). Furthermore, the role of GIS and remote sensing in monitoring and managing urban biodiversity, showing how data from these technologies support evidence-based decision-making for urban ecosystem conservation (Oppong et al., 2023). These case studies collectively underscore the vital role of GIS and remote sensing in advancing urban ecological projects, ensuring that urban planning is both sustainable and ecologically compatible.

The adaptability and precision of GIS and remote sensing technologies are particularly valuable in assessing and responding to environmental challenges within urban settings. The hierarchical integration of these tools to handle diverse, yet complementary spatial data, enhancing analytical flexibility and inventory management in urban systems (Mesev, 1997). Additionally, the contribution of remote sensing to ecologically oriented urban planning, despite the noted challenges such as data repetition and compounded errors, highlighting the critical need for structured integration to truly benefit urban policy and planning (Wellmann et al., 2020).

In practical applications, the impact of remote sensing and GIS is evident in specific urban development projects. For example, a new remote sensing-based ecological index that successfully assessed urban ecological quality, revealing patterns of ecological deterioration and providing crucial insights for urban management (Hu & Xu, 2018). Similarly, urban sprawl and its effects, employing remote sensing and GIS to quantify urban form and its dynamics over time, offering a detailed understanding of the spatial and temporal variability of urban expansion (Jat et al., 2008).

Moreover, the integration of these technologies not only supports ecological assessments but also facilitates the planning and implementation of urban projects. For instance, the use of GIS and remote sensing in designing urban information systems demonstrated how high-resolution imagery and automated data extraction can enhance urban planning processes, providing a robust framework for managing the complexities of urban growth (Alkan & Bulut, 2010). This comprehensive approach ensures that urban planners have accurate, timely, and relevant data to make informed decisions that respect and enhance the urban ecosystem.

7.4.2 Discussion of challenges and solutions in integrating these technologies into urban ecological practices

The integration of GIS and remote sensing technologies into urban ecological practices presents numerous challenges that must be navigated to harness the full potential of these tools. Below, we discuss several key challenges identified in the process, along with proposed solutions that aim to address these obstacles effectively (Table 7.1).

TABLE 7.1 Challenges and solutions in integrating geographic information systems (GIS) and remote sensing in urban planning.

Category	Issue	Description	Impact	Solution	References
Data management	Complexity and errors	Data repetition, compounded errors.	Compromises data integrity.	Advanced management systems.	Mesev (1997), Himanen et al. (2019), Yalcin et al. (2022), Zhao and Zhang (2020)
Technological access	Technology disparity	Limited technology in developing regions.	Hinders GIS and remote sensing use.	Capacity building and training.	Noor et al. (2014), Rezvani et al. (2023), Awan et al. (2021), Viqueira et al. (2020)
System integration	Poor integration	New data poorly integrated with existing systems.	Underutilized potential.	Interdisciplinary collaboration.	Wellmann et al. (2020), Yu and Fang (2023), Deng et al. (2022), Simon (2024)
Policy development	Lack of supportive policies	Absence of frameworks supporting GIS integration.	Misalignment with urban goals.	Develop supportive policies.	Arimjaya and Dimyati (2022), Badach et al. (2020), Son et al. (2023)

7.5 Challenges and limitations

7.5.1 Data quality and availability

Data quality and availability are critical issues in the field of remote sensing and GIS, impacting everything from urban planning to ecological monitoring (Liu et al., 2016). The resolution, accuracy, and accessibility of remote sensing data determine how effectively these technologies can be applied to real-world problems (Musse et al., 2018). Here is an in-depth look at each of these issues (Table 7.2).

7.5.2 Technical and skill gaps

The effective utilization of GIS and remote sensing technologies requires a deep pool of specialized skills and knowledge, which can highlight significant technical and skill gaps, particularly in regions or organizations that lack robust educational and training infrastructures (Kerski, 2015). Addressing these gaps is vital for harnessing the full potential of these technologies in diverse applications like urban planning and environmental monitoring (de Klerk & Buchanan, 2017). The technical complexity of GIS and remote sensing is substantial as it encompasses various processes such as data collection, processing, and interpretation, each demanding a comprehensive understanding of the technology and the specific application area (Cao & Lam, 2023). For instance, data collection involves understanding the specifications of different remote sensing platforms, like satellites and drones, and their sensors, including optical, radar, and LiDAR, which is essential for gathering high-quality data suited to particular needs (Jafarbiglu & Pourreza, 2022). Furthermore, data processing requires significant preprocessing to rectify issues related to atmospheric, sensor, and angular distortions using sophisticated algorithms and software (Kumar et al., 2024; Tetarwal & Kumar, 2024). Lastly, data analysis and interpretation require advanced capabilities in image processing and geographic information analysis techniques such as classification, regression, and change detection, alongside a thorough comprehension of the ecological or urban processes under study to ensure accurate data interpretation (Li et al., 2024).

Skill gaps in GIS and remote sensing span several critical areas, each demanding specialized capability. Technical skills, such as data processing and proficiency in GIS software, are essential and typically obtained through dedicated education and training (Lindner et al., 2022). Additionally, professionals must possess robust analytical skills to effectively analyze and integrate spatial data, which require strong analytical thinking and problem-solving abilities (Sivarajah et al., 2017). Interdisciplinary knowledge is also crucial, particularly in applications like urban ecology, where understanding relevant domains such as urban planning and environmental science is necessary (Wellmann et al., 2020). Furthermore, expertise in specific software and programming skills for GIS and remote sensing, such as ArcGIS or Python, is increasingly important for conducting sophisticated analyses and automating processes (White & Powell, 2019). These varied skills collectively enhance a professional's ability to effectively utilize GIS and remote sensing technologies in their work.

TABLE 7.2 Key issues in remote sensing and geographic information systems: impacts on data quality and utility.

Category	Subcategory	Issue description	Impact	Reference
Resolution	Spatial	Size of the smallest detectable object by a sensor. High requirements in urban settings.	Needs more storage and processing.	Wellmann et al. (2020)
	Temporal	Frequency of data collection over the same area.	Lower frequency may miss rapid changes.	Pazúr et al. (2021)
	Spectral	Ability to distinguish fine wavelength intervals.	Higher resolution increases data complexity and processing needs.	Aasen et al. (2018)
Accuracy	Geometric	Precision in matching imagery location to real-world coordinates.	Errors can affect mapping and analysis significantly.	Deliry and Avdan (2021)
	Radiometric	Precision of electromagnetic spectrum measurements.	Affects interpretation of material properties and conditions.	Abdulraheem et al. (2023)
	Classification	Accuracy in categorizing land cover or other features.	Misclassification leads to incorrect decisions and analyses.	Ebrahimy et al. (2021)
Accessibility	Cost	Expense of high-resolution and high-frequency data.	Limits access for smaller entities. Some data behind paywalls.	Nichols et al. (2019)
	Availability	Variability in the availability of recent and relevant data across different regions.	Coverage gaps in some areas.	Sheffield et al. (2018)
	Usability	Requirement of specialized knowledge and software for data processing and analysis.	Technical barriers for organizations without necessary capabilities.	Paz and Pow-Sang (2016)
Data integration		Challenges in integrating data from various sources due to discrepancies in formats, resolutions, and coordinates.	Standardization efforts ongoing but not fully resolved.	Li et al. (2017)
Interoperability		Aligning datasets from different GIS platforms and remote sensing products.	Often requires extensive data manipulation or transformation for combined analysis.	Bill et al. (2022)

To effectively address skill gaps in GIS and remote sensing, a comprehensive approach involving multiple strategies is essential. Increasing access to specialized education through university courses, vocational training, online platforms, and workshops is fundamental in cultivating a knowledgeable workforce (Bauer et al., 2021). Furthermore, ongoing professional development is crucial; certifications, advanced courses, and seminars enable current professionals to keep pace with evolving technologies and methodologies (Webster-Wright, 2009). Additionally, interdisciplinary programs that blend GIS and remote sensing with domain-specific education can produce adept professionals capable of leveraging these technologies in various fields (Bowlick et al., 2017). Partnerships and collaborations between academia, industry, and government are also vital, as they facilitate knowledge exchange and offer practical experiences through internships and projects. Lastly, democratizing technology by enhancing the usability of GIS and remote sensing tools through open-source software and community platforms can significantly lower barriers to entry, broadening access to these critical technologies (Kumar et al., 2017).

7.5.3 Integration with other data sources

Integrating GIS and remote sensing data with socio-economic datasets is a powerful approach for holistic urban and environmental analysis (Hamud et al., 2019). However, this integration presents several challenges that can affect the accuracy and usability of the combined insights. Addressing these challenges requires careful consideration of data compatibility, scale, and privacy issues (Table 7.3).

7.6 Future directions

7.6.1 Technological advancements

Emerging technologies in GIS and remote sensing are rapidly evolving, offering new and improved ways to enhance urban ecological research (Wellmann et al., 2020). These advancements are not only refining existing methodologies but also opening new avenues for deeper insights and more efficient management of urban environments (Javed et al., 2022). Here are some of the key emerging technologies and their potential impacts:

LiDAR: LiDAR technology has transformed topographical mapping and vegetation analysis. Unlike traditional imaging, LiDAR can penetrate through canopy layers to provide accurate 3D data about ground surfaces and vegetation structure (Eitel et al., 2016). This is crucial for urban ecology in mapping and analyzing urban forestry, detailed flood

TABLE 7.3 Challenges and solutions in integrating geographic information systems, remote sensing, and socio-economic data.

Category	Challenge	Description	Solution	References
Data compatibility	Format issues	Difficulties in integrating raster-based remote sensing data with vector-based socio-economic data.	Standardization and conversion of data formats.	Radočaj et al. (2020), Mekuriaw et al. (2021), Giuliani et al. (2017), Blazquez and Domenech (2018)
	Metadata issues	Variations or absence of metadata can complicate data integration.	Ensuring compatible, comprehensive metadata.	Ulrich et al. (2022)
Scale and resolution	Spatial discrepancies	Misalignment in spatial resolution between data types.	Spatial aggregation or disaggregation to align datasets.	Avtar et al. (2020)
	Temporal discrepancies	Frequency of data collection varies significantly.	Adjusting analysis techniques to accommodate temporal mismatches.	Bennett and Smith (2017)
Privacy and ethics	Data sensitivity	Socio-economic data may include sensitive personal information.	Anonymization and aggregation of sensitive data.	Akindote et al. (2023), Majeed et al. (2022)
	Regulatory compliance	Legal requirements for data protection must be met.	Adherence to legal standards like GDPR or HIPAA.	Bakare et al. (2024)
Data quality and reliability	Bias and error	Socio-economic data may be subjective and prone to errors.	Validation and cross-verification of datasets.	Black et al. (2017), Greganti et al. (2021)
	Consistency and continuity	Need for consistent data collection methodologies over time.	Standardizing data collection procedures.	Tuthill et al. (2020)
Technical challenges	Expertise and infrastructure	Specialized skills and robust infrastructure are required for data integration.	Technical training and investment in technological infrastructure.	Asch et al. (2018), Blišťan et al. (2015), Rao et al. (2019)
Overcoming challenges	Organizational and policy support	Need for professional development and supportive policy frameworks.	Professional development initiatives, advanced tools, open data initiatives, and data sharing frameworks.	Shahin et al. (2017), Akingbemisilu (2024)

risk modeling due to terrain variations, and assessing biomass. Urban planners use LiDAR to create more accurate and detailed green infrastructure plans, enhancing ecological networks within cities (Li, Liu, et al., 2020).

Hyperspectral imaging: Hyperspectral sensors collect information across hundreds of narrow spectral bands, unlike the few broad bands used in conventional remote sensing. This technology provides detailed spectral signatures that can differentiate between species of plants, stages of plant health, and even subtle changes in water quality (Ravikanth et al., 2017). For urban ecology, hyperspectral imaging enables more nuanced studies of urban biodiversity and can help monitor pollution levels across large areas, improving interventions in urban water systems and green spaces (Zhu et al., 2019).

Real-time GIS: Real-time GIS integrates live data directly into GIS systems, allowing for dynamic mapping and analysis. This technology is particularly beneficial for monitoring urban environments where conditions change rapidly, such as traffic flows, air quality levels, and public space usage (Wang, Zhong, et al., 2019). Real-time GIS enables urban ecologists to study how urban life interacts with the environment in real-time, leading to better-informed decisions about urban planning and emergency responses (Rezvani et al., 2023).

AI and machine learning: AI and machine learning are playing increasingly significant roles in processing and interpreting the large datasets generated by remote sensing and GIS. These technologies are used to automate complex processes such as pattern recognition, change detection, and predictive modeling (Yang et al., 2022). In urban ecology, AI can predict urban growth patterns, simulate environmental impacts, and optimize resource management strategies, making ecological and conservation efforts more proactive and responsive (Shaikh & Birajdar, 2024).

UAV drones with advanced sensors: The use of UAVs equipped with advanced sensors continues to rise, offering a cost-effective method for collecting high-resolution data (Mohsan et al., 2022). Drones can be deployed to monitor hard-to-reach urban areas and provide frequent updates on various ecological parameters. They are particularly useful for detailed studies on urban wildlife, vegetation health monitoring, and management of micro-climate conditions in urban heat islands (Zhang & Zhu, 2023).

Integration with Internet of Things (IoT): The integration of GIS and remote sensing with IoT devices is a significant advancement (Safari Bazargani et al., 2021). Sensors placed throughout urban areas can continuously collect data on various environmental parameters, such as temperature, humidity, air pollutants, and noise levels. This sensor data can be visualized and analyzed within GIS platforms, providing a comprehensive tool for monitoring and managing urban ecological systems in real time (Li & Wang, 2022).

7.6.2 Policy and planning

The advancements in GIS and remote sensing technologies are set to have a profound impact on urban planning and policy-making, enabling more informed, efficient, and dynamic urban development strategies (Kuffer et al., 2021). These technologies offer detailed insights and data-driven approaches that can guide decision-makers in crafting policies that not only foster urban growth but also ensure environmental sustainability and resilience (Bell et al., 2023). Here is how these advancements could shape future urban planning initiatives:

Enhanced spatial planning: With high-resolution remote sensing and advanced GIS analytics, urban planners can access detailed maps and models of urban areas (Eilola et al., 2021). This allows for precise zoning, land-use planning, and infrastructure development that are aligned with environmental conservation goals. Planners can identify optimal locations for green spaces, such as parks and urban forests, which are essential for biodiversity, reducing urban heat island effects, and improving air quality (Semeraro et al., 2021). These technologies also make it easier to integrate natural features and ecosystems into urban layouts, promoting a balance between built environments and natural landscapes (Wang, Pan, et al., 2019).

Data-driven urban growth management: Future urban planning can utilize predictive analytics from GIS and remote sensing to forecast urban growth patterns and the consequent environmental impacts (Harb et al., 2020). By modeling different development scenarios, planners can visualize the outcomes of various approaches, choosing paths that minimize ecological disruption and maximize sustainability (Bibri, 2023). For example, tools like scenario analysis help in understanding the implications of expanding public transit networks versus increasing road infrastructure, allowing cities to plan growth in a way that prioritizes environmental and social benefits (Tehrani et al., 2024).

Real-time monitoring and adaptive management: Advancements in real-time GIS and integration with the IoT enable continuous monitoring of urban conditions such as traffic, air quality, and energy usage (Kaginalkar et al., 2021). This information can be used to implement adaptive urban management strategies that respond to current conditions rather than rely solely on static, long-term plans. For instance, dynamic traffic management systems, informed by real-time data, can reduce congestion and pollution levels, enhancing urban livability and sustainability (Zeng et al., 2024).

Climate resilience and disaster preparedness: The ability of remote sensing and GIS to monitor climate variables and model disaster scenarios greatly enhances urban resilience planning (Twumasi et al., 2019). Urban areas can be designed or modified to better withstand climatic events such as floods, heatwaves, and hurricanes. Detailed risk assessment maps can guide infrastructure investments, ensuring that vulnerable areas are prioritized and that developments are built to resist expected changes in climate conditions (Rezvani et al., 2023).

Public participation and engagement: Future urban planning can benefit from the democratization of GIS and remote sensing data, making these tools accessible to the public through interactive platforms and applications (Brovelli et al., 2016). This can enhance public participation in planning processes, allowing residents to contribute data, share their preferences, and have a more direct influence on decisions that affect their communities (Karagiannopoulou et al., 2022). For example, participatory GIS applications can involve community members in identifying areas in need of restoration or protection, fostering a more inclusive and community-oriented approach to urban development (Jiménez-Caldera et al., 2024).

Regulatory compliance and environmental monitoring: Emerging technologies enable more effective monitoring of environmental compliance and the impacts of urban policies (Glicksman et al., 2017). For instance, continuous remote sensing surveillance can detect unauthorized land-use changes or identify areas where pollution exceeds regulatory limits, prompting timely interventions. This capability ensures that policies are effectively implemented and adjusted as necessary to meet environmental and public health goals (Oguejiofor et al., 2023).

7.6.3 Public participation geographic information systems

PPGIS represents a transformative approach in urban planning and policy-making by incorporating public input directly into GIS (Kahila-Tani, 2015). This integration significantly enhances community engagement and democratizes data-driven decision-making processes. The potential for increased public engagement through accessible GIS applications is substantial, facilitating broader involvement in urban development and environmental management (Bastos et al., 2022).

Enhanced accessibility and inclusivity: PPGIS makes GIS technology accessible to the general public, breaking down the barriers traditionally associated with complex GIS tools. By providing user-friendly interfaces and simplifying the data presentation, PPGIS enables individuals without technical backgrounds to understand and contribute to spatial decision-making processes (Kantola & Tuulentie, 2020). This inclusivity ensures that a wider range of perspectives and local knowledge is incorporated into planning efforts, which can lead to more equitable and effective outcomes (Aranda et al., 2023).

Improved communication and transparency: PPGIS promotes transparency in planning processes by making the same information available to both decision-makers and the public. This open access to data helps build trust between communities and local governments (Kahila-Tani et al., 2019). Furthermore, PPGIS facilitates better communication by allowing the public to visualize the potential impacts of proposed projects or policies on their communities through maps and spatial analyses. This visual representation helps nonexperts grasp complex issues and engages them more effectively in discussions and decision-making (Aranda et al., 2023).

Empowerment and local knowledge integration: By enabling public contributions to GIS databases, PPGIS empowers community members to share their local knowledge and insights. Residents can identify and map important community resources, environmental hazards, or areas in need of development or conservation (Zaman et al., 2022). This locally sourced data can be invaluable, often revealing patterns and priorities that may not be visible through conventional data collection methods. For instance, in environmental conservation projects, local inputs can help identify critical habitats or pollution sources that are not documented in official data sets (Ghose & Welcenbach, 2018).

Facilitating collaborative problem solving: PPGIS can serve as a platform for collaborative problem-solving, where community members, stakeholders, and planners come together to address urban and environmental challenges (Robinson et al., 2017). Through interactive GIS applications, multiple stakeholders can propose, visualize, and modify solutions in real-time, fostering a collaborative approach to urban planning. This process not only enhances the quality of the solutions by incorporating diverse viewpoints but also increases public commitment to the resulting plans (Babelon et al., 2021).

Enhanced monitoring and feedback: Accessible GIS applications allow the public to participate in monitoring and providing feedback on ongoing projects and policies (Kar et al., 2016). For instance, community members can use mobile GIS applications to report issues, track progress, or suggest improvements for urban development projects. This ongoing engagement keeps projects aligned with community needs and allows for adaptive management based on real-time public feedback (Wilson et al., 2019).

Policy development and civic education: PPGIS is also a powerful tool for civic education, helping citizens understand the complexities of urban and environmental planning (Kantola & Tuulentie, 2020). By engaging with GIS applications, the public can learn about the challenges and considerations involved in planning decisions, fostering a more informed and active citizenry. Moreover, this engagement can lead to more informed policy development, as public understanding and support for initiatives increase (Jankowski et al., 2022).

7.7 Conclusion

The application of GIS and remote sensing technologies in urban ecology is not only transforming how we understand and manage urban environments but is also pivotal in shaping sustainable cities of the future. These technologies offer unparalleled abilities to collect, analyze, and visualize spatial and temporal data, enabling more precise and effective interventions in urban planning, environmental management, and policy formulation. From enhancing land use and land cover change detection to improving biodiversity and habitat analysis, and from monitoring urban climates to managing water resources, the benefits of GIS and remote sensing are vast. They provide crucial data that helps in the management of urban growth, the conservation of natural resources, and the mitigation of environmental impacts. Moreover, the integration of these technologies with emerging tools like LiDAR, hyperspectral imaging, and real-time GIS further amplifies their potential, opening new avenues for advanced urban ecological research. However, the effective implementation of GIS and remote sensing technologies also presents challenges, including issues related to data quality, resolution, privacy, and the need for specialized skills. Overcoming these challenges requires ongoing advancements in technology, better policies for data access and sharing, and enhanced training and education in GIS and remote sensing applications. Looking ahead, the continued evolution and integration of GIS and remote sensing with other technological innovations promise to enhance our ability to design more resilient and sustainable urban environments. As urban areas continue to grow, leveraging these tools will be crucial in ensuring that urban development is balanced with ecological and social sustainability. In summary, the strategic application of GIS and remote sensing is essential for the future of urban ecology, enabling cities to become more adaptable, sustainable, and livable for future generations.

References

Aasen, H., Honkavaara, E., Lucieer, A., & Zarco-Tejada, P. J. (2018). Quantitative remote sensing at ultra-high resolution with UAV spectroscopy: A review of sensor technology, measurement procedures, and data correctionworkflows. *Remote Sensing*, *10*(7). Available from https://doi.org/10.3390/rs10071091, https://res.mdpi.com.

Abd-Elhamid, H. F., Zeleňáková, M., Barańczuk, J., Gergelova, M. B., & Mahdy, M. (2023). Historical trend analysis and forecasting of shoreline change at the nile delta using RS data and GIS with the DSAS tool. *Remote Sensing*, *15*(7). Available from https://doi.org/10.3390/rs15071737, http://www.mdpi.com/journal/remotesensing.

Abdulraheem, M. I., Zhang, W., Li, S., Moshayedi, A. J., Farooque, A. A., & Hu, J. (2023). Advancement of remote sensing for soil measurements and applications: A comprehensive review. *Sustainability*, *15*(21), 15444. Available from https://doi.org/10.3390/su152115444.

Abebe, G., Getachew, D., & Ewunetu, A. (2022). Analysing land use/land cover changes and its dynamics using remote sensing and GIS in Gubalafito district, Northeastern Ethiopia. *SN Applied Sciences*, *4*(1), 30.

Abid, S. K., Sulaiman, N., Chan, S. W., Nazir, U., Abid, M., Han, H., Ariza-Montes, A., & Vega-Muñoz, A. (2021). Toward an integrated disaster management approach: How artificial intelligence can boost disaster management. *Sustainability*, *13*(22), 12560. Available from https://doi.org/10.3390/su132212560.

Aboelnour, M., & Engel, B. A. (2018). Application of remote sensing techniques and geographic information systems to analyze land surface temperature in response to land use/land cover change in greater Cairo Region, Egypt. *Journal of Geographic Information System*, *10*(01), 57–88. Available from https://doi.org/10.4236/jgis.2018.101003.

Adjovu, G. E., Stephen, H., James, D., & Ahmad, S. (2023). Overview of the application of remote sensing in effective monitoring of water quality parameters. *Remote Sensing*, *15*(7), 1938.

Ahmad, A. Y., & Al-Ghouti, M. A. (2020). Approaches to achieve sustainable use and management of groundwater resources in Qatar: A review. *Groundwater for Sustainable Development*, *11*. Available from https://doi.org/10.1016/j.gsd.2020.100367, http://www.journals.elsevier.com/groundwater-for-sustainable-development/.

Akindote, O. J., Adegbite, A. O., Dawodu, S. O., Omotosho, A., Anyanwu, A., & Maduka, C. P. (2023). Comparative review of big data analytics and GIS in healthcare decision-making. *World Journal of Advanced Research and Reviews*, *20*(3), 1293–1302. Available from https://doi.org/10.30574/wjarr.2023.20.3.2589.

Akingbemisilu, T. H. (2024). A critical evaluation of government role in spatial data infrastructures for healthcare decision-making. *Journal of Public Policy and Administration*, *8*(1), 14–23. Available from https://doi.org/10.11648/j.jppa.20240801.13.

Alaimo, L. S., Ivaldi, E., Landi, S., & Maggino, F. (2022). Measuring and evaluating socio-economic inequality in small areas: An application to the urban units of the Municipality of Genoa. *Socio-economic Planning Sciences, 83*. Available from https://doi.org/10.1016/j.seps.2021.101170, http://www.elsevier.com/inca/publications/store/3/1/7/.

Alamgir, M., Hashim, M., Uddin, M. N., Mahmud, M. R., Islam, A. R. M. T., & Shahid, S. (2024). Enhancing water governance and water resources management in Bangladesh. *International Journal of River Basin Management*. Available from https://doi.org/10.1080/15715124.2024.2356219, http://www.tandfonline.com/loi/trbm20.

Alberti, M., Marzluff, J., & Hunt, V. M. (2017). Urban driven phenotypic changes: empirical observations and theoretical implications for eco-evolutionary feedback. *Philosophical Transactions of the Royal Society B: Biological Sciences, 372*(1712), 20160029.

Alganci, U., Soydas, M., & Sertel, E. (2020). Comparative research on deep learning approaches for airplane detection from very high-resolution satellite images. *Remote Sensing, 12*(3), 458.

Alipouri, E., Nami, M. H., & Naderi, M. (2024). An overview of the application of remote sensing technologies in disaster management (with an emphasis on natural hazards). *Disaster Prevention and Management Knowledge, 13*(4), 490−507.

Alkan, M., & Bulut, G. (2010). GIS and remote-sensing-based urban-information system design and development: A case study for Kozlu, Zonguldak. *Scientific Research and Essays, 5*(19), 2889−2899. Available from http://www.academicjournals.org/sre/PDF/pdf2010/4Oct/Alkan%20and%20Bulut.pdf, Turkey.

Almeida, C. R. D., Teodoro, A. C., & Gonçalves, A. (2021). Study of the urban heat island (UHI) using remote sensing data/techniques: A systematic review. *Environments, 8*(10), 105.

Angelidou, M., Psaltoglou, A., Komninos, N., Kakderi, C., Tsarchopoulos, P., & Panori, A. (2018). Enhancing sustainable urban development through smart city applications. *Journal of Science and Technology Policy Management, 9*(2), 146−169. Available from https://doi.org/10.1108/JSTPM-05-2017-0016, http://www.emeraldinsight.com/loi/jstpm.

Apostolaki, S. (2024). *Nature-based solutions for sustainable stormwater management as means to increase resilience to climate change, promote circularity and improve city aesthetics* (pp. 315−331). Springer Science and Business Media LLC. Available from 10.1007/978-3-031-50725-0_18.

Aranda, N. R., De Waegemaeker, J., & Van de Weghe, N. (2023). The evolution of public participation GIS (PPGIS) barriers in spatial planning practice. *Applied Geography, 155*, 102940.

Arif, M., Sengupta, S., Mohinuddin, S. K., & Gupta, K. (2023). Dynamics of land use and land cover change in peri urban area of Burdwan city, India: A remote sensing and GIS based approach. *GeoJournal, 88*(4), 4189−4213. Available from https://doi.org/10.1007/s10708-023-10860-3.

Arimjaya, I. W. G. K., & Dimyati, M. (2022). Remote sensing and geographic information systems technics for spatial-based development planning and policy. *International Journal of Electrical and Computer Engineering, 12*(5), 5073. Available from https://doi.org/10.11591/ijece.v12i5.pp5073-5083.

Asch, M., Moore, T., Badia, R., Beck, M., Beckman, P., Bidot, T., Bodin, F., Cappello, F., Choudhary, A., de Supinski, B., Deelman, E., Dongarra, J., Dubey, A., Fox, G., Fu, H., Girona, S., Gropp, W., Heroux, M., Ishikawa, Y., . . . Zacharov, I. (2018). Big data and extreme-scale computing: Pathways to convergence-toward a shaping strategy for a future software and data ecosystem for scientific inquiry. *International Journal of High Performance Computing Applications, 32*(4), 435−479. Available from https://doi.org/10.1177/1094342018778123.

Ashkezari, A. D., Hosseinzadeh, N., Chebli, A., & Albadi, M. (2018). Development of an enterprise geographic information system (GIS) integrated with smart grid. *Sustainable Energy, Grids and Networks, 14*, 25−34. Available from https://doi.org/10.1016/j.segan.2018.02.001, http://www.journals.elsevier.com/sustainable-energy-grids-and-networks/.

Avtar, R., Komolafe, A. A., Kouser, A., Singh, D., Yunus, A. P., Dou, J., Kumar, P., Gupta, R. D., Johnson, B. A., Thu Minh, H. V., Aggarwal, A. K., & Kurniawan, T. A. (2020). Assessing sustainable development prospects through remote sensing: A review. *Remote Sensing Applications: Society and Environment, 20*. Available from https://doi.org/10.1016/j.rsase.2020.100402, http://www.journals.elsevier.com/remote-sensing-applications-society-and-environment/.

Awan, U., Arnold, M. G., & Gölgeci, I. (2021). Enhancing green product and process innovation: Towards an integrative framework of knowledge acquisition and environmental investment. *Business Strategy and the Environment, 30*(2), 1283−1295.

AzariJafari, H., Xu, X., Gregory, J., & Kirchain, R. (2021). Urban-scale evaluation of cool pavement impacts on the urban heat island effect and climate change. *Environmental Science and Technology, 55*(17), 11501−11510. Available from https://doi.org/10.1021/acs.est.1c00664, http://pubs.acs.org/journal/esthag.

Babelon, I., Pánek, J., Falco, E., Kleinhans, R., & Charlton, J. (2021). Between consultation and collaboration: Self-reported objectives for 25 web-based geoparticipation projects in urban planning. *ISPRS International Journal of Geo-Information, 10*(11), 783. Available from https://doi.org/10.3390/ijgi10110783.

Badach, J., Voordeckers, D., Nyka, L., & Van Acker, M. (2020). A framework for air quality management zones - useful GIS-based tool for urban planning: Case studies in Antwerp and Gdańsk. *Building and Environment, 174*, 106743. Available from https://doi.org/10.1016/j.buildenv.2020.106743.

Bai, H., Li, Z., Guo, H., Chen, H., & Luo, P. (2022). Urban green space planning based on remote sensing and geographic information systems. *Remote Sensing, 14*(17), 4213. Available from https://doi.org/10.3390/rs14174213.

Bakare, S. S., Adeniyi, A. O., Akpuokwe, C. U., & Eneh, N. E. (2024). Data privacy laws and compliance: A comparative review of the EU GDPR and USA regulations. *Computer Science and IT Research Journal, 5*(3), 528−543. Available from https://doi.org/10.51594/csitrj.v5i3.859.

Balla, D., Zichar, M., Tóth, R., Kiss, E., Karancsi, G., & Mester, T. (2020). Geovisualization techniques of spatial environmental data using different visualization tools. *Applied Sciences, 10*(19), 6701. Available from https://doi.org/10.3390/app10196701.

Barot, S., Abbadie, L., Auclerc, A., Barthélémy, C., Bérille, E., Billet, P., Clergeau, P., Consales, J. N., Deschamp-Cottin, M., David, A., Devigne, C., Dham, V., Dusza, Y., Gaillard, A., Gonzalez, E., Hédont, M., Labarraque, D., Le Bastard, A. M., Morel, J. L., ... Veyrières, M. (2019). Urban ecology, stakeholders and the future of ecology. *Science of the Total Environment*, *667*, 475–484. Available from https://doi.org/10.1016/j.scitotenv.2019.02.410, http://www.elsevier.com/locate/scitotenv.

Bastos, D., Fernández-Caballero, A., Pereira, A., & Rocha, N. P. (2022). Smart city applications to promote citizen participation in city management and governance: A systematic review. *Informatics*, *9*(4). Available from https://doi.org/10.3390/informatics9040089, http://www.mdpi.com/journal/informatics.

Batista e Silva, F., Freire, S., Schiavina, M., Rosina, K., Marín-Herrera, M. A., Ziemba, L., Craglia, M., Koomen, E., & Lavalle, C. (2020). Uncovering temporal changes in Europe's population density patterns using a data fusion approach. *Nature Communications*, *11*(1). Available from https://doi.org/10.1038/s41467-020-18344-5, http://www.nature.com/ncomms/index.html.

Batubara, B., Kooy, M., & Zwarteveen, M. (2023). Politicising land subsidence in Jakarta: How land subsidence is the outcome of uneven sociospatial and socionatural processes of capitalist urbanization. *Geoforum; Journal of Physical, Human, and Regional Geosciences*, *139*, 103689.

Bauer, T., Immitzer, M., Mansberger, R., Vuolo, F., Márkus, B., Wojtaszek, M. V., Földváry, L., Szablowska-Midor, A., Kozak, J., Oliveira, I., van Lieshout, A., Vekerdy, Z., Ninsawat, S., & Mozumder, C. (2021). The making of a joint e-learning platform for remote sensing education: Experiences and lessons learned. *Remote Sensing*, *13*(9), 1718. Available from https://doi.org/10.3390/rs13091718.

Bayomi, N., & Fernandez, J. E. (2023). Eyes in the sky: Drones applications in the built environment under climate change challenges. *Drones*, *7*(10). Available from https://doi.org/10.3390/drones7100637, http://www.mdpi.com/journal/drones.

Bayulken, B., Huisingh, D., & Fisher, P. M. (2021). How are nature based solutions helping in the greening of cities in the context of crises such as climate change and pandemics? A comprehensive review. *Journal of Cleaner Production*, *288*, 125569.

Bell, A., Klein, D., Rieser, J., Kraus, T., Thiel, M., & Dech, S. (2023). Scientific evidence from space—a review of spaceborne remote sensing applications at the science–policy interface. *Remote Sensing*, *15*(4).

Bennett, M. M., & Smith, L. C. (2017). Advances in using multitemporal night-time lights satellite imagery to detect, estimate, and monitor socioeconomic dynamics. *Remote Sensing of Environment*, *192*, 176–197. Available from https://doi.org/10.1016/j.rse.2017.01.005, http://www.elsevier.com/inca/publications/store/5/0/5/7/3/3.

Bibri, S. E. (2021). Data-driven smart sustainable cities of the future: An evidence synthesis approach to a comprehensive state-of-the-art literature review. *Sustainable Futures*, *3*.

Bibri, S. E. (2022). Eco-districts and data-driven smart eco-cities: Emerging approaches to strategic planning by design and spatial scaling and evaluation by technology. *Land Use Policy*, *113*, 105830. Available from https://doi.org/10.1016/j.landusepol.2021.105830.

Bibri, S. E. (2023). Data-driven smart eco-cities of the future: An empirically informed integrated model for strategic sustainable urban development. *World Futures*, *79*(7-8), 703–746. Available from https://doi.org/10.1080/02604027.2021.1969877, http://www.tandfonline.com/toc/gwof20/current.

Bibri, S. E., Krogstie, J., Kaboli, A., & Alahi, A. (2024). Smarter eco-cities and their leading-edge artificial intelligence of things solutions for environmental sustainability: A comprehensive systematic review. *Environmental Science and Ecotechnology*, *19*. Available from https://doi.org/10.1016/j.ese.2023.100330, http://www.journals.elsevier.com/environmental-science-and-ecotechnology.

Bikis, A., & Wang, Q. (2023). Urban air pollution and greenness in relation to public health. *Journal of Environmental and Public Health*, *2023*, 1–18. Available from https://doi.org/10.1155/2023/8516622.

Bilaşco, Ş., Hognogi, G. G., Roşca, S., Pop, A. M., Iuliu, V., Fodorean, I., Marian-Potra, A. C., & Sestras, P. (2022). Flash flood risk assessment and mitigation in digital-era governance using unmanned aerial vehicle and GIS spatial analyses case study: Small river basins. *Remote Sensing*, *14*(10). Available from https://doi.org/10.3390/rs14102481, https://www.mdpi.com/2072-4292/14/10/2481/pdf?version = 1653205173.

Bill, R., Blankenbach, J., Breunig, M., Haunert, J. H., Heipke, C., Herle, S., Maas, H. G., Mayer, H., Meng, L., Rottensteiner, F., Schiewe, J., Sester, M., Sörgel, U., & Werner, M. (2022). Geospatial information research: State of the art, case studies and future perspectives. *Journal of Photogrammetry, Remote Sensing and Geoinformation Science*, *90*(4), 349–389. Available from https://doi.org/10.1007/s41064-022-00217-9, https://rd.springer.com/journal/volumesAndIssues/41064.

Bingqing, L. U., Zhang, X., & Jin, W. E. N. (2020). Real world effectiveness of information and communication technologies in disaster relief: A systematic review. *Iranian Journal of Public Health*, *49*(10), 1813.

Black, N., Johnston, D. W., & Suziedelyte, A. (2017). Justification bias in self-reported disability: New evidence from panel data. *Journal of Health Economics*, *54*, 124–134.

Blazquez, D., & Domenech, J. (2018). Big Data sources and methods for social and economic analyses. *Technological Forecasting and Social Change*, *130*, 99–113. Available from https://doi.org/10.1016/j.techfore.2017.07.027.

Blištan, P., Kovanič, Ľ., & Kovaničová, M. (2015). The importance of geographic information systems education at universities in the process of building a european knowledge-based society. *Procedia-Social and Behavioral Sciences*, *191*, 2458–2462.

Bo, Y., Mesner, N., Drew, M. K., & Durfee, D. (2018). Integrated teaching and practice: green infrastructure planning and green roof performance in a semi-arid campus environment, USA. *Landscape Architecture Frontiers*, *6*(5), 44. Available from https://doi.org/10.15302/j-laf-20180504.

Bokefode, J. D., Bhise, A. S., Satarkar, P. A., & Modani, D. G. (2016). Developing a secure cloud storage system for storing IoT data by applying role based encryption. *Procedia Computer Science*, *89*(43), 50. Available from http://www.sciencedirect.com/science/journal/18770509, 10.1016/j.procs.2016.06.007.

Bowes, B. D., Wang, C., Ercan, M. B., Culver, T. B., Beling, P. A., & Goodall, J. L. (2022). Reinforcement learning-based real-time control of coastal urban stormwater systems to mitigate flooding and improve water quality. *Environmental Science: Water Research and Technology*, *8*(10), 2065–2086. Available from https://doi.org/10.1039/d1ew00582k, http://pubs.rsc.org/en/journals/journal/ew.

Bowlick, F. J., Goldberg, D. W., & Bednarz, S. W. (2017). Computer science and programming courses in geography departments in the United States. *The Professional Geographer*, *69*(1), 138–150.

Brook, J. R., Setton, E. M., Seed, E., Shooshtari, M., Doiron, D., Awadalla, P., Brauer, M., Hu, H., McGrail, K., Stieb, D., Subarrao, P., Demers, P., Manuel, D., McLaughlin, J., Carlsten, C., Azad, M., Atkinson, S., Burnett, R., Lou, W., ... Widener, M. (2018). The Canadian urban environmental health research consortium - a protocol for building a national environmental exposure data platform for integrated analyses of urban form and health. *BMC Public Health*, *18*(1). Available from https://doi.org/10.1186/s12889-017-5001-5, http://www.biomedcentral.com/bmcpublichealth.

Brovelli, M. A., Minghini, M., & Zamboni, G. (2016). Public participation in GIS via mobile applications. *ISPRS Journal of Photogrammetry and Remote Sensing*, *114*, 306–315.

Bukvic, A., Mitchell, A., Shao, Y., & Irish, J. L. (2023). Spatiotemporal implications of flooding on relocation risk in rural and urban coastal municipalities. *Land Use Policy*, *132*, 106754. Available from https://doi.org/10.1016/j.landusepol.2023.106754.

Burke, M., Driscoll, A., Lobell, D. B., & Ermon, S. (2021). Using satellite imagery to understand and promote sustainable development. *Science*, *371* (6535). Available from https://doi.org/10.1126/science.abe8628, https://science.sciencemag.org/content/371/6535/eabe8628.

Cai, Y., Chen, Y., & Tong, C. (2019). Spatiotemporal evolution of urban green space and its impact on the urban thermal environment based on remote sensing data: A case study of Fuzhou City, China. *Urban Forestry and Urban Greening*, *41*, 333–343.

Cao, C., & Lam, N. S. N. (2023). *Understanding the scale and resolution effects in remote sensing and GIS. Scale in remote sensing and GIS* (pp. 57–72). United States: CRC Press. Available from http://www.tandfebooks.com/doi/book/9781351417624, 10.1201/9780203740170-4.

Cao, Y., Xu, C., Aziz, N. M., & Kamaruzzaman, S. N. (2023). BIM–GIS integrated utilization in urban disaster management: The contributions, challenges, and future directions. *Remote Sensing*, *15*(5).

Caro-Borrero, A., Carmona-Jiménez, J., Rivera-Ramírez, K., & Bieber, K. (2021). The effects of urbanization on aquatic ecosystems in peri-urban protected areas of Mexico City: The contradictory discourse of conservation amid expansion of informal settlements. *Land Use Policy*, *102*, 105226. Available from https://doi.org/10.1016/j.landusepol.2020.105226.

Carrasco, C. A., Lombillo, I., Sánchez-Espeso, J. M., & Balbás, F. J. (2022). Quantitative and qualitative analysis on the integration of geographic information systems and building information modeling for the generation and management of 3D Models. *Buildings*, *12*(10). Available from https://doi.org/10.3390/buildings12101672, http://www.mdpi.com/journal/buildings.

Castro, C. V., & Rifai, H. S. (2021). Development and assessment of a web-based national spatial data infrastructure for nature-based solutions and their social, hydrological, ecological, and environmental co-benefits. *Sustainability*, *13*(19), 11018. Available from https://doi.org/10.3390/su131911018.

Chan, C. S., Nozu, K., & Cheung, T. O. L. (2020). Tourism and natural disaster management process: perception of tourism stakeholders in the case of Kumamoto earthquake in Japan. *Current Issues in Tourism*, *23*(15), 1864–1885.

Chan, F. K. S., Chuah, C. J., Ziegler, A. D., Dąbrowski, M., & Varis, O. (2018). Towards resilient flood risk management for Asian coastal cities: Lessons learned from Hong Kong and Singapore. *Journal of Cleaner Production*, *187*, 576–589. Available from https://doi.org/10.1016/j.jclepro.2018.03.217, https://www.journals.elsevier.com/journal-of-cleaner-production.

Chang, S., Su, K., Jiang, X., You, Y., Li, C., & Wang, L. (2023). Impacts and predictions of urban expansion on habitat connectivity networks: A multi-scenario simulation approach. *Forests*, *14*(11), 2187. Available from https://doi.org/10.3390/f14112187.

Chatterjee, R. S., Singha, S., Aggarwal, A., Sharma, V., Pranjal, P., Karunakalage, A., Jain, P. K., Nagar, A., Mitra, D. S., Kumar, D., Patel, N. R., & Chauhan, P. (2023). Reconnaissance to characterisation of land subsidence due to groundwater overdraft and oil extraction in and around Mehsana City, Gujarat, India by long-term hybrid differential interferometric SAR technique. *Journal of Hydrology*, *627*, 130441. Available from https://doi.org/10.1016/j.jhydrol.2023.130441.

Chen, Y., Gu, P., Schulte, N., Zhou, X., Mara, S., Croes, B. E., Herner, J. D., & Vijayan, A. (2022). A new mobile monitoring approach to characterize community-scale air pollution patterns and identify local high pollution zones. *Atmospheric Environment*, *272*. Available from https://doi.org/10.1016/j.atmosenv.2022.118936, http://www.elsevier.com/locate/atmosenv.

Cheng, L., Wang, L., Feng, R., & Yan, J. (2021). Remote sensing and social sensing data fusion for fine-resolution population mapping with a multimodel neural network. *IEEE Journal of Selected Topics in Applied Earth Observations and Remote Sensing*, *14*, 5973–5987. Available from https://doi.org/10.1109/JSTARS.2021.3086139, http://ieeexplore.ieee.org/xpl/RecentIssue.jsp?punumber = 4609443.

Chrysoulakis, N., Somarakis, G., Stagakis, S., Mitraka, Z., Wong, M.-S., & Ho, H.-C. (2021). Monitoring and evaluating nature-based solutions implementation in urban areas by means of earth observation. *Remote Sensing*, *13*(8), 1503. Available from https://doi.org/10.3390/rs13081503.

Clarke, B. J., Otto, F. E., & Jones, R. G. (2021). Inventories of extreme weather events and impacts: Implications for loss and damage from and adaptation to climate extremes. *Climate Risk Management*, *32*, 100285.

Coleman, N., Esmalian, A., & Mostafavi, A. (2020). Equitable resilience in infrastructure systems: Empirical assessment of disparities in hardship experiences of vulnerable populations during service disruptions. *Natural Hazards Review*, *21*(4), 04020034.

Contreras-Alonso, M. a R., Ezquerra-Canalejo, A., Pérez-Martín, E., Herrero-Tejedor, T. R., López-Cuervo Medina, S., & Ribeiro da Silva, C. (2020). Environmental assessment of obstacle limitation surfaces (OLS) in airports using geographic information technologies. *PLoS ONE*, *15*(2), e0229378. Available from https://doi.org/10.1371/journal.pone.0229378.

Damaševičius, R., Bacanin, N., & Misra, S. (2023). From sensors to safety: Internet of emergency services (IoES) for emergency response and disaster management. *Journal of Sensor and Actuator Networks*, *12*(3), 41.

Dandge, K. P., & Patil, S. S. (2022). Spatial distribution of ground water quality index using remote sensing and GIS techniques. *Applied Water Science*, *12*(1). Available from https://doi.org/10.1007/s13201-021-01546-7.

de Klerk, H. M., & Buchanan, G. (2017). Remote sensing training in African conservation. *Remote Sensing in Ecology and Conservation*, *3*(1), 7–20. Available from https://doi.org/10.1002/rse2.36, http://onlinelibrary.wiley.com/journal/10.1002/(ISSN)2056-3485.

Deliry, S. I., & Avdan, U. (2021). Accuracy of unmanned aerial systems photogrammetry and structure from motion in surveying and mapping: A review. *Journal of the Indian Society of Remote Sensing*, *49*(8), 1997–2017. Available from https://doi.org/10.1007/s12524-021-01366-x, http://www.springerlink.com/content/0255-660X.

Deng, X., Liu, P., Liu, X., Wang, R., Zhang, Y., He, J., & Yao, Y. (2019). Geospatial big data: New paradigm of remote sensing applications. *IEEE Journal of Selected Topics in Applied Earth Observations and Remote Sensing*, *12*(10), 3841–3851. Available from https://doi.org/10.1109/JSTARS.2019.2944952, http://ieeexplore.ieee.org/xpl/RecentIssue.jsp?punumber = 4609443.

Deng, Y., Xie, L., Xing, C., & Cai, L. (2022). Digital city landscape planning and design based on spatial information technology. *Neural Computing and Applications*, *34*(12), 9429–9440. Available from https://doi.org/10.1007/s00521-021-06377-w, https://www.springer.com/journal/521.

Diener, A., & Mudu, P. (2021). How can vegetation protect us from air pollution? A critical review on green spaces' mitigation abilities for air-borne particles from a public health perspective - with implications for urban planning. *Science of the Total Environment*, *796*, 148605. Available from https://doi.org/10.1016/j.scitotenv.2021.148605.

Dobbs, C., Hernández, A., de la Barrera, F., Miranda, M. D., & Paecke, S. R. (2018). Integrating urban biodiversity mapping, citizen science and technology. In A. Ossola, & J. Niemelä (Eds.), *Urban biodiversity: From research to practice* (pp. 236–247).

Dronova, I., Kislik, C., Dinh, Z., & Kelly, M. (2021). A review of unoccupied aerial vehicle use in wetland applications: Emerging opportunities in approach, technology, and data. *Drones*, *5*(2). Available from https://doi.org/10.3390/drones5020045, https://www.mdpi.com/2504-446X/5/2/45/pdf.

Du, P., Xia, J., Du, Q., Luo, Y., & Tan, K. (2013). Evaluation of the spatio-temporal pattern of urban ecological security using remote sensing and GIS. *International Journal of Remote Sensing*, *34*(3), 848–863. Available from https://doi.org/10.1080/01431161.2012.714503, https://www.tandfonline.com/loi/tres20.

Duan, H., Cao, Z., Shen, M., Liu, D., & Xiao, Q. (2019). Detection of illicit sand mining and the associated environmental effects in China's fourth largest freshwater lake using daytime and nighttime satellite images. *Science of the Total Environment*, *647*, 606–618. Available from https://doi.org/10.1016/j.scitotenv.2018.07.359, http://www.elsevier.com/locate/scitotenv.

Dube, E. (2020). The build-back-better concept as a disaster risk reduction strategy for positive reconstruction and sustainable development in Zimbabwe: A literature study. *International Journal of Disaster Risk Reduction*, *43*, 101401. Available from https://doi.org/10.1016/j.ijdrr.2019.101401.

Díaz, C. G., Zambrana-Vasquez, D., & Bartolomé, C. (2024). Building resilient cities: A comprehensive review of climate change adaptation indicators for urban design. *Energies*, *17*(8), 1959.

Ebrahimy, H., Mirbagheri, B., Matkan, A. A., & Azadbakht, M. (2021). Per-pixel land cover accuracy prediction: A random forest-based method with limited reference sample data. *ISPRS Journal of Photogrammetry and Remote Sensing*, *172*, 17–27. Available from https://doi.org/10.1016/j.isprsjprs.2020.11.024, http://www.elsevier.com/inca/publications/store/5/0/3/3/4/0.

Eilola, S., Käyhkö, N., & Fagerholm, N. (2021). Lessons learned from participatory land use planning with high-resolution remote sensing images in Tanzania: Practitioners' and participants' perspectives. *Land Use Policy*, *109*, 105649.

Eilola, S., Käyhkö, N., Ferdinands, A., & Fagerholm, N. (2019). A bird's eye view of my village—developing participatory geospatial methodology for local level land use planning in the Southern Highlands of Tanzania. *Landscape and Urban Planning*, *190*, 103596.

Eitel, J. U. H., Höfle, B., Vierling, L. A., Abellán, A., Asner, G. P., Deems, J. S., Glennie, C. L., Joerg, P. C., LeWinter, A. L., Magney, T. S., Mandlburger, G., Morton, D. C., Müller, J., & Vierling, K. T. (2016). Beyond 3-D: The new spectrum of lidar applications for earth and ecological sciences. *Remote Sensing of Environment*, *186*, 372–392. Available from https://doi.org/10.1016/j.rse.2016.08.018, http://www.elsevier.com/inca/publications/store/5/0/5/7/3/3.

Elshorbany, Y. F., Kapper, H. C., Ziemke, J. R., & Parr, S. A. (2021). The status of air quality in the united states during the covid-19 pandemic: A remote sensing perspective. *Remote Sensing*, *13*(3), 1–23. Available from https://doi.org/10.3390/rs13030369, https://www.mdpi.com/2072-4292/13/3/369/pdf.

Fang, Z., Yue, P., Zhang, M., Xie, J., Wu, D., & Jiang, L. (2023). A service-oriented collaborative approach to disaster decision support by integrating geospatial resources and task chain. *International Journal of Applied Earth Observation and Geoinformation*, *117*, 103217. Available from https://doi.org/10.1016/j.jag.2023.103217.

Farrell, C., Livesley, S. J., Arndt, S. K., Beaumont, L., Burley, H., Ellsworth, D., Esperon-Rodriguez, M., Fletcher, T. D., Gallagher, R., Ossola, A., Power, S. A., Marchin, R., Rayner, J. P., Rymer, P. D., Staas, L., Szota, C., Williams, N. S. G., & Leishman, M. (2022). Can we integrate ecological approaches to improve plant selection for green infrastructure? *Urban Forestry and Urban Greening*, *76*, 127732. Available from https://doi.org/10.1016/j.ufug.2022.127732.

Feizizadeh, B., Ronagh, Z., Pourmoradian, S., Gheshlaghi, H. A., Lakes, T., & Blaschke, T. (2021). An efficient GIS-based approach for sustainability assessment of urban drinking water consumption patterns: A study in Tabriz city, Iran. *Sustainable Cities and Society*, *64*. Available from https://doi.org/10.1016/j.scs.2020.102584, http://www.elsevier.com/wps/find/journaldescription.cws_home/724360/description#description.

Fernández-Nóvoa, D., González-Cao, J., & García-Feal, O. (2024). Enhancing flood risk management: A comprehensive review on flood early warning systems with emphasis on numerical modeling. *Water*, *16*(10), 1408.

Fisher, R. A., & Koven, C. D. (2020). Perspectives on the future of land surface models and the challenges of representing complex terrestrial systems. *Journal of Advances in Modeling Earth Systems*, *12*(4). Available from https://doi.org/10.1029/2018MS001453, http://agupubs.onlinelibrary.wiley.com/agu/journal/10.1002/(ISSN)1942-2466/.

Foody, G., Long, G., Schultz, M., & Olteanu-Raimond, A. M. (2024). Assuring the quality of VGI on land use and land cover: Experiences and learnings from the LandSense project. *Geo-Spatial Information Science*, *27*(1), 16–37. Available from https://doi.org/10.1080/10095020.2022.2100285, http://www.tandfonline.com/loi/tgsi20.

Francesco, F. D., Leopold, L., & Tosun, J. (2020). Distinguishing policy surveillance from policy tracking: Transnational municipal networks in climate and energy governance. *Journal of Environmental Policy and Planning*, *22*(6), 857–869.

Fu, Y., Li, J., Weng, Q., Zheng, Q., Li, L., Dai, S., & Guo, B. (2019). Characterizing the spatial pattern of annual urban growth by using time series Landsat imagery. *Science of the Total Environment*, *666*, 274–284. Available from https://doi.org/10.1016/j.scitotenv.2019.02.178, http://www.elsevier.com/locate/scitotenv.

Gallotti, R., Sacco, P., & De Domenico, M. (2021). Complex urban systems: challenges and integrated solutions for the sustainability and resilience of cities. *Complexity*, *2021*(1), 1782354.

Galán-Madruga, D. (2023). Urban air quality changes resulting from the lockdown period due to the COVID-19 pandemic. *International Journal of Environmental Science and Technology*, *20*(7), 7083–7098. Available from https://doi.org/10.1007/s13762-022-04464-6.

Ghalehteimouri, K. J., Ros, F. C., & Rambat, S. (2024). Flood risk assessment through rapid urbanization LULC change with destruction of urban green infrastructures based on NASA Landsat time series data: A case of study Kuala Lumpur between 1990–2021. *Ecological Frontiers*, *44*(2), 289–306.

Ghamisi, P., Rasti, B., Yokoya, N., Wang, Q., Hofle, B., Bruzzone, L., Bovolo, F., Chi, M., Anders, K., Gloaguen, R., Atkinson, P. M., & Benediktsson, J. A. (2019). Multisource and multitemporal data fusion in remote sensing: A comprehensive review of the state of the art. *IEEE Geoscience and Remote Sensing Magazine*, *7*(1), 6–39. Available from https://doi.org/10.1109/MGRS.2018.2890023, http://ieeexplore.ieee.org/xpl/RecentIssue.jsp?punumber = 6245518.

Ghose, R., & Welcenbach, T. (2018). "Power to the people": Contesting urban poverty and power inequities through open GIS. *Canadian Geographer*, *62*(1), 67–80. Available from https://doi.org/10.1111/cag.12442.

Giuliani, G., Nativi, S., Obregon, A., Beniston, M., & Lehmann, A. (2017). Spatially enabling the global framework for climate services: Reviewing geospatial solutions to efficiently share and integrate climate data & information. *Climate Services*, *8*, 44–58. Available from https://doi.org/10.1016/j.cliser.2017.08.003.

Glicksman, R. L., Markell, D. L., & Monteleoni, C. (2017). Technological innovation, data analytics, and environmental enforcement. *Ecology LQ*, *44*, 41.

Gomes, V. C., Queiroz, G. R., & Ferreira, K. R. (2020). An overview of platforms for big earth observation data management and analysis. *Remote Sensing*, *12*(8), 1253.

Gonzalez-Trevizo, M. E., Martinez-Torres, K. E., Armendariz-Lopez, J. F., Santamouris, M., Bojorquez-Morales, G., & Luna-Leon, A. (2021). Research trends on environmental, energy and vulnerability impacts of Urban Heat Islands: An overview. *Energy and Buildings*, *246*, 111051. Available from https://doi.org/10.1016/j.enbuild.2021.111051.

Greganti, C., Demarie, T. F., Ringbauer, M., Jones, J. A., Saggio, V., Calafell, I. A., Rozema, L. A., Erhard, A., Meth, M., Postler, L., Stricker, R., Schindler, P., Blatt, R., Monz, T., Walther, P., & Fitzsimons, J. F. (2021). Cross-verification of independent quantum devices. *Physical Review X*, *11*(3). Available from https://doi.org/10.1103/PhysRevX.11.031049, https://journals.aps.org/prx/abstract/10.1103/PhysRevX.11.031049.

Gu, Z., & Zeng, M. (2024). The use of artificial intelligence and satellite remote sensing in land cover change detection: Review and perspectives. *Sustainability*, *16*(1), 274. Available from https://doi.org/10.3390/su16010274.

Guha, S., & Kumar, S. (2018). Emergence of big data research in operations management, information systems, and healthcare: Past contributions and future roadmap. *Production and Operations Management*, *27*(9), 1724–1735. Available from https://doi.org/10.1111/poms.12833, http://onlinelibrary.wiley.com/journal/10.1111/(ISSN)1937-5956.

Guimarães, N., Pádua, L., Marques, P., Silva, N., Peres, E., & Sousa, J. J. (2020). Forestry remote sensing from unmanned aerial vehicles: A review focusing on the data, processing and potentialities. *Remote Sensing*, *12*(6). Available from https://doi.org/10.3390/rs12061046, https://res.mdpi.com/d_attachment/remotesensing/remotesensing-12-01046/article_deploy/remotesensing-12-01046.pdf.

Guo, Y., Zhao, S., Zhao, X., Wang, H., & Shi, W. (2023). Evaluation of the spatiotemporal change of ecological quality under the context of urban expansion—a case study of typical urban agglomerations in China. *Remote Sensing*, *16*(1), 45.

Halder, B., Bandyopadhyay, J., & Banik, P. (2021). Monitoring the effect of urban development on urban heat island based on remote sensing and geo-spatial approach in Kolkata and adjacent areas, India. *Sustainable Cities and Society*, *74*, 103186.

Halder, B., Bandyopadhyay, J., Khedher, K. M., Fai, C. M., Tangang, F., & Yaseen, Z. M. (2022). Delineation of urban expansion influences urban heat islands and natural environment using remote sensing and GIS-based in industrial area. *Environmental Science and Pollution Research*, *29*(48), 73147–73170. Available from https://doi.org/10.1007/s11356-022-20821-x, https://link.springer.com/journal/11356.

Hall, J., & Hossain, A. A. (1980). Mapping urbanization and evaluating its possible impacts on stream water quality in Chattanooga, Tennessee, using GIS and remote sensing. *Sustainability*, *12*(5).

Hamud, A. M., Prince, H. M., & Shafri, H. Z. (2019). Landuse/landcover mapping and monitoring using remote sensing and GIS with environmental integration. *IOP Conference Series: Earth and Environmental Science*, *357*(1), 012038.

Hao, H., & Wang, Y. (2021). Assessing disaster impact in real time: Data-driven system integrating humans, hazards, and the built environment. *Journal of Computing in Civil Engineering*, *35*(5). Available from https://doi.org/10.1061/(ASCE)CP.1943-5487.0000970, http://ascelibrary.org/cpo/resource/1/jccee5.

Harb, M., Garschagen, M., Cotti, D., Krätzschmar, E., Baccouche, H., Khaled, K. B., Bellert, F., Chebil, B., Fredj, A. B., Ayed, S., Shekhar, H., & Hagenlocher, M. (2020). Integrating data-driven and participatory modeling to simulate future urban growth scenarios: Findings from Monastir, Tunisia. *Urban Science*, *4*(1), 10. Available from https://doi.org/10.3390/urbansci4010010.

Hawken, S., Rahmat, H., Sepasgozar, S. M. E., & Zhang, K. (2021). The SDGs, ecosystem services and cities: A network analysis of current research innovation for implementing urban sustainability. *Sustainability*, *13*(24), 14057. Available from https://doi.org/10.3390/su132414057.

Haworth, B. T., Bruce, E., Whittaker, J., & Read, R. (2018). The good, the bad, and the uncertain: Contributions of volunteered geographic information to community disaster resilience. *Frontiers in Earth Science*, *6*. Available from https://doi.org/10.3389/feart.2018.00183, https://www.frontiersin.org/articles/10.3389/feart.2018.00183/pdf.

Heinzlef, C., Becue, V., & Serre, D. (2020). A spatial decision support system for enhancing resilience to floods: Bridging resilience modelling and geovisualization techniques. *Natural Hazards and Earth System Sciences*, *20*(4), 1049–1068.

Herath, H. M. M. S. D., Fujino, T., & Senavirathna, M. D. H. J. (2023). A review of emerging scientific discussions on green infrastructure (GI)-prospects towards effective use of urban flood plains. *Sustainability*, *15*(2), 1227.

Himanen, L., Geurts, A., Foster, A. S., & Rinke, P. (2019). Data-driven materials science: Status, challenges, and perspectives. *Advanced Science*, *6* (21). Available from https://doi.org/10.1002/advs.201900808, http://onlinelibrary.wiley.com/journal/10.1002/(ISSN)2198-3844.

Hu, X., & Xu, H. (2018). A new remote sensing index for assessing the spatial heterogeneity in urban ecological quality: A case from Fuzhou City, China. *Ecological Indicators*, *89*, 11–21. Available from https://doi.org/10.1016/j.ecolind.2018.02.006.

Hui, R. H., Abimbola, O., & Walter, S. (2023). Galen, Geodesign for multi-scalar consensus: Lessons from flood adaptation pathways planning. *Landscape Research Record*, *11*.

Hussain, M. A., Chen, Z., Shoaib, M., Shah, S. U., Khan, J., & Ying, Z. (2022). Sentinel-1A for monitoring land subsidence of coastal city of Pakistan using Persistent Scatterers In-SAR technique. *Nature Research, China Scientific Reports*, *12*(1). Available from https://doi.org/10.1038/s41598-022-09359-7, http://www.nature.com/srep/index.html.

Huylenbroeck, L., Laslier, M., Dufour, S., Georges, B., Lejeune, P., & Michez, A. (2020). Using remote sensing to characterize riparian vegetation: A review of available tools and perspectives for managers. *Journal of Environmental Management*, *267*, 110652. Available from https://doi.org/10.1016/j.jenvman.2020.110652.

Ishtiaq, M., Khan, M. J., Khan, S. A., Ghani, J., Ullah, Z., Nawab, J., Alrefaei, A. F., Almutairi, M. H., & Alharbi, S. N. (2024). Potentially harmful elements and health risk assessment in groundwater of urban industrial areas. *Frontiers in Environmental Science*, *12*. Available from https://doi.org/10.3389/fenvs.2024.1332965, journal.frontiersin.org/journal/environmental-science.

Jafarbiglu, H., & Pourreza, A. (2022). A comprehensive review of remote sensing platforms, sensors, and applications in nut crops. *Computers and Electronics in Agriculture*, *197*, 106844. Available from https://doi.org/10.1016/j.compag.2022.106844.

Jankowski, P., Forss, K., Czepkiewicz, M., Saarikoski, H., & Kahila, M. (2022). Assessing impacts of PPGIS on urban land use planning: Evidence from Finland and Poland. *European Planning Studies*, *30*(8), 1529–1548. Available from https://doi.org/10.1080/09654313.2021.1882393.

Jat, M. K., Garg, P. K., & Khare, D. (2008). Monitoring and modelling of urban sprawl using remote sensing and GIS techniques. *International Journal of Applied Earth Observation and Geoinformation*, *10*(1), 26–43.

Javed, A. R., Shahzad, F., Rehman, S. ur, Zikria, Y. B., Razzak, I., Jalil, Z., & Xu, G. (2022). Future smart cities: Requirements, emerging technologies, applications, challenges, and future aspects. *Cities*, *129*, 103794. Available from https://doi.org/10.1016/j.cities.2022.103794.

Jayasooriya, V. M., Ng, A. W. M., Muthukumaran, S., & Perera, C. B. J. (2020). Optimization of green infrastructure practices in industrial areas for runoff management: A review on issues, challenges and opportunities. *Water*, *12*(4). Available from https://doi.org/10.3390/W12041024, http://www.mdpi.com/2073-4441/12/4/1024.

Jiménez-Caldera, J., Durango-Severiche, G. Y., Pérez-Arévalo, R., Serrano-Montes, J. L., Rodrigo-Comino, J., & Caballero-Calvo, A. (2024). Methodological proposal for the inclusion of citizen participation in the management and planning of urban public spaces. *Cities*, *150*. Available from https://doi.org/10.1016/j.cities.2024.105008, https://www.sciencedirect.com/science/journal/02642751.

Josset, L., Lall, U., Prakash, D., & Dinar, A. (2024). 13.

Kabisch, N., Selsam, P., Kirsten, T., Lausch, A., & Bumberger, J. (2019). A multi-sensor and multi-temporal remote sensing approach to detect land cover change dynamics in heterogeneous urban landscapes. *Ecological Indicators*, *99*, 273–282. Available from https://doi.org/10.1016/j.ecolind.2018.12.033, http://www.elsevier.com/locate/ecolind.

Kaginalkar, A., Kumar, S., Gargava, P., & Niyogi, D. (2021). Review of urban computing in air quality management as smart city service: An integrated IoT, AI, and cloud technology perspective. *Urban Climate*, *39*, 100972. Available from https://doi.org/10.1016/j.uclim.2021.100972.

Kahila-Tani, M. (2015). Reshaping the planning process using local experiences: Utilising PPGIS in participatory urban planning.

Kahila-Tani, M., Kytta, M., & Geertman, S. (2019). Does mapping improve public participation? Exploring the pros and cons of using public participation GIS in urban planning practices. *Landscape and Urban Planning*, *186*, 45–55.

Kamarulzaman., Jaafar., Said., Saad, S. N. M., & Mohan. (2023). UAV implementations in urban planning and related sectors of rapidly developing nations: A review and future perspectives for Malaysia. *Remote Sensing*, *15*(11).

Kamran, M., & Yamamoto, K. (2023). Evolution and use of remote sensing in ecological vulnerability assessment: A review. *Ecological Indicators*, *148*, 110099. Available from https://doi.org/10.1016/j.ecolind.2023.110099.

Kantola, S., & Tuulentie, S. (2020). Participation in a large Arctic city – the possibilities of PPGIS for improving interaction. *Polar Geography*, *43* (4), 295–312. Available from https://doi.org/10.1080/1088937x.2020.1767709.

Kar, B., Sieber, R., Haklay, M., & Ghose, R. (2016). Public participation GIS and participatory GIS in the era of GeoWeb. *The Cartographic Journal*, *53*(4), 296–299. Available from https://doi.org/10.1080/00087041.2016.1256963.

Karagiannopoulou, A., Tsertou, A., Tsimiklis, G., & Amditis, A. (2022). Data fusion in earth observation and the role of citizen as a sensor: A scoping review of applications, methods and future trends. *Remote Sensing*, *14*(5), 1263. Available from https://doi.org/10.3390/rs14051263.

Karimi, F., & Sultana, S. (2024). Urban expansion prediction and land use/land cover change modeling for sustainable urban development. *Sustainability*, *16*(6), 2285. Available from https://doi.org/10.3390/su16062285.

Kaur, G., & Afaq, Y. (2024). Developments in deep learning for change detection in remote sensing: A review. *Transactions in GIS*, *28*(2), 223–257. Available from https://doi.org/10.1111/tgis.13133, http://onlinelibrary.wiley.com/journal/10.1111/(ISSN)1467-9671.

Kerski, J. J. (2015). *Opportunities and challenges in using geospatial technologies for education. Geospatial technologies and geography education in a changing world: Geospatial practices and lessons learned* (pp. 183–194). United States: Springer Japan. Available from http://www.springer.com/in/book/9784431555186, 10.1007/978-4-431-55519-3_15.

Koehler, J., & Kuenzer, C. (2020). Forecasting spatio-temporal dynamics on the land surface using earth observation data—a review. *Remote Sensing*, *12*(21), 3513. Available from https://doi.org/10.3390/rs12213513.

Kuffer, M., Pfeffer, K., & Persello, C. (2021). Special issue "remote-sensing-based urban planning indicators". *Remote Sensing*, *13*(7), 1264.

Kumar, M., Khamis, K., Stevens, R., Hannah, D. M., & Bradley, C. (2024). In-situ optical water quality monitoring sensors—applications, challenges, and future opportunities. *Frontiers in Water*, *6*. Available from https://doi.org/10.3389/frwa.2024.1380133, http://www.frontiersin.org/journals/water.

Kumar, P., Gupta, K., Karnatak, H. C., Siddiqui, A., & Senthil Kumar, A. (2017). Geo-enabled e-democracy tools and services for smart cities. *Advances in 21st Century Human Settlements*, 391–440. Available from https://www.springer.com/series/13196, 10.1007/978-981-10-4035-1_13.

Lacerda, L. Id. A., da Silveira, J. A. R., Santos, C. A. G., da Silva, R. M., Silva, A. M., do Nascimento, T. V. M., Ribeiro, E. L., & de Freitas, P. V. N. (2021). Urban forest loss using a GIS-based approach and instruments for integrated urban planning: A case study of João Pessoa, Brazil. *Journal of Geographical Sciences*, *31*(10), 1529–1553. Available from https://doi.org/10.1007/s11442-021-1910-4, http://www.springer.com/sgw/cda/frontpage/0,11855,1-40391-70-66229542-0,00.html?changeHeader = true.

Lassalle, G. (2021). Monitoring natural and anthropogenic plant stressors by hyperspectral remote sensing: Recommendations and guidelines based on a meta-review. *Science of the Total Environment*, *788*, 147758. Available from https://doi.org/10.1016/j.scitotenv.2021.147758.

Le, T., Hassan, F., Le, C., & Jeong, H. D. (2022). Understanding dynamic data interaction between civil integrated management technologies: A review of use cases and enabling techniques. *International Journal of Construction Management*, *22*(6), 1011–1032. Available from https://doi.org/10.1080/15623599.2019.1678863, http://www.tandfonline.com/loi/tjcm20.

Li, B., Xie, D., Wu, Y., Zheng, L., Xu, C., Zhou, Y., Fu, Y., Wang, C., Liu, B., & Zuo, X. (2024). Synthesis and detection algorithms for oblique stripe noise of space-borne remote sensing images. *IEEE Transactions on Geoscience and Remote Sensing*, *62*, 1–14. Available from https://doi.org/10.1109/tgrs.2024.3360268.

Li, F., Liu, X., Zhang, X., Zhao, D., Liu, H., Zhou, C., & Wang, R. (2017). Urban ecological infrastructure: an integrated network for ecosystem services and sustainable urban systems. *Journal of Cleaner Production*, *163*, S12. Available from https://doi.org/10.1016/j.jclepro.2016.02.079.

Li, J., Pei, Y., Zhao, S., Xiao, R., Sang, X., & Zhang, C. (2020). A review of remote sensing for environmental monitoring in China. *Remote Sensing*, *12*(7), 1130.

Li, J., & Wang, Y. (2022). Characteristic analysis and integration method of urban planning data based on GIS of internet of things. *Sustainable Computing: Informatics and Systems*, *36*, 100801. Available from https://doi.org/10.1016/j.suscom.2022.100801.

Li, X., Liu, C., Wang, Z., Xie, X., Li, D., & Xu, L. (2020). Airborne LiDAR: State-of-the-art of system design, technology and application. *Measurement Science and Technology*, *32*(3). Available from https://doi.org/10.1088/1361-6501/abc867, https://iopscience.iop.org/article/10.1088/1361-6501/abc867.

Lindner, C., Rienow, A., Otto, K. H., & Juergens, C. (2022). Development of an app and teaching concept for implementation of hyperspectral remote sensing data into school lessons using augmented reality. *Remote Sensing*, *14*(3). Available from https://doi.org/10.3390/rs14030791, https://www.mdpi.com/2072-4292/14/3/791/pdf.

Liu, J., Li, J., Li, W., & Wu, J. (2016). Rethinking big data: A review on the data quality and usage issues. *ISPRS Journal of Photogrammetry and Remote Sensing*, *115*, 134–142. Available from https://doi.org/10.1016/j.isprsjprs.2015.11.006.

Liu, Y., Song, W., & Deng, X. (2019). Understanding the spatiotemporal variation of urban land expansion in oasis cities by integrating remote sensing and multi-dimensional DPSIR-based indicators. *Ecological Indicators*, *96*, 23–37.

Liu, Y., Zuo, R., & Dong, Y. (2021). Analysis of temporal and spatial characteristics of urban expansion in Xiaonan District from 1990 to 2020 using time series Landsat imagery. *Remote Sensing*, *13*(21), 4299.

Lloyd, C. T., Chamberlain, H., Kerr, D., Yetman, G., Pistolesi, L., Stevens, F. R., Gaughan, A. E., Nieves, J. J., Hornby, G., MacManus, K., Sinha, P., Bondarenko, M., Sorichetta, A., & Tatem, A. J. (2019). Global spatio-temporally harmonised datasets for producing high-resolution gridded population distribution datasets. *Big Earth Data*, *3*(2), 108–139. Available from https://doi.org/10.1080/20964471.2019.1625151, https://www.tandfonline.com/action/journalInformation?journalCode = tbed20.

Loidl, M., Wallentin, G., Cyganski, R., Graser, A., Scholz, J., & Haslauer, E. (2016). GIS and transport modeling-strengthening the spatial perspective. *ISPRS International Journal of Geo-Information*, *5*(6). Available from https://doi.org/10.3390/ijgi5060084, http://www.mdpi.com/2220-9964/5/6/84.

Lopez, B. E., Magliocca, N. R., & Crooks, A. T. (2019). Challenges and opportunities of social media data for socio-environmental systems research. *Land*, *8*(7), 107.

Lopez, T., Al Bitar, A., Biancamaria, S., Güntner, A., & Jäggi, A. (2020). On the use of satellite remote sensing to detect floods and droughts at large scales. *Surveys in Geophysics*, *41*(6), 1461–1487. Available from https://doi.org/10.1007/s10712-020-09618-0.

Lourdes, K. T., Hamel, P., Gibbins, C. N., Sanusi, R., Azhar, B., & Lechner, A. M. (2022). Planning for green infrastructure using multiple urban ecosystem service models and multicriteria analysis. *Landscape and Urban Planning*, *226*. Available from https://doi.org/10.1016/j.landurbplan.2022.104500, http://www.elsevier.com/inca/publications/store/5/0/3/3/4/7.

De Luca, C., Langemeyer, J., Vaňo, S., Baró, F., & Andersson, E. (2021). Adaptive resilience of and through urban ecosystem services: A transdisciplinary approach to sustainability in Barcelona. *Ecology and Society*, *26*(4). Available from https://doi.org/10.5751/ES-12535-260438, http://www.ecologyandsociety.org.

Ma, Yf, & You, Xy (2022). A sustainable conservation strategy of wildlife in urban ecosystems: Case of Gallinula chloropus in Beijing-Tianjin-Hebei region. *Ecological Informatics*, *68*. Available from https://doi.org/10.1016/j.ecoinf.2022.101571, http://www.elsevier.com/wps/find/journaldescription.cws_home/705192/description#description.

Mahabir, R., Croitoru, A., Crooks, A., Agouris, P., & Stefanidis, A. (2018). A critical review of high and very high-resolution remote sensing approaches for detecting and mapping slums: Trends, challenges and emerging opportunities. *Urban Science*, *2*(1), 8. Available from https://doi.org/10.3390/urbansci2010008.

Majeed, A., Khan, S., & Hwang, S. O. (2022). Toward privacy preservation using clustering based anonymization: Recent advances and future research outlook. *IEEE Access*, *10*, 53066–53097.

Makanda, K., Nzama, S., & Kanyerere, T. (2022). Assessing the role of water resources protection practice for sustainable water resources management: A review. *Water*, *14*(19), 3153.

Marchante, E., López-Núñez, F. A., Duarte, L. N., & Marchante, H. (2024). *The role of citizen science in biodiversity monitoring: when invasive species and insects meet. Biological invasions and global insect decline* (pp. 291–314). Academic Press.

Martínez-Graña, A. M., & Valdés Rodríguez, V. (2016). Remote sensing and GIS applied to the landscape for the environmental restoration of urbanizations by means of 3D virtual reconstruction and visualization (Salamanca, Spain). *ISPRS International Journal of Geo-information*, *5*(1), 2.

Mathew, A., Arunab, K. S., & Sharma, A. K. (2024). Revealing the urban heat Island: Investigating spatiotemporal surface temperature dynamics, modeling, and interactions with controllable and non-controllable factors. *Remote Sensing Applications: Society and Environment*, *35*, 101219.

Mauree, D., Naboni, E., Coccolo, S., Perera, A. T. D., Nik, V. M., & Scartezzini, J. L. (2019). A review of assessment methods for the urban environment and its energy sustainability to guarantee climate adaptation of future cities. *Renewable and Sustainable Energy Reviews*, *112*, 733–746. Available from https://doi.org/10.1016/j.rser.2019.06.005, https://www.journals.elsevier.com/renewable-and-sustainable-energy-reviews.

Mavrakou, T., Polydoros, A., Cartalis, C., & Santamouris, M. (2018). Recognition of thermal hot and cold spots in urban areas in support of mitigation plans to counteract overheating: Application for Athens. *Climate*, *6*(1). Available from https://doi.org/10.3390/cli6010016, http://www.mdpi.com/2225-1154/6/1/16/pdf.

Mazzanti, P., Scancella, S., Virelli, M., Frittelli, S., Nocente, V., & Lombardo, F. (2022). Assessing the performance of multi-resolution satellite SAR images for post-earthquake damage detection and mapping aimed at emergency response management. *Remote Sensing*, *14*(9), 2210. Available from https://doi.org/10.3390/rs14092210.

Mbuh, M., Metzger, P., Brandt, P., Fika, K., & Slinkey, M. (2019). Application of real-time GIS analytics to support spatial intelligent decision-making in the era of big data for smart cities. *EAI Endorsed Transactions on Smart Cities*, *4*(9).

McHale, M. R., Beck, S. M., Pickett, S. T. A., Childers, D. L., Cadenasso, M. L., Rivers, L., Swemmer, L., Ebersohn, L., Twine, W., & Bunn, D. N. (2018). Democratization of ecosystem services—a radical approach for assessing nature's benefits in the face of urbanization. *Ecosystem Health and Sustainability*, *4*(5), 115–131. Available from https://doi.org/10.1080/20964129.2018.1480905, https://spj.science.org/journal/ehs.

McKenzie, Z., Kumler, M. P., Ma, R., Williams, K., & Hayes, W. K. (2023). Eyes from the sky: Application of satellite-based indices to assess vegetation casualty on Grand Bahama Island one year post-Hurricane Dorian. *Remote Sensing Applications: Society and Environment*, *32*. Available from https://doi.org/10.1016/j.rsase.2023.101044, https://www.sciencedirect.com/science/journal/23529385.

Mehryar, S., Sasson, I., & Surminski, S. (2022). Supporting urban adaptation to climate change: What role can resilience measurement tools play? *Urban Climate*, *41*, 101047.

Mekuriaw, A., Sewnet, H., Wubie, A. M., & Amsalu, T. (2021). Analysis of land data acquisition, management, and their challenges in Ethiopia. *African Journal on Land Policy and Geospatial Sciences e-ISSN*, *4*(2).

Mesev, V. (1997). Remote sensing of urban systems: Hierarchical integration with GIS. *Computers, Environment and Urban Systems*, *21*(3-4), 175–187. Available from https://doi.org/10.1016/s0198-9715(97)10003-5.

Mohsan, S. A. H., Khan, M. A., Noor, F., Ullah, I., & Alsharif, M. H. (2022). Towards the unmanned aerial vehicles (UAVs): A comprehensive review. *Drones*, *6*(6). Available from https://doi.org/10.3390/drones6060147, https://www.mdpi.com/2504-446X/6/6/147/pdf?version = 1655300789.

Mortula, M., Ali, T., Bachir, A., Elaksher, A., & Abouleish, M. (2020). Towards monitoring of nutrient pollution in coastal lake using remote sensing and regression analysis. *Water*, *12*(7), 1954.

Muhamad Khair, N. K., Lee, K. E., & Mokhtar, M. (2020). Sustainable city and community empowerment through the implementation of community-based monitoring: A conceptual approach. *Sustainability*, *12*(22), 9583.

Mukonza, S. S., & Chiang, J. L. (2022). Satellite sensors as an emerging technique for monitoring macro- and microplastics in aquatic ecosystems. *Water Emerging Contaminants and Nanoplastics*, *1*(4). Available from https://doi.org/10.20517/wecn.2022.12, https://f.oaes.cc/xmlpdf/80dbcbd1-b821-40ff-ac8c-d950f370d651/5203_down.pdf?v = 2.

Munawar, H. S., Hammad, A. W., & Waller, S. T. (2022). Remote sensing methods for flood prediction: A review. *Sensors*, *22*(3), 960.

Musse, M. A., Barona, D. A., & Rodriguez, L. M. S. (2018). Urban environmental quality assessment using remote sensing and census data. *International journal of applied earth observation and geoinformation*, *71*, 95–108.

Naeini, A. M., Tabesh, M., & Soltaninia, S. (2024). Modeling the effect of land use change to design a suitable low impact development (LID) system to control surface water pollutants. *Science of the Total Environment*, *932*, 172756.

Nagendra, N. P., Narayanamurthy, G., & Moser, R. (2022). Management of humanitarian relief operations using satellite big data analytics: The case of Kerala floods. *Annals of Operations Research*, *319*(1), 885–910.

Nedd, R., Light, K., Owens, M., James, N., Johnson, E., & Anandhi, A. (2021). A synthesis of land use/land cover studies: Definitions, classification systems, meta-studies, challenges and knowledge gaps on a global landscape. *Land*, *10*(9), 994. Available from https://doi.org/10.3390/land10090994.

Neyns, R., & Canters, F. (2022). Mapping of urban vegetation with high-resolution remote sensing: A review. *Remote Sensing*, *14*(4), 1031. Available from https://doi.org/10.3390/rs14041031.

Nichols, C. R., Wright, L. D., Bainbridge, S. J., Cosby, A., Hénaff, A., Loftis, J. D., Cocquempot, L., Katragadda, S., Mendez, G. R., Letortu, P., Le Dantec, N. L., Resio, D., & Zarillo, G. (2019). Collaborative science to enhance coastal resilience and adaptation. *Frontiers in Marine Science*, *6*. Available from https://doi.org/10.3389/fmars.2019.00404, https://www.frontiersin.org/articles/10.3389/fmars.2019.00404.

Nitoslawski, S. A., Galle, N. J., van den Bosc, C. K., & Steenberg, J. W. N. (2019). Smarter ecosystems for smarter cities? A review of trends, technologies, and turning points for smart urban forestry. *Sustainable Cities and Society*, *51*. Available from https://doi.org/10.1016/j.scs.2019.101770, http://www.elsevier.com/wps/find/journaldescription.cws_home/724360/description#description.

Noor, N. M., Abdullah, A., & Rosni, N. A. (2014). Leveraging of remote sensing and GIS on mapping in urban and regional planning applications. *IOP Conference Series: Earth and Environmental Science*, *20*(1), 012004.

Noori, A., Bonakdari, H., Hassaninia, M., Morovati, K., Khorshidi, I., Noori, A., & Gharabaghi, B. (2022). A reliable GIS-based FAHP-FTOPSIS model to prioritize urban water supply management scenarios: A case study in semi-arid climate. *Sustainable Cities and Society*, *81*, 103846. Available from https://doi.org/10.1016/j.scs.2022.103846.

Nova, K. (2023). AI-enabled water management systems: An analysis of system components and interdependencies for water conservation. *Eigenpub Review of Science and Technology*, *7*(1), 105–124.

Odongo, R. (2023). Remote sensing applications in environmental monitoring. *European Journal of Natural Sciences*, *1*, 1–12.

Oguejiofor, B. B., Omotosho, A., Abioye, K. M., Alabi, A. M., Oguntoyinbo, F. N., Daraojimba, A. I., & Daraojimba, C. (2023). A review on data-driven regulatory compliance in Nigeria. *International Journal of Applied Research in Social Sciences*, *5*(8), 231–243. Available from https://doi.org/10.51594/ijarss.v5i8.571.

Olds, H. T., Corsi, S. R., Dila, D. K., Halmo, K. M., Bootsma, M. J., & McLellan, S. L. (2018). High levels of sewage contamination released from urban areas after storm events: A quantitative survey with sewage specific bacterial indicators. *PLoS Medicine*, *15*(7). Available from https://doi.org/10.1371/journal.pmed.1002614, http://medicine.plosjournals.org/perlserv/?request = index-html&issn = 1549-1676.

Olson, D., & Anderson, J. (2021). Review on unmanned aerial vehicles, remote sensors, imagery processing, and their applications in agriculture. *Agronomy Journal*, *113*(2), 971–992. Available from https://doi.org/10.1002/agj2.20595, https://acsess.onlinelibrary.wiley.com/journal/14350645.

Oppong, J., Ning, Z. H., Twumasi, Y., Antwi, R. A., Anokye, M., Ahoma, G., Annan, J., Namwamba, J. B., Loh, P., & Akinrinwoye, C. (2023). The integration of remote sensing and geographic information system (GIS) in managing urban ecosystems. *The International Archives of the Photogrammetry, Remote Sensing and Spatial Information Sciences*, 169–175. Available from https://doi.org/10.5194/isprs-archives-xlviii-m-3-2023-169-2023.

Ouchra, H., Belangour, A., & Erraissi, A. (2022). A comprehensive study of using remote sensing and geographical information systems for urban planning. *Internetworking Indonesia Journal*, *12*, 15–20.

Ouma, Y. O. (2016). Advancements in medium and high resolution Earth observation for land-surface imaging: Evolutions, future trends and contributions to sustainable development. *Advances in Space Research*, *57*(1), 110–126. Available from https://doi.org/10.1016/j.asr.2015.10.038, http://www.journals.elsevier.com/advances-in-space-research/.

Pandit, A., Minné, E. A., Li, F., Brown, H., Jeong, H., James, J. A. C., Newell, J. P., Weissburg, M., Chang, M. E., Xu, M., Yang, P., Wang, R., Thomas, V. M., Yu, X., Lu, Z., & Crittenden, J. C. (2017). Infrastructure ecology: An evolving paradigm for sustainable urban development. *Journal of Cleaner Production*, *163*, S19. Available from https://doi.org/10.1016/j.jclepro.2015.09.010.

Pascapurnama, D. N., Murakami, A., Chagan-Yasutan, H., Hattori, T., Sasaki, H., & Egawa, S. (2018). Integrated health education in disaster risk reduction: Lesson learned from disease outbreak following natural disasters in Indonesia. *International Journal of Disaster Risk Reduction*, *29*, 94–102. Available from https://doi.org/10.1016/j.ijdrr.2017.07.013, http://www.journals.elsevier.com/international-journal-of-disaster-risk-reduction/.

Paz, F., & Pow-Sang, J. A. (2016). A systematic mapping review of usability evaluation methods for software development process. *International Journal of Software Engineering and Its Applications*, *10*(1), 165–178.

Pazúr, R., Price, B., & Atkinson, P. M. (2021). Fine temporal resolution satellite sensors with global coverage: An opportunity for landscape ecologists. *Landscape Ecology*, *36*, 2199–2213.

Pei, T., Xu, J., Liu, Y., Huang, X., Zhang, L., Dong, W., Qin, C., Song, C., Gong, J., & Zhou, C. (2021). GIScience and remote sensing in natural resource and environmental research: Status quo and future perspectives. *Geography and Sustainability*, *2*(3), 207–215. Available from https://doi.org/10.1016/j.geosus.2021.08.004, http://www.journals.elsevier.com/geography-and-sustainability.

Penteado, H. M. (2023). Effects of open space configurations and development patterns on future urban wildlife habitats and populations. *City and Environment Interactions*, *19*, 100106. Available from https://doi.org/10.1016/j.cacint.2023.100106.

Perrelet, K., Moretti, M., Dietzel, A., Altermatt, F., & Cook, L. M. (2024). Engineering blue-green infrastructure for and with biodiversity in cities. *NPJ Urban Sustainability*, *4*(1). Available from https://doi.org/10.1038/s42949-024-00163-y, https://www.nature.com/npjurbansustain/.

Pfadenhauer, L. M., Gerhardus, A., Mozygemba, K., Lysdahl, K. B., Booth, A., Hofmann, B., Wahlster, P., Polus, S., Burns, J., Brereton, L., & Rehfuess, E. (2017). Making sense of complexity in context and implementation: The context and implementation of complex interventions (CICI) framework. *Implementation Science*, *12*(1). Available from https://doi.org/10.1186/s13012-017-0552-5, http://www.implementationscience.com/.

Pickett, S. T. A., Cadenasso, M. L., Baker, M. E., Band, L. E., Boone, C. G., Buckley, G. L., Groffman, P. M., Grove, J. M., Irwin, E. G., Kaushal, S. S., Ladeau, S. L., Miller, A. J., Nilon, C. H., Romolini, M., Rosi, E. J., Swan, C. M., & Szlavecz, K. (2020). Theoretical perspectives of the Baltimore ecosystem study: Conceptual evolution in a social-ecological research project. *Bioscience*, *70*(4), 297–314. Available from https://doi.org/10.1093/biosci/biz166, http://bioscience.oxfordjournals.org/.

Pineo, H. (2022). Towards healthy urbanism: Inclusive, equitable and sustainable (THRIVES) – an urban design and planning framework from theory to praxis. *Cities and Health*, *6*(5), 974–992. Available from https://doi.org/10.1080/23748834.2020.1769527.

Pretzsch, H., Moser-Reischl, A., Rahman, M. A., Pauleit, S., & Rötzer, T. (2023). Towards sustainable management of the stock and ecosystem services of urban trees. From theory to model and application. *Trees*, *37*(1), 177–196. Available from https://doi.org/10.1007/s00468-021-02100-3.

Pérez-González, C. J., Colebrook, M., Roda-García, J. L., & Rosa-Remedios, C. B. (2019). Developing a data analytics platform to support decision making in emergency and security management. *Expert Systems with Applications*, *120*, 167–184. Available from https://doi.org/10.1016/j.eswa.2018.11.023.

Quamar, M. M., Al-Ramadan, B., Khan, K., Shafiullah, M., & El Ferik, S. (2023). Advancements and applications of drone-integrated geographic information system technology—a review. *Remote Sensing*, *15*(20). Available from https://doi.org/10.3390/rs15205039, http://www.mdpi.com/journal/remotesensing.

Radočaj, D., Obhođaš, J., Jurišić, M., & Gašparović, M. (2020). Global open data remote sensing satellite missions for land monitoring and conservation: A review. *Land*, *9*(11), 402. Available from https://doi.org/10.3390/land9110402.

Rahaman, M. H., Masroor, M., & Sajjad, H. (2023). Integrating remote sensing derived indices and machine learning algorithms for precise extraction of small surface water bodies in the lower Thoubal river watershed, India. *Journal of Cleaner Production*, *422*, 138563.

Rahman, M. S., Di, L., Yu, E., Lin, L., & Yu, Z. (2021). Remote sensing based rapid assessment of flood crop damage using novel disaster vegetation damage index (DVDI). *International Journal of Disaster Risk Science*, *12*(1), 90–110. Available from https://doi.org/10.1007/s13753-020-00305-7, http://www.springer.com/earth + sciences + and + geography/natural + hazards/journal/13753.

Rao, T. R., Mitra, P., Bhatt, R., & Goswami, A. (2019). The big data system, components, tools, and technologies: A survey. *Knowledge and Information Systems*, *60*(3), 1165–1245. Available from https://doi.org/10.1007/s10115-018-1248-0, https://www.springer.com/journal/10115.

Ravanelli, R., Nascetti, A., Cirigliano, R. V., Di Rico, C., Leuzzi, G., Monti, P., & Crespi, M. (2018). Monitoring the impact of land cover change on surface urban heat island through Google Earth Engine: Proposal of a global methodology, first applications and problems. *Remote Sensing*, *10*(9). Available from https://doi.org/10.3390/rs10091488, https://res.mdpi.com/remotesensing/remotesensing-10-01488/article_deploy/remotesensing-10-01488.pdf?filename = &attachment = 1.

Ravikanth, L., Jayas, D. S., White, N. D. G., Fields, P. G., & Sun, D. W. (2017). Extraction of spectral information from hyperspectral data and application of hyperspectral imaging for food and agricultural products. *Food and Bioprocess Technology*, *10*(1). Available from https://doi.org/10.1007/s11947-016-1817-8, http://www.springer.com/life + sci/food + science/journal/11947.

Razzaghi Asl, S., & Pearsall, H. (2023). How do spatial factors of green spaces contribute to flood regulation in urban areas? A systematic mapping approach. *Progress in Physical Geography: Earth and Environment*, *47*(5), 702–720. Available from https://doi.org/10.1177/03091333231156511.

Reed, M. S., Ferré, M., Martin-Ortega, J., Blanche, R., Lawford-Rolfe, R., Dallimer, M., & Holden, J. (2021). Evaluating impact from research: A methodological framework. *Research Policy*, *50*(4), 104147. Available from https://doi.org/10.1016/j.respol.2020.104147.

Rezvani, S. M. H. S., Falcão, M. J., Komljenovic, D., & de Almeida, N. M. (2023). A systematic literature review on urban resilience enabled with asset and disaster risk management approaches and GIS-based decision support tools. *Applied Sciences*, *13*(4), 2223. Available from https://doi.org/10.3390/app13042223.

Robinson, J. A., Block, D., & Rees, A. (2017). Community geography: Addressing barriers in public participation GIS. *The Cartographic Journal*, *54*(1), 5–13.

Rodríguez, L., García, J. J., Carreño, F., & Martínez, B. (2019). Integration of physiological knowledge into hybrid species distribution modelling to improve forecast of distributional shifts of tropical corals. *Diversity and Distributions*, *25*(5), 715–728. Available from https://doi.org/10.1111/ddi.12883, http://onlinelibrary.wiley.com/journal/10.1111/(ISSN)1472-4642.

Rodríguez-López, L., Duran-Llacer, I., Bravo Alvarez, L., Lami, A., & Urrutia, R. (2023). Recovery of water quality and detection of Algal Blooms in Lake Villarrica through landsat satellite images and monitoring data. *Remote Sensing*, *15*(7), 1929.

Rossetto, R., De Filippis, G., Borsi, I., Foglia, L., Cannata, M., Criollo, R., & Vázquez-Suñé, E. (2018). Integrating free and open source tools and distributed modelling codes in GIS environment for data-based groundwater management. *Environmental Modelling and Software*, *107*, 210–230. Available from https://doi.org/10.1016/j.envsoft.2018.06.007, http://www.elsevier.com/inca/publications/store/4/2/2/9/2/1.

Roth, R. E. (2021). Cartographic design as visual storytelling: Synthesis and review of map-based narratives, genres, and tropes. *Cartographic Journal*, *58*(1), 83–114. Available from https://doi.org/10.1080/00087041.2019.1633103, http://www.tandfonline.com/loi/ycaj20#.VwHa901f1Qs.

Safaei, M., Bashari, H., Kleinebecker, T., Fakheran, S., Jafari, R., & Große-Stoltenberg, A. (2023). Mapping terrestrial ecosystem health in drylands: Comparison of field-based information with remotely sensed data at watershed level. *Landscape Ecology*, *38*(3), 705–724. Available from https://doi.org/10.1007/s10980-022-01454-4, https://www.springer.com/journal/10980.

Safari Bazargani, J., Sadeghi-Niaraki, A., & Choi, S. M. (2021). A survey of GIS and IoT integration: Applications and architecture. *Applied Sciences*, *11*(21), 10365.

Saha, P., & Paul, B. (2021). Identification of potential strategic sites for city planning based on water quality through GIS-AHP-integrated model. *Environmental Science and Pollution Research*, *28*(18), 23073–23086. Available from https://doi.org/10.1007/s11356-020-12292-9, https://link.springer.com/journal/11356.

Samarakkody, A., Amaratunga, D., & Haigh, R. (2023). Technological innovations for enhancing disaster resilience in smart cities: A comprehensive urban Scholar's analysis. *Sustainability*, *15*(15), 12036.

Santos, E., Carvalho, M., & Martins, S. (2023). Sustainable water management: Understanding the socioeconomic and cultural dimensions. *Sustainability*, *15*(17), 13074.

Santos, J., Yip, C., Thekdi, S., & Pagsuyoin, S. (2020). Workforce/population, economy, infrastructure, geography, hierarchy, and time (WEIGHT): Reflections on the plural dimensions of disaster resilience. *Risk Analysis*, *40*(1), 43–67. Available from https://doi.org/10.1111/risa.13186, http://onlinelibrary.wiley.com/journal/10.1111/(ISSN)1539-6924.

Saravani, M. J., Saadatpour, M., & Shahvaran, A. R. (2024). A web GIS based integrated water resources assessment tool for Javeh Reservoir. *Expert Systems with Applications*, *252*, 124198.

Sarkissian., Sayah, M. J., Abdallah, C., Zaninetti, J. M., & Nedjai, R. (2022). Land use planning to reduce flood risk: opportunities, challenges and uncertainties in developing countries. *Sensors*, *22*(18).

Schmelz, W. J., & Psuty, N. P. (2022). Application of geomorphological maps and LiDAR to volumetrically measure coastal geomorphological change from Hurricane Sandy at Fire Island National Seashore. *Geomorphology*, *408*. Available from https://doi.org/10.1016/j.geomorph.2022.108262, http://www.elsevier.com/inca/publications/store/5/0/3/3/3/4/.

Semeraro, T., Scarano, A., Buccolieri, R., Santino, A., & Aarrevaara, E. (2021). Planning of urban green spaces: An ecological perspective on human benefits. *Land*, *10*(2), 105. Available from https://doi.org/10.3390/land10020105.

Sepasgozar, S. M. E., Hawken, S., Sargolzaei, S., & Foroozanfa, M. (2019). Implementing citizen centric technology in developing smart cities: A model for predicting the acceptance of urban technologies. *Technological Forecasting and Social Change*, *142*, 105–116. Available from https://doi.org/10.1016/j.techfore.2018.09.012, https://www.journals.elsevier.com/technological-forecasting-and-social-change.

Shafapourtehrany, M., Batur, M., Shabani, F., Pradhan, B., Kalantar, B., & Özener, H. (2023). A comprehensive review of geospatial technology applications in earthquake preparedness, emergency management, and damage assessment. *Remote Sensing*, *15*(7), 1939.

Shahin, M., Babar, M. A., & Zhu, L. (2017). Continuous integration, delivery and deployment: A systematic review on approaches, tools, challenges and practices. *IEEE Access*, *5*, 3909–3943.

Shaikh, M., & Birajdar, F. (2024). Advancements in remote sensing and GIS for sustainable groundwater monitoring: Applications, challenges, and future directions. *International Journal of Research in Engineering*, *7*(3), 16–24.

Shariat, R., Roozbahani, A., & Ebrahimian, A. (2019). Risk analysis of urban storm water infrastructure systems using fuzzy spatial multi-criteria decision making. *Science of the Total Environment*, *647*, 1468–1477.

Sharma, R., Kamble, S. S., & Gunasekaran, A. (2018). Big GIS analytics framework for agriculture supply chains: A literature review identifying the current trends and future perspectives. *Computers and Electronics in Agriculture*, *155*, 103–120.

Shawaqfah, M., Almomani, F., & Al-Rousan, T. (2021). Potential use of treated wastewater as groundwater recharge using GIS techniques and modeling tools in Dhuleil-Halabat well-field/Jordan. *Water*, *13*(11), 1581.

Sheffield, J., Wood, E. F., Pan, M., Beck, H., Coccia, G., Serrat-Capdevila, A., & Verbist, K. (2018). Satellite remote sensing for water resources management: potential for supporting sustainable development in data-poor regions. *Water Resources Research*, *54*(12), 9724–9758. Available from https://doi.org/10.1029/2017wr022437.

Shen, Y., Morsy, M. M., Huxley, C., Tahvildari, N., & Goodall, J. L. (2019). Flood risk assessment and increased resilience for coastal urban watersheds under the combined impact of storm tide and heavy rainfall. *Journal of Hydrology*, *579*. Available from https://doi.org/10.1016/j.jhydrol.2019.124159, http://www.elsevier.com/inca/publications/store/5/0/3/3/4/3.

Shi, H., Xian, G., Auch, R., Gallo, K., & Zhou, Q. (2021). Urban heat island and its regional impacts using remotely sensed thermal data—a review of recent developments and methodology. *Land*, *10*(8), 867. Available from https://doi.org/10.3390/land10080867.

Shinde, S. P., Barai, V. N., Bansod, R. D., Atre, A. A., Gavit, B. K., & Kadam, S. A. (2023). A review on water resource planning and management with special reference to groundwater using remote sensing and GIS techniques. *International Journal of Economic Plants*, *10*(May, 2), 083–091. Available from https://doi.org/10.23910/2/2022.0516b.

Shreenithi, M., Sujithra, M., Senthilkumar, B., Shanmugapriyaa, D., & Rizanuma, T. (2024). *Spatial data visualization with python: Techniques, tools, and real-world applications. Geospatial application development using python programming* (pp. 123–162). India: IGI Global. Available from https://www.igi-global.com/book/geospatial-application-development-using-python/329970, 10.4018/979-8-3693-1754-9.ch005.

Shrestha, S., Poudyal, K. N., Bhattarai, N., Dangi, M. B., & Boland, J. J. (2022). An assessment of the impact of land use and land cover change on the degradation of ecosystem service values in kathmandu valley using remote sensing and GIS. *Sustainability*, *14*(23), 15739. Available from https://doi.org/10.3390/su142315739.

Silveira, C., Dias, A. T. C., Amaral, F. G., de Gois, G., & Pistón, N. (2024). The importance of private gardens and their spatial composition and configuration to urban heat island mitigation. *Sustainable Cities and Society*, *112*. Available from https://doi.org/10.1016/j.scs.2024.105589, https://www.sciencedirect.com/science/journal/22106707.

Simkin, R. D., Seto, K. C., McDonald, R. I., & Jetz, W. (2022). Biodiversity impacts and conservation implications of urban land expansion projected to 2050. *Proceedings of the National Academy of Sciences of the United States of America*, *119*(12). Available from https://www.pnas.org/doi/10.1073/pnas.2117297119, 10.1073/pnas.2117297119.

Simon, S. (2024). A participatory urban governance framework for transitioning cities towards sustainability: Linking the protection of urban ecosystems functions and circularity principles. *Advances in Urban Regional Development and Planning*, *1*(1), 1–17.

Singleton, A., & Arribas-Bel, D. (2021). Geographic data science. *Geographical Analysis*, *53*(1), 61–75. Available from https://doi.org/10.1111/gean.12194, http://onlinelibrary.wiley.com/journal/10.1111/(ISSN)1538-4632.

Sivarajah, U., Kamal, M. M., Irani, Z., & Weerakkody, V. (2017). Critical analysis of Big Data challenges and analytical methods. *Journal of Business Research*, *70*, 263–286. Available from https://doi.org/10.1016/j.jbusres.2016.08.001, http://www.elsevier.com/locate/jbusres.

Sodiq, A., Baloch, A. A. B., Khan, S. A., Sezer, N., Mahmoud, S., Jama, M., & Abdelaal, A. (2019). Towards modern sustainable cities: Review of sustainability principles and trends. *Journal of Cleaner Production*, *227*, 972–1001. Available from https://doi.org/10.1016/j.jclepro.2019.04.106, https://www.journals.elsevier.com/journal-of-cleaner-production.

Son, T. H., Weedon, Z., Yigitcanlar, T., Sanchez, T., Corchado, J. M., & Mehmood, R. (2023). Algorithmic urban planning for smart and sustainable development: Systematic review of the literature. *Sustainable Cities and Society*, *94*. Available from https://doi.org/10.1016/j.scs.2023.104562, http://www.elsevier.com/wps/find/journaldescription.cws_home/724360/description#description.

Song, J., Zhao, C., Lin, T., Li, X., & Prishchepov, A. V. (2019). Spatio-temporal patterns of traffic-related air pollutant emissions in different urban functional zones estimated by real-time video and deep learning technique. *Journal of Cleaner Production*, *238*. Available from https://doi.org/10.1016/j.jclepro.2019.117881, https://www.journals.elsevier.com/journal-of-cleaner-production.

Song, Y., Wang, X., Tan, Y., Wu, P., Sutrisna, M., Cheng, J. C. P., & Hampson, K. (2017). Trends and opportunities of BIM-GIS integration in the architecture, engineering and construction industry: A review from a spatio-temporal statistical perspective. *ISPRS International Journal of Geo-Information*, *6*(12), 397. Available from https://doi.org/10.3390/ijgi6120397.

Southerland, V. A., Anenberg, S. C., Harris, M., Apte, J., Hystad, P., van Donkelaar, A., Martin, R. V., Beyers, M., & Roy, A. (2021). Assessing the distribution of air pollution health risks within cities: A neighborhood-scale analysis leveraging high-resolution data sets in the bay area, california. *Environmental Health Perspectives*, *129*(3). Available from https://doi.org/10.1289/EHP7679, https://ehp.niehs.nih.gov/doi/pdf/10.1289/EHP7679.

Staniszewska, S., Denegri, S., Matthews, R., & Minogue, V. (2018). Reviewing progress in public involvement in NIHR research: Developing and implementing a new vision for the future. *BMJ Open*, *8*(7), e017124. Available from https://doi.org/10.1136/bmjopen-2017-017124.

Stavropoulos, S., Zaimes, G. N., Filippidis, E., Diaconu, D. C., & Emmanouloudis, D. (2020). Mitigating flash floods with the use of new technologies: A multi-criteria decision analysis to map flood susceptibility for Zakynthos Island, Greece. *Journal of Urban and Regional Analysis*, *12*(2), 233–248. Available from https://doi.org/10.37043/JURA.2020.12.2.7, http://www.jurareview.ro/resources/pdf/volume_30_mitigating_flash_-floods_with_the_use_of_new_technologies:_a_%CE%9Culti-criteria_decision_analysis_to_map_flood_susceptibility_for_zakynthos_island_-greece:abstract.pdf.

Stoian, A., Poulain, V., Inglada, J., Poughon, V., & Derksen, D. (2019). Land cover maps production with high resolution satellite image time series and convolutional neural networks: Adaptations and limits for operational systems. *Remote Sensing*, *11*(17), 1986.

Štrbac, S., Kašanin-Grubin, M., Pezo, L., Stojić, N., Lončar, B., Ćurčić, L., & Pucarević, M. (2023). Green infrastructure designed through nature-based solutions for sustainable urban development. *International Journal of Environmental Research and Public Health*, *20*(2), 1102. Available from https://doi.org/10.3390/ijerph20021102.

Sumadi, H. (2024). Causes of groundwater resource decline in the city of bandung perspective of Islamic economic law. *Suhuf*, *36*(1), 66–77. Available from https://doi.org/10.23917/suhuf.v36i1.4414.

Sun, A. Y., & Scanlon, B. R. (2019). How can Big Data and machine learning benefit environment and water management: A survey of methods, applications, and future directions. *Environmental Research Letters*, *14*(7). Available from https://doi.org/10.1088/1748-9326/ab1b7d, https://iopscience.iop.org/article/10.1088/1748-9326/ab1b7d/pdf.

Sunny, M. A. U. (2024). Unveiling spatial insights: Navigating the parameters of dynamic geographic information systems (GIS) analysis. *International Journal of Science and Research Archive*, *11*(2), 1976–1985. Available from https://doi.org/10.30574/ijsra.2024.11.2.0690.

Surya, B., Salim, A., Hernita, H., Suriani, S., Menne, F., & Rasyidi, E. S. (2021). Land use change, urban agglomeration, and urban sprawl: A sustainable development perspective of makassar city, indonesia. *Land*, *10*(6). Available from https://doi.org/10.3390/land10060556, https://www.mdpi.com/2073-445X/10/6/556/pdf.

Swan, C. M., Brown, B., Borowy, D., Cavender-Bares, J., Jeliazkov, A., Knapp, S., Lososová, Z., Padullés Cubino, J., Pavoine, S., Ricotta, C., & Sol, D. (2021). A framework for understanding how biodiversity patterns unfold across multiple spatial scales in urban ecosystems. *Ecosphere*, *12*(7). Available from https://doi.org/10.1002/ecs2.3650, http://esajournals.onlinelibrary.wiley.com/hub/journal/10.1002/(ISSN)2150-8925/.

Tan, P. Y., Zhang, J., Masoudi, M., Alemu, J. B., Edwards, P. J., Grêt-Regamey, A., Richards, D. R., Saunders, J., Song, X. P., & Wong, L. W. (2020). A conceptual framework to untangle the concept of urban ecosystem services. *Landscape and Urban Planning*, *200*, 103837.

Tang, J., Liang, J., Yang, Y., Zhang, S., Hou, H., & Zhu, X. (2022). Revealing the structure and composition of the restored vegetation cover in semi-arid mine dumps based on LiDAR and hyperspectral images. *Remote Sensing*, *14*(4), 978. Available from https://doi.org/10.3390/rs14040978.

Tang, J., Liu, D., Shang, C., & Niu, J. (2024). Impacts of land use change on surface infiltration capacity and urban flood risk in a representative karst mountain city over the last two decades. *Journal of Cleaner Production*, *454*, 142196. Available from https://doi.org/10.1016/j.jclepro.2024.142196.

Tang, P., Huang, J., Zhou, H., Fang, C., Zhan, Y., & Huang, W. (2021). Local and tele coupling coordination degree model of urbanization and the eco-environment based on RS and GIS: A case study in the Wuhan urban agglomeration. *Sustainable Cities and Society*, *75*.

Tang, X., & Adesina, J. A. (2022). Integrated watershed management framework and groundwater resources in Africa—a review of West Africa sub-region. *Water*, *14*(3), 288. Available from https://doi.org/10.3390/w14030288.

Tehrani, A. A., Veisi, O., Fakhr, B. V., & Du, D. (2024). Predicting solar radiation in the urban area: A data-driven analysis for sustainable city planning using artificial neural networking. *Sustainable Cities and Society, 100*. Available from https://doi.org/10.1016/j.scs.2023.105042, http://www.elsevier.com/wps/find/journaldescription.cws_home/724360/description#description.

Tetarwal, V., & Kumar, S. (2024). A comprehensive review on computer vision analysis of aerial data. arXiv:2402.09781.

Timilsina, S., Aryal, J., & Kirkpatrick, J. B. (2020). Mapping urban tree cover changes using object-based convolution neural network (OB-CNN). *Remote Sensing, 12*(18), 3017.

Tomar, P., Singh, S. K., Kanga, S., & Pattanaik, A. (2023). Water bodies mapping and monitoring using high-resolution satellite images. *Sustainability, Agri, Food and Environmental Research.*

Tomar, P., Singh, S. K., Kanga, S., Pattanaik, A., & Meraj, G. (2024). Use of geospatial technique in urban flood hazard management. *Disaster Management and Environmental Sustainability*, 155–167.

Tran, T. J., Helmus, M. R., & Behm, J. E. (2020). Green infrastructure space and traits (GIST) model: Integrating green infrastructure spatial placement and plant traits to maximize multifunctionality. *Urban Forestry and Urban Greening, 49*, 126635.

Trinder, J., & Liu, Q. (2020). Assessing environmental impacts of urban growth using remote sensing. *Geo-Spatial Information Science, 23*(1), 20–39.

Tsatsaris, A., Kalogeropoulos, K., Stathopoulos, N., Louka, P., Tsanakas, K., Tsesmelis, D. E., Krassanakis, V., Petropoulos, G. P., Pappas, V., & Chalkias, C. (2021). Geoinformation technologies in support of environmental hazards monitoring under climate change: An extensive review. *ISPRS International Journal of Geo-Information, 10*(2). Available from https://doi.org/10.3390/ijgi10020094, https://www.mdpi.com/2220-9964/10/2/94.

Tuthill, E. L., Maltby, A. E., DiClemente, K., & Pellowski, J. A. (2020). Longitudinal qualitative methods in health behavior and nursing research: Assumptions, design, analysis and lessons learned. *International Journal of Qualitative Methods, 19*. Available from https://doi.org/10.1177/1609406920965799, http://ijq.sagepub.com/content/by/year.

Twumasi, N. Y. D., Shao, Z., & Orhan, A. (2019). Remote sensing and gis methods in urban disaster monitoring and management—an overview. *International Journal of Trend in Scientific Research and Development, 3*, 918–926.

Ulrich, H., Kock-Schoppenhauer, A.-K., Deppenwiese, N., Gött, R., Kern, J., Lablans, M., Majeed, R. W., Stöhr, M. R., Stausberg, J., Varghese, J., Dugas, M., & Ingenerf, J. (2022). Understanding the nature of metadata: Systematic review. *Journal of Medical Internet Research, 24*(1), e25440. Available from https://doi.org/10.2196/25440.

Uysal, Ç., Yılmaz, Ö., Koç, E., Yöntem, E., & Koppelaar, R. (2021). An environmental evaluation framework for planning and monitoring of nature based solutions for sustainable urban management. *Open Research Europe, 1*, 97. Available from https://doi.org/10.12688/openreseurope.13746.1.

Vaitis, M., Feidas, H., Symeonidis, P., Kopsachilis, V., Dalaperas, D., Koukourouvli, N., Simos, D., & Taskaris, S. (2019). Development of a spatial database and web-GIS for the climate of Greece. *Earth Science Informatics, 12*(1), 97–115. Available from https://doi.org/10.1007/s12145-018-0351-9, http://www.springer.com/geosciences/computer + %26 + mathematical + applications/journal/12145.

Ventura, D., Grosso, L., Pensa, D., Casoli, E., Mancini, G., Valente, T., Scardi, M., & Rakaj, A. (2023). Coastal benthic habitat mapping and monitoring by integrating aerial and water surface low-cost drones. *Frontiers in Marine Science, 9*, 1096594.

Viqueira, J. R. R., Villarroya, S., Mera, D., & Taboada, J. A. (2020). Smart environmental data infrastructures: Bridging the gap between earth sciences and citizens. *Applied Sciences, 10*(3), 856. Available from https://doi.org/10.3390/app10030856.

Wachendorf, M., Fricke, T., & Möckel, T. (2018). Remote sensing as a tool to assess botanical composition, structure, quantity and quality of temperate grasslands. *Grass and Forage Science, 73*(1), 1–14.

Wachtel, G., Schmöcker, J. D., Hadas, Y., Gao, Y., Nahum, O. E., & Ben-Moshe, B. (2021). Planning for tourist urban evacuation routes: A framework for improving the data collection and evacuation processes. *Environment and Planning B: Urban Analytics and City Science, 48*(5), 1108–1125. Available from https://doi.org/10.1177/2399808321994575, http://journals.sagepub.com/loi/epbb.

Wang, H., Pan, Y., & Luo, X. (2019). Integration of BIM and GIS in sustainable built environment: A review and bibliometric analysis. *Automation in Construction, 103*, 41–52.

Wang, S., Cao, J., & Philip, S. Y. (2020). Deep learning for spatio-temporal data mining: A survey. *IEEE Transactions on Knowledge and Data Engineering, 34*(8), 3681–3700.

Wang, S., Zhong, Y., & Wang, E. (2019). An integrated GIS platform architecture for spatiotemporal big data. *Future Generation Computer Systems, 94*, 160–172.

Wang, X., & Xie, H. (2018). A review on applications of remote sensing and geographic information systems (GIS) in water resources and flood risk management. *Water, 10*(5), 608. Available from https://doi.org/10.3390/w10050608.

Webster-Wright, A. (2009). Reframing professional development through understanding authentic professional learning. *Review of Educational Research, 79*(2), 702–739. Available from https://doi.org/10.3102/0034654308330970.

Wellmann, T., Lausch, A., Andersson, E., Knapp, S., Cortinovis, C., Jache, J., Scheuer, S., Kremer, P., Mascarenhas, A., Kraemer, R., Haase, A., Schug, F., & Haase, D. (2020). Remote sensing in urban planning: Contributions towards ecologically sound policies. *Landscape and Urban Planning, 204*, 103921. Available from https://doi.org/10.1016/j.landurbplan.2020.103921.

White, P., & Powell, S. (2019). Code-literacy for GIS librarians: A discussion of languages, use cases, and competencies. *Journal of Map and Geography Libraries, 15*(1), 45–67. Available from https://doi.org/10.1080/15420353.2019.1660754.

Wilson, A., Tewdwr-Jones, M., & Comber, R. (2019). Urban planning, public participation and digital technology: App development as a method of generating citizen involvement in local planning processes. *Environment and Planning B: Urban Analytics and City Science, 46*(2), 286–302.

Wu, A. N., & Biljecki, F. (2022). GANmapper: Geographical data translation. *International Journal of Geographical Information Science*, *36*(7), 1394–1422. Available from https://doi.org/10.1080/13658816.2022.2041643, http://www.tandf.co.uk/journals/titles/13658816.asp.

Xia, C., Zhang, A., Wang, H., & Liu, J. (2020). Delineating early warning zones in rapidly growing metropolitan areas by integrating a multiscale urban growth model with biogeography-based optimization. *Land Use Policy*, *90*, 104332. Available from https://doi.org/10.1016/j.landusepol.2019.104332.

Yalcin, A. S., Kilic, H. S., & Delen, D. (2022). The use of multi-criteria decision-making methods in business analytics: A comprehensive literature review. *Technological Forecasting and Social Change*, *174*, 121193.

Yang, K., Han, Q., & de Vries, B. (2024). Urbanization effects on the food-water-energy nexus within ecosystem services: A case study of the Beijing-Tianjin-Hebei urban agglomeration in China. *Ecological Indicators*, *160*, 111845.

Yang, L., Driscol, J., Sarigai, S., Wu, Q., Chen, H., & Lippitt, C. D. (2022). Google earth engine and artificial intelligence (AI): A comprehensive review. *Remote Sensing*, *14*(14). Available from https://doi.org/10.3390/rs14143253, https://www.mdpi.com/2072-4292/14/14/3253/pdf?version = 1657102306.

Yang, X., Sanesi, G., Li, X., Chen, W. Y., & Lafortezza, R. (2021). *Remote sensing and urban green infrastructure. Urban remote sensing: Monitoring, synthesis, and modeling in the urban environment* (pp. 447–468). Wiley. Available from 10.1002/9781119625865.ch20.

Yang, Y., Javanroodi, K., & Nik, V. M. (2021). Climate change and energy performance of European residential building stocks—a comprehensive impact assessment using climate big data from the coordinated regional climate downscaling experiment. *Applied Energy*, *298*, 117246.

Yu, D., & Fang, C. (2023). Urban remote sensing with spatial big data: A review and renewed perspective of urban studies in recent decades. *Remote Sensing*, *15*(5), 1307.

Yu, D., Yin, J., Wilby, R. L., Lane, S. N., Aerts, J. C. J. H., Lin, N., Liu, M., Yuan, H., Chen, J., Prudhomme, C., Guan, M., Baruch, A., Johnson, C. W. D., Tang, X., Yu, L., & Xu, S. (2020). Disruption of emergency response to vulnerable populations during floods. *Nature Sustainability*, *3*(9), 728–736. Available from https://doi.org/10.1038/s41893-020-0516-7, http://www.nature.com/natsustain/.

Yu, J., Li, X., Guan, X., & Shen, H. (2024). A remote sensing assessment index for urban ecological livability and its application. *Geo-Spatial Information Science*, *27*(2), 289–310. Available from https://doi.org/10.1080/10095020.2022.2072775, http://www.tandfonline.com/loi/tgsi20.

Zaman, S., Korpilo, S., Horcea-Milcu, A. I., & Raymond, C. (2022). Associations between landscape values, self-reported knowledge, and land-use: A public participation GIS assessment. *Ecosystems and People*, *18*(1), 212–225. Available from https://doi.org/10.1080/26395916.2022.2052749, https://tandfonline.com/loi/tbsm22.

Zeng, F., Pang, C., & Tang, H. (2024). Sensors on internet of things systems for the sustainable development of smart cities: A systematic literature review. *Sensors*, *24*(7), 2074.

Zhang, B., Wu, Y., Zhao, B., Chanussot, J., Hong, D., Yao, J., & Gao, L. (2022). Progress and challenges in intelligent remote sensing satellite systems. *IEEE Journal of Selected Topics in Applied Earth Observations and Remote Sensing*, *15*, 1814–1822. Available from https://doi.org/10.1109/JSTARS.2022.3148139, http://ieeexplore.ieee.org/xpl/RecentIssue.jsp?punumber = 4609443.

Zhang, Z., Arshad, A., Zhang, C., Hussain, S., & Li, W. (2020). Unprecedented temporary reduction in global air pollution associated with COVID-19 forced confinement: A continental and city scale analysis. *Remote Sensing*, *12*(15), 2420. Available from https://doi.org/10.3390/rs12152420.

Zhang, Z., & Zhu, L. (2023). A review on unmanned aerial vehicle remote sensing: Platforms, sensors, data processing methods, and applications. *Drones*, *7*(6), 398. Available from https://doi.org/10.3390/drones7060398.

Zhao, Z., & Zhang, Y. (2020). Impact of smart city planning and construction on economic and social benefits based on big data analysis. *Complexity*, *2020*(1), 8879132.

Zhou, D., Xiao, J., Bonafoni, S., Berger, C., Deilami, K., Zhou, Y., Frolking, S., Yao, R., Qiao, Z., & Sobrino, J. A. (2018). Satellite remote sensing of surface urban heat islands: Progress, challenges, and perspectives. *Remote Sensing*, *11*(1), 48.

Zhou, W., Pickett, S. T., & Cadenasso, M. L. (2017). Shifting concepts of urban spatial heterogeneity and their implications for sustainability. *Landscape Ecology*, *32*, 15–30.

Zhu, J., & Wu, P. (2021). Towards effective BIM/GIS data integration for smart city by integrating computer graphics technique. *Remote Sensing*, *13*(10), 1889. Available from https://doi.org/10.3390/rs13101889.

Zhu, S., Yang, Y., Yan, Y., Causone, F., Jin, X., Zhou, X., & Shi, X. (2022). An evidence-based framework for designing urban green infrastructure morphology to reduce urban building energy use in a hot-humid climate. *Building and Environment*, *219*, 109181. Available from https://doi.org/10.1016/j.buildenv.2022.109181.

Zhu, Z., Qiu, S., & Ye, S. (2022). Remote sensing of land change: A multifaceted perspective. *Remote Sensing of Environment*, *282*, 113266.

Zhu, Z., Zhou, Y., Seto, K. C., Stokes, E. C., Deng, C., Pickett, S. T. A., & Taubenböck, H. (2019). Understanding an urbanizing planet: Strategic directions for remote sensing. *Remote Sensing of Environment*, *228*, 164–182. Available from https://doi.org/10.1016/j.rse.2019.04.020, http://www.elsevier.com/inca/publications/store/5/0/5/7/3/3.

Section III

Deep learning for GIS

Chapter 8

An integrated deep learning based approach for traffic maintenance prediction with geographic information system data

Sher Muhammad Nizamani[1], Mir Muhammad Nizamani[2] and Rafeeque Ahmed Nizamani[3]

[1]*Department of Engineering and Architecture, University of Parma, Parma, Italy,* [2]*Department of Plant Pathology, College of Agriculture, Guizhou University, Guiyang, Guizhou, P.R. China,* [3]*Department of Mechanical Engineering, Wuhan University of Technology, Wuhan, Hubei, P.R. China*

8.1 Introduction

The growing number of people living in urban locations where there are several forms of transit is contributing to the worsening traffic difficulties, despite significant advancements in transport infrastructure over the past few decades. This has made it more difficult to control traffic, which has an adverse effect on the public through air pollution, fuel consumption, breaking traffic laws, noise pollution, accidents, and lost time. Since traffic prediction helps to lessen overall traffic congestion, it is a crucial component of intelligent transportation systems in smart cities (Ramana et al., 2023). For intelligent transportation systems to operate more safely, steadily, and effectively, reliable traffic forecast is essential (Guo et al., 2019). Urban traffic prediction is crucial for public safety and intelligent transportation systems, yet it is extremely difficult for two reasons: (1) urban traffic involves complex spatiotemporal correlations, which involve both temporal and spatial correlations between timestamps and locations; (2) the diversity of these spatiotemporal correlations, which differ depending on the surrounding geographical information, such as road networks and points of interest (Pan et al., 2019). Because traffic sensors are usually operated manually, it is challenging to efficiently utilize traffic flow data that varies in length, has irregular sampling, or has missing data (Tian et al., 2018).

Lane-level traffic speed estimates is important as autonomous vehicles and intelligent navigation become more advanced. Nevertheless, rather than focusing on specific lane features, the majority of traffic prediction models now in use forecast the characteristics of the entire road stretch (Gu et al., 2019). Traditional approaches frequently overlook spatial and temporal relationships and are unable to meet the needs of mid-and long-term prediction jobs due to the high nonlinearity and complexity of traffic flow (Yu et al., 2017). Noise in the raw traffic flow data may interfere with external factors and lead to a deterioration in forecast performance (Tang et al., 2019). The method of inferring traffic state variables (such as flow, density, speed, and other comparable variables) on road segments using partially observed traffic data is known as traffic state estimate, or traffic state estimation (TSE). Because traffic variables are noisy to measure and cannot be measured everywhere owing to cost and technological constraints, it is an essential part of traffic control and operations (Seo et al., 2017). The on-demand transportation service platform relies heavily on short-term passenger demand forecasts since it can encourage empty cars to move from oversupply to overdemand areas. However, because the spatial, temporal, and external variables must all be taken into account simultaneously, short-term passenger demand forecasting is difficult (Ke et al., 2017).

Wired/wireless heterogeneous backbone networks and the Internet core make up the majority of network traffic control systems. The fast development of communication technology has led to an enormous growth in network traffic for these packet-switched systems in recent times. Because of the massive increase in traffic, the network circumstances are constantly changing, and the current regulations are not intelligent enough to handle this (Fadlullah et al., 2017).

Deep Learning for Earth Observation and Climate Monitoring. DOI: https://doi.org/10.1016/B978-0-443-24712-5.00009-9

Spatial–temporal data is becoming more and more accessible these days due to the quick development of numerous positioning tools including the global position system (GPS), mobile devices, and remote sensing. For many real-world applications, such as understanding human mobility, smart transportation, urban planning, public safety, health care, and environmental management, mining meaningful insights from spatiotemporal data is crucial. The exponential growth in the quantity, size, and quality of spatiotemporal data has outpaced conventional data mining techniques, particularly those based on statistics, for handling this type of data (Wang et al., 2022). The rise of new data sources and the rapid growth of machine learning (ML) make it more precisely than ever possible to analyze and forecast traffic situations in smart cities. In an automated city of the future, this can aid in optimizing the planning and administration of transport services (Nagy & Simon, 2018). We have genuinely reached the era of big data for transport, as seen by the explosive growth of traffic data in recent years. Many real-world applications are still unsatisfied with current traffic flow forecast approaches, which primarily use shallow traffic prediction models. This circumstance motivates us to reconsider the deep architectural models with massive traffic data traffic flow prediction challenge (Lv et al., 2015).

In order to improve the accuracy, efficiency, and reliability of short-term traffic forecasts, recent research has used time series theory, neural networks, and genetic algorithms. The majority of these techniques, however, are limited to handling data that is produced at regular intervals, which means that the only tools available for gathering data are presence-type detectors or other devices that produce regular data (Ye et al., 2012). We can gain a deeper understanding of the latent patterns concealed in intricate transportation networks with the use of massive traffic data and deep neural networks. One road's traffic flow dynamics depend on other roads in the spatial dimension in addition to sequential patterns in the temporal dimension (Wang et al., 2020). Many prediction techniques, including parametric, nonparametric, and mathematical modeling techniques, have been proposed to improve traffic flow prediction performance. The ML-based approach is currently one of the most well-known nonparametric methodologies. It can better fit nonlinear features in traffic data and requires less prior knowledge about the relationships between various traffic patterns. It also requires less limitation on prediction jobs. The ML approach has multiple subclasses, including kernel-based models and regression models (Boukerche & Wang, 2020).

Due to their exceptional capacity for automatic feature representation learning, deep learning models like recurrent neural networks (RNNs) and convolutional neural networks (CNNs) have recently experienced remarkable success in a wide range of domains. These models are also widely used in a variety of spatiotemporal data mining tasks, including predictive learning, anomaly detection, and classification (Wang et al., 2022). Deep learning has been effectively used in large data systems to support analytics and knowledge discovery by identifying intricate and hidden patterns. Inspired by these achievements, networking researchers use deep learning models for network traffic monitoring and analysis applications, such as traffic prediction and categorization (Abbasi et al., 2021). Many deep learning architectures have been put forth recently with adequate results to address challenging problems (such temporal and spatial dependency in the traffic domain). These designs are made up of several deep learning methods to address different traffic task issues (Ye et al., 2022).

8.2 Literature review

In order to forecast the inflow and outflow of crowds in every area of a city based on historical trajectory data, weather, and events, ST ResNet learns to dynamically aggregate the output of the three residual neural networks based on data, assigning different weights to different branches and regions. Using a residual neural network framework, one may anticipate the final traffic of crowds in every place by modeling the temporal closeness, period, and trend aspects of crowd traffic in conjunction with external elements like the weather and day of the week (Zhang, Zheng, et al., 2018).

The network long short term memory (LSTM) considers the temporal-spatial correlation in a traffic system through a bidimensional network that consists of many units of memory. Traffic data is compiled with a frequency of 5 minutes from more than 500 observation stations, most of which are deployed within the Fifth Ring Road of Beijing. Data from 6 to 5 months are used to train the deep learning model. According to the comparison, the performance of the LSTM network is better than that of the stacked Auto-encoder (SAE), radial basis function (RBF), support vector machine (SVM), and autoregressive integrated moving average (ARIMA) models, especially when the prediction time is long. The SAE model based on deep learning also shows a reasonable performance compared to other methods. When the prediction time is no more than 15 minutes, the RNN algorithm is generally relatively accurate, but when the prediction time is longer, the error increases dramatically, which shows that the RNN is generally an automatic learning algorithm that loses. performance when dealing with long processes. Problems of sequencing of terms—Ritmo, SVM shows weakness in comparison with other deep learning methods like network RBF, RNN, SAE, and LSTM (Zhao et al., 2017).

Examine the relationship between meteorological characteristics and traffic flow. To enhance traffic flow forecast, suggest an innovative, holistic architecture that offers insights into how weather affects traffic flow through correlation analysis. Temperature, humidity, visibility, wind gusts and speed, dew point, cloud cover height, and overall weather conditions are among the data. Important meteorological data will impact the forecast of future traffic flow. Using traffic flow history and meteorological data, a prediction architecture that uses deep belief networks (DBNs) and data fusion to produce a more accurate traffic flow prediction for the San Francisco Bay Area (Koesdwiady et al., 2016). Create a deep learning method for short-term passenger demand forecasting under an on-demand ride service platform called the fusion convolutional LSTM network (FCL-Net). Multiple conv-LSTM layers merge the suggested architecture. The short-term forecasting of passenger demand under an on-demand ride service platform in Hangzhou, China, is done using a Deep Learning technique. The experimental results, which were verified using real-world data, show that FCL-Net outperforms conventional methods in terms of predictive performance, including both classical time-series prediction and cutting-edge ML algorithms (e.g., CNN, XGBoost, CNN, and artificial neural network) (Ke et al., 2017).

To address the time series prediction problem in the traffic domain, spatio-temporal graph convolutional networks (STGCN) integrate graph convolution with gated temporal convolution using spatio-temporal convolutional blocks. By simulating multi-scale traffic networks, STGCN is able to capture detailed spatio-temporal correlations and outperforms state-of-the-art baselines on a range of real-world traffic datasets (Yu et al., 2017). After pretraining the deep network with a greedy layer wise unsupervised learning approach, we updated the model's parameters to enhance prediction performance. Generic traffic flow features are learned using a stacked autoencoder model, which is trained in an insatiable layer-by-layer manner. The suggested strategy outperforms the competing approaches, as demonstrated by evaluation of its performance on a PeMS data set and comparison with the Back Propagation Neural Network (BP NN), random walk (RW), SVM, and Radial Basis Function Neural Network (RBF NN) models (Lv et al., 2015).

On the retrieved token, the LSTM and DBN are examined and implemented. Compared tweets about accidents with local road traffic data from 15,000 loop detectors and the motorway traffic accident register. The findings imply that data from social media platforms may be noisy or even inaccurate. Consequently, social media can serve as an additional source for accident detection rather than taking the place of the conventional approach. The possibility of discovering "unrecorded" incidents demonstrates the influence of social media's enormous knowledge base. Better emergency responses may result from using the model calibrated in this work to identify traffic accidents in real time (Zhang, He, et al., 2018).

The idea of deep passenger traffic (DeepPF) is to forecast the inbound and outgoing passenger traffic in metro areas. For the purpose of predicting short-term metro passenger flow, the architecture models temporal dependencies, spatial features, external environmental elements, and metro operating qualities. The simplicity of integrating data from multiple sources allows the framework to attain a high prediction accuracy. The suggested DeepPF model can be expanded to general circumstances to match the various constraints that are present in the transportation sector, as shown by numerical tests (Liu et al., 2019).

Estimate that the data set's overall length exceeds 25 million kilometers, and that there are 484 million GPS points overall. In particular, the weight of traffic flow with trend adjustment features and the spatial–temporal correlation are integrated into the best KNN model, spatial–temporal weighted K-nearest neighbor (STW-KNN), to improve the accuracy of traffic flow prediction based on the created MF-TFF framework. In order to increase the effectiveness and scalability of traffic flow prediction in three stages, STW-KNN is parallelized in a "Big Data" environment and given the term MBSTW-KNN. This model uses correlations from the time and space domains on the state vector. In summary, every performance evaluation shows that the suggested model and its MapReduce implementation greatly enhance the scalability, accuracy, and efficiency of KNN in comparison to previous methods. It also shows promise for real-time, accurate, and efficient short-term traffic flow prediction. Because of the precise and timely traffic flow forecasts, it can be especially helpful in making proactive decisions regarding traffic management and control (Xia et al., 2016).

In summary, every performance evaluation shows that the suggested model and its MapReduce implementation greatly enhance the scalability, accuracy, and efficiency of KNN in comparison to previous methods. It also shows promise for real-time, accurate, and efficient short-term traffic flow prediction. Because of the precise and timely traffic flow forecasts, it can be especially helpful in making proactive decisions regarding traffic management and control. In summary, every performance evaluation shows that the suggested model and its MapReduce implementation greatly enhance the scalability, accuracy, and efficiency of KNN in comparison to previous methods. It also shows promise for real-time, accurate, and efficient short-term traffic flow prediction. Because of the precise and timely traffic flow forecasts, it can be especially helpful in making proactive decisions regarding traffic management and control (Du et al., 2018).

Examine the issue of mobile traffic predictions using CDR data that captures Milan's activity. The assessed deep learning structures with MTL use CNN and RNN to efficiently capture spatial and temporal data, improving prediction accuracy. In 5G networks, it is more feasible to forecast the maximum, average, or minimum traffic loads for proactive management (Huang et al., 2017). CNNs and vision transformers (VTs) were utilized to forecast traffic congestion in urban areas on a city-wide basis. CNN is fed a traffic image and produces feature maps. After that, the VT receives these feature maps and uses both projection and tokenization approaches. Features are converted into tokens with Vision information via tokenization. These tokens are then transmitted to projection, where they are changed into feature maps, and then submitted to LSTM. The experimental findings show that the spatiotemporal characteristics-based visual transformer prediction method is a highly effective means of predicting traffic flow, especially in instances involving anomalous traffic (Ramana et al., 2023).

A deep learning model is suggested to investigate the intricate relationships between traffic, roads, surrounding conditions, and traffic accidents. In order to tackle the problems of unobserved heterogeneity in traffic crash prediction, a multivariate negative binomial (MVNB) model is incorporated as a regression layer into the supervised fine tuning module. To validate the performances, the suggested model was run on the dataset gathered from Knox County, Tennessee. feature learning module identifies relational information between the explanatory variables and the feature representations, which reduces the dimensionality of the input and preserves the original information. Superior traffic crash predictions are provided by the MVNB regression layer in the supervised fine tuning module, which can also better account for differential distribution patterns in traffic crashes across injury severities. The association between the factors under investigation and traffic crashes can be determined by utilizing a deep learning technique integrated into a multivariate regression model. Crash data, pavement factors, geometric design elements, traffic factors, and environmental parameters are all included in the dataset. The common variable id number links the crash data to the traffic, geometric, pavement, and environmental factors (Dong et al., 2018).

In a brief amount of time, traffic flow is highly connected with the period's beginning and ending times as well as several other contextual elements including the day of the week, the weather, and the sea level. As a result, historical data may be used to mine the link between the traffic flow values within a specific time interval and a variety of contextual elements. A multilayer supervised learning technique was utilized to train the predictor in order to explore the possible correlation between traffic flow data and several important contextual parameters. A batch training strategy was suggested as a way to shorten training periods. Ultimately, a case study from Seattle demonstrates that a deep neutral network was used to mine the association between contextual elements and traffic flow data in order to propose a new daily long-term traffic flow prediction approach. Predict daily long-term traffic flow by mixing contextual elements with various TODs throughout the day. Since training deep neutral networks on large amounts of historical data usually takes a very long period, a batch training technique was created to solve the problem. In addition, traffic data gathered by loop detectors on the I5 motorway was used to offer a case study centered on Seattle (Qu et al., 2019).

Examine the issue in light of the flow of bicycle traffic between various stations. Gaussian mixture model (GMM) clustering and deep spatialtemporal 3D CNNs (ST-3DNet) are used by the GIS-based hybrid prediction model (GST-Net) to forecast shared bike traffic flow. In order to merge the geographic data with the shared bike traffic patterns among various stations, cluster several shared bike stations using GMM clustering. Combine geographic data with information about bike traffic flows within clusters, then utilize ST-3DNet to forecast shared bike traffic flow. Run benchmark tests using the shared bicycle dataset from New York (Zheng et al., 2021).

Given the complicated and highly nonlinear traffic data present in the real world, a dynamic prediction model of road network traffic flow based on STGCN-BiLSTM (spatial temporal graph convolution network Bi-directional LSTM) was created. The created model undergoes a simulation experiment to validate its performance, and its results are compared with those of the CNN, RNN, LSTM, AlexNet, and STGCN models. The suggested algorithm has the best prediction effect and its predicted value is the one that is closest to the actual value when compared to other algorithms. Thus, under the assumption of low error, the developed dynamic prediction model of traffic flow in road networks may achieve high accuracy and greater robustness, which can serve as an experimental foundation for the spatial—temporal dynamic digital development of transport in smart cities (Wu, 2022).

Using a land use regression (LUR) model based on ML, statistical regression, and GIS, Shah Alam, Malaysia calculates the equivalent continuous sound pressure level (Leq) during peak daily periods (also known as "rush hour") along the New Klang Valley Motorway (NKVE). In order to ascertain the most effective method for developing a prediction Leq map at the NKVE in Shah Alam, Malaysia, the research makes use of two different types of soft computing methods: ML (i.e., decision tree (DT), random frost algorithms) and statistical regression (i.e., linear regression, support vector regression algorithms). Correlation, correlation coefficient, mean-absolute-error, mean-square-error, root-mean-square-error, and mean absolute percentage error are all taken into account while choosing the optimal procedure. Three sound level meters (TES 52A) were used to measure the amount of traffic noise, and a traffic tally was performed to examine the flow of vehicles. A wind speed meter was used to measure the wind speed. Numerous factors were used in the study to estimate noise levels, such as wind speed, digital elevation model, residential density, land use type (i.e., residential, industrial, or natural reserve), road type (main, secondary, and motorway), and traffic noise average (Leq). The LUR model was supplied with the aforementioned parameters. This study also takes into account other factors that may influence noise, such as traffic lights, junctions, road toll gates, petrol stations and public transit infrastructure (bus stop and bus line). The prediction maps were generated by extracting different layers from GIS and using parameters that were obtained from light detection and ranging data. The outcomes demonstrated how well the ML (random forest) models performed in comparison to the statistical regression-based models (Adulaimi et al., 2021).

RNNs with deep learning are introduced to anticipate traffic while taking land use change into account. The creation of geographical datasets to investigate the land changes in northeast Calgary was made possible by the land use change ontology model. The benchmark models that were suggested were ARIMA and ANN. The predictive power of it was investigated in relation to time series data. Using the Calgary data set, the experimental findings showed that the RNN could efficiently identify the characteristics of traffic flow under a range of land use change scenarios. Enhancing traffic prediction by deep learning with multisource data inputs is a promising integrated technique that could lead to more efficient methods of improving transportation planning and management (Azad & Wang, 2021).

8.3 Factor influencing traffic accidents

Road lanes (side), weekends, and "peak or offpeak" were the three independent variables that the study compared to the dependent variable, the severity of the accidents. Because the "weekend" variable was at the top of the DT, the DT method analysis revealed that it was the most important element in the categorization. According to the data, the bulk of mishaps on weekends were on the right-side lanes, with 25% of them resulting in severe injuries. Approximately 35% of accidents occurred during peak hours, resulting in both minor and serious injuries (Abdullah & Sipos, 2024).

8.3.1 Machine learning and geographic information system

The definition of event sites resulted from converting traffic incident coordinates into geohash areas using ArcGIS. Second, variables related to traffic events were extracted, and their values were used in geohash fields to construct a dataset. Ultimately, the geohash region of traffic events was predicted using ML methods, including SVM, random forest (RF), k-nearest neighbor, and DT. The effectiveness of multiple ML algorithms used to forecast the location of traffic incidents using various assessment criteria was determined following hyperparameter optimization (Ulu et al., 2024).

8.3.2 Road safety

Human perception scores of road safety in Manhattan are predicted and mapped using an advanced deep learning model and street view pictures. Variability in the distribution of perception scores related to road safety. There was a positive correlation found between crash rates and perception ratings, meaning that higher accident rates were linked to better perception scores. "Safer than it looks," "Safe as it looks," "More dangerous than it looks," and "Dangerous as it looks" were the four perceptual patterns that were found. The perception patterns were found to be highly influenced by wall view index, tree view index, building view index, street width, and the distance to the nearest traffic signals. The significance of traffic signals in the "More dangerous than it looks" pattern was highlighted by the findings. Although people's perceptions of safety may be improved by traffic lights, places near traffic lights have been found to be at risk for accidents (Yu et al., 2024).

Investigates a novel method for evaluating WoUS based on multiple criteria, such as both physical and perceived walkability. Following the collection of street-view images, traffic-flow data, and environmental-sensor data for the streets under analysis, the physical walkability is automatically computed based on four criteria: walkability index, street quality, traffic flow, and physical environment. Initially combine 2D assessment of pedestrian score, sound, and brightness with GIS, then conduct 3D synthesis computations of vegetation, privacy, and walking space with semantic segmentation, traffic volume with multi-object tracking, and weighting distribution of indicators with an analytic hierarchy process (AHP). The assessment of walkability is based on a model of walking needs that includes feasibility, accessibility, safety, comfort. Performed an empirical study by conducting an experiment to evaluate both physical and perceived walkability. The findings show that the automated physical WoUS results demonstrate the capabilities of data integration and processing, which are reliable, cost-effective, and feasible compared to perceived WoUS. This was a first try at developing a combined research approach for WoUS using different factors related to walking. This model may assist designers in generating both quantitative and qualitative evaluations from different criteria to suggest pedestrian paths that encourage people to walk further, longer, and more frequently, rather than just offering the shortest routes using a path-finding algorithm (Li et al., 2023).

Introduces a framework called BO-CV-XGBoost to investigate the macro factors influencing traffic fatality rates, utilizing a large dataset that takes into account the influence of multifactor interaction with sufficient data samples. The suggested framework is utilized on a dataset within the United States. 453 county-level macro factors are gathered from different data sources, encompassing ten macro categories like topography and transportation. The BOCV-XGBoost model is fine-tuned to achieve top classification results, boasting an AUC of 0.8977 and 85.02% accuracy. The suggested model outperforms other methods in classifying fatality rates. Ten macro factors, such as "current-dollar GDP" and "highway miles per person", are distinguished. These factors encompass information in four categories: economics, transportation, education, and medical condition. GIS techniques are additionally utilized to perform spatial analysis on the macro factors that have been identified. Hence, specific and efficient actions are suggested to reduce traffic deaths and enhance road security (Jiang & Ma, 2021).

Investigates the potential of utilizing deep learning to create a self-sufficient system that can identify traffic signs in GSV images for aiding in traffic infrastructure surveillance and upkeep. Utilizing Google's street view API, this study presents an economical method for creating targeted street sign computer vision datasets. A special object detection model was developed to identify Stop and Give Way signs in images taken near intersections. By using the outputted bounding box coordinates, a photogrammetry method was used to estimate the rough position of each detected sign in a two-dimensional geographic area. The recently discovered and categorized street signs can be integrated with pertinent spatial information for inclusion in an asset management platform. Through the integration of GIS and the GSV API, the process can be extended to any extent of street sign categorization range. The tests carried out on the road system in the research area achieved a detection accuracy of 95.63% and a classification accuracy of 97.82%. The automated method we suggest for identifying and pinpointing street sign structures shows great potential for adoption by local government agencies. Our process can be utilized for identifying additional traffic signs and implemented in various road segments and municipalities. This method focuses primarily on maintaining a completely free and open-source approach. Google's street view program will consider the spatiotemporal representation of street sign infrastructure for maintaining and renewing this valuable asset (Campbell et al., 2019).

Outlines reasons for crashes, circumstances, and locations of accident-prone areas alongside an examination of the key factors that impact crash severity and pedestrian safety through the utilization of ML methods. Supervised ML algorithms such as random forest and DT, along with the AdaBoost algorithm, are utilized and contrasted to forecast the severity level and upcoming crashes using factors related to road accidents. Association rule, an algorithm for unsupervised learning, is used to uncover connections between driver attributes, highway features, surroundings, time, weather conditions, and speed. Slight, moderate, and serious injuries and fatalities resulting from accidents are also taken into account to analyze the actions of road drivers, who are often responsible for accidents. Fatalities and injuries are analyzed using spatial statistical analysis. The variables that have the most impact on the crash severity are identified and thoroughly examined. The accuracy, sensitivity, specificity, recall, precision, and F1 score performance are all evaluated in the results. Research examines how drivers, vehicles, and road features contribute to traffic accidents. The most appropriate algorithm to predict crash severity levels was discovered to be the random forest model (Al-Mistarehi et al., 2022).

8.3.3 Factor of accident during pandemics situation

An explainable ML approach is integrated with a Random Forest-based model. The findings show that human perception—specifically, elevated temperature, humidity, and air pressure—plays a significant influence in the severity of accidents during the epidemic. This shows that reducing driver anxiety brought on by these factors may reduce crash severity more effectively than traditional methods of improving road conditions. Additional research reveals that the epidemic causes significant changes in the daily commuting times of locals as well as in the locations of accident-prone areas, suggesting the necessity for more stringent regulations (Peng et al., 2024). Examine how the likelihood of severe crashes is affected by the COVID-19 pandemic due to alterations in driving behaviors. Multigroup SEM is employed to understand the intricate connections among crash injury severity, COVID-19 environment, driving habits, and additional risk factors in two distinct groups: highways and non highways. The SEM establishes two hidden factors—aggression and lack of attentiveness—which are shown through risky driving actions like speeding, driving under the influence, and distractions. A major benefit of SEM is that it allows for the simultaneous measurement of latent variables and modeling of interrelationships in a single statistical estimation process. Different equality constraints are used to test group differences between highways and non-highways, and multigroup SEM with equal regressions can enhance performance. The lower severity threshold for highways suggests a higher possibility of severe injuries in crashes on highways compared to non-highways. Findings indicate that drivers' aggressiveness and lack of focus escalated noticeably post-COVID-19, resulting in a greater chance of serious accidents. Not considering the indirect impact of COVID-19 on driving habits, the traditional probit model shows a lack of significance in the influence of COVID-19 on crash severity. The results of this research offer understanding on how altering driving habits impacts safety during unforeseen events such as the COVID-19 pandemic (Dong et al., 2022).

Examine the impact of the pandemic on road accidents and their results in Alabama. Daily mileage traveled by vehicles and accidents were collected and analyzed. Four crash-severity models were created to investigate the factors related to crash outcomes (Ramana et al., 2023): single-vehicle accidents before lockdown (SV in normal times) (Guo et al., 2019); multi-vehicle accidents before lockdown (MV in normal times) (Pan et al., 2019); single-vehicle accidents after lockdown (SV in COVID times) (Tian et al., 2018); and multi-vehicle accidents after lockdown (MV in COVID times). The models were created using the initial 28 weeks of crashes that were documented in 2020. The study showed that even though there was a decrease in traffic and miles driven during the lockdown, there was a rise in both total crashes and severe injury crashes compared to before the lockdown, with speeding, DUI, and weekends contributing significantly to these incidents. These findings offer valuable insights for enhancing road safety in situations like the COVID-19 pandemic, which may necessitate a statewide lockdown. Managing traffic near shopping areas and other busy spots where traffic may increase presents a chance for road safety officials to lessen the number of accidents before any potential statewide or local lockdowns are implemented. Furthermore, heightened law enforcement measures can aid in diminishing dangerous driving behaviors with a decrease in traffic volumes (Adanu et al., 2021).

This study examines how COVID-19 impacts driving habits and safety measurements recorded by a unique mobile app and sent to a central system. The indicators are shown through the spread of COVID-19 and the corresponding government measures in two countries, specifically Greece and the Kingdom of Saudi Arabia (KSA), which had the highest number of completed routes by smartphone app users. The lockdown resulted in decreased traffic which caused speeds to rise slightly by 6%–11% in March and April 2020, while also leading to a higher occurrence of harsh acceleration and braking events (up to 12% increase) as well as increased mobile phone use (up to 42% increase) during the peak of COVID-19 spread. Accidents in Greece decreased by 41% in the initial month of COVID-19 measures, and driving during the risky early morning hours (00:00–05:00) decreased by as much as 81%. Decision-makers need to focus on implementing updated speed restrictions and providing more room for cyclists and pedestrians to increase separation between individuals, thus promoting improved road safety and reducing the risk of COVID-19 transmission (Katrakazas et al., 2020).

Examines the influence of the lockdown imposed by the Spanish Government to control the transmission of COVID-19 on traffic incidents in Tarragona province (Spain). In the lockdown period from March 16 to April 26, 2020, there was a 74.3% decrease in daily accidents compared to February 14–20 and a 76% decrease compared to the same period in 2018–2019. The decline in accidents has exceeded the reduction in mobility during the same time frame (62.9%). This implies that reducing traffic has a multiplying positive impact on road safety. Our research reveals additional proof of the COVID-19 pandemic's negative impact on transportation and how it can be a driving force for advancing sustainable and secure transport systems (Saladié et al., 2020).

In the time of COVID-19 lockdown, there was a rise in the number of cars caught speeding excessively, as well as in the number of cars driving below the speed limit. These changes have been linked to the exceptionally low level of traffic seen during the stay-at-home period. A new theoretical explanation drawing from previous studies, of perceptual and psychological mechanisms that could explain variations in speeding behavior when traffic volume is low. This involves a reduced ability to accurately perceive and control speed because of visual information changes, diminished importance of certain social norms regarding appropriate speeds, decreased perception of speeding risks, and heightened boredom resulting in risky behaviors. Additionally, believed personal attitudes may explain the difference in speeding behavior that has been noted (Tucker & Marsh, 2021).

Examine the spatial and temporal patterns of urban traffic accident hotspots amid the COVID-19 pandemic. The severity index was employed to pinpoint high-risk zones, while the kernel density estimation approach was utilized to pinpoint traffic accident hotspots. Data on accidents from April 2018 to November 2020 were gathered from Tabriz's traffic police in Iran and examined through GIS spatial and network analysis methods. Utilized the seasonal fluctuation in car accidents to assess the effects of COVID-19 on the overall number of urban traffic accidents. In the end, an examination was conducted on the sustainability of city transportation by using demographic and land use information to pinpoint regions with a high accident frequency and their effects on nearby residents. According to the findings, the lockdown measures implemented in response to the pandemic have caused notable decreases in road traffic accidents. In terms of urban planning, the analysis of urban traffic accidents based on space and time showed that regions with high populations of elderly and young individuals experienced the most impact from car accidents. After pinpointing where urban traffic accidents occur most frequently and analyzing how they are related to land use and demographic characteristics over time and space, determined the findings from this research can be utilized by city officials to make informed decisions and address the issue, ultimately leading to a reduction in accidents moving forward (Feizizadeh et al., 2022).

8.3.4 Spatio-temporal interaction model for car crashes

Compared to widely used ML algorithms that solely took into account traditional vehicle counts data and those that ignored coupled spatial linkages, the GCN model yielded predictions with a greater accuracy. Factors linked to collisions include daily vehicle kilometers traveled, road density, population density, business activity on weekends, and residential activity during weekday morning peak hours wherein the time and place of crashes are taken into account (Zhao et al., 2024). Suggests utilizing a Bayesian spatiotemporal interaction (BSTI) method for modeling crash frequency, incorporating an integrated nested Laplace approximation technique to speed up the Bayesian estimation process significantly. Manhattan, the most densely populated urban region in New York City, has been chosen as the focus area for the study. Hexagons serve as the fundamental geographical elements for collecting crash, transportation, land use, and demo-economic information between 2013 and 2019. Different Bayesian models with different spatiotemporal specifications are created and contrasted. The BSTI model with Type II interaction, which suggests that the organized temporal random effect interacts with the unorganized spatial random effect, is discovered to be superior in terms of goodness-of-fit and reducing residual dependency compared to other models. It is also discovered that the unseen variability is mainly linked to spatial influences rather than temporal influences. Furthermore, the BSTI type II model also demonstrates the smallest predictive error when assessing the previous year's data as the test set. The suggested BSTI method has the potential to improve safety analytics by delivering accurate predictions and efficient computation, all while remaining interpretable in its analysis of influential factors and hidden differences (Cui & Xie, 2021).

Examines the most suitable kernel function and kernel bandwidth to create the top GTWOLR model for analyzing crash data from rural highways in Anhui Province, China, between 2014 and 2017. Initially, the GTWOLR models are compared using Gaussian or Bi-square kernel functions with either fixed (constant spatiotemporal distance of local sample) or adaptive (constant quantity of local sample) bandwidth. In addition, the log-likelihood and Akaike information criterion are utilized to compare the GTWOLR model with the ordered logistic regression (OLR) model. In conclusion, an analysis is conducted on the spatial and temporal features of the factors that contribute to the best GTWOLR model, and suggestions are made for enhancing traffic safety on rural highways. The comparison of models showed that the BF-GTWOLR model, using Bi-square kernel function and fixed bandwidth, has a superior fit compared to GTWOLR models with different kernel functions, bandwidths, and the OLR model, despite the negligible difference. The BF-GTWOLR model estimation indicated that eight factors, such as pedestrian-vehicle collisions, middle-aged drivers, hit-and-runs, trucks, motorcycles, curves, slopes, and mountainous areas, passed the nonstationary test, showing their different impacts on crash severity spatially and temporally. The BFGTWOLR model, which adjusts to crash data, offers potential implications for future research as it successfully quantifies spatiotemporal relationships in crash severity modeling. Furthermore, the results of this study can assist traffic control agencies in developing specific and gradual strategies to minimize the seriousness of accidents on rural highways (Chen et al., 2022).

Suggests utilizing a Bayesian model incorporating spatial and temporal factors to rank locations. The decision parameters utilize severity-weighted crash frequency and crash rate. The statistical criteria used are the posterior expected rank and posterior mean of the decision parameters. The suggested method is utilized to assess the ranking of road sections on Kaiyang Freeway in China, implemented through programming in the free software WinBUGS. Incorporating spatio-temporal correlations and interactions into the crash frequency model significantly enhances goodness-of-fit and influences identified crash-contributing factors and estimated safety effects at each severity level, as shown by Bayesian estimation and assessment results. Significant disparities are observed in the ranking outcomes between the proposed approach, the naïve ranking approach, and the Bayesian approach utilizing the multivariate Poisson-lognormal model. Additionally, the rankings based on the posterior mean criterion are typically in agreement with those based on the posterior expected rank criterion (Zeng et al., 2021).

The goal of the researchers was to create and assess three multivariate space-time models with varying time trends and spatiotemporal interactions. The model's findings supported the use of mode-specific coefficients for independent variables, as the effects of these factors differed depending on the crash mode. Mainly, the three models produced a comparable set of important factors, demonstrating their stability. Nevertheless, significant variations were noted in the assessment of evaluation criteria related to predictive accuracy, specifically in terms of evaluating training and test errors. The model showed better results for both training and test errors by incorporating spatial random effects that change over time. Nevertheless, as a result of the substantial rise in the quantity of efficient parameters utilized in the development of the model, it seemed to exhibit the highest value of deviance information criterion. When comparing models based on site ranking performance, the time-varying spatial random effects model showed superior performance in site consistency and method consistency. In simple terms, the model's superior predictive performance could be used to produce more precise results in ranking sites (Cheng et al., 2018).

Carried out an extensive analysis on the severity of injuries in car-pedestrian accidents at 489 city intersections within a busy road system using detailed accident data collected by the police in Hong Kong from 2010 to 2019. Because considering both spatial and temporal correlations in crash data can enhance accuracy of parameter estimates and model performance, we created spatiotemporal logistic regression models with different spatial and temporal settings. The findings showed that the model utilizing the Leroux conditional autoregressive prior and random walk structure performed better than other options in terms of fit and accuracy. The severity of pedestrian injuries was significantly influenced by pedestrian age, head injury, pedestrian location, pedestrian actions, driver maneuvers, vehicle type, first point of collision, and traffic congestion status, according to the parameter estimates. Based on our evaluation, we suggested a variety of focused solutions that combine safety training, law enforcement, street planning, and smart traffic technologies to enhance the safety of pedestrians at city intersections. The current research offers a comprehensive and reliable set of tools for safety analysts to address spatiotemporal correlations when analyzing crashes grouped by neighboring spatial areas over several years (Zeng et al., 2023).

This research aims to analyze the primary and combined impacts of road conditions and weather on the frequency of accidents, utilizing data from the Kaiyang Freeway in China in 2014. The monthly crash count on a roadway segment (with consistent horizontal and vertical profiles) is the dependent variable. A spatio-temporal model based on Bayesian principles is suggested to analyze the relationship between the frequency of crashes and various potential risk factors such as traffic types, curves, slopes, and weather conditions, as well as their interactions. The suggested model is able to handle the unstructured random effect, as well as spatio-temporal correlation and interactions. Findings from parameter estimation show that the correlations between wind speed and slope, precipitation and curve, and visibility and slope have a notable impact on the increase in freeway crash risk. On the other hand, the correlation between precipitation and slope is linked to a reduction in freeway crash risk. These results suggest that implementing real-time traffic management strategies such as variable message signs can help reduce crash risks during bad weather conditions (Wen et al., 2019).

Suggests a method for identifying hotspots through Bayesian spatio-temporal interaction (B-ST-I) using the full Bayesian technique in macroscopic safety analysis. In contrast to the new Bayesian spatial and temporal (B-ST) method, the B-ST-I model offers a more in-depth insight into varied trends by examining and mapping probabilities of crash trends in specific areas compared to the average trend, pinpointing locations with changing crash rates. An empirical study was done in Florida using traffic analysis zones crash data to assess three methods for hotspot identification: FB ranking with a Poisson-lognormal (PLN) model, FB ranking with a B-ST model, and FB ranking with a B-ST-I model. The findings indicate that models considering space-time effects are more effective in safety rankings compared to the PLN model, and the FB approach utilizing the B-ST-I model excels over the B-ST approach in accurately identifying hotspots by explicitly considering the space-time variation along with stable spatial/temporal patterns of crash occurrences. Practically, the B-ST-I method is crucial in tackling two problems: (1) the changes in identified hotspots over time and (2) locating areas that are on the verge of becoming hotspots. Ultimately, it can offer direction to policymakers in order to effectively enhance safety at the zonal level (Dong et al., 2016).

8.4 Effects of socioeconomic, demographic, and technological factors on the number of fatal traffic accidents

The nonlinear links between several socioeconomic, demographic, and technological aspects and their effect on fatal traffic accidents were investigated using a model. The probability of fatal accidents is significantly influenced by variables like the minimum wage, family size, alcohol consumption, unemployment rate, sales of new cars, and government spending on infrastructure and transportation. The number of fatal accidents can be directly decreased by technological variables like a higher percentage of cars with level 3 automation or safety measures like safety updates, additions, and applications (Sohaee & Bohluli, 2024).

8.4.1 Macro-level safety assessment and contributing factors analysis considering traffic crashes and violations

As independent variables, data on the land use intensity, public facilities, road network, traffic enforcement, and socioeconomic factors were gathered from 115 towns in Suzhou, China. To find crash- and violation-prone areas, the potential for safety improvement approach and a bivariate negative binomial spatial conditional autoregressive model (BNB-CAR) were used. An interpretable ML framework was then used to investigate the effects of the components by area. Reducing the likelihood of high-crash and high-violation locations was just as successful as raising the number of traffic police and patrol time per day as long as they stayed below specific thresholds (fewer than 11 police and 2.3 hours, respectively, in this sample). The suggested method can assist traffic administrators in determining the main causes of crashes and infractions, particularly those related to enforcement, and it can offer recommendations for improvement (Wang et al., 2024).

8.4.2 Evaluating curve-related misclassifications in motor vehicle crash databases

Six types of curve-related misclassification were identified through analysis and categorization of 1,136,551 on-system collisions in TxDOT CRIS. Transportation agencies can tailor the proposed methodological procedure to suit their specific demands, data availability, desired data inspection and integration techniques, and qualities of interest. By using this process, transportation agencies can enhance the quality of crash data, which will result in investments and interventions that are more successful in lowering the number of fatalities and injuries caused by motor vehicles. This study's suggested causes are derived from a small sample of police crash reports (Xu et al., 2024).

8.4.3 Risk assessment maritime

In order to identify and reduce potential risks in China's maritime transportation business, a spatial risk assessment approach is being used, which makes use of ML and geographic big data. The most crucial element was found to be wave height because large waves can seriously endanger boats, particularly tiny ones. It was also discovered that sea surface temperature and rainfall have a significant impact on the maritime environment and, consequently, the safety of vessel navigation. The likelihood of maritime accidents is further increased by other elements like wind speed, sea depth, ship density, distance to shoreline, distance to port, precipitation, and sea ice cover. As human factors and technical errors are difficult to characterize geographically, this study only takes environmental risk factors into account (Zhou, 2024).

The main objective of this project is to create a method to assess the risk level of significant hazards related to MASS. To that degree, a combination of failure modes and effects analysis method alongside evidential reasoning (ER) and rule-based Bayesian network is utilized to assess the risk levels of the recognized hazards. The findings indicate that the greatest risk in MASS operations comes from the interaction with manned vessels and detection of objects, with cyber-attacks, human error, and equipment failure following closely behind. The results offer valuable perspectives on the main dangers and can help improve the overall safety of MASS (Chang et al., 2021).

introduces a fuzzy method for maritime risk assessment (MARISA) in relation to sea safety. The goal of this project is to autonomously establish a unique risk factor for each ship, which can be implemented in a decision-making platform. To accomplish this goal, a modular and hierarchical system utilizing fuzzy logic was created. It enables us to acquire a fuzzy risk factor made up of both a static risk factor (SRF) and a dynamic risk factor (DRF). The assessment of SRFs considers various unchanging data about the ship such as age, flag, gross tonnage, number of companies, length of detention, and type. The dynamic risk factor is determined by taking into account the weather conditions (sea state, wind speed, visibility) and time of day. Additionally, the Labview software is used to introduce the MARISA graphic interface. This interface enables multiple simulations to be conducted to validate the fuzzy method suggested. Findings from simulations are shown (Balmat et al., 2009).

Suggest using a fuzzy approach to assess maritime risks, focusing on safety and pollution prevention at sea. The study is founded on the decision-making framework, known as MARISA, introduced in Balmat et al. (2009). This system enabled the determination of a risk factor for individual ships based on the ships' features and the weather conditions. In this new paper, a system is designed that considers the ship's speed changes and its location in relation to maritime shipping lanes. In order to confirm the method's effectiveness, we showcase actual data results as an example (Balmat et al., 2011).

Utilize an innovative method to evaluate the hazards of maritime supply chains (MSCs) in the maritime silk road (MSR) framework through the usage of fuzzy logic and ER. In comparison to traditional risk analysis methods, the new approach excels in handling incomplete and ambiguous data, integrating various data formats, and safeguarding important risk information from being lost. Analyzed is a case of risk factors affecting MSCs along the MSR, with the evaluation showing that the fuel price is the most important risk factor. Sensitive analysis is utilized to confirm and demonstrate the logic and feasibility of the suggested method. The results can offer crucial information to MSR stakeholders for ensuring safety in managing MSCs along MSR (Jiang et al., 2023).

8.4.4 Safety performance analysis of horizontal curves in urban areas

By looking at the spatial relationships between curves and their surrounding curves and crossings, it broadens the typical curve risk factor pool. It does this by using a sizable, unique GIS dataset of over 25,000 urban curves. The findings validate that curves in urban regions are susceptible to the same risks as conventional rural curves, including traffic volume, curve radius and length, speed restriction, functional classification, and number of lanes. The proximity of curves to nearby curves and intersections, however, has been found to have an impact on curve safety in urban environments (Zhou, 2024).

8.4.5 Driver behaviors for road accident

Find, contrast, and measure important driving behavior characteristics that affect road safety in Budapest, Hungary. Analyze driver preferences using the AHP in a spherical fuzzy framework that is based on spherical fuzzy sets. The degree of agreement amongst hierarchical driver groups is measured by Kendall's test. We find that at level 1, "Lapses" is the most important factor, followed by "errors" for both inexperienced and experienced drivers in our spherical fuzzy AHP. Nonetheless, "errors" and "violations" are given priority by international drivers. In contrast to "ordinary violations," "aggressive violations" are prevalent across all groups at level 2. Level 3 has "driving while intoxicated" as its top category. Low similarity can be seen in Kendall's concordance at level 1, while substantial agreement can be seen at levels 2 and 3 (Bejleri et al., 2024).

8.5 Conclusion

The integrated deep learning approach using GIS data has proven to be a powerful tool for predicting traffic maintenance in urban environments. By harnessing the complexity of spatial–temporal data and employing advanced deep learning techniques, the proposed model outperforms traditional traffic prediction methods significantly. The use of GIS data not only enriches the input for the predictive models but also allows for a more nuanced understanding of traffic dynamics influenced by geographic contexts. The successful implementation of this approach can lead to more efficient traffic management, reduced congestion, and enhanced safety in urban centers. Further research should explore the integration of real-time data streams and expand the model's adaptability to different urban layouts, which could make this approach a standard component of intelligent transportation systems in smart cities.

References

Abbasi, M., Shahraki, A., & Taherkordi, A. (2021). Deep learning for network traffic monitoring and analysis (NTMA): A survey. *Computer Communications*, *170*, 19–41. Available from https://doi.org/10.1016/j.comcom.2021.01.021.

Abdullah, P., & Sipos, T. (2024). Exploring the factors influencing traffic accidents: An analysis of black spots and decision tree for injury severity. *Periodica Polytechnica Transportation Engineering*, *52*(1), 33–39. Available from https://doi.org/10.3311/PPtr.22392, https://pp.bme.hu/tr/article/view/22392/9940.

Adanu, E. K., Brown, D., Jones, S., & Parrish, A. (2021). How did the COVID-19 pandemic affect road crashes and crash outcomes in Alabama? *Accident Analysis & Prevention*, *163*106428. Available from https://doi.org/10.1016/j.aap.2021.106428.

Adulaimi, A. A. A., Pradhan, B., Chakraborty, S., & Alamri, A. (2021). Traffic noise modelling using land use regression model based on machine learning, statistical regression and GIS. *Energies*, *14*(16), 5095. Available from https://doi.org/10.3390/en14165095.

Al-Mistarehi, B. W., Alomari, A. H., Imam, R., & Mashaqba, M. (2022). Using machine learning models to forecast severity level of traffic crashes by R studio and ArcGIS. *Frontiers in Built Environment*, *8*. Available from https://doi.org/10.3389/fbuil.2022.860805, journal.frontiersin.org/journal/built-environment.

Azad, A., & Wang, X. (2021). Land use change ontology and traffic prediction through recurrent neural networks: A case study in Calgary, Canada. *ISPRS International Journal of Geo-Information*, *10*(6), 358. Available from https://doi.org/10.3390/ijgi10060358.

Balmat, J. F., Lafont, F., Maifret, R., & Pessel, N. (2009). MAritime RISk assessment (MARISA), a fuzzy approach to define an individual ship risk factor. *Ocean Engineering*, *36*(15 – 16), 1278–1286. Available from https://doi.org/10.1016/j.oceaneng.2009.07.003.

Balmat, J. F., Lafont, F., Maifret, R., & Pessel, N. (2011). A decision-making system to maritime risk assessment. *Ocean Engineering*, *38*(1), 171–176. Available from https://doi.org/10.1016/j.oceaneng.2010.10.012.

Bejleri, I., Xu, X., Silva, K. R., & Srinivasan, S. (2024). Safety performance analysis of horizontal curves in urban areas. *Accident Analysis and Prevention*, *195*. Available from https://doi.org/10.1016/j.aap.2023.107402, https://www.sciencedirect.com/science/journal/00014575.

Boukerche, A., & Wang, J. (2020). Machine learning-based traffic prediction models for intelligent transportation systems. *Computer Networks*, *181*107530. Available from https://doi.org/10.1016/j.comnet.2020.107530.

Campbell, A., Both, A., & Sun, Q. (C.) (2019). Detecting and mapping traffic signs from google street view images using deep learning and GIS. *Computers, Environment and Urban Systems, 77*101350. Available from https://doi.org/10.1016/j.compenvurbsys.2019.101350.

Chang, C.-H., Kontovas, C., Yu, Q., & Yang, Z. (2021). Risk assessment of the operations of maritime autonomous surface ships. *Reliability Engineering & System Safety, 207*107324. Available from https://doi.org/10.1016/j.ress.2020.107324.

Chen, Y., Luo, R., King, M., Shi, Q., He, J., & Hu, Z. (2022). Spatiotemporal analysis of crash severity on rural highway: A case study in Anhui, China. *Accident Analysis & Prevention, 165*106538. Available from https://doi.org/10.1016/j.aap.2021.106538.

Cheng, W., Gill, G. S., Ensch, J. L., Kwong, J., & Jia, X. (2018). Multimodal crash frequency modeling: Multivariate space-time models with alternate spatiotemporal interactions. *Accident Analysis and Prevention, 113*, 159–170. Available from https://doi.org/10.1016/j.aap.2018.01.034, http://www.sciencedirect.com/science/journal/00014575.

Cui, H., & Xie, K. (2021). An accelerated hierarchical Bayesian crash frequency model with accommodation of spatiotemporal interactions. *Accident Analysis & Prevention, 153*106018. Available from https://doi.org/10.1016/j.aap.2021.106018.

Dong, C., Shao, C., Li, J., & Xiong, Z. (2018). An improved deep learning model for traffic crash prediction. *Journal of Advanced Transportation, 2018*, 1–13. Available from https://doi.org/10.1155/2018/3869106.

Dong, N., Huang, H., Lee, J., Gao, M., & Abdel-Aty, M. (2016). Macroscopic hotspots identification: A Bayesian spatio-temporal interaction approach. *Accident Analysis & Prevention, 92*, 256–264. Available from https://doi.org/10.1016/j.aap.2016.04.001.

Dong, X., Xie, K., & Yang, H. (2022). How did COVID-19 impact driving behaviors and crash Severity? A multigroup structural equation modeling. *Accident Analysis & Prevention, 172*106687. Available from https://doi.org/10.1016/j.aap.2022.106687.

Du, S., Li, T., Gong, X., & Horng, S. J. (2018). A hybrid method for traffic flow forecasting using multimodal deep learning. *arXiv*. Available from https://doi.org/10.48550/arxiv.1803.02099, https://arxiv.org.

Fadlullah, Z. M., Tang, F., Mao, B., Kato, N., Akashi, O., Inoue, T., & Mizutani, K. (2017). State-of-the-art deep learning: Evolving machine intelligence toward tomorrow's intelligent network traffic control systems. *IEEE Communications Surveys & Tutorials, 19*(4), 2432–2455. Available from https://doi.org/10.1109/comst.2017.2707140.

Feizizadeh, B., Omarzadeh, D., Sharifi, A., Rahmani, A., Lakes, T., & Blaschke, T. (2022). A GIS-based spatiotemporal modelling of urban traffic accidents in tabriz city during the COVID-19 pandemic. *Sustainability, 14*(12), 7468. Available from https://doi.org/10.3390/su14127468.

Gu, Y., Lu, W., Qin, L., Li, M., & Shao, Z. (2019). Short-term prediction of lane-level traffic speeds: A fusion deep learning model. *Transportation Research Part C: Emerging Technologies, 106*, 1–16. Available from https://doi.org/10.1016/j.trc.2019.07.003.

Guo, S., Lin, Y., Li, S., Chen, Z., & Wan, H. (2019). Deep spatial–temporal 3D convolutional neural networks for traffic data forecasting. *IEEE Transactions on Intelligent Transportation Systems, 20*(10), 3913–3926. Available from https://doi.org/10.1109/tits.2019.2906365.

Huang, C.W., Chiang, C.T., & Li, Q. (2017). A study of deep learning networks on mobile traffic forecasting. IEEE International Symposium on Personal, Indoor and Mobile Radio Communications, PIMRC 2017-16 Institute of Electrical and Electronics Engineers Inc. Taiwan. Available from https://doi.org/10.1109/PIMRC.2017.8292737.

Jiang, F., & Ma, J. (2021). A comprehensive study of macro factors related to traffic fatality rates by XGBoost-based model and GIS techniques. *Accident Analysis & Prevention, 163*106431. Available from https://doi.org/10.1016/j.aap.2021.106431.

Jiang, M., Liu, Y., Lu, J., Qu, Z., & Yang, Z. (2023). Risk assessment of maritime supply chains within the context of the Maritime Silk Road. *Ocean & Coastal Management, 231*106380. Available from https://doi.org/10.1016/j.ocecoaman.2022.106380.

Katrakazas, C., Michelaraki, E., Sekadakis, M., & Yannis, G. (2020). A descriptive analysis of the effect of the COVID-19 pandemic on driving behavior and road safety. *Transportation Research Interdisciplinary Perspectives, 7*100186. Available from https://doi.org/10.1016/j.trip.2020.100186.

Ke, J., Zheng, H., Yang, H., & Chen, X. (M.) (2017). Short-term forecasting of passenger demand under on-demand ride services: A spatio-temporal deep learning approach. *Transportation Research Part C: Emerging Technologies, 85*, 591–608. Available from https://doi.org/10.1016/j.trc.2017.10.016.

Koesdwiady, A., Soua, R., & Karray, F. (2016). Improving traffic flow prediction with weather information in connected cars: A deep learning approach. *IEEE Transactions on Vehicular Technology, 65*(12), 9508–9517. Available from https://doi.org/10.1109/tVT.2016.2585575.

Li, Y., Yabuki, N., & Fukuda, T. (2023). Integrating GIS, deep learning, and environmental sensors for multicriteria evaluation of urban street walkability. *Landscape and Urban Planning, 230*104603. Available from https://doi.org/10.1016/j.landurbplan.2022.104603.

Liu, Y., Liu, Z., & Jia, R. (2019). DeepPF: A deep learning based architecture for metro passenger flow prediction. *Transportation Research Part C: Emerging Technologies, 101*, 18–34. Available from https://doi.org/10.1016/j.trc.2019.01.027.

Lv, Y., Duan, Y., Kang, W., Li, Z., & Wang, F. Y. (2015). Traffic flow prediction with big data: A deep learning approach. *IEEE Transactions on Intelligent Transportation Systems, 16*(2), 865–873. Available from https://doi.org/10.1109/TITS.2014.2345663, http://ieeexplore.ieee.org/xpl/RecentIssue.jsp?punumber = 6979.

Nagy, A. M., & Simon, V. (2018). Survey on traffic prediction in smart cities. *Pervasive and Mobile Computing, 50*, 148–163. Available from https://doi.org/10.1016/j.pmcj.2018.07.004.

Pan, Z., Liang, Y., Wang, W., Yu, Y., Zheng, Y., & Zhang, J. (2019). Urban traffic prediction from spatio-temporal data using deep meta learning. Proceedings of the ACM SIGKDD International Conference on Knowledge Discovery and Data Mining 9781450362016 1720 1730 Association for Computing Machinery China. 10.1145/3292500.3330884.

Peng, Q., Bakkar, Y., Wu, L., Liu, W., Kou, R., & Liu, K. (2024). Transportation resilience under Covid-19 uncertainty: A traffic severity analysis. *Transportation Research Part A: Policy and Practice, 179*103947. Available from https://doi.org/10.1016/j.tra.2023.103947.

Qu, L., Li, W., Li, W., Ma, D., & Wang, Y. (2019). Daily long-term traffic flow forecasting based on a deep neural network. *Expert Systems with Applications, 121*, 304–312. Available from https://doi.org/10.1016/j.eswa.2018.12.031.

Ramana, K., Srivastava, G., Kumar, M. R., Gadekallu, T. R., Lin, J. C.-W., Alazab, M., & Iwendi, C. (2023). A vision transformer approach for traffic congestion prediction in urban areas. *IEEE Transactions on Intelligent Transportation Systems*, *24*(4), 3922–3934. Available from https://doi.org/10.1109/tits.2022.3233801.

Saladié, Ò., Bustamante, E., & Gutiérrez, A. (2020). COVID-19 lockdown and reduction of traffic accidents in Tarragona province, Spain. *Transportation Research Interdisciplinary Perspectives*, *8*100218. Available from https://doi.org/10.1016/j.trip.2020.100218.

Seo, T., Bayen, A. M., Kusakabe, T., & Asakura, Y. (2017). Traffic state estimation on highway: A comprehensive survey. *Annual Reviews in Control*, *43*, 128–151. Available from https://doi.org/10.1016/j.arcontrol.2017.03.005.

Sohaee, N., & Bohluli, S. (2024). Nonlinear analysis of the effects of socioeconomic, demographic, and technological factors on the number of fatal traffic accidents. *Safety*, *10*(1), 11. Available from https://doi.org/10.3390/safety10010011.

Tang, J., Chen, X., Hu, Z., Zong, F., Han, C., & Li, L. (2019). Traffic flow prediction based on combination of support vector machine and data denoising schemes. *Physica A: Statistical Mechanics and its Applications*, *534*120642. Available from https://doi.org/10.1016/j.physa.2019.03.007.

Tian, Y., Zhang, K., Li, J., Lin, X., & Yang, B. (2018). LSTM-based traffic flow prediction with missing data. *Neurocomputing*, *318*, 297–305. Available from https://doi.org/10.1016/j.neucom.2018.08.067.

Tucker, A., & Marsh, K. L. (2021). Speeding through the pandemic: Perceptual and psychological factors associated with speeding during the COVID-19 stay-at-home period. *Accident Analysis & Prevention*, *159*106225. Available from https://doi.org/10.1016/j.aap.2021.106225.

Ulu, M., Kilic, E., & Türkan, Y. S. (2024). Prediction of traffic incident locations with a geohash-based model using machine learning algorithms. *Applied Sciences*, *14*(2), 725. Available from https://doi.org/10.3390/app14020725.

Wang, S., Cao, J., & Yu, P. S. (2022). Deep learning for spatio-temporal data mining: A survey. *IEEE Transactions on Knowledge and Data Engineering*, *34*(8), 3681–3700. Available from https://doi.org/10.1109/tkde.2020.3025580.

Wang, X., Ma, Y., Wang, Y., Jin, W., Wang, X., Tang, J., Jia, C., & Yu, J. (2020). Traffic flow prediction via spatial temporal graph neural network. The web conference 2020 – Proceedings of the world wide web conference, WWW 2020 9781450370233 1082 1092 Association for Computing Machinery, Inc., China. http://dl.acm.org/citation.cfm?id = 3366423, 10.1145/3366423.3380186.

Wang, X., Zhang, X., & Pei, Y. (2024). A systematic approach to macro-level safety assessment and contributing factors analysis considering traffic crashes and violations. *Accident Analysis & Prevention*, *194*107323. Available from https://doi.org/10.1016/j.aap.2023.107323.

Wen, H., Zhang, X., Zeng, Q., & Sze, N. N. (2019). Bayesian spatial–temporal model for the main and interaction effects of roadway and weather characteristics on freeway crash incidence. *Accident Analysis & Prevention*, *132*105249. Available from https://doi.org/10.1016/j.aap.2019.07.025.

Wu, S. (2022). Spatiotemporal dynamic forecasting and analysis of regional traffic flow in urban road networks using deep learning convolutional neural network. *IEEE Transactions on Intelligent Transportation Systems*, *23*(2), 1607–1615. Available from https://doi.org/10.1109/tits.2021.3098461.

Xia, D., Wang, B., Li, H., Li, Y., & Zhang, Z. (2016). A distributed spatial–temporal weighted model on MapReduce for short-term traffic flow forecasting. *Neurocomputing*, *179*, 246–263. Available from https://doi.org/10.1016/j.neucom.2015.12.013.

Xu, Y., Han, Z., Zhang, Z., & Murphy, M. (2024). A methodological procedure for evaluating curve-related misclassifications in motor vehicle crash databases. *Research in Transportation Economics*, *103*101389. Available from https://doi.org/10.1016/j.retrec.2023.101389.

Ye, J., Zhao, J., Ye, K., & Xu, C. (2022). How to build a graph-based deep learning architecture in traffic domain: A survey. *IEEE Transactions on Intelligent Transportation Systems*, *23*(5), 3904–3924. Available from https://doi.org/10.1109/tits.2020.3043250.

Ye, Q., Szeto, W. Y., & Wong, S. C. (2012). Short-term traffic speed forecasting based on data recorded at irregular intervals. *IEEE Transactions on Intelligent Transportation Systems*, *13*(4), 1727–1737. Available from https://doi.org/10.1109/TITS.2012.2203122.

Yu, B., Yin, H., & Zhu, Z. (2017). Spatio-temporal graph convolutional networks: A deep learning framework for traffic forecasting. *arXiv*. Available from https://doi.org/10.48550/arxiv.1709.04875, https://arxiv.org.

Yu, X., Ma, J., Tang, Y., Yang, T., & Jiang, F. (2024). Can we trust our eyes? Interpreting the misperception of road safety from street view images and deep learning. *Accident Analysis & Prevention*, *197*107455. Available from https://doi.org/10.1016/j.aap.2023.107455.

Zeng, Q., Wang, Q., Zhang, K., Wong, S. C., & Xu, P. (2023). Analysis of the injury severity of motor vehicle–pedestrian crashes at urban intersections using spatiotemporal logistic regression models. *Accident Analysis & Prevention*, *189*107119. Available from https://doi.org/10.1016/j.aap.2023.107119.

Zeng, Q., Xu, P., Wang, X., Wen, H., & Hao, W. (2021). Applying a Bayesian multivariate spatio-temporal interaction model based approach to rank sites with promise using severity-weighted decision parameters. *Accident Analysis & Prevention*, *157*106190. Available from https://doi.org/10.1016/j.aap.2021.106190.

Zhang, J., Zheng, Y., Qi, D., Li, R., Yi, X., & Li, T. (2018). Predicting citywide crowd flows using deep spatio-temporal residual networks. *Artificial Intelligence*, *259*, 147–166. Available from https://doi.org/10.1016/j.artint.2018.03.002.

Zhang, Z., He, Q., Gao, J., & Ni, M. (2018). A deep learning approach for detecting traffic accidents from social media data. *Transportation Research Part C: Emerging Technologies*, *86*, 580–596. Available from https://doi.org/10.1016/j.trc.2017.11.027.

Zhao, J., Liu, P., & Li, Z. (2024). Exploring the impact of trip patterns on spatially aggregated crashes using floating vehicle trajectory data and graph convolutional networks. *Accident Analysis & Prevention*, *194*107340. Available from https://doi.org/10.1016/j.aap.2023.107340.

Zhao, Z., Chen, W., Wu, X., Chen, P. C. Y., & Liu, J. (2017). LSTM network: A deep learning approach for short-term traffic forecast. *IET Intelligent Transport Systems*, *11*(2), 68–75. Available from https://doi.org/10.1049/iet-its.2016.0208.

Zheng, W., Deng, H., & Han, F. (2021). GST-Net: A GIS-based hybrid prediction model for shared bike traffic flow. *Proceedings – 2021 21st international conference on software quality, reliability and security companion*, QRS-C 2021 9781665478366 941 946 Institute of Electrical and Electronics Engineers Inc., China. Available from http://ieeexplore.ieee.org/xpl/mostRecentIssue.jsp?punumber = 9741747, https://doi.org/10.1109/QRS-C55045.2021.00142.

Zhou, X. (2024). Spatial risk assessment of maritime transportation in offshore waters of China using machine learning and geospatial big data. *Ocean & Coastal Management*, *247*106934. Available from https://doi.org/10.1016/j.ocecoaman.2023.106934.

Chapter 9

Enriching the metadata of map images: a deep learning approach with geographic information systems-based data augmentation

Entaj Tarafder[1,*], Sharmili Khatun[2], Muhammad Awais[3], Muhammad Qayyum[4], Mir Muhammad Nizamani[1,*] and Yong Wang[1]

[1]*Department of Plant Pathology, College of Agriculture, Guizhou University, Guiyang, Guizhou, P.R. China,* [2]*Department of Geography, Ranaghat College, Kalyani University, Ranaghat, West Bengal, India,* [3]*Institute of Plant Sciences, Faculty of Natural Sciences, University of Sindh, Jamshoro, Sindh, Pakistan,* [4]*School of Economics and Statistics, Guangzhou University, Guangzhou, P.R. China*

9.1 Introduction

In the intricate tapestry of the digital age, map images serve as indispensable tools for navigation, urban planning, environmental studies, and a myriad of other applications. These maps, whether physical or digital, have always been enriched by metadata—data about data. Metadata provides critical context, aiding in the accurate interpretation and effective utilization of the map (Raes et al., 2021). Traditionally, metadata might include details such as scale, projection, source, or date of creation. These basic elements have long been sufficient to provide users with a foundational understanding of the map's content and purpose. However, as we move further into an era defined by automation, vast datasets, and intelligent systems, the requirement for richer, more detailed metadata has become increasingly evident (Ohm et al., 2023). Modern applications of maps demand more sophisticated metadata to enhance their functionality and precision. For instance, advanced metadata can include real-time updates, user-generated data, and machine-readable information, which collectively empower more dynamic and interactive uses of maps. This evolution in metadata sophistication ensures that maps are not merely static representations but dynamic tools capable of adapting to the complex needs of contemporary users (Gao et al., 2017). The progression toward more detailed metadata is driven by the burgeoning needs of various sectors that rely heavily on accurate and up-to-date spatial information. In urban planning, for example, enriched metadata can support complex simulations and predictive modeling, aiding planners in making informed decisions about infrastructure development and resource allocation (Jobst & Gartner, 2019). Similarly, in environmental studies, detailed metadata allows researchers to track changes over time with greater precision, facilitating more effective conservation efforts and policy-making. The push for more comprehensive metadata is thus a response to the growing demands of an increasingly data-driven world, where the ability to interpret and utilize spatial information with accuracy and efficiency is paramount (Newman et al., 2017).

Deep learning, a subfield of machine learning, has recently revolutionized various sectors, including image processing, natural language understanding, and recommendation systems, to name a few (Cheng et al., 2023; Zhang et al., 2019). Its potential to extract intricate patterns and features from images makes it a promising approach for automating and enhancing metadata generation for map images. By leveraging deep learning algorithms, it is possible to analyze and interpret complex visual data with unprecedented accuracy and efficiency, facilitating the creation of detailed and dynamic metadata that can significantly improve the usability and functionality of maps. However, map images pose

*Entaj Tarafder and Mir Muhammad Nizamani contributed equally to the work.

Deep Learning for Earth Observation and Climate Monitoring. DOI: https://doi.org/10.1016/B978-0-443-24712-5.00010-5

unique challenges due to their diverse nature, the importance of geographical context, and their inherent complexities (Khelifi & Mignotte, 2020; Liu et al., 2021). Unlike typical images used in other domains, map images contain layers of information that must be accurately parsed and understood. This includes geographical features, political boundaries, infrastructure details, and more. The need to maintain spatial accuracy and contextual relevance adds another layer of complexity to the task of metadata generation. Traditional image processing techniques often fall short in addressing these unique requirements, necessitating the development of specialized deep learning models tailored to the nuances of map imagery. Furthermore, the diverse nature of map images means that a one-size-fits-all approach is seldom effective. Different types of maps—such as topographic, thematic, and satellite maps—each have their own set of characteristics and requirements for metadata (Kuźma & Mościcka, 2020). Deep learning models must be trained to recognize and interpret these varying elements accurately. The importance of geographical context also cannot be overstated, as it is crucial for ensuring that the metadata generated is relevant and meaningful. This requires integrating geographic information systems (GIS) data and other contextual datasets into the deep learning process, further complicating the task (Brodeur et al., 2019).

This is where GIS come into play. GIS, with its capacity to manage, analyze, and display geospatial data, presents a unique opportunity to augment the capabilities of deep learning in the context of map image metadata enrichment (Huang et al., 2021). By integrating GIS with deep learning, it becomes possible to harness the strengths of both technologies, leading to more sophisticated and accurate metadata generation. GIS provides the geographical precision and contextual awareness needed to interpret complex map images accurately, while deep learning offers the predictive power to analyze and extract meaningful patterns and features from these images. Marrying the predictive prowess of deep learning with the geographical precision of GIS can pave the way for an advanced approach to metadata creation—one that is more automated, context-aware, and enriched (Bhatti, Bazai, et al., 2023; Bhatti, Huang, et al., 2023; Bhatti, Song, et al., 2023; Kerski, 2022). This integration enables the development of systems that can automatically generate detailed and contextually relevant metadata for a wide range of map images. For example, a deep learning model can be trained to recognize and classify various geographical features in a satellite image, while GIS can provide the necessary contextual information to ensure that these features are accurately represented in the metadata (Veenendaal et al., 2017). This synergy between deep learning and GIS not only enhances the quality and accuracy of the metadata but also significantly reduces the time and effort required for its generation (Kucklick et al., 2021). Moreover, the combination of deep learning and GIS can lead to the creation of dynamic and interactive maps that can adapt to changing conditions and user needs. For instance, in urban planning, such systems can provide real-time updates and predictive insights based on current geographical data, helping planners make more informed decisions (Bentivoglio et al., 2022). In environmental studies, they can enable more precise tracking and analysis of environmental changes over time. The enriched metadata generated through this integrated approach can also facilitate more effective data sharing and collaboration across various sectors, promoting better decision-making and resource management (Keskin & Kettunen, 2023).

This chapter seeks to unravel the convergence of deep learning and GIS in the realm of map metadata enrichment. We will explore the challenges of current metadata enrichment practices, delve into the intricacies of GIS-based data augmentation, and illustrate the transformative potential of combining these technologies (Hochschild et al., 2020). Through real-world case studies, results, and evaluations, we aim to offer a comprehensive perspective on this emerging interdisciplinary field, setting the stage for future explorations and innovations (Huang et al., 2021).

9.2 Background

The intertwining relationship between maps, metadata, and technological advancements is rooted deeply within the annals of history. To better appreciate the contemporary intersection of deep learning, GIS, and map metadata, we must first journey through the evolution of these domains and understand their historical significance.

9.2.1 The evolution of map images and their metadata

Maps, in their essence, have been humanity's navigational companions for millennia. From rudimentary sketches on cave walls and parchment scrolls to sophisticated digital renderings, maps have charted our understanding of the world (Small, 2023). These early maps, though primitive, served a crucial purpose in helping people navigate their surroundings, plan hunts, and later, conduct trade and exploration. As human civilization advanced, so did the complexity and accuracy of maps (Aamir et al., 2023). The development of cartography saw the transition from simple, hand-drawn depictions to detailed and accurate representations of geographical regions, aided by advances in technology and understanding. Alongside this evolution, the accompanying metadata has transformed from simple legends and annotations to intricate, layered information databases (Finger et al., 2017). In the past, map metadata was often limited to basic information such as place names, scale, and directional indicators. These annotations were sufficient for the needs of the time but lacked the depth and breadth required for more complex analyses and applications (Bhatti, Bazai, et al., 2023; Bhatti, Huang, et al., 2023; Bhatti, Song, et al., 2023). However, as the use of maps expanded beyond basic navigation to include scientific research, urban planning, and environmental monitoring, the need for more detailed and comprehensive metadata became apparent. This led to the development of more sophisticated metadata systems that could capture a wide range of information about the maps and their content. The advent of the digital age further facilitated the transition from static, isolated metadata to dynamic, interconnected datasets that can be updated and accessed in real-time (Bhatti et al., 2021; Cerchione et al., 2023). Digital maps brought about a revolution in how maps are created, used, and shared. Unlike their static predecessors, digital maps can be easily updated to reflect the latest information, ensuring that users always have access to current data (Batty et al., 2010). This real-time capability is particularly valuable in applications such as disaster response, where timely and accurate information can save lives (Shah et al., 2019). Furthermore, digital maps can be integrated with other data sources, creating interconnected datasets that provide richer, more contextualized information. In this digital landscape, metadata has evolved to include a wide array of data points, from geographical coordinates and elevation levels to demographic statistics and environmental conditions (Guo et al., 2020). This enriched metadata allows for more precise and nuanced analyses, supporting a variety of applications in different fields. For instance, urban planners can use this data to design more efficient and sustainable cities, while environmental scientists can monitor and model changes in ecosystems with greater accuracy. The ability to access and analyze this data in real-time further enhances its value, enabling more informed decision-making and more effective responses to emerging challenges (Brumana et al., 2018).

9.2.2 Deep learning and image analysis

Deep learning emerged from the broader field of artificial intelligence (Fig. 9.1), specifically inspired by neural networks and our understanding of the human brain's architecture (Liu et al., 2024). It has since evolved into a computational powerhouse, capable of recognizing intricate patterns within vast amounts of data (Alhatami et al., 2024; Mehonic et al., 2020). The fundamental concept behind deep learning is its ability to mimic the neural structures of the brain, enabling machines to learn and make decisions in a manner similar to humans (Huang et al., 2024). This has opened up new possibilities across various domains, with deep learning revolutionizing the way we process and analyze information (Saxe et al., 2021). In image processing, deep learning models, particularly convolutional neural networks (CNNs), have displayed unparalleled prowess in feature extraction, object detection, and semantic segmentation. CNNs are designed to automatically and adaptively learn spatial hierarchies of features from input images, making them exceptionally effective for tasks involving visual data. Their layered structure allows them to detect low-level features

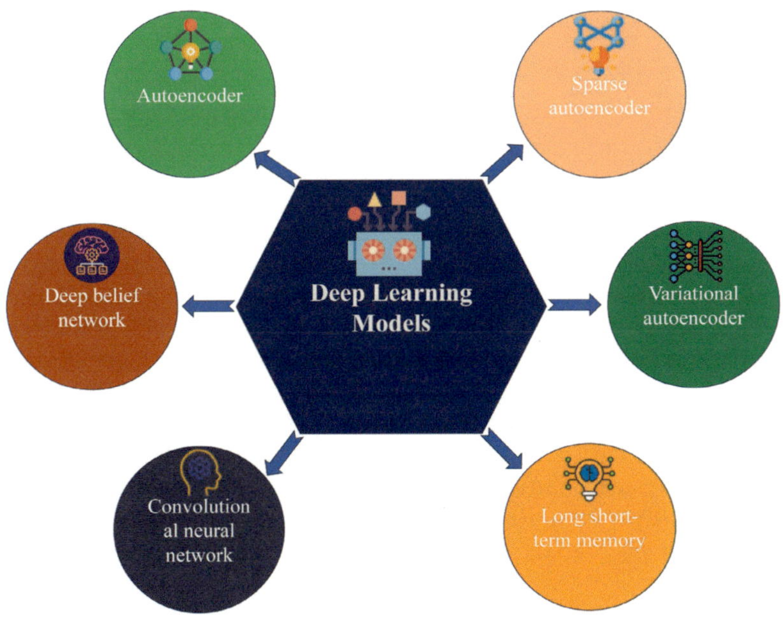

FIGURE 9.1 Deep learning emerged from the broader field of artificial intelligence. Overview of various deep learning models. The central hexagon represents "Deep Learning Models," which include Autoencoder, Sparse Autoencoder, Variational Autoencoder, Long Short-Term Memory (LSTM), Convolutional Neural Network (CNN), and Deep Belief Network (DBN). Each model is depicted in a circular shape connected to the central hexagon, illustrating the interconnections among these deep learning techniques.

such as edges and textures in the initial layers and progressively more complex patterns in deeper layers (Archana & Jeevaraj, 2024). This hierarchical learning process enables CNNs to achieve remarkable accuracy in recognizing and interpreting visual information. The ability of deep learning models to "learn" from vast datasets allows for granular insights, making them prime candidates for the nuanced world of map images (Sani-Mohammed et al., 2022). In the context of maps, the application of CNNs can significantly enhance the accuracy and efficiency of metadata generation (Nawaz et al., 2024). For example, CNNs can be trained to identify and categorize different geographical features such as roads, rivers, buildings, and vegetation (Zohuri & Rahmani, 2024). They can also detect changes over time, making them invaluable for tasks like monitoring urban development or environmental changes. Moreover, the deep learning approach is highly adaptable, allowing models to be fine-tuned for specific types of maps or particular use cases (Younesi et al., 2024). This flexibility is crucial given the diverse nature of map images, which can range from detailed urban layouts to vast, sparsely populated rural areas (Wu, Qiu, et al., 2023; Wu, Ren, et al., 2023). By training deep learning models on diverse datasets, it is possible to develop systems capable of handling a wide variety of map-related tasks with high precision (Hosain et al., 2024).

9.2.3 The emergence and significance of geographic information systems

GIS have been integral to the modern understanding and manipulation of spatial data. Originating in the 1960s as rudimentary computer-aided cartography tools, GIS has evolved to become a multidisciplinary domain that integrates spatial data with analytical capabilities. Initially, GIS provided a basic platform for digitizing and visualizing maps, but its capabilities have expanded exponentially over the decades (Wang et al., 2019). Today, GIS is not merely a tool for creating maps but a sophisticated system for analyzing spatial relationships and patterns, enabling users to gain deeper insights into geographical data. Beyond just visual representation, GIS enables the layering of data, spatial analysis, and complex geospatial modeling (Heitzler et al., 2017). This functionality allows for the integration of various types of data—such as demographic information, environmental conditions, and infrastructure details—into a single, coherent framework. The ability to overlay multiple layers of data on a single map facilitates comprehensive analysis and decision-making (Janssen et al., 2017). For instance, urban planners can overlay population density maps with transportation networks to identify optimal locations for new infrastructure. Similarly, environmental scientists can combine satellite imagery with climate data to monitor and predict ecological changes. GIS's significance in urban planning, environmental studies, disaster management, and various other fields underscores its versatility and relevance in today's world. In urban planning, GIS helps planners design cities that are more efficient, sustainable, and livable by providing detailed spatial data on land use, traffic patterns, and public services (Sarker et al., 2020). In environmental studies, GIS is used to track changes in ecosystems, analyze the impact of human activities on natural habitats, and develop conservation strategies. During disaster management, GIS provides critical information for emergency response, such as identifying affected areas, planning evacuation routes, and coordinating rescue operations (Pricope et al., 2019). Moreover, GIS's role extends to fields such as public health, where it is used to map disease outbreaks and plan healthcare services, and in business, where companies use GIS for market analysis and logistics planning. The versatility of GIS is further enhanced by its integration with other technologies such as remote sensing, GPS, and now, increasingly, with artificial intelligence and deep learning (Paul et al., 2020).

9.2.4 The need for richer metadata in a digital age

As digital transformation permeates every facet of modern society, the sheer volume and complexity of data available have grown exponentially. This rapid expansion has touched all domains, including geospatial data, where map images have become increasingly detailed. With the advent of advanced tools such as satellite imagery, drone captures, and other sophisticated technologies, the complexity and richness of map images have surged (Giordan et al., 2020; Nex et al., 2022). These advancements provide a wealth of information but also pose significant challenges in terms of data management and utilization. The need for richer metadata, therefore, is not just desirable but essential. Detailed metadata plays a crucial role in several key areas. Firstly, it aids in faster and more accurate image retrieval (Jane et al., 2014). With comprehensive metadata, users can efficiently search and access specific map images from vast datasets based on various criteria such as location, time, or content type. This capability is particularly important in applications that require timely and precise information, such as disaster response and urban planning (Fan et al., 2021). Moreover, enriched metadata supports machine learning models in deciphering context. In the realm of deep learning, the ability of models to interpret and analyze images is greatly enhanced when they are provided with contextual information. Metadata that includes details about the geographical features, environmental conditions, and temporal changes within an image can significantly improve the performance of these models (Sun et al., 2020). This is especially relevant in tasks such as object detection, classification, and segmentation, where understanding the context can lead to more accurate and meaningful results. Furthermore, sophisticated applications, from autonomous driving to geospatial analytics, rely heavily on detailed metadata to function optimally (Ahmad et al., 2022). In autonomous driving, for instance, vehicles equipped with advanced sensors and cameras generate and rely on vast amounts of spatial data to navigate safely. Accurate and up-to-date metadata is crucial for these systems to interpret their surroundings correctly and make informed decisions. Similarly, in geospatial analytics, detailed metadata allows for more precise modeling and analysis of spatial phenomena, leading to better insights and decision-making (Iyer et al., 2023).

9.3 The significance of rich metadata

In an ever-evolving digital landscape, metadata has transitioned from a supplementary component to an essential pillar of information management and interpretation. Rich metadata, which refers to metadata that is detailed, comprehensive, and context-aware, plays a pivotal role in various applications and domains, especially in the context of map images (Guerra-Manzanares & Bahsi, 2022). This evolution signifies a shift in how we perceive and utilize metadata, recognizing its integral role in enhancing the value and usability of data. In the realm of map images, rich metadata is indispensable. These images, generated from a variety of advanced tools such as satellites and drones, contain a wealth of information that can only be fully exploited through detailed and context-aware metadata (Mayernik, 2016). For example, metadata that includes information on the time and date of capture, geographical coordinates, environmental conditions, and even the specific sensors used can greatly enhance the utility of map images. Such detailed metadata allows users to interpret the images accurately, ensuring that they can be effectively used in applications ranging from environmental monitoring to urban planning (Hu et al., 2022). Moreover, rich metadata facilitates more efficient and accurate data retrieval. In large geospatial databases, searching for specific map images can be like finding a needle in a haystack without adequate metadata. Detailed descriptions and contextual information allow for more precise search queries, enabling users to quickly locate the images they need based on specific criteria (Ma et al., 2021). This efficiency is crucial in fields where timely access to accurate information can make a significant difference, such as in disaster response or real-time traffic management. Additionally, the importance of rich metadata extends to the realm of machine learning and artificial intelligence. Machine learning models, especially those used in image processing and analysis, rely heavily on high-quality metadata to train effectively (Chen et al., 2016). Metadata provides these models with the necessary context to understand and interpret complex visual data. For instance, in autonomous driving, metadata can help machine learning algorithms distinguish between different types of roads, recognize traffic signs, and understand various environmental conditions, thereby improving the safety and reliability of these systems (Guo et al., 2019). Furthermore, rich metadata is vital for ensuring interoperability and data integration across different systems and platforms. As digital transformation continues to drive the integration of diverse data sources, having standardized and comprehensive metadata ensures that data from different origins can be seamlessly combined and analyzed. This interoperability is particularly important in collaborative projects and initiatives that span multiple organizations or even countries (Blasch et al., 2021).

9.3.1 Enhancing data interpretability

Imagine embarking on a journey without a compass or understanding of the terrain. Such is the ordeal when deciphering intricate map images sans rich metadata. This metadata serves as a guiding light, ensuring neither humans nor machines stray in their interpretations. It chronicles vital elements, from the map's historical essence to its intricate geographical narratives. Like a seasoned guide, enriched metadata ensures the map's stories are told accurately and wholly (Stray et al., 2022). Without rich metadata, interpreting complex map images can be a daunting task. Maps often contain layers of information, including topographical details, infrastructure, and environmental features. Rich metadata provides the necessary context to navigate these layers effectively. For humans, this means being able to understand the nuances of the terrain, historical landmarks, and current developments with clarity. For machines, rich metadata equips algorithms with the information needed to analyze and process map images accurately, leading to more reliable outputs in applications such as autonomous navigation and geographic information systems (Sun & Scanlon, 2019). The value of rich metadata extends beyond mere guidance. It also preserves the historical and geographical integrity of map images. By documenting the origins, sources, and transformations of the data, metadata ensures that the historical context of a map is not lost. This is crucial for researchers and historians who rely on accurate records to trace the development of geographical regions over time (Urdu et al., 2023). Enriched metadata captures these details, enabling a comprehensive understanding of the map's significance and evolution. Moreover, rich metadata facilitates the communication of complex geographical narratives. Maps are powerful storytelling tools, conveying information about the physical world and human interactions within it (Duckham et al., 2023). Enriched metadata acts as a storyteller, elucidating the relationships between different elements on the map. It explains why certain features are present, how they interact, and what implications they might have. This deeper level of understanding is essential for fields like urban planning, environmental management, and disaster response, where decisions depend on precise and contextually aware geographical information. In a digital age where data are abundant but often overwhelming, rich metadata stands out as a beacon of clarity (Burns et al., 2024). It transforms map images from static representations into dynamic, interactive resources that can be explored and understood in depth. By providing detailed annotations and contextual information, enriched metadata makes maps more accessible and useful for a wide range of applications. Whether for a researcher delving into historical landscapes, a city planner designing future infrastructure, or an AI system navigating autonomous vehicles, rich metadata ensures that the journey through map images is informed, accurate, and meaningful (Iapaolo, 2023).

9.3.2 Improved search and retrieval

In the sprawling universe of geospatial data, finding the right map image can be akin to spotting a needle in a haystack. Yet, with rich metadata, this task becomes infinitely more manageable. This metadata does not just recognize keywords; it grasps the context, the story, the intent behind a query. It ensures searches are not just accurate; they are intuitive, aligning seamlessly with the user's unique needs and intentions (Kerr et al., 2016). Rich metadata transforms the search process by providing a depth of understanding that goes beyond basic keyword matching. Traditional search methods often rely on simple tags or descriptions, which can lead to results that are too broad or irrelevant (Bogers & Petras, 2017). In contrast, enriched metadata encompasses detailed information about the map image, such as its geographical scope, the specific features it highlights, the time period it represents, and the purpose for which it was created. This level of detail allows search algorithms to parse queries more effectively and deliver results that are highly relevant and specific to the user's needs (Chapman et al., 2020). Moreover, rich metadata enables searches that are not only precise but also intuitive. When a user searches for a map image, they often have a particular context or intent in mind, such as planning a hiking trip, researching historical land use, or analyzing environmental changes. Rich metadata captures this context by including information on how the map image can be used, its relevance to different types of queries, and its connections to other datasets. As a result, search results are not just a random collection of images but a curated selection that aligns closely with the user's intentions (Löffler et al., 2021). This intuitive search capability is particularly valuable in professional fields such as urban planning, environmental science, and disaster management, where the ability to quickly find and interpret relevant map images can have significant real-world implications. For instance, an urban planner looking for maps of a city's transportation infrastructure can benefit from metadata that highlights recent updates, ongoing construction projects, and future development plans (Mosannenzadeh et al., 2017). Similarly, an environmental scientist studying deforestation patterns can use metadata that includes temporal data, types of vegetation, and satellite imaging details to track changes over time. In addition to improving search accuracy and intuition, rich metadata also enhances the overall user experience by making interactions with geospatial data more efficient and rewarding (Santos et al., 2019). Users spend less time sifting through irrelevant results and more time engaging with the precise information they need. This efficiency is crucial in high-stakes situations, such as emergency response, where timely access to accurate geospatial data can facilitate quicker and more effective decision-making (Matricardi et al., 2020).

9.3.3 Automation and scalability

As we usher in an era dominated by automation, rich metadata emerges as the crucial cog in this machinery. For deep learning models, it is akin to a gourmet meal, enriching their data diet and amplifying their analytical prowess. Metadata, in its detailed and contextualized form, provides deep learning algorithms with the nuanced information necessary to make accurate predictions and decisions (Keeler, 2020). This enrichment enables models to discern subtle patterns and relationships within the data, leading to more sophisticated and reliable outputs. Furthermore, when faced with the Herculean task of processing vast datasets, enriched metadata streamlines and accelerates the process, ensuring scalability never comes at the cost of precision (Mancuso et al., 2023). In the world of big data, the volume and complexity of information can be overwhelming. However, rich metadata acts as an organizational framework, categorizing and indexing data in a way that enhances accessibility and usability. This structured approach allows deep learning models to efficiently navigate through large datasets, reducing processing times and improving overall performance

(Newman & Witsell, 2021). The benefits of rich metadata in automation extend beyond mere efficiency. By providing comprehensive context, it enhances the interpretability of model outputs. For example, in the realm of autonomous vehicles, rich metadata can include details about road conditions, weather patterns, and traffic signals. This contextual information helps the vehicle's AI system make more informed decisions, such as adjusting speed in response to road hazards or predicting the behavior of other drivers (Iapaolo, 2023). Similarly, in healthcare, enriched metadata can incorporate patient history, treatment protocols, and diagnostic results, enabling more accurate and personalized medical recommendations from AI systems. Moreover, rich metadata supports the continuous improvement and refinement of deep learning models. As models process data and generate outputs, metadata can capture feedback and performance metrics, providing valuable insights for iterative training and optimization (Burns et al., 2024). This feedback loop ensures that models remain adaptive and responsive to new information, continuously enhancing their accuracy and effectiveness over time. In addition to enhancing individual models, rich metadata facilitates the integration and interoperability of multiple AI systems. In complex environments, such as smart cities or interconnected industrial operations, various AI systems must work together seamlessly. Metadata serves as a common language, enabling different systems to share and interpret data consistently. This interoperability is crucial for creating cohesive and efficient automated solutions that can address multifaceted challenges in real-time (Puliafito et al., 2021).

9.3.4 Facilitating interoperability

In today's digital ecosystem, collaboration is key. Map images, often needing to dance across platforms, find their rhythm thanks to enriched metadata. This metadata bridges gaps, ensuring a GIS system's intricacies align harmoniously with a digital library's framework. In doing so, it champions a standardized, unified approach to map management, fostering seamless integrations across diverse platforms (Aristidou et al., 2019). The ability to share and integrate map images across various systems is crucial in an era where digital collaboration and data interoperability are essential. Enriched metadata serves as a common language that different platforms can understand and utilize effectively. For instance, a GIS system might generate detailed spatial data and analysis, while a digital library might focus on cataloging and providing access to a wide range of digital resources (Pricope et al., 2019). Enriched metadata ensures that the information from these diverse sources can be accurately aligned and integrated, facilitating a cohesive and comprehensive data environment. This standardized approach to map management not only improves efficiency but also enhances the quality of data analysis and decision-making. When map images are enriched with detailed metadata, they become more than just visual representations of geographical information. They carry with them a wealth of contextual details that can be leveraged for more in-depth analysis. For example, a GIS system using enriched metadata can better understand the temporal and spatial relationships between different datasets, leading to more accurate predictions and insights (Paul et al., 2020). Moreover, the interoperability facilitated by enriched metadata is vital for collaborative projects that span multiple organizations and disciplines. In environmental research, for example, scientists from various institutions might need to share and analyze data on climate change impacts. Enriched metadata ensures that the datasets from different sources are compatible and can be seamlessly integrated into a unified framework, enabling comprehensive analysis and fostering collaboration (Trani et al., 2018). Similarly, in urban planning, different departments—such as transportation, public works, and housing—can work together more effectively by sharing map images and data through a common metadata standard. Enriched metadata also enhances the user experience by making it easier to find and access relevant map images. In a digital library setting, metadata can provide detailed descriptions and classifications, allowing users to perform precise searches and quickly locate the specific maps they need (Chassanoff & Altman, 2020). This capability is particularly important for researchers and professionals who rely on accurate and timely information to make informed decisions. Furthermore, enriched metadata supports the long-term preservation and usability of map images. By documenting the origins, transformations, and context of map data, metadata ensures that future users can understand and trust the information, even as technology and platforms evolve. This preservation aspect is critical for maintaining the integrity and value of digital archives and repositories (Remy et al., 2019).

9.3.5 Enhancing user experience

Beyond mere visualization, rich metadata invites users into an interactive realm. It beckons them to delve deeper, to explore the myriad layers and stories embedded within map images. Rich metadata transforms static maps into dynamic, interactive experiences, allowing users to engage with the data in meaningful ways. They can uncover historical contexts, analyze geographical changes over time, and understand complex spatial relationships (Brumana et al., 2018). This interactive engagement fosters a deeper connection between users and the map images, transforming passive observation into active exploration. Moreover, with this granular understanding, platforms can anticipate user needs, curating experiences that resonate on a personal level. Enriched metadata provides the contextual clues necessary for platforms to understand user preferences and behaviors (Veenendaal et al., 2017). By analyzing how users interact with different layers of data, platforms can predict what information might be most relevant or interesting to them. This anticipatory capability allows for the creation of personalized experiences that not only provide the information users seek but also introduce them to new insights and perspectives they might not have considered. A symbiotic relationship between users and map data, marked by curiosity, exploration, and tailored interactions (Van Rijmenam, 2022). Users are no longer mere recipients of information; they become active participants in the data discovery process. This relationship is mutually beneficial: users gain a richer understanding of the world through interactive and personalized map experiences, while platforms gather valuable insights into user behavior and preferences, which can be used to further refine and enhance the services provided. For instance, in educational settings, students can use interactive maps with enriched metadata to conduct detailed geographical studies, exploring various aspects of a region's physical and cultural landscape (McDaniel, 2022). This hands-on approach not only makes learning more engaging but also helps students develop critical thinking and analytical skills. In tourism, interactive maps can provide travelers with personalized recommendations based on their interests and past behaviors, enhancing their travel experience by highlighting attractions and activities that are most likely to appeal to them. In urban planning and public policy, interactive maps with rich metadata can facilitate community engagement and participatory decision-making (McDaniel & Ingram, 2023). Residents can explore proposed developments, understand their potential impacts, and provide feedback through the same platform. This transparency and interactivity can lead to more informed and inclusive planning processes, where the voices of the community are heard and considered (Kerski & Baker, 2019). Furthermore, businesses can leverage interactive maps to gain deeper insights into market trends and consumer behavior. By analyzing geographical data alongside sales and demographic information, companies can identify new market opportunities, optimize supply chains, and tailor marketing strategies to specific regions. The interactivity and personalization enabled by rich metadata ensure that these insights are both actionable and relevant (Li et al., 2018).

9.3.6 Strengthening data security and management

In the intricate dance of data exchange, trust is paramount. Here, enriched metadata plays the dual role of chronicler and gatekeeper. It meticulously records every move—be it an edit, access, or transfer—ensuring transparency and accountability. By maintaining a comprehensive log of all interactions with the map data, enriched metadata provides a clear audit trail that can be used to track changes, identify responsible parties, and verify the integrity of the data (Anderson, 2018). This level of transparency is crucial in fostering trust among users, as it assures them that the data they are accessing is accurate, reliable, and well-maintained. Simultaneously, enriched metadata enforces boundaries, delineating who can view or modify the map data. This role as a gatekeeper involves setting permissions and access controls that define the scope of interaction for different users (van Gasselt & Naß, 2024). For instance, sensitive geographical information might be restricted to authorized personnel, while other data could be made publicly accessible. By implementing these controls, enriched metadata ensures that only those with the appropriate credentials can access or alter the data, thereby protecting it from unauthorized use or tampering. In a world teeming with data breaches and misuse, rich metadata stands as the sentinel, ensuring map images are shielded, secure, and leveraged responsibly (Rehberger, 2020). The security protocols embedded within the metadata framework can include encryption standards, user authentication processes, and compliance with data protection regulations. These measures help safeguard the data against cyber threats and ensure that it is used ethically and legally. For organizations handling large volumes of geospatial data, this security is essential not only for protecting their assets but also for maintaining their reputation and trustworthiness (Wu et al., 2024). Moreover, the accountability provided by enriched metadata extends to data sharing and collaboration. When multiple stakeholders are involved in a project, knowing who has accessed or modified the data, and when, becomes critical (Luo et al., 2021). Enriched metadata facilitates this by offering detailed records of user interactions, which can be reviewed and audited as necessary. This capability is particularly valuable in collaborative environments, such as urban planning or environmental research, where multiple parties need to coordinate and trust the data they are working with (Subramaniam et al., 2021). Enriched metadata also supports the responsible use of data by embedding ethical guidelines and usage policies directly within the metadata. These guidelines can inform users about the intended use of the data, any restrictions on its application, and the consequences of misuse. By providing this context, enriched metadata helps ensure that users are aware of their responsibilities and the ethical considerations associated with handling sensitive geospatial information (Pacheco et al., 2023).

9.4 Challenges in metadata enrichment

While the potential benefits of rich metadata for map images are substantial, achieving this enriched state is not without its challenges. The sheer volume and diversity of data present significant obstacles, as map images vary widely in resolution, scale, and the types of information they contain. Integrating these diverse datasets into a cohesive metadata framework requires sophisticated algorithms and extensive computational resources (Alma'aitah et al., 2020). The dynamic nature of geospatial data adds another layer of complexity, necessitating real-time updates to reflect changes in the physical world, such as new infrastructure developments or natural disasters. Keeping metadata current in such a rapidly evolving landscape is a continual challenge (Hu et al., 2022). Technical challenges also arise from integrating deep learning models with GIS systems, requiring large training datasets and seamless interoperability. Despite the power of these models, acquiring and curating the necessary training data can be time-consuming and costly. Human expertise remains crucial to validate metadata accuracy and provide nuanced understanding, necessitating interdisciplinary collaboration among cartographers, geographers, and data scientists (Trani et al., 2018). Additionally, privacy and ethical concerns must be addressed, ensuring that detailed metadata is used responsibly and does not infringe on privacy rights. Developing robust data governance frameworks and adhering to ethical standards are essential steps in mitigating these risks. Overcoming these challenges involves a combination of advanced technology, interdisciplinary collaboration, and robust governance, ultimately enabling the full realization of enriched metadata's promise in enhancing the utility and accuracy of map images (Rehberger, 2020).

9.4.1 Diversity and complexity of map images

Map images spring forth from a myriad of sources, encompassing satellite visuals, intricate topographical surveys, and treasured historical archives, each imprinting unique metadata requirements (Capello & Rosenbaum, 2020). The vast landscape of map images showcases a kaleidoscope of scales and resolutions, from detailed city blocks to expansive continents. This variability necessitates enrichment techniques that are robust and highly adaptive, ensuring each map's uniqueness is accurately captured and conveyed (Akbulut et al., 2023). Enrichment processes must handle high-resolution satellite imagery with the same finesse as low-resolution historical maps, preserving essential details and context. Adaptable metadata frameworks are crucial, accommodating diverse data sources and types, and ensuring users can quickly access relevant information regardless of the map's origin or scale (Zhao et al., 2023). Standardizing metadata across different maps enhances usability and interoperability, facilitating complex geospatial analyses and seamless data sharing across disciplines and organizations. By addressing these challenges, enriched metadata transforms map images into dynamic tools for insight and discovery, unlocking new possibilities for analysis and application across various fields (Ullah et al., 2024).

9.4.2 Limitations of traditional methods

Traditional methods of metadata enrichment often tread the labor-intensive path of manual annotations. While they offer a human touch, these methods are not immune to errors and can be excessively time-consuming (Gutiérrez, 2015). The meticulous effort required to manually annotate map images means that scaling these methods up presents a Herculean challenge, often proving impractical for large datasets. Additionally, the inconsistent tapestry of standards woven by different sectors or organizations adds to the complexity, leading to disparities in metadata depth, format, and granularity (Moratón Gutiérrez, 2016). These inconsistencies can hamper the integration and usability of map data across different platforms and applications, highlighting the need for more automated, standardized, and scalable solutions in metadata enrichment (Ghoussaini et al., 2021).

9.4.3 Computational challenges

The universe of map images is vast, and enriching their metadata through computational methods, especially deep learning, requires significant firepower. Large datasets are both a boon and a bane, offering richness but demanding substantial computational prowess. Training deep learning models adds another layer of complexity (Zhang et al., 2021). Procuring labeled data is a herculean task, often requiring extensive manual effort and expertise to ensure accuracy and relevance. Additionally, balancing these models to prevent overfitting or underfitting is paramount, especially for nuanced tasks like curating metadata. These challenges necessitate advanced algorithms, robust computational resources, and meticulous validation processes to achieve high-quality, scalable metadata enrichment for diverse and intricate map images (de-Lima-Santos & Salaverría, 2021).

9.4.4 Integrating geographic information systems-based augmentation

Merging the worlds of GIS data and map images for enrichment is akin to walking a tightrope. This fusion brings forth challenges in data compatibility, the intricacies of integration techniques, and the relentless pursuit of geographical precision (Paul et al., 2020). Ensuring that diverse data types from various sources can seamlessly interact requires sophisticated algorithms and robust data management frameworks. Moreover, the ever-evolving nature of geographical terrains and land utilizations means the GIS data must always be in sync with current realities, ensuring map images and their augmentations are time-relevant (Mihai et al., 2022). This necessitates continuous updates and real-time data processing capabilities to maintain the accuracy and relevance of the enriched metadata. The integration process must also handle discrepancies in scale, resolution, and data formats, demanding adaptable and scalable solutions. Consequently, the task of harmonizing GIS data with map images not only requires advanced technological solutions but also a deep understanding of geographical data dynamics to achieve precise and meaningful enrichment (Hu et al., 2022).

9.4.5 Ensuring quality and accuracy

Quality assurance in metadata enrichment is a nonnegotiable criterion. Validating the accuracy of metadata, especially when crafted through automated avenues, demands stringent evaluation frameworks (Chandak et al., 2024). Automated systems, while efficient, can occasionally misinterpret or overlook subtle features within map images, necessitating robust quality checks to ensure reliability. Maps can sometimes be cryptic, harboring features that defy singular interpretations, adding a layer of complexity to the enrichment process (Flemotomos et al., 2021). This ambiguity poses a challenge, compelling metadata creators to tread cautiously, ensuring clarity and precision in their descriptors. To address this, metadata enrichment must incorporate both automated precision and human oversight, where experts review and refine the data to capture nuanced details accurately. Implementing comprehensive validation protocols, such as cross-referencing with established databases and utilizing feedback loops for continuous improvement, is essential to uphold the integrity and usability of enriched metadata (Chandak et al., 2024).

9.4.6 Ethical and privacy concerns

The road to metadata enrichment is fraught with ethical landmines. While the aim is to provide clarity, there is a risk of spotlighting sensitive or protected zones, inadvertently breaching privacy protocols (Garrett, 2021). Enriched metadata can unintentionally expose vulnerable areas, such as wildlife reserves or private properties, leading to potential misuse or exploitation. Safeguarding against such breaches requires implementing strict data governance policies and ethical guidelines to ensure that sensitive information is adequately protected. Furthermore, the training data, foundational for deep learning models or GIS augmentations, might harbor biases (Paul et al., 2020). These biases can arise from various sources, such as the selective inclusion of certain types of data, historical inaccuracies, or socio-political influences embedded within the data. If unchecked, these biases can weave their way into the metadata, potentially skewing perceptions and interpretations (Ahmad et al., 2020). For example, biased training data might emphasize certain geographical features while neglecting others, leading to an imbalanced representation of the area. To mitigate these risks, it is crucial to employ diverse and representative training datasets, along with rigorous bias detection and correction mechanisms. Continuous monitoring and assessment of the metadata enrichment process can help identify and address any emerging biases. Additionally, involving a multidisciplinary team in the development and validation stages can provide a broader perspective and help ensure that ethical considerations are thoroughly integrated into the process (Mihai et al., 2022).

9.5 Deep learning: a new frontier for metadata enrichment

The emergence of deep learning has transformed many areas of technological development, particularly in fields related to data interpretation and pattern recognition. Its intersection with the realm of map metadata enrichment presents a promising pathway to overcome previously mentioned challenges and open new horizons (Atitallah et al., 2020). Deep learning algorithms, especially CNNs, excel at extracting intricate features from complex datasets, making them ideal for analyzing and enriching map images (Nawaz et al., 2024). By automating the detection and annotation of geographical features, deep learning can significantly reduce the time and effort required for manual metadata creation. Furthermore, deep learning models can continuously learn and improve from new data inputs, ensuring that metadata remains current and accurate in the face of changing geographical landscapes. This adaptive capability addresses the challenge of keeping GIS data synchronized with real-world changes, providing real-time updates and maintaining the relevance of enriched metadata (Pricope et al., 2019). Additionally, deep learning's ability to handle large and diverse datasets helps in standardizing metadata across different sources, enhancing interoperability and integration (Nawaz et al., 2024). Moreover, deep learning can mitigate some ethical concerns associated with metadata enrichment. For instance, advanced bias detection algorithms can be integrated into the deep learning process to identify and correct biases in training data, promoting fair and balanced metadata generation. Automated systems can also be programmed to recognize and flag sensitive areas, ensuring that privacy and security protocols are upheld (Bhatti et al., 2024).

9.5.1 The basics of deep learning

Deep learning draws its inspiration from the labyrinth of neural networks in the human brain, giving birth to artificial neural networks. These computational models are adept at processing intricate patterns in vast datasets (Stanley et al., 2019). As these networks grew in depth, adding layers upon layers, their ability to recognize and decipher advanced patterns evolved, culminating in the concept now known as "deep learning." This profound evolution in neural networks has enabled deep learning to offer unparalleled capabilities in pattern recognition and interpretation (Matsuo et al., 2022). Deep learning models, particularly deep neural networks (DNNs) and CNNs, excel in processing complex, high-dimensional data. Each layer in a deep learning model extracts specific features from the input data, gradually building up a detailed and nuanced understanding of patterns and relationships (Wu, Qiu et al., 2023; Wu, Ren et al., 2023). This hierarchical learning approach allows deep learning algorithms to perform tasks that were previously unattainable with traditional machine learning methods, such as recognizing objects in images, understanding natural language, and making accurate predictions based on large datasets. In the context of map metadata enrichment, deep learning's advanced pattern recognition capabilities are particularly valuable (Hosain et al., 2024). These models can automatically identify and annotate geographical features, such as roads, rivers, and buildings, within map images. By leveraging vast amounts of training data, deep learning algorithms can learn to distinguish between different types of terrain and land use, providing detailed and accurate metadata annotations. This automation reduces the reliance on manual metadata creation, saving time and resources while improving the consistency and quality of the data. Furthermore, deep learning's ability to adapt and improve over time is crucial for maintaining up-to-date and relevant metadata (Yao et al., 2017). As new data becomes available, deep learning models can be retrained to incorporate the latest information, ensuring that the metadata reflects current geographical conditions. This dynamic updating process is essential for applications that depend on accurate and timely geospatial data, such as urban planning, disaster response, and environmental monitoring. In addition to enhancing the accuracy and richness of metadata, deep learning can also help address ethical and privacy concerns. Advanced algorithms can be designed to detect and mitigate biases in the training data, ensuring that the generated metadata is fair and unbiased. Additionally, deep learning models can incorporate privacy-preserving techniques, such as differential privacy, to protect sensitive information while still providing valuable insights (Akgun & Greenhow, 2022).

9.5.2 Convolutional neural networks and map images

CNNs are a specialized avatar of neural networks, meticulously crafted for image data. Their architecture boasts of convolutional layers that are adept at spatial data processing (Peng et al., 2024). These networks excel in their ability to automatically and adaptively discern spatial hierarchies of features from map images. In simple terms, CNNs can dissect an image, identifying and learning the significance of each pixel and its relation to its neighbors (Luo et al., 2023). The strength of CNNs lies in their unique structure, which includes convolutional layers, pooling layers, and fully connected layers. Convolutional layers apply filters to the input image to create feature maps, capturing essential patterns such as edges, textures, and shapes. This process of applying filters is akin to sliding a magnifying glass over different parts of the image to identify and highlight key features. Pooling layers then reduce the dimensionality of these feature maps, retaining the most critical information while discarding less important details. This reduction helps in managing computational complexity and preventing overfitting. Through multiple layers of convolution and pooling, CNNs build a hierarchical understanding of the image (Wu, Qiu et al., 2023; Wu, Ren et al., 2023). The initial layers might capture simple features like edges and corners, while deeper layers recognize more complex patterns and structures, such as roads, buildings, and natural formations in map images (Luo et al., 2023). This layered approach allows CNNs to effectively learn and represent the spatial relationships and hierarchies within the data. One of the remarkable capabilities of CNNs is their adaptability. As they are trained on large datasets, they learn to fine-tune their filters and weights to optimize feature detection for the specific type of imagery they are analyzing. This adaptability is particularly beneficial in the context of map images, where variations in scale, resolution, and content can be significant. By learning from diverse training data, CNNs become proficient at generalizing their knowledge to new, unseen images, ensuring robust performance across different mapping scenarios (Luo et al., 2023). Moreover, CNNs can handle various image processing tasks crucial for map metadata enrichment, such as object detection, image segmentation, and classification. Object detection involves identifying and locating specific features within an image, such as identifying all buildings in a satellite image. Image segmentation partitions the image into segments, or regions, that represent different objects or features, providing a detailed map of what is present in each part of the image. Classification assigns labels to whole images or specific regions, categorizing them based on the features detected (Chengping Zhang et al., 2024).

9.5.3 Deep learning for automated metadata generation

Deep learning's prowess lies in its pattern recognition capabilities. It can seamlessly identify recurring patterns, flag anomalies, and spotlight unique facets in map images. This ability to discern complex patterns and structures within vast datasets is what makes deep learning particularly powerful for geospatial analysis (Klinger et al., 2018). By leveraging sophisticated algorithms, deep learning models can process and interpret the intricate details of map images, revealing insights that might be overlooked by traditional methods. Moreover, semantic segmentation techniques in deep learning take this capability a step further by meticulously labeling various components of a map. These techniques can distinguish and categorize different elements within an image, such as meandering roads, vast water bodies, and bustling urban sprawls (Webb, 2018). Semantic segmentation works by assigning a class label to each pixel in an image, effectively partitioning the map into meaningful regions. This level of granularity allows for a comprehensive understanding of the spatial relationships and characteristics of the various features depicted in the map. Such granularity in understanding adds invaluable layers of context to metadata, making it more insightful (Lucieri et al., 2020). When map images are enriched with detailed, context-aware metadata, they become far more than simple visual representations. The enriched metadata can provide information about the types of terrain, the density of urban areas, the flow of rivers, and the presence of infrastructure, among other things. This detailed metadata enhances the utility of map images for a wide range of applications, from urban planning and environmental monitoring to navigation and disaster management (Guerguiev et al., 2017). For instance, in urban planning, having access to detailed metadata about road networks, building locations, and land use patterns can significantly improve decision-making processes. Planners can use this information to design more efficient transportation systems, optimize land use, and develop sustainable urban environments. Similarly, in environmental monitoring, enriched metadata can help track changes in natural landscapes, monitor the health of ecosystems, and predict the impact of human activities on the environment (Ha et al., 2020). Additionally, the ability of deep learning models to flag anomalies and highlight unique facets in map images is particularly valuable for identifying areas of interest or concern. For example, in disaster management, these models can quickly detect changes in the landscape caused by natural disasters, such as floods, earthquakes, or wildfires. This rapid identification enables emergency responders to allocate resources more effectively and mitigate the impact of such events (Venigandla & Tatikonda, 2021).

9.5.4 Transfer learning and pretrained models

Transfer learning stands as a testament to the efficiency of deep learning. It revolves around the idea of harnessing pretrained models—models trained on one task—and repurposing them for another related task. This approach economizes on computational time and resources. Instead of starting from scratch, transfer learning enables the use of existing models that have already learned relevant features from extensive datasets (Bugliarello et al., 2022). These models can then be fine-tuned for specific applications, significantly speeding up the training process and reducing the need for vast amounts of new data. In the realm of map metadata enrichment, transfer learning can be a game-changer. It offers a shortcut, allowing models to use knowledge acquired from related tasks or domains and apply it to map images (Belgacem et al., 2020). For example, a model trained on general image recognition tasks can be repurposed to identify geographical features in map images. The layers of the pretrained model that capture basic patterns such as edges, textures, and shapes can be retained, while the final layers can be adjusted to focus on the specific details relevant to map data, such as roads, buildings, and natural formations. This method not only saves time and computational power but also leverages the robust feature extraction capabilities of established models. The initial training phase, which is often the most resource-intensive part, has already been completed (Ma et al., 2020). By building on this foundation, transfer learning allows for quicker adaptation to new, specialized tasks with improved performance. This is particularly advantageous for tasks like semantic segmentation, where models must label various components of a map with high precision. Moreover, transfer learning can help overcome the challenge of limited labeled data in specialized fields. In many cases, acquiring a large, annotated dataset for a specific type of map can be difficult and costly. Transfer learning mitigates this issue by enabling models to generalize from the knowledge embedded in pretrained networks, thus requiring fewer labeled examples to achieve high accuracy. In practical terms, implementing transfer learning in map metadata enrichment can lead to more accurate and detailed annotations (Nkwentsha, 2022). For instance, a model pretrained on satellite imagery can be fine-tuned to recognize specific urban structures or natural features unique to a particular region. This adaptability enhances the quality and relevance of the metadata, making it more useful for various applications such as urban planning, environmental monitoring, and disaster response (Baena, 2023). Furthermore, the ability of transfer learning to integrate knowledge from multiple domains can foster innovative approaches to geospatial analysis. By combining insights from different types of imagery and related tasks, models can develop a more comprehensive understanding of complex landscapes. This holistic perspective is crucial for addressing multifaceted challenges that span various disciplines and data sources (Girelli Consolaro et al., 2023).

9.5.5 Challenges and limitations of deep learning

Deep learning, for all its brilliance, is not without its challenges. A primary concern is its voracious appetite for data. Effective training demands extensive, labeled datasets, which can be both time-consuming and costly to procure (Suleyman, 2023). The quality of outcomes hinges not only on the quantity but also on the representativeness and bias-free nature of this data. Biased datasets can lead to skewed results, perpetuating inaccuracies and potentially causing significant issues in applications reliant on precise and fair data interpretations (Chamorro-Premuzic, 2023). Ensuring the datasets are comprehensive and unbiased is essential to the integrity and reliability of deep learning models. Additionally, deep learning models, especially the intricate ones, have garnered a reputation as "black boxes." Their decision-making intricacies often remain shrouded in mystery, challenging our understanding of their inner workings (Crawford & Paglen, 2021). This lack of transparency poses significant concerns, particularly in critical applications where understanding the reasoning behind a model's output is crucial. The opacity of these models can hinder trust and acceptance among users and stakeholders who need to ensure that the decisions made by the models are sound and justifiable. To mitigate these challenges, several approaches are being explored (Hosseini et al., 2018). For instance, techniques such as transfer learning and data augmentation can help address the data hunger by making better use of existing datasets and artificially expanding them. Moreover, efforts to improve the interpretability of deep learning models are ongoing (Voinov et al., 2016). Techniques such as attention mechanisms, visualization tools, and explainable AI (XAI) frameworks aim to shed light on the decision-making processes of these models. These methods strive to make the models more transparent and their outputs more understandable, thereby increasing trust and usability. Furthermore, establishing rigorous validation and testing protocols is vital to identify and correct biases within the training data (Süsser et al., 2022). Ensuring diversity in the datasets and continually monitoring the models for biased outcomes can help mitigate the risks associated with biased training data. Collaboration between data scientists, domain experts, and ethicists can foster the development of fairer and more accountable models (Casacuberta et al., 2024).

9.5.6 Integrating geographic information systems with deep learning

GIS data act as the compass for deep learning models, guiding them with vital spatial context. This context augments the model's predictive prowess, especially when generating metadata (Boyko et al., 2023). By providing geographical and spatial information, GIS data help deep learning models understand the relationships and patterns within map images more accurately. This added layer of context ensures that the generated metadata is not only more precise but also more relevant to the geographical nuances of the area being analyzed (Chandak et al., 2024). Additionally, GIS data can play a pivotal role in data augmentation. It can be amalgamated with map images to enhance the richness and variability of training datasets (Boyko et al., 2023). Data augmentation techniques, such as adding noise, rotating images, or changing scales, benefit significantly from the integration of GIS data, as it introduces realistic variations that reflect actual geographical changes. This symbiotic integration of GIS and deep learning ensures that models are exposed to a diverse set of scenarios, improving their robustness and generalization capabilities (Hosain et al., 2024). For example, combining satellite images with corresponding GIS layers, such as topographical maps, land use data, or population density maps, provides a multifaceted training set that deep learning models can learn from (Chandak et al., 2024). This enriched dataset enables the models to detect and understand complex spatial patterns, such as urban sprawl, deforestation, or water body changes, with greater accuracy. Consequently, the metadata generated from these models is more comprehensive and insightful, capturing the intricate details of the geographical features (Yue & Shyu, 2024). Furthermore, GIS data can enhance the interpretability of deep learning models. By aligning the model outputs with spatial data, it becomes easier to visualize and understand the patterns that the model has learned (Usama, 2023). This spatial alignment can help validate the model's predictions and provide a clear, contextual explanation of its decision-making process. For instance, if a model predicts changes in land use, GIS data can help map these predictions onto actual geographical locations, making the outcomes more tangible and actionable (Zou et al., 2024).

9.6 Geographic information systems in the world of deep learning

GIS have long stood as powerful tools for the management, visualization, and analysis of spatial data. When combined with the capabilities of deep learning, GIS can be harnessed in innovative ways, offering enhanced data interpretations, richer visualizations, and more robust spatial decision-making processes (Goodchild, 2009). GIS offer a dynamic framework to capture, store, and analyze spatial data (Nizamani et al., 2021). These data come in varied forms, from vector representations like points, lines, and polygons that depict specific locations or paths, to raster data which capture the Earth's surface in pixelated detail, as seen in satellite images (Nizamani et al., 2023). The core of GIS revolves around operations like geoprocessing, which manipulates this spatial data; spatial querying, which retrieves specific data subsets; and spatial analysis, which delves deep to extract insights and patterns from the data (Kokla & Guilbert, 2020). GIS stands as a robust tool in the arsenal of deep learning, enhancing its training datasets. By introducing spatial transformations, variations, or even synthetic data generation, GIS ensures richer, more varied data pools ready for model training. Furthermore, GIS's strength lies in spatial contextualization. It sews in spatial references and geospatial attributes, allowing deep learning models to produce predictions that are not just accurate but also spatially aware and context-rich (Klein et al., 2023). Deep learning, coupled with GIS, automates the complex task of land use classification. Whether it is identifying dense urban conglomerates, vast water expanses, or serene forests, the synergy ensures precise categorization. Beyond land types, this duo shines in object detection within map data—pinpointing buildings that punctuate a skyline, vehicles that traverse roads, or integral infrastructure elements. By leveraging GIS data, this process gains a heightened degree of accuracy (Wang et al., 2022). The merging of time and space gives rise to spatiotemporal analyses. Through a marriage of GIS and deep learning, one can track the morphing face of our planet—be it the sprawling growth of cities, the receding boundaries of glaciers, or the vanishing green of forests. The insights gleaned from past spatial data also empower predictive modeling, forecasting events like potential flood zones or the trajectory of urban sprawl, offering a foresight that can shape our response (Kankanamge et al., 2020). Melding GIS with deep learning is not without its hurdles. Data integration, for instance, demands meticulous attention, ensuring that traditional GIS data melds seamlessly with data primed for deep learning, upholding both compatibility and geographic accuracy. With vast, high-resolution geospatial datasets, scalability becomes a concern, striving to maintain computational efficiency. Moreover, a pivotal aspect remains geographical accuracy, ensuring that the outputs and predictions of deep learning mirror the real-world geographical nuances (Tagde et al., 2021). The real-world reverberations of the GIS and deep learning alliance are profound. Urban planning becomes a calculated affair, with data-driven decisions shaping infrastructure placements, urban growth trajectories, and zoning rules. The environmental domain too benefits, monitoring shifts in our natural world, predicting potential calamities, or evaluating habitat changes. In the bustling lanes of transportation and logistics, these technologies come together to carve optimal routes, dissect traffic flows, and blueprint transportation systems—all powered by enhanced spatial data (Bibri et al., 2022).

9.7 Proposed methodology

Building on the knowledge of the potentialities of deep learning and GIS, this chapter outlines a comprehensive methodology to enrich the metadata of map images (Lefevre et al., 2017). The objective is to combine the strengths of both domains to create a system that automates metadata generation, ensuring accuracy, richness, and relevance. The first step in map metadata enrichment requires sourcing map images from a blend of historical treasure troves and cutting-edge satellite technologies. Once amassed, GIS datasets are seamlessly integrated, ensuring that map images have the correct spatial references. But data are not always pristine. The crucial task of data cleaning ensues, scrubbing away anomalies and inaccuracies. Ultimately, the goal is to achieve data standardization, morphing diverse data into a consistent format, scale, and resolution, primed for subsequent processing (Lefevre et al., 2017). Deep learning offers a plethora of architectures, each with its merits. The key is in picking the right one, with a leaning toward CNNs for image-related tasks. But the architecture is just the skeleton; the muscles are the hyperparameters. By fine-tuning elements like the learning rate, batch size, or the number of layers, the model's performance can be optimized to its peak potential (Zhao et al., 2018). Data augmentation is like adding colors to an artist's palette. Using GIS, map images undergo spatial transformations; they are rotated, translated, scaled, offering a richer dataset. But why stop at reality? Synthetic data generation steps in, painting GIS-informed imaginary map scenarios, further bolstering the training dataset (Cho et al., 2017). Training a deep learning model is akin to schooling. The vast dataset is split into classrooms—the training, validation, and test sets. Using batch processing, the model is fed data in digestible chunks, learning from each. As it learns, continuous validation acts as the teacher's oversight, ensuring the model does not just memorize but truly understands, preventing overfitting (Yang, Song, et al., 2023; Yang, Vining, et al., 2023). Post training, the model turns into a map detective. Through feature extraction, it identifies and categorizes unique characteristics within map images (Tanhua et al., 2019). Going deeper, it delves into semantic understanding, interpreting the context and meaning of these features. This treasure trove of information is then systematically compiled, crafting a comprehensive metadata structure (Ali et al., 2023). GIS is not done yet. In the postprocessing phase, it refines the metadata, ensuring pixel-perfect alignment with geographical truths. Beyond mere refinement, it amplifies the metadata, infusing it with enriched geospatial insights, making the metadata not just accurate but insightful (Crooks et al., 2018). No process is truly complete without evaluation. The generated metadata undergoes rigorous quantitative assessment, juxtaposed against ground truths or expert annotations. But in this digital age, users play a pivotal role. By incorporating user feedback mechanisms, the process remains dynamic, always evolving. This feedback, paired with evaluation insights, feeds back into the system, refining models and methodologies in a perpetual cycle of improvement (Mach & Field, 2017).

9.8 Case studies

The potential of combining deep learning with GIS for enriching metadata can be best understood through real-world applications. This chapter delves into a selection of case studies that exemplify the practicalities, challenges, and successes of the proposed methodology. Deep within the archives, historical maps whisper tales of bygone eras. The primary aim here is to amplify these whispers, making historical geographies resonate with today's audience. By deploying deep learning, we unearth old landmarks, trace long-forgotten transport routes, and sketch ancient terrains. When juxtaposed with contemporary GIS data, we not only see the past but also how it morphs into our present. The resulting tapestry reveals urban evolutions, ancient landmarks in today's contexts, and deepens our connection with history (Wu, Qiu et al., 2023; Wu, Ren et al., 2023). As cities pulse and grow, understanding their beat becomes crucial. The focus is on automating the categorization of urban areas—from residential cocoons to bustling commercial zones, tranquil parks, and industrious regions. With deep learning models trained on the bird's-eye views from satellites and enhanced by GIS data, we gain an accurate insight into the city's heartbeat. Such knowledge paves the way for well-informed urban planning, resource allocations, and ensures the city's evolution remains on the right track (Liu et al., 2020). Wetlands, the earth's ecological treasure troves, need vigilant guardians. The goal is clear: monitor their health and territorial spread. Harnessing deep learning, satellite images reveal the wetlands in their entirety, allowing us to witness their shifts and murmurs over time. GIS data, bearing notes on local flora, fauna, and the waters' purity, further accentuates this understanding. With these tools, we can leap into action when needed, ensuring these wetlands continue to thrive and serve as ecological havens (Yang, Song, et al., 2023; Yang, Vining, et al., 2023). In the game of urban development, every move counts. Here, the objective is to foresee the ripples of infrastructural changes before they touch the ground. Through GIS simulations, proposed developments rise in virtual spaces (Barghchi et al., 2009). Deep learning models then step in as oracles, predicting how these changes would stir the surroundings. Such foresight empowers urban developers, ensuring that growth does not come at nature's expense and that infrastructures weave seamlessly into the existing urban fabric (Qadir & Fatah, 2023). In the sprawling fields, crops sway with secrets of their health and yield. The mission is to decode these secrets from the skies. Multispectral satellite images, processed by deep learning models, reveal the tales of different crops and their vitality. GIS data, bearing the memories of the soil, rains, and man-made rivers, adds depth to this narrative. Farmers, armed with these insights, can then dance in tandem with nature, ensuring bountiful harvests and securing our food chains (Puzzarini & Barone, 2020).

9.9 Results and evaluations

Assessing the effectiveness of enriched metadata through well-defined metrics and visual representations is paramount to ensure the methodology's success. This chapter elucidates the specific metrics utilized and provides a comprehensive visualization and discussion of the results. When we venture into evaluating the enriched metadata's robustness, several yardsticks come into play. We start with accuracy—a simple yet profound measure of how often the enriched metadata hits the bull's-eye against a validated reference. Then, precision enters, dissecting how many of the model's confident claims (like identified landmarks or terrains) genuinely hold water. But it is not just about what is claimed; recall probes into what is missed, gauging how many actual map features the model successfully identifies (Wu et al., 2019). When balancing precision and recall becomes a tightrope act, the F1-Score offers a harmonious solution, especially when data classes are skewed. Yet, metrics are not solely about numbers; the User Satisfaction Score introduces a human touch, gauging user sentiments on the metadata's quality. Finally, we cannot overlook efficiency. Processing time becomes our watchful sentinel, clocking the speed of metadata enrichment, hinting at the methodology's agility (Di Lascio et al., 2018). Though this text cannot illustrate, imagine vivid visual aids decoding results. A bar graph stands tall, each bar representing the model's performance—accuracy, precision, recall—across varied map types, facilitating a direct comparison (Ragni et al., 2020). Then, a colorful pie chart emerges, its slices revealing user sentiments—from utmost satisfaction to indifference or discontent. For a spatial touch, a heat map paints a picture, its gradient hues spotlighting areas in map images where the model either shone brightly or dimmed. And as we ponder on scalability, a line graph plots the journey, drawing connections between map complexities and the time taken for metadata enrichment (Al-Shdeifat & Al-Jamal, 2023). Deep dives follow the visuals. In performance analysis, we dissect the model's feats and flaws, pondering upon reasons for discrepancies in accuracy across map genres. User feedback, a goldmine of insights, undergoes scrutiny (Ragni et al., 2020). Echoed praises or recurring grievances provide a compass, guiding toward the enriched metadata's practical implications. And as we stand proud of our methodology, we also turn to our peers, juxtaposing our techniques against the traditional, weighing pros and cons. Visuals, though silent, speak volumes. From the bar graphs to heat maps, they whisper tales of patterns, strengths, and growth avenues, providing a holistic view of the methodology's prowess (Winter & Davidson, 2022).

9.10 Limitations and potential solutions

While the convergence of GIS and deep learning offers transformative potential for map metadata enrichment, it also presents its own set of challenges. Acknowledging these limitations and devising strategies to overcome them is vital for the continual evolution of the methodology. Marrying GIS with deep learning, while promising, is not devoid of challenges. High on this list is the computational weight deep learning brings, particularly when parsing spatial intricacies. GIS, with its voluminous spatial data, only compounds this, often stretching computational muscles to their limits (Qadir & Fatah, 2023). But it is not just about crunching numbers; the aligning of stars, or rather data, becomes pivotal. A tiny misalignment between GIS data and images can snowball into glaring metadata errors. And while deep learning models can be experts in their trained niches, asking them to wear many hats, to generalize across varied map genres, can be a tall order. Add the exponentially ballooning GIS datasets to the mix, and scalability becomes a daunting tightrope act. Lastly, time waits for none, and the ever-shifting terrains, evolving urban landscapes, and changing land uses turn temporal dynamics into a perpetual game of catch-up for models (Paul et al., 2020). But where challenges arise, solutions beckon. Facing towering computational demands head-on, one can tap into the might of distributed computing, with graphics processing units (GPUs) leading the charge. With platforms like TensorFlow and PyTorch rallying behind GPU acceleration, processing prowess can get a significant boost. The alignment ordeal has its saviors too. A meticulous preprocessing regimen, aided by stalwarts like Geospatial Data Abstraction Library (GDAL) and Quantum Geographic Information System (QGIS), can align spatial data with precision, avoiding downstream errors (O'Neill et al., 2021). And as we seek versatility in our models, transfer learning emerges as a beacon, allowing them to learn from diverse terrains and fine-tune their expertise. As datasets burgeon, optimized databases like PostGIS step in, ensuring data are not just stored, but also retrieved with finesse. Finally, to dance to the evolving tunes of time, models can embrace continuous learning. By frequently updating their knowledge from the latest spatial narratives, they stay relevant, informed, and ready for the future (Haslip & Gullo, 2018).

9.11 Future prospects

As technology advances and the spheres of GIS and deep learning evolve, the landscape of map metadata enrichment will undoubtedly experience profound shifts. Exploring these forthcoming trends and identifying potential areas of research can provide direction and inspiration for future endeavors. We stand at the cusp of a transformative era where GIS, deep learning, and map metadata are converging in unprecedented ways (O'Neill et al., 2021). In the offing are systems capable of real-time metadata enrichment, a game-changer for dynamic, fast-evolving landscapes. But as data multiply, the luxury of manually labeled data wanes, propelling methods like semisupervised and unsupervised learning to the forefront (Qadir & Fatah, 2023). The future also hints at integrative multimodal systems, where a symphony of text, images, audio, and sensor data offers a richer geospatial tapestry. Adaptive learning models, agile enough to dance to the shifting beats of terrains without extensive retraining, are another promising frontier. And as the boundaries between the virtual and real blur, enriched metadata might soon be the heartbeat of immersive augmented reality (AR) and virtual reality (VR) geospatial odysseys (Winter & Davidson, 2022). The potential of enriched metadata is akin to a key unlocking myriad domains. Foremost are smart cities, where this metadata could sculpt urban landscapes with intelligence, ensuring they are not only connected but also sustainable. In the verdant realm of environmental conservation, such detailed metadata could be the eyes monitoring delicate ecosystems, guiding conservation actions with precision (Qadir & Fatah, 2023). When calamities strike, real-time enriched metadata can be a beacon for disaster management, optimizing life-saving responses. In the echoing corridors of our past, enriched metadata can breathe life into historical and archeological narratives, aiding preservation and enlightening the public (Guo et al., 2019). The roads we travel, too, could be reshaped, with transportation and logistics sectors benefitting from better planning and fluid operations. And for the wanderlust-filled souls, augmented tourism beckons, where journeys are enriched with layers of insights, making every step an informed exploration (Liu et al., 2020).

9.12 Conclusion

The endeavor to enrich map metadata through the fusion of GIS and deep learning represents a transformative step in the way we perceive, understand, and interact with geospatial data. This journey, from recognizing the importance of rich metadata to harnessing advanced computational methods, has showcased both the potential and challenges of this interdisciplinary approach. We have witnessed the power of deep learning, traditionally seen in areas like image recognition and natural language processing, make significant inroads into the realm of GIS. Its ability to sift through complex datasets, detect patterns, and generate enriched metadata has reshaped expectations and set new benchmarks for quality and depth. Yet, this journey has not been without its challenges. The integration of vast GIS datasets with intricate deep learning models has tested computational capacities, demanded robust preprocessing tools, and necessitated continuous adaptation to changing geospatial realities. Recognizing these challenges and addressing them proactively has been integral to the methodology's evolution. Looking ahead, the horizon is replete with opportunities. Emerging trends, from real-time metadata enrichment to the integration of AR and VR, are pushing boundaries. There is immense potential waiting to be tapped, from smart urban planning to innovative tourism experiences. The future beckons researchers, developers, and practitioners to explore, innovate, and elevate the state of map metadata enrichment to unprecedented heights. In closing, the endeavor to enhance map metadata is more than a mere technical exercise; it is a pathway to a richer, more informed understanding of our world. By leveraging the strengths of both GIS and deep learning, we are not just adding layers of data to maps; we are adding layers of context, understanding, and insight. The journey has only just begun, and the map to the future is rich with promise.

References

Aamir, M., Rahman, Z., Abro, W. A., Bhatti, U. A., Dayo, Z. A., & Ishfaq, M. (2023). Brain tumor classification utilizing deep features derived from high-quality regions in MRI images. *Biomedical Signal Processing and Control, 85,* 104988.

Ahmad, R.M., Yao, X., Nawaz, S.A., Bhatti, U.A., Mehmood, A., Bhatti, M.A., & Shaukat, M.U. (2020). Robust image watermarking method in wavelet domain based on sift features. In *Proceedings of the 2020 third international conference on artificial intelligence and pattern recognition* (pp. 180–185).

Ahmad, T., Madonski, R., Zhang, D., Huang, C., & Mujeeb, A. (2022). Data-driven probabilistic machine learning in sustainable smart energy/smart energy systems: Key developments, challenges, and future research opportunities in the context of smart grid paradigm. *Renewable and Sustainable Energy Reviews, 160,* 112128.

Akbulut, U., Cifci, M. A., & Aslan, Z. (2023). Hybrid modeling for stream flow estimation: Integrating machine learning and federated learning. *Applied Sciences, 13*(18), 10203.

Akgun, S., & Greenhow, C. (2022). Artificial intelligence in education: Addressing ethical challenges in K-12 settings. *AI and Ethics, 2*(3), 431–440.

Alhatami, E., Huang, M., & Bhatti, U. A. (2024). *Image fusion techniques and applications for remote sensing and medical images. Deep learning for multimedia processing applications* (pp. 154–175). CRC Press.

Ali, A., Li, J., Chen, H., Bhatti, U. A., & Khan, A. (2023). Real-time spammers detection based on metadata features with machine learning. *Intelligent Automation and Soft Computing, 38*(3).

Alma'aitah, W. Z. A., Talib, A. Z., & Osman, M. A. (2020). Opportunities and challenges in enhancing access to metadata of cultural heritage collections: A survey. *Artificial Intelligence Review, 53,* 3621–3646.

Al-Shdeifat, Z., & Al-Jamal, D. (2023). Visual thinking and reading comprehension: Foreign language setting as an example. *International Journal of Educational Methodology*, *9*(2), 367–385.

Anderson, J. (2018). Securing, standardizing, and simplifying electronic health record audit logs through permissioned blockchain technology. *Dartmouth College Undergraduate Theses*, 135.

Archana, R., & Jeevaraj, P. E. (2024). Deep learning models for digital image processing: A review. *Artificial Intelligence Review*, *57*(1), 11.

Aristidou, A., Shamir, A., & Chrysanthou, Y. (2019). Digital dance ethnography: Organizing large dance collections. *Journal on Computing and Cultural Heritage*, *12*(4), 1–27.

Atitallah, S. B., Driss, M., Boulila, W., & Ghézala, H. B. (2020). Leveraging deep learning and IoT big data analytics to support the smart cities development: Review and future directions. *Computer Science Review*, *38*, 100303.

Baena, R. (2023). Beyond the training task in classification: looking at extension of the notion of generalization, Doctoral dissertation, Ecole nationale supérieure Mines-Télécom Atlantique.

Barghchi, M., Omar, D., & Aman, M. S. (2009). Cities, sports facilities development, and hosting events. *European Journal of Social Sciences*, *10*(2), 185–195.

Batty, M., Crooks, A., Hudson-Smith, A., Milton, R., Anand, S., Jackson, M., & Morley, J. (2010). *Data mash-ups and the future of mapping*. Bristol: JISC.

Belgacem, A., Beghdad-Bey, K., Nacer, H., & Bouznad, S. (2020). Efficient dynamic resource allocation method for cloud computing environment. *Cluster Computing*, *23*(4), 2871–2889.

Bentivoglio, R., Isufi, E., Jonkman, S. N., & Taormina, R. (2022). Deep learning methods for flood mapping: A review of existing applications and future research directions. *Hydrology and Earth System Sciences*, *26*(16), 4345–4378.

Bhatti, G., Mohan, H., & Singh, R. R. (2021). Towards the future of smart electric vehicles: Digital twin technology. *Renewable and Sustainable Energy Reviews*, *141*, 110801.

Bhatti, M. A., Song, Z., Bhatti, U. A., & Ahmad, N. (2023). Predicting the impact of change in air quality patterns due to COVID-19 lockdown policies in multiple urban cities of henan: A deep learning approach. *Atmosphere*, *14*(5), 902.

Bhatti, U. A., Bazai, S. U., Hussain, S., Fakhar, S., Ku, C. S., Marjan, S., & Jing, L. (2023). Deep learning-based trees disease recognition and classification using hyperspectral data. *Computers, Materials and Continua*, *77*(1).

Bhatti, U. A., Huang, M., Neira-Molina, H., Marjan, S., Baryalai, M., Tang, H., & Bazai, S. U. (2023). MFFCG–multi feature fusion for hyperspectral image classification using graph attention network. *Expert Systems with Applications*, *229*, 120496.

Bhatti, U. A, Tang, H, Khan, A, Ghadi, Y. Y, Bhatti, M. A, & Khan, K. A (2024). Investigating the nexus between energy, socio-economic factors and environmental pollution: A geo-spatial multi regression approach. *Gondwana Research*, *130*, 308–325.

Bibri, S. E., Allam, Z., & Krogstie, J. (2022). The metaverse as a virtual form of data-driven smart urbanism: Platformization and its underlying processes, institutional dimensions, and disruptive impacts. *Computational Urban Science*, *2*(1), 24.

Blasch, E., Pham, T., Chong, C. Y., Koch, W., Leung, H., Braines, D., & Abdelzaher, T. (2021). Machine learning/artificial intelligence for sensor data fusion–opportunities and challenges. *IEEE Aerospace and Electronic Systems Magazine*, *36*(7), 80–93.

Bogers, T., & Petras, V. (2017). Supporting book search: A comprehensive comparison of tags vs. controlled vocabulary metadata. *Data and Information Management*, *1*(1), 17–34.

Boyko, J., Cohen, J., Fox, N., Veiga, M.H., Li, J.I., Liu, J., & Xie, X. (2023). An interdisciplinary outlook on large language models for scientific research. arXiv preprint arXiv:2311.04929.

Brodeur, J., Coetzee, S., Danko, D., Garcia, S., & Hjelmager, J. (2019). Geographic information metadata—an outlook from the international standardization perspective. *ISPRS International Journal of Geo-Information*, *8*(6), 280.

Brumana, R., Oreni, D., Caspani, S., & Previtali, M. (2018). Virtual museums and built environment: narratives and immersive experience via multitemporal geodata hub. *Virtual Archaeology Review*, *9*(19), 34–49.

Bugliarello, E., Liu, F., Pfeiffer, J., Reddy, S., Elliott, D., Ponti, E.M., & Vulić, I. (2022). IGLUE: A benchmark for transfer learning across modalities, tasks, and languages. In *International conference on machine learning* (pp. 2370–2392). PMLR.

Burns, A., Lee, C., On, T., Xiong, C., Peck, E., & Mahyar, N. (2024). From invisible to visible: Impacts of metadata in communicative data visualization. *IEEE Transactions on Visualization and Computer Graphics*, *30*(7), 3427–3443.

Capello, E., & Rosenbaum, J. B. (Eds.), (2020). *Cartographic expeditions and visual culture in the nineteenth-century Americas*. Routledge.

Casacuberta, D., Guersenzvaig, A., & Moyano-Fernández, C. (2024). Justificatory explanations in machine learning: For increased transparency through documenting how key concepts drive and underpin design and engineering decisions. *AI and Society*, *39*(1), 279–293.

Cerchione, R., Centobelli, P., Riccio, E., Abbate, S., & Oropallo, E. (2023). Blockchain's coming to hospital to digitalize healthcare services: Designing a distributed electronic health record ecosystem. *Technovation*, *120*102480.

Chamorro-Premuzic, T. (2023). *I, human: AI, automation, and the quest to reclaim what makes us unique*. Harvard Business Press.

Chandak, T., Jayashree, J., & Vijayashree, J. (2024). *Trends and advancements of AI and XAI in drug discovery. Explainable AI (XAI) for sustainable development: Trends and applications* (pp. 233–255). Taylor & Francis.

Chapman, A., Simperl, E., Koesten, L., Konstantinidis, G., Ibáñez, L. D., Kacprzak, E., & Groth, P. (2020). Dataset search: A survey. *The VLDB Journal*, *29*(1), 251–272.

Chassanoff, A., & Altman, M. (2020). Curation as "interoperability with the future": Preserving scholarly research software in academic libraries. *Journal of the Association for Information Science and Technology*, *71*(3), 325–337.

Chen, J., Dowman, I., Li, S., Li, Z., Madden, M., Mills, J., & Heipke, C. (2016). Information from imagery: ISPRS scientific vision and research agenda. *ISPRS Journal of Photogrammetry and Remote Sensing*, *115*, 3–21.

Cheng, M., Tang, H., Khan, A., Melethil Sethumadhavan, S., Assam, M., Li, D., & Bhatti, U. A. (2023). A digital twin-based visual servoing with extreme learning machine and differential evolution. *International Journal of Intelligent Systems, 2023.*

Chengping Zhang, M. A., Guan, Y., Al-Razgan, M., Awwad, E. M., Ullah, R., Bhatti, U. A., & Ghadi, Y. Y. (2024). Enhancing lung cancer diagnosis with data fusion and mobile edge computing using DenseNet and CNN. *Journal of Cloud Computing, 13(1)*, 91.

Cho, J., Yun, S., Mu Lee, K., & Young Choi, J. (2017). Palettenet: Image recolorization with given color palette. In *Proceedings of the IEEE conference on computer vision and pattern recognition workshops* (pp. 62–70).

Crawford, K., & Paglen, T. (2021). Excavating AI: The politics of images in machine learning training sets. *AI and Society*, *36*(4), 1105–1116.

Crooks, A., Malleson, N., Malleson, N., Manley, E., & Heppenstall, A. (2018). *Agent-based modelling and geographical information systems: A practical primer.* Sage.

de-Lima-Santos, M. F., & Salaverría, R. (2021). From data journalism to artificial intelligence: Challenges faced by La Nación in implementing computer vision in news reporting. *Palabra Clave*, *24*(3), e2437.

Di Lascio, E., Gashi, S., & Santini, S. (2018). Unobtrusive assessment of students' emotional engagement during lectures using electrodermal activity sensors. *Proceedings of the ACM on Interactive, Mobile, Wearable and Ubiquitous Technologies*, *2*(3), 1–21.

Duckham, M., Sun, Q. C., & Worboys, M. F. (2023). *GIS: A computing perspective*. CRC Press.

Fan, C., Zhang, C., Yahja, A., & Mostafavi, A. (2021). Disaster city digital twin: A vision for integrating artificial and human intelligence for disaster management. *International Journal of Information Management*, *56,* 102049.

Finger, J. H., Smith, C. M., Hayamizu, T. F., McCright, I. J., Xu, J., Law, M., & Ringwald, M. (2017). The mouse gene expression database (GXD): 2017 update. *Nucleic Acids Research*, *45*(D1), D730–D736.

Flemotomos, N., Martinez, V. R., Chen, Z., Creed, T. A., Atkins, D. C., & Narayanan, S. (2021). Automated quality assessment of cognitive behavioral therapy sessions through highly contextualized language representations. *PLoS ONE*, *16*(10), e0258639.

Gao, S., Li, L., Li, W., Janowicz, K., & Zhang, Y. (2017). Constructing gazetteers from volunteered big geo-data based on Hadoop. *Computers, Environment and Urban Systems*, *61*, 172–186.

Garrett, P. M. (2021). *Dissenting social work: Critical theory, resistance and pandemic*. Routledge.

Ghoussaini, M., Mountjoy, E., Carmona, M., Peat, G., Schmidt, E. M., Hercules, A., & Dunham, I. (2021). Open targets genetics: Systematic identification of trait-associated genes using large-scale genetics and functional genomics. *Nucleic Acids Research*, *49*(D1), D1311–D1320.

Giordan, D., Adams, M. S., Aicardi, I., Alicandro, M., Allasia, P., Baldo, M., & Troilo, F. (2020). The use of unmanned aerial vehicles (UAVs) for engineering geology applications. *Bulletin of Engineering Geology and the Environment*, *79*, 3437–3481.

Girelli Consolaro, N., Shinde, S. S., Naseh, D., & Tarchi, D. (2023). Analysis and performance evaluation of transfer learning algorithms for 6G wireless networks. *Electronics*, *12*(15), 3327.

Goodchild, M. F. (2009). Geographic information systems and science: Today and tomorrow. *Annals of GIS*, *15*(1), 3–9.

Guerguiev, J., Lillicrap, T. P., & Richards, B. A. (2017). Towards deep learning with segregated dendrites. *Elife*, *6,* e22901.

Guerra-Manzanares, A., & Bahsi, H. (2022). On the relativity of time: Implications and challenges of data drift on long-term effective android malware detection. *Computers and Security*, *122,* 102835.

Guo, H., Nativi, S., Liang, D., Craglia, M., Wang, L., Schade, S., & Annoni, A. (2020). Big Earth Data science: An information framework for a sustainable planet. *International Journal of Digital Earth*, *13*(7), 743–767.

Guo, J., Kurup, U., & Shah, M. (2019). Is it safe to drive? An overview of factors, metrics, and datasets for driveability assessment in autonomous driving. *IEEE Transactions on Intelligent Transportation Systems*, *21*(8), 3135–3151.

Gutiérrez, L.M. (2015). Thematic patterning in English and Spanish: Contrastive annotation of a bilingual newspaper corpus for linguistic and computational applications, Doctoral dissertation, Universidad Complutense De Madrid.

Ha, N., Xu, K., Ren, G., Mitchell, A., & Ou, J. Z. (2020). Machine learning-enabled smart sensor systems. *Advanced Intelligent Systems*, *2*(9), 2000063.

Haslip, M. J., & Gullo, D. F. (2018). The changing landscape of early childhood education: Implications for policy and practice. *Early Childhood Education Journal*, *46*, 249–264.

Heitzler, M., Lam, J. C., Hackl, J., Adey, B. T., & Hurni, L. (2017). GPU-accelerated rendering methods to visually analyze large-scale disaster simulation data. *Journal of Geovisualization and Spatial Analysis*, *1*, 1–18.

Hochschild, V., Braun, A., Sommer, C., Warth, G., & Omran, A. (2020). Visualizing landscapes by geospatial techniques. In *Modern approaches to the visualization of landscapes*, (pp. 47–78). Springer.

Hosain, M. T., Zaman, A., Abir, M. R., Akter, S., Mursalin, S., & Khan, S. S. (2024). Synchronizing object detection: Applications, advancements and existing challenges. *IEEE Access*.

Hosseini, M., Shahri, A., Phalp, K., & Ali, R. (2018). Four reference models for transparency requirements in information systems. *Requirements Engineering*, *23*, 251–275.

Hu, Y., Gui, Z., Wang, J., & Li, M. (2022). Enriching the metadata of map images: A deep learning approach with GIS-based data augmentation. *International Journal of Geographical Information Science*, *36*(4), 799–821.

Huang, M. Q., Ninić, J., & Zhang, Q. (2021). BIM, machine learning and computer vision techniques in underground construction: Current status and future perspectives. *Tunnelling and Underground Space Technology*, *108,* 103677.

Huang, M., Zhang, X. S., Bhatti, U. A., Wu, Y., Zhang, Y., & Ghadi, Y. Y. (2024). An interpretable approach using hybrid graph networks and explainable AI for intelligent diagnosis recommendations in chronic disease care. *Biomedical Signal Processing and Control*, *91,* 105913.

Iapaolo, F. (2023). The system of autono-mobility: computer vision and urban complexity—reflections on artificial intelligence at urban scale. *AI and Society*, *38*(3), 1111–1122.

Iyer, C. K., Ganguli, S., & Pandey, V. (2023). *Perspectives on geospatial artificial intelligence platforms for multimodal spatiotemporal datasets. Advances in scalable and intelligent geospatial analytics* (pp. 17–63). Taylor & Francis.

Jane, G., White, H. C., Carrier, S., & Scherle, R. (2014). *A metadata best practice for a scientific data repository. Metadata best practices and guidelines* (pp. 28–46). Routledge.

Janssen, S. J., Porter, C. H., Moore, A. D., Athanasiadis, I. N., Foster, I., Jones, J. W., & Antle, J. M. (2017). Towards a new generation of agricultural system data, models and knowledge products: Information and communication technology. *Agricultural Systems*, *155*, 200–212.

Jobst, M., & Gartner, G. (2019). Changing paradigm in map production and geoinformation management—an introduction. In *Service-oriented mapping: Changing paradigm in map production and geoinformation management*, (pp. 3–42). Springer.

Kankanamge, N., Yigitcanlar, T., Goonetilleke, A., & Kamruzzaman, M. (2020). Determining disaster severity through social media analysis: Testing the methodology with South East Queensland Flood tweets. *International Journal of Disaster Risk Reduction*, *42*, 101360.

Keeler, J. K. (2020). *Feasting on Words: What university students learn when they study food writing and food media*. University of South Florida.

Kerr, D. W., Hogle, I. B., Ort, B. S., & Thornton, W. J. (2016). A review of 15 years of Spartina management in the San Francisco Estuary. *Biological Invasions*, *18*(8), 2247–2266.

Kerski, J. J. (2022). Online, engaged instruction in geography and GIS using IoT feeds, web mapping services, and field tools within a spatial thinking framework. *The Geography Teacher*, *19*(3), 93–101.

Kerski, J. J., & Baker, T. R. (2019). Infusing educational practice with Web GIS. In *Geospatial technologies in geography education*, (pp. 3–19). Springer International Publishing.

Keskin, M., & Kettunen, P. (2023). Potential of eye-tracking for interactive geovisual exploration aided by machine learning. *International Journal of Cartography*, *9*(2), 150–172.

Khelifi, L., & Mignotte, M. (2020). Deep learning for change detection in remote sensing images: Comprehensive review and meta-analysis. *IEEE Access*, *8*, 126385–126400.

Klein, L., Seidl, D., Fulneček, J., Prokop, L., Mišák, S., & Dvorský, J. (2023). Antenna contactless partial discharges detection in covered conductors using ensemble stacking neural networks. *Expert Systems with Applications*, *213*, 118910.

Klinger, J., Mateos-Garcia, J., & Stathoulopoulos, K. (2018). Deep learning, deep change? Mapping the development of the artificial intelligence general purpose technology. arXiv preprint arXiv:1808.06355.

Kokla, M., & Guilbert, E. (2020). A review of geospatial semantic information modeling and elicitation approaches. *ISPRS International Journal of Geo-Information*, *9*(3), 146.

Kucklick, J.P., Müller, J., Beverungen, D., & Müller, O. (2021). Quantifying the impact of location data for real estate appraisal-a GIS-based deep learning approach. In *ECIS*.

Kuźma, M., & Mościcka, A. (2020). Evaluation of metadata describing topographic maps in a National Library. *Heritage Science*, *8*(1), 113.

Lefevre, S., Tuia, D., Wegner, J. D., Produit, T., & Nassar, A. S. (2017). Toward seamless multiview scene analysis from satellite to street level. *Proceedings of the IEEE*, *105*(10), 1884–1899.

Li, L., Chi, T., Hao, T., & Yu, T. (2018). Customer demand analysis of the electronic commerce supply chain using Big Data. *Annals of Operations Research*, *268*, 113–128.

Liu, R., Nageotte, F., Zanne, P., de Mathelin, M., & Dresp-Langley, B. (2021). Deep reinforcement learning for the control of robotic manipulation: A focussed mini-review. *Robotics*, *10*(1), 22.

Liu, Y., Wang, R., Lu, Y., Li, Z., Chen, H., Cao, M., & Song, Y. (2020). Natural outdoor environment, neighbourhood social cohesion and mental health: Using multilevel structural equation modelling, streetscape and remote-sensing metrics. *Urban Forestry and Urban Greening*, *48*, 126576.

Liu, Z., Li, J., Ashraf, M., Syam, M. S., Asif, M., Awwad, E. M., & Bhatti, U. A. (2024). Remote sensing-enhanced transfer learning approach for agricultural damage and change detection: A deep learning perspective. *Big Data Research*, *36*, 100449.

Löffler, F., Wesp, V., König-Ries, B., & Klan, F. (2021). Dataset search in biodiversity research: Do metadata in data repositories reflect scholarly information needs? *PLoS ONE*, *16*(3), e0246099.

Lucieri, A., Bajwa, M.N., Braun, S.A., Malik, M.I., Dengel, A., & Ahmed, S. (2020). On interpretability of deep learning based skin lesion classifiers using concept activation vectors. In *2020 international joint conference on neural networks (IJCNN)*, IEEE (pp. 1–10).

Luo, N., Pritoni, M., & Hong, T. (2021). An overview of data tools for representing and managing building information and performance data. *Renewable and Sustainable Energy Reviews*, *147*, 111224.

Luo, Z., Cai, S., Dong, J., Ming, R., Qiu, L., Zhan, X., & Han, X. (2023). RaBit: Parametric modeling of 3D biped cartoon characters with a topological-consistent dataset. In *Proceedings of the IEEE/CVF conference on computer vision and pattern recognition* (pp. 12825–12835).

Ma, J. W., Czerniawski, T., & Leite, F. (2021). An application of metadata-based image retrieval system for facility management. *Advanced Engineering Informatics*, *50*, 101417.

Ma, Y., Fan, A., He, J., Nelakurthi, A. R., & Maciejewski, R. (2020). A visual analytics framework for explaining and diagnosing transfer learning processes. *IEEE Transactions on Visualization and Computer Graphics*, *27*(2), 1385–1395.

Mach, K. J., & Field, C. B. (2017). Toward the next generation of assessment. *Annual Review of Environment and Resources*, *42*, 569–597.

Mancuso, I., Petruzzelli, A. M., & Panniello, U. (2023). Innovating agri-food business models after the Covid-19 pandemic: The impact of digital technologies on the value creation and value capture mechanisms. *Technological Forecasting and Social Change*, *190*, 122404.

Matricardi, E. A. T., Skole, D. L., Costa, O. B., Pedlowski, M. A., Samek, J. H., & Miguel, E. P. (2020). Long-term forest degradation surpasses deforestation in the Brazilian Amazon. *Science*, *369*(6509), 1378–1382.

Matsuo, Y., LeCun, Y., Sahani, M., Precup, D., Silver, D., Sugiyama, M., & Morimoto, J. (2022). Deep learning, reinforcement learning, and world models. *Neural Networks*, *152*, 267–275.

Mayernik, M. S. (2016). Research data and metadata curation as institutional issues. *Journal of the Association for Information Science and Technology*, *67*(4), 973–993.

McDaniel, P. N. (2022). Teaching, learning, and exploring the geography of North America with virtual globes and geovisual narratives. *Journal of Geography*, *121*(4), 125–140.

McDaniel, P. N., & Ingram, U. (2023). Integrating ArcGIS Online and digital story mapping for active learning in systematic geography courses. *Journal of Geography in Higher Education*, 1–31.

Mehonic, A., Sebastian, A., Rajendran, B., Simeone, O., Vasilaki, E., & Kenyon, A. J. (2020). Memristors—from in-memory computing, deep learning acceleration, and spiking neural networks to the future of neuromorphic and bio-inspired computing. *Advanced Intelligent Systems*, *2*(11) 2000085.

Mihai, S., Yaqoob, M., Hung, D. V., Davis, W., Towakel, P., Raza, M., & Nguyen, H. X. (2022). Digital twins: A survey on enabling technologies, challenges, trends and future prospects. *IEEE Communications Surveys and Tutorials*.

Moratón Gutiérrez, L. (2016). *Thematic patterning in English and Spanish: contrastive annotation of a bilingual newspaper corpus for liguistic and computational applications. Doctoral Thesis*. Departamento de Filología Inglesa Facultad de Filología Universidad Complutense de Madrid Madrid.

Mosannenzadeh, F., Bisello, A., Vaccaro, R., D'Alonzo, V., Hunter, G. W., & Vettorato, D. (2017). Smart energy city development: A story told by urban planners. *Cities*, *64*, 54–65.

Nawaz, S. A., Li, J., Bhatti, U. A., Shoukat, M. U., & Ahmad, R. M. (2024). *Deep learning applications in digital image security: Latest methods and techniques. Deep learning for multimedia processing applications* (pp. 110–153). Taylor & Francis.

Newman, E. L., & Witsell, E. (Eds.), (2021). *The food network recipe: Essays on cooking, celebrity and competition*. McFarland.

Newman, G., Chandler, M., Clyde, M., McGreavy, B., Haklay, M., Ballard, H., & Gallo, J. (2017). Leveraging the power of place in citizen science for effective conservation decision making. *Biological Conservation*, *208*, 55–64.

Nex, F., Armenakis, C., Cramer, M., Cucci, D. A., Gerke, M., Honkavaara, E., & Skaloud, J. (2022). UAV in the advent of the twenties: Where we stand and what is next. *ISPRS Journal of Photogrammetry and Remote Sensing*, *184*, 215–242.

Nizamani, M. M., Bhatti, U. A., Cheng, X. L., Nizamani, F. G., Rind, R. A., Khokhar, A. A., & Yang, D. M. (2021). The connections between aboveground biomass and plant diversity of roadside trees, density and diversity on different types of roads in Karachi. *Polish Journal of Environmental Studies*, *30*(3), 2691–2700.

Nizamani, M. M., Zhang, Q., Muhae-Ud-Din, G., Awais, M., Qayyum, M., Farhan, M., & Wang, Y. (2023). *Application of GIS and remote-sensing technology in ecosystem services and biodiversity conservation. Deep learning for multimedia processing applications* (pp. 284–321). CRC Press.

Nkwentsha, X. (2022). Semi-supervised transfer learning for medical images as an alternative to ImageNet transfer learning. Masters Thesis. Faculty of Engineering and the Built Environment, Department of Electrical Engineering.

Ohm, T., Solà, M.C., Karjus, A., & Schich, M. (2023). Collection space navigator: An interactive visualization interface for multidimensional datasets. arXiv preprint arXiv:2305.06809.

O'Neill, L., Anantharama, N., Buntine, W., & Angus, S. D. (2021). *Quantitative discourse analysis at scale-AI, NLP and the transformer revolution (No. 2021-12)*. Monash University, SoDa Laboratories.

Pacheco, A., Da Silva, C. G., & De Freitas, M. C. V. (2023). A metadata model for authenticity in digital archival descriptions. *Archival Science*, *23*(4), 629–673.

Paul, P., Aithal, P. S., Bhimali, A., Kalishankar, T., Saavedra, M. R., & Aremu, P. S. B. (2020). Geo information systems and remote sensing: Applications in environmental systems and management. *International Journal of Management, Technology, and Social Sciences*, *5*(2), 11–18.

Peng, M., Liu, Y., Khan, A., Ahmed, B., Sarker, S. K., Ghadi, Y. Y., & Ali, Y. A. (2024). Crop monitoring using remote sensing land use and land change data: Comparative analysis of deep learning methods using pre-trained CNN models. *Big Data Research*, *36*, 100448.

Pricope, N. G., Mapes, K. L., & Woodward, K. D. (2019). Remote sensing of human–environment interactions in global change research: A review of advances, challenges and future directions. *Remote Sensing*, *11*(23), 2783.

Puliafito, A., Tricomi, G., Zafeiropoulos, A., & Papavassiliou, S. (2021). Smart cities of the future as cyber physical systems: Challenges and enabling technologies. *Sensors*, *21*(10), 3349.

Puzzarini, C., & Barone, V. (2020). A never-ending story in the sky: The secrets of chemical evolution. *Physics of Life Reviews*, *32*, 59–94.

Qadir, A. M. A., & Fatah, A. O. (2023). Platformization and the metaverse: Opportunities and challenges for urban sustainability and economic development. *EAI Endorsed Transactions on Energy Web*, *10*(1).

Raes, L., Michiels, P., Adolphi, T., Tampere, C., Dalianis, A., McAleer, S., & Kogut, P. (2021). DUET: A framework for building interoperable and trusted digital twins of smart cities. *IEEE Internet Computing*, *26*(3), 43–50.

Ragni, F., Tucciarelli, R., Andersson, P., & Lingnau, A. (2020). Decoding stimulus identity in occipital, parietal and inferotemporal cortices during visual mental imagery. *Cortex*, *127*, 371–387.

Rehberger, J. (2020). *Cybersecurity attacks–red team strategies: A practical guide to building a penetration testing program having homefield advantage*. Packt Publishing Ltd.

Remy, L., Ivanović, D., Theodoridou, M., Kritsotaki, A., Martin, P., Bailo, D., & Jeffery, K. (2019). Building an integrated enhanced virtual research environment metadata catalogue. *The Electronic Library*, *37*(6), 929–951.

Sani-Mohammed, A., Yao, W., & Heurich, M. (2022). Instance segmentation of standing dead trees in dense forest from aerial imagery using deep learning. *ISPRS Open Journal of Photogrammetry and Remote Sensing*, *6*, 100024.

Santos, F., Meneses, P., & Hostert, P. (2019). Monitoring long-term forest dynamics with scarce data: A multi-date classification implementation in the Ecuadorian Amazon. *European Journal of Remote Sensing*, *52*(sup1), 62–78.

Sarker, M. N. I., Peng, Y., Yiran, C., & Shouse, R. C. (2020). Disaster resilience through big data: Way to environmental sustainability. *International Journal of Disaster Risk Reduction*, *51,* 101769.

Saxe, A., Nelli, S., & Summerfield, C. (2021). If deep learning is the answer, what is the question? *Nature Reviews. Neuroscience*, *22*(1), 55–67.

Shah, S. A., Seker, D. Z., Hameed, S., & Draheim, D. (2019). The rising role of big data analytics and IoT in disaster management: Recent advances, taxonomy and prospects. *IEEE Access*, *7*, 54595–54614.

Small, M. F. (2023). *Here begins the dark sea: Venice, a medieval monk, and the creation of the most accurate map of the world.* Simon and Schuster.

Stanley, K. O., Clune, J., Lehman, J., & Miikkulainen, R. (2019). Designing neural networks through neuroevolution. *Nature Machine Intelligence*, *1*(1), 24–35.

Stray, J., Halevy, A., Assar, P., Hadfield-Menell, D., Boutilier, C., Ashar, A., & Vasan, N. (2022). Building human values into recommender systems: An interdisciplinary synthesis. arXiv preprint arXiv:2207.10192.

Subramaniam, P., Ma, Y., Li, C., Mohanty, I., & Fernandez, R.C. (2021). Comprehensive and comprehensible data catalogs: The what, who, where, when, why, and how of metadata management. arXiv preprint arXiv:2103.07532.

Suleyman, M. (2023). *The coming wave: Technology, power, and the twenty-first century's greatest dilemma.* Crown.

Sun, A. Y., & Scanlon, B. R. (2019). How can Big Data and machine learning benefit environment and water management: A survey of methods, applications, and future directions. *Environmental Research Letters*, *14*(7), 073001.

Sun, W., Bocchini, P., & Davison, B. D. (2020). Applications of artificial intelligence for disaster management. *Natural Hazards*, *103*(3), 2631–2689.

Süsser, D., Gaschnig, H., Ceglarz, A., Stavrakas, V., Flamos, A., & Lilliestam, J. (2022). Better suited or just more complex? On the fit between user needs and modeller-driven improvements of energy system models. *Energy*, *239,* 121909.

Tagde, P., Tagde, S., Bhattacharya, T., Tagde, P., Chopra, H., Akter, R., & Rahman, M. H. (2021). Blockchain and artificial intelligence technology in e-Health. *Environmental Science and Pollution Research*, *28*, 52810–52831.

Tanhua, T., Pouliquen, S., Hausman, J., O'brien, K., Bricher, P., De Bruin, T., & Zhao, Z. (2019). Ocean FAIR data services. *Frontiers in Marine Science*, *6*, 440.

Trani, L., Atkinson, M., Bailo, D., Paciello, R., & Filgueira, R. (2018). Establishing core concepts for information-powered collaborations. *Future Generation Computer Systems*, *89*, 421–437.

Ullah, F., Ullah, I., Khan, R. U., Khan, S., Khan, K., & Pau, G. (2024). Conventional to deep ensemble methods for hyperspectral image classification: A comprehensive survey. *IEEE Journal of Selected Topics in Applied Earth Observations and Remote Sensing*.

Urdu, D., Goense, D., Booij, J., Graumans, C., & Kempenaar, C. (2023). Analysis of metadata standards for the exchange of image datasets and algorithms in the agricultural domain: *A metadata-oriented approach to identify minimum interoperability mechanisms for image data and deep learning algorithms that is used for vision-based applications in agriculture.* Sprint Robotics Project PL4. 0 WP7 (No. WPR-OT-1019). Wageningen Plant Research.

Usama, M. (2023). Application of machine learning techniques for traffic state estimation, pattern recognition, and crash detection. The University of Alabama in Huntsville.

van Gasselt, S., & Naß, A. (2024). Cartographic metadata for improving accessibility and facilitating knowledge extraction and validation in planetary mapping based on remote-sensing observations. *ISPRS International Journal of Geo-Information*, *13*(3), 69.

Van Rijmenam, M. (2022). *Step into the metaverse: How the immersive internet will unlock a trillion-dollar social economy*. John Wiley & Sons.

Veenendaal, B., Brovelli, M. A., & Li, S. (2017). Review of web mapping: Eras, trends and directions. *ISPRS International Journal of Geo-Information*, *6*(10), 317.

Venigandla, K., & Tatikonda, V. M. (2021). Improving diagnostic imaging analysis with RPA and deep learning technologies. *Power System Technology*, *45*(4).

Voinov, A., Kolagani, N., McCall, M. K., Glynn, P. D., Kragt, M. E., Ostermann, F. O., & Ramu, P. (2016). Modelling with stakeholders—next generation. *Environmental Modelling and Software*, *77*, 196–220.

Wang, J., Bretz, M., Dewan, M. A. A., & Delavar, M. A. (2022). Machine learning in modelling land-use and land cover-change (LULCC): Current status, challenges and prospects. *Science of the Total Environment*, *822,* 153559.

Wang, S., Zhong, Y., & Wang, E. (2019). An integrated GIS platform architecture for spatiotemporal big data. *Future Generation Computer Systems*, *94*, 160–172.

Webb, S. (2018). Deep learning for biology. *Nature*, *554*(7693)), 555–557.

Winter, J. S., & Davidson, E. (2022). Harmonizing regulatory regimes for the governance of patient-generated health data. *Telecommunications Policy*, *46*(5)102285.

Wu, J., Orlandi, F., O'Sullivan, D., & Dev, S. (2024). LinkedGeoClimate: An interoperable platform for climate data access within geographical context. *IEEE Journal of Selected Topics in Applied Earth Observations and Remote Sensing.*

Wu, W., Ren, C., Yin, A., & Zhang, X. (2023). MixerNet-SAGA a novel deep learning architecture for superior road extraction in high-resolution remote sensing imagery. *Applied Sciences*, *13*(18), 10067.

Wu, X., Li, J., Bhatti, U.A., & Chen, Y.W. (2019). Logistic map and contourlet-based robust zero watermark for medical images. In *Innovation in medicine and healthcare systems, and multimedia: Proceedings of KES-InMed-19 and KES-IIMSS-19 conferences* (pp. 115–123), Springer Singapore.

Wu, Y., Qiu, Y., & Wang, H. (2023). Urban cartography and spatial politics: Ethnic tensions, westernization, and decolonization in Hangzhou's transformation within the global south. *Journal of Geography and Cartography*, *6*(2), 2853.

Yang, J., Vining, N., Kheradmand, S., Carr, N., Sigal, L., & Sheffer, A. (2023). Subpixel deblurring of anti-aliased raster clip-art. *Computer Graphics Forum*, *42*(2), 61–76.

Yang, T., Song, W., Huang, Z., & Wu, J. (2023). The values of biological diversity: A perspective from China. *Biological Conservation*, *287*110341.

Yao, Y., Li, X., Liu, X., Liu, P., Liang, Z., Zhang, J., & Mai, K. (2017). Sensing spatial distribution of urban land use by integrating points-of-interest and Google Word2Vec model. *International Journal of Geographical Information Science*, *31*(4), 825–848.

Younesi, A., Ansari, M., Fazli, M., Ejlali, A., Shafique, M., & Henkel, J. (2024). A comprehensive survey of convolutions in deep learning: Applications, challenges, and future trends. arXiv preprint arXiv:2402.15490.

Yue, Y., & Shyu, J. Z. (2024). A paradigm shift in crisis management: The nexus of AGI-driven intelligence fusion networks and blockchain trustworthiness. *Journal of Contingencies and Crisis Management*, *32*(1), e12541.

Zhang, C., Bengio, S., Hardt, M., Recht, B., & Vinyals, O. (2021). Understanding deep learning (still) requires rethinking generalization. *Communications of the ACM*, *64*(3), 107–115.

Zhang, S., Yao, L., Sun, A., & Tay, Y. (2019). Deep learning-based recommender system: A survey and new perspectives. *ACM computing surveys*, *52*(1), 1–38.

Zhao, J., Zhao, W., Deng, B., Wang, Z., Zhang, F., Zheng, W., & Burke, A. F. (2023). Autonomous driving system: A comprehensive survey. *Expert Systems with Applications*, 122836.

Zhao, X., Liu, L., Qi, S., Teng, Y., Li, J., & Qian, W. (2018). Agile convolutional neural network for pulmonary nodule classification using CT images. *International Journal of Computer Assisted Radiology and Surgery*, *13*, 585–595.

Zohuri, B., & Rahmani, F. M. (2024). Computational intelligence for signal and image processing unleashing the power of algorithms. *Clinical Case Reports and Studies*, *5*(2).

Zou, X., Yan, Y., Hao, X., Hu, Y., Wen, H., Liu, E., & Liang, Y. (2024). Deep learning for cross-domain data fusion in urban computing: Taxonomy, advances, and outlook. arXiv preprint arXiv:2402.19348.

Chapter 10

Enhancing geospatial insights: a data-driven approach to multi-source remote sensing fusion

Mughair Aslam Bhatti and Xin Sun
School of Information and Communication Engineering, Hainan University, Haikou, P.R. China

10.1 Introduction

Remote-sensing satellite observation of the Earth's surface from space attributes characteristics and changes rules of time and space. Its application potential depends on the sensor's ability to detect the intensity of electromagnetic energy reflected or emitted by ground objects in different spectrums and to monitor changes over space and time. Typically, image spatial resolution, temporal resolution, and spectral resolution characterize this ability. Obtaining high-resolution remote-sensing data has been identified as one of the three major challenges by the International Society of Photogrammetry and Remote Sensing and is also listed as one of the 16 major scientific and technological projects in China. However, due to constraints imposed by satellite orbits, observation modes, and sensor performance, there is a mutual restriction between space, time, and spectral resolution, making it impossible to achieve "clear vision," "fast measurement," and "accurate calculation" simultaneously. As a result, it is challenging to finely describe the spatial pattern and evolution process of complex geological phenomena.

Multi-source remote-sensing information fusion technology can achieve comprehensive processing and application of complementary information from multiplatform and multimodal observation data, generating high "time-space-spectrum" seamless observation data across large scenes. Traditional approaches to solving this problem have mainly expanded from the perspective of model-driven methods, using degradation processes, physical mechanisms, and prior knowledge to build functional models of specific targets, obtaining the optimal estimate through function solving (Irani & Peleg, 1991; Lertrattanapanich & Bose, 2002). Significant progress has been made in improving the spatial, spectral, and temporal indexes of remote-sensing data, focusing on the following four aspects:

1. Multi-temporal/multiangle fusion: This approach utilizes redundant spatio-temporal information caused by subpixel motion among multiple images, fusing and complementing them to produce high-quality, high-resolution images. Traditional methods include subpixel migration based on the frequency domain, nonuniform interpolation, iterative back projection, convex set projection, regularization, and sparse coding based on the spatial domain (Farsiu et al., 2004; Stark & Oskoui, 1989; Yang et al., 2010).
2. Space-spectrum fusion: This method alleviates the mutual restriction between spatial resolution and spectral resolution, leveraging the space-spectrum complementarity between multisource image data to generate remote-sensing images with both high spatial and high spectral resolution. Techniques include panchromatic/multispectral image fusion and panchromatic (or multispectral)/hyperspectral image fusion, with challenges arising due to large spectral differences. Mainstream methods include component substitution, multiresolution analysis, and variational optimization-based fusion methods (1980) (Cliché et al., 1985; Hallada & Cox, 1983).
3. Time-space fusion: This method addresses the trade-off between temporal and spatial resolution, generating time-continuous, high spatial resolution remote-sensing images. Methods include fusion based on single-phase image pair assistance and multitemporal image pair assistance, with algorithms like STARFM and ESTARFM being widely used in applications such as surface reflectance and surface temperature (Li et al., 2018; Niu, 2012; Shen et al., 2013).

Deep Learning for Earth Observation and Climate Monitoring. DOI: https://doi.org/10.1016/B978-0-443-24712-5.00020-8

4. Point-surface data fusion: This approach combines high-accuracy, sparse point-like data from ground stations with wide-coverage, continuous planar data from satellite observations. The goal is to obtain quantitative products with high accuracy, wide coverage, and spatial continuity. Point-surface fusion has been applied in geosciences, including snow depth inversion models based on deep belief networks, hourly PM2.5 concentration inversion models based on LSTM networks, and precipitation inversion models based on random forests (Xue et al., 2017).

Overall, while model-driven approaches have stronger inclusiveness and can depict relationships between time, space, and spectrum degradation, they often assume linear regression between multisource data. This limitation hinders the fusion accuracy when dealing with nonlinear characteristics of amplitude space, space spectrum, spatio-temporal, and anisotropic relations.

In recent years, the rapid development of computing power and artificial intelligence algorithms has led to the emergence of data-driven multisource remote-sensing information fusion technology. This approach does not require assumptions or prior knowledge; instead, it mines large data samples to construct feature relationships between input and output data pairs, extracting rich statistical features and achieving superior data fusion results. This paper attempts to summarize data-driven multisource remote-sensing information fusion methods, focusing on anisotropic and isotropic remote-sensing data fusion as well as quantitative inversion-oriented information fusion, while identifying current challenges and potential development directions in this field.

10.2 Isotropic remote-sensing data fusion

Isotropic remote-sensing data fusion refers to the fusion of observed data under the same imaging method, among which the data fusion between optical data in the visible-near infrared band imaging is the most common. The main purpose of this kind of technology is to relieve the mutual restriction of spatial resolution, temporal resolution, spectral resolution, and other observation indexes of single remote-sensing observation data caused by the technical limitations of the sensor system and the influence of space-space-earth imaging conditions to obtain remote-sensing data with high quality in time, space, and spectral resolution. It mainly includes multitemporal/multiangle remote ssensing information fusion, remote sensing information space-spectrum fusion, and remote sensing information time-space fusion.

10.2.1 Multitemporal/multiangle remote-sensing information fusion

Multi-temporal/multiangle remote-sensing information fusion mainly refers to the process of mining and fusing complementary information from multiple remote-sensing images with subpixel displacement and scene aliasing to reconstruct one or more high-quality remote-sensing images. Typical application scenarios include multitemporal image fusion and cloud removal and superresolution reconstruction. Among them, superresolution reconstruction can be traced back to the subpixel migration method based on the frequency domain, which proves that the registration and fusion of multiple images provide the possibility of resolution improvement. Due to the limitation of the frequency domain method in integrating prior information from high-resolution images, the traditional method has gradually transitioned to the spatial domain-based method. Typical representatives include nonuniform interpolation, iterative back projection, convex set projection, regularization, and sparse coding (Bhatti et al., 2020; Nawaz et al., 2022).

With the rise of deep learning, data-driven superresolution methods show stronger performance in solving complex nonlinear inverse problems. Thanks to the nonlinear feature expression ability of deep neural networks, the deep representation of superresolution problems can be realized. At the same time, the emergence of massive remote-sensing data also provides training data support for model learning. At present, multitemporal superresolution is mainly improved by image registration and fusion reconstruction. Among them, image registration is divided into the optical flow-based method and the kernel-based method. Fusion is mainly divided into direct fusion (Bhatti et al., 2020; Bhatti, Bazai, et al., 2023), attention-based fusion, and multistep fusion. Registration based on optical flow occurs explicitly at the image level. By estimating the optical flow between the image to be registered and the reference image, the motion relationship between the images due to pixel movement is encoded in the optical flow. The image to be registered is then distorted to the reference image using the optical flow map. Kernel-based registration occurs implicitly on the feature layer surface, and the convolution processing of the registered features is carried out adaptively by a learnable convolution kernel. The representative methods include kernel regression, deformable convolution, nonlocal modules, etc. Kernel regression and variable convolution focus on the local motion and migration of the image and capture or track the pixel movement by convolution so that the local receptive field of the convolution can always learn the redundant information for reconstruction. The nonlocal module focuses on the global response of pixels and allocates weights to different regions in the global through nonlocal attention. Subsequent fusion is responsible for aggregating the spatio-temporal information after registration. In direct fusion, simple convolution is often used to map nonlinear features after series. Attention-based fusion can modulate the contribution ratio of features in fusion, focus on useful information, and suppress useless information. Multi-step fusion integrates the information from multitemporal images into the spatial information of the reference images in a multistage or cyclic manner to ensure that the reconstruction results are highly consistent with the reference targets. Fig. 10.1 shows the current superresolution reconstruction process of multitemporal/multiangle remote-sensing images based on deep learning. It can be seen that compared with traditional methods, the data-driven model can reconstruct more textures and details.

10.2.2 Spatial-spectral fusion of remote-sensing information

Spatial-spectral remote-sensing information fusion refers to the use of the complementarity between spatial and spectral information of different remote-sensing image data to generate high spatial resolution and high spectral resolution remote-sensing images and solve constraint environments, such as imaging systems for sensor devices in space remote-sensing image details and the spectrum of the problem of constrained fineness. Common space-spectrum fusion methods mainly involve three types of shadow data: panchromatic image, multispectral image, and hyperspectral image. The spatial resolution of these three types of data decreases gradually, while the spectral resolution increases gradually.

Panchromatic multispectral fusion, which means obtaining high-quality multispectral remote-sensing data by fusing panchromatic and multispectral images, is the most classical means of space-spectrum fusion of remote-sensing information, which was proposed as early as the 1980s. With the vigorous development of spaceborne remote-sensing technology, the pattern of multispace and multispectral joint earth observation tends to take shape. Panchromatic and multispectral fusion technology has developed rapidly, and a large number of methods have been put forward continuously. In general, the current traditional panchromatic multispectral fusion algorithms can be divided into four categories: fusion methods based on component replacement, fusion methods based on multiresolution analysis, fusion methods based on model optimization, and fusion methods based on machine learning. Among them, the method based

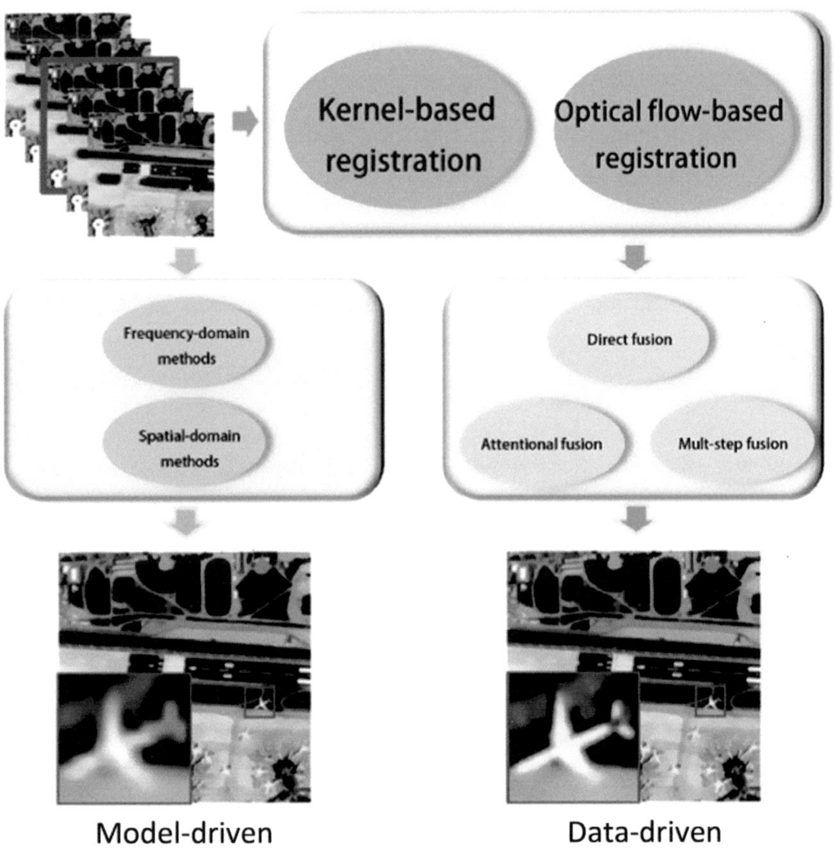

FIGURE 10.1 The multiview super resolution fusion.

on component replacement has better spatial details and serious spectral distortion. However, the method based on multiresolution analysis has better spectral fidelity but less spatial detail. Considering the advantages of the two types of fusion methods, the literature proposed a fusion framework combining component replacement and multiresolution analysis. An algorithm based on the model optimization problem as an inverse problem is to fusion, to exert different prior constraints and solve the inverse problem, compared with the fusion method based on component replacement and multiresolution analysis, and its mathematical model is more rigorous, with higher precision, but its accuracy depends heavily on the accuracy of prior knowledge, and to solve the complex, operation time is higher. The earliest fusion method based on machine learning is based on sparse expression or dictionary learning. The idea is to learn a complete spectral dictionary through a large number of samples, then calculate the dictionary and the corresponding sparse coefficients on panchromatic and multispectral images, and study the relationship between the dictionary and the corresponding coefficients to achieve the fusion of space-spectral information.

Since 2015, deep learning has gradually gained attention in panchromatic and multispectral fusion, as shown in Fig. 10.2. The main idea is to directly learn the nonlinear relationship between low and high-spatial-resolution remote-sensing data by building different networks. The panchromatic multispectral fusion algorithm based on deep learning was first proposed by scholars in 2015 when a simple deep neural network with a sparse denoising autoencoder was used. Then, a convolutional neural network, which is more suitable for image processing, was used in panchromatic multispectral fusion. With the rise of residual networks, some scholars added residual connections based on convolutional neural networks or reduced residual learning. However, a single convolution cannot fully mine the multiscale information in remote sensing images. In reference, multiscale convolution is introduced into the deep residual network to better increase the feature mining ability of the model. Although residual networks with multiscale convolution have been able to obtain better details than traditional algorithms, some scholars have introduced adversarial learning in the field of image generation into convolutional neural networks to further improve image details. In addition, considering the interpretability of deep learning and the difficulty of solving the optimization model, scholars proposed a panchromatic multispectral fusion algorithm combining the optimization model and deep learning. To help the network better learn the nonlinear relationship between high and low spatial resolution remote-sensing images, in recent years, scholars have paid more attention to the relationship between remote-sensing images and blocks and used different modules to solve it, such as attention mechanisms, graph convolution, and transformer with multihead attention. Fig. 10.3 shows the difference in the panchromatic multispectral fusion effect between the deep learning algorithm and the traditional model-based algorithm. The data-driven fusion algorithm has more advantages in detail injection than the model-based algorithm.

In addition to panchromatic multispectral fusion, panchromatic hyperspectral fusion and multispectral hyperspectral fusion have also attracted more and more attention. Most algorithms for these two types of fusion are derived from panchromatic multispectral fusion algorithms. In addition, scholars also used 3D convolution for hyperspectral image feature mining to improve the fusion accuracy, and some scholars combined the Laplacian method to enhance spatial details. Compared with panchromatic multispectral fusion, the fusion of remote-sensing image data involving hyperspectral is more difficult because the spatial resolution difference between the panchromatic image and the hyperspectral image is greater, and the spectral scale difference is also increased several times. Therefore, how to alleviate the

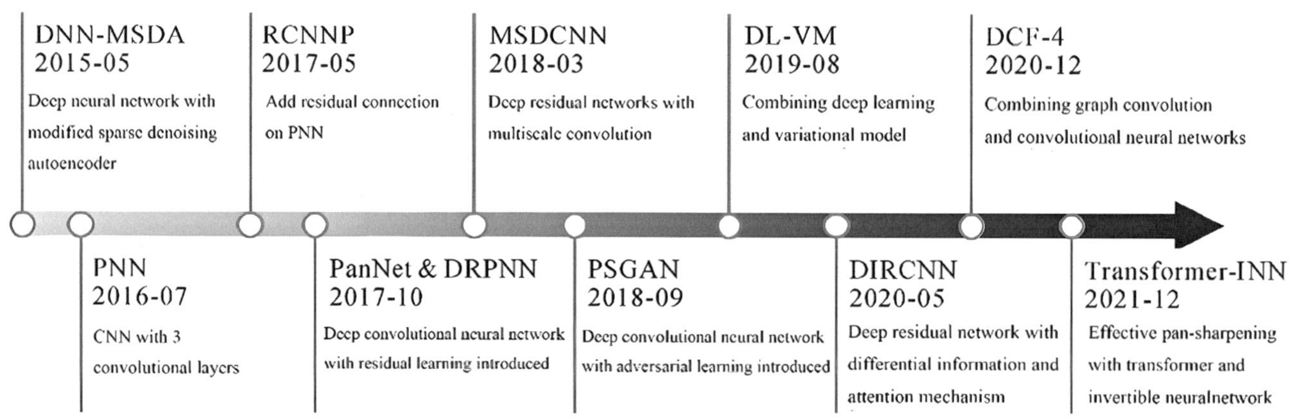

FIGURE 10.2 Development of deep learning in pan-sharpening.

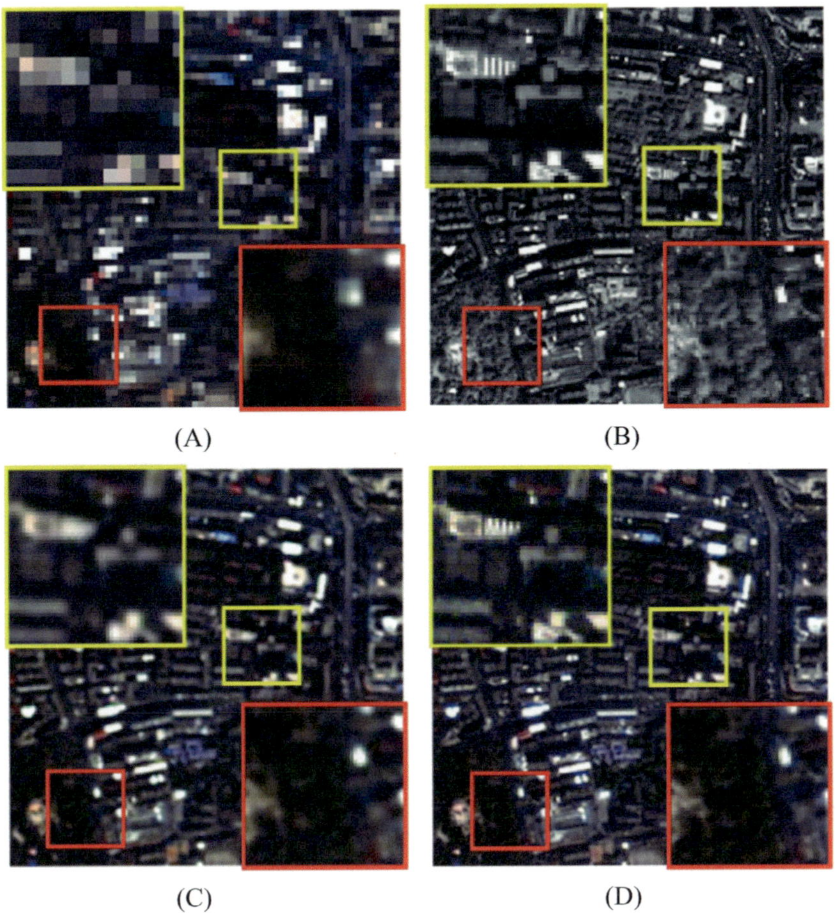

FIGURE 10.3 Spatial-spectral fusion. (A) Multispectral image (resampled). (B) Panchromatic images. (C) Fusion results of traditional methods. (D) Fusion results of deep learning.

scale gap of direct fusion between the two through effective means or auxiliary data is still worth pondering by the majority of researchers (Bhatti et al., 2020; Bhatti, Huang, et al., 2023).

10.2.3 Time-space fusion of remote-sensing information

Time-space fusion of remote-sensing information refers to the fusion of a low spatial resolution image with high temporal resolution and a high spatial resolution image with a low temporal resolution to generate an image sequence with sufficient spatial and temporal resolution. Over the past 10 years, space remote-sensing information fusion has attracted a lot of attention. Scholars have developed many space-time fusion methods, which can mainly be classified into the following categories: fusion method based on space-time filtering, the fusion method based on the decomposition of mixed pixels, based on the study of fusion method, the fusion method based on bayesian. Among them, methods based on spatio-temporal filtering are widely used in time-space fusion methods. Literature first used the spatio-temporal adaptive reflection fusion model (STARFM), which has been continuously developed by scholars. For example, the spike cap transform was introduced to calculate the conversion coefficient between high- and low-resolution images, and joint regression. The Kriging method added bilateral filtering to determine the weight; kernel drive was used for weighted combination; and nonlocal filtering was used. The time-space fusion method based on mixed pixel decomposition was first proposed by literature in 1999. After scholars continuously introduced new constraints and auxiliary information, a robust time-space fusion system based on mixed pixel decomposition was gradually formed. Literature used sparse representation to solve the time-space fusion of remote-sensing information, which opened the prelude of the time-space fusion method based on learning. Subsequent scholars kept exploring to improve the completeness and accuracy of the learned dictionary and derived a series of time-space fusion algorithms based on dictionary learning.

In addition, some learning-based fusion algorithms do not rely on dictionary learning but use deep learning to solve the time-space fusion of remote-sensing information. The last one is to transform the fusion problem into the optimal estimation problem by constructing the inverse fusion problem and solving the problem by using the Bayesian maximum a posteriori estimation. The difference between each algorithm lies in the different emphasis of the inverse problem model and the different optimization algorithms used. In general, learning-based and Bayes-based algorithms are the up-and-comers. Although they can achieve good results at present, it still needs to be verified by a lot of practice whether they can maintain robustness in long-time series remote-sensing data (Bhatti et al., 2022).

10.3 Anisotropic observation data fusion

Anisotropic observation data fusion is one of the research hotspots in recent years. Its main purpose is to make a more accurate interpretation of earth observation data by using multidimensional feature information provided by anisotropic observation data with different acquisition methods. This section will introduce the fusion of anisotropic remote-sensing images and spatio-temporal big data information in detail.

10.3.1 Multisource anisotropic remote-sensing data fusion

Different remote-sensing data are imaged in different ways. For example, SAR images and LiDAR images are imaged by active remote sensing, while optical remote-sensing images are imaged by passive remote sensing. The same ground object behaves differently in remote-sensing images under different imaging modes, thus forming more dimensional data. Multi-source Anisotropic remote-sensing data fusion is mainly to integrate the multidimensional features of Anisotropic remote-sensing data to make more full use of the information in the image and more accurate interpretation.

In the research of SAR and optical remote-sensing image fusion, SAR image-assisted optical data restoration is mainly represented. Some scholars took the lead in using multifrequency SAR images to repair the cloud area of optical images in the same area by pixel-level replacement algorithm, which effectively improved the utilization efficiency of optical images and the accuracy of ground object information interpretation. Furthermore, the cloud area of the Sentinel 2 image was repaired by using a single-frequency SAR image and the corresponding Landsat image through sparse expression. In addition, scholars also proposed a repair process of using the generative adversarial network to first complete the transformation from SAR to optics and then fill the cloud area, followed by several related works. The U-net structured convolutional neural network was used to introduce the multitemporal Sentinel 1 and Sentinel 2 data into the repair task of target Sentinel 2 data, and more accurate repair results were obtained. There were also subsequent related works. It is worth mentioning that the network of the above deep learning algorithm is trained with the data of a specific region at a specific time, which cannot be applied at other times and scenes. To make the pretrained model have a stronger generalization ability, some scholars published a group of large-scale single-phase SAR-optical data sets. Some scholars introduced Landsat auxiliary data and published triplet datasets, each triplet including paired Sentinel 1, Sentinel 2, and Landsat-8 data. The multitemporal Sentinel 2 data was further introduced as an auxiliary, and the paired multitemporal Sentinel 1 and Sentinel 2 data sets were published to complete the optical image restoration. Fig. 10.4 shows the general development of optical data restoration using SAR data as a reference.

Fig. 10.5 shows the cloud removal results of optical images using SAR images as a reference. It can be seen that SAR images are sufficient to complete the repair of optical information.

The fusion of more accurate interpretation of image information is mainly represented by the task of ground object classification. Based on optical data classification, more accurate ground object classification can be achieved by supplementing the features of the data obtained by active remote sensing. Firstly, scholars extracted ground object

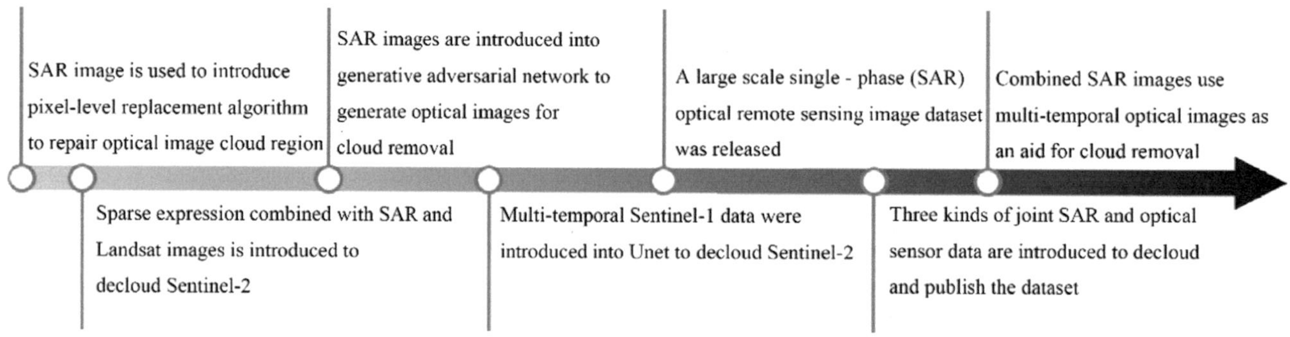

FIGURE 10.4 Development of optical data recovery assisted by SAR data.

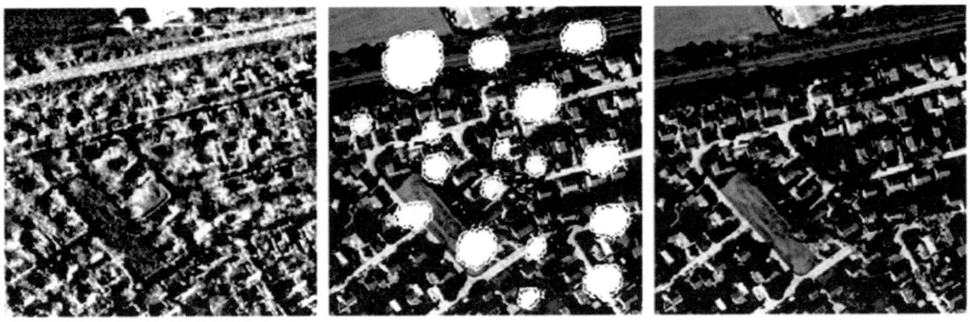
(A) (B) (C)

FIGURE 10.5 The example of optical data recovery with SAR data. (A) SAR image, (B) target image, and (C) repair results.

information from Landsat TM optical data, ALOS visible light and near-infrared data, and PALSAR data through the traditional maximum likelihood method and artificial neural network and obtained more accurate classification results than using optical images alone. Then, scholars used IHS transform and wavelet transforms to fuse MODIS data and RADASAT-2 data, and the fusion results were easier to be classified by traditional ISODATA and other algorithms. Then, some scholars used the random forests to extract planting areas from Sentinel-1 SAR data and Landsat-8 optical data and found that the classification results using anisotropic data were much higher than the classification results using single data. There were also many similar machine-learning algorithms, such as the support vector machine and the Chu. In addition, scholars also proposed a stacked and sparse autoencoder model based on deep learning to estimate biomass using LiDAR data, optical data, and SAR data. The results are more accurate than traditional algorithms and machine learning algorithms. Subsequent work includes the use of convolutional neural networks, U-net networks, deep isotropic feature fusion, etc. At the same time, the image classification results obtained by combining anisotropic data are much better than those obtained by using optical data alone.

10.3.2 Remote-sensing observation and spatio-temporal big data fusion

Due to the long period of remote-sensing data acquisition, remote-sensing data alone cannot achieve the purpose of real-time monitoring of ground events. Spatiotemporal big data, including social media data, can effectively reflect reality in real-time. The main purpose of the fusion of remote-sensing observation and spatio-temporal big data is to make full use of the large-scale characteristics of remote-sensing data and the real-time characteristics of social media data to complete the large-scale real-time monitoring of ground events. In recent years, the fusion of remote-sensing observation and spatio-temporal big data has been studied in depth (Liu et al., 2019).

Based on the decision tree algorithm, accurate and effective monitoring of floods and other disasters was realized by collecting the hot content of Twitter and matching the remote-sensing image of the corresponding region. The subsequent work added topographic data and wiki data based on the above data. Later, some scholars found that when using deep neural networks to classify urban areas of remote-sensing images, adding social media data matching with pixels as an auxiliary would obtain higher accuracy classification results, and there are still similar works in the future. Deep learning was used for the fusion analysis of remote-sensing images and social media data to accurately estimate the division of urban functional areas. Through the analysis of motility-light remote-sensing data and social media data, the dynamic changes of the urban population on a daily scale were obtained, and the social and economic situation of the city was further analyzed in the subsequent work.

10.4 Information fusion for quantitative inversion

The transformation of satellite remote-sensing observation into surface parameter information requires quantitative inversion. However, it is affected by observation technology (optics, radar, lidar, etc.) and observation environment (light source, cloud, aerosol, etc.), and other factors. The quantitative products derived from the inversion of a single remote-sensing data source often have great limitations. Therefore, it is of great significance to explore the complementary characteristics of anisotropic data, integrate multisource information, and improve the quality of quantitative inversion products (Cheng et al., 2023).

10.4.1 Basic principle and framework of point-surface data fusion

Traditional earth environment monitoring relies on the construction of ground observation stations, such as meteorological stations, ocean observation stations, ecological stations, atmospheric environment monitoring stations, etc. Ground station data is a kind of point-like data, which has the characteristics of high accuracy, high timeliness, and good stability, but sparse distribution, limited monitoring range, and high construction cost. Satellite remote-sensing, as a new observation technology, uses the method of Earth imaging to observe the Earth's environment and can get planar observation data. It has the advantages of large areas, wide coverage, and continuous observation, and forms complementary advantages with site observation. Therefore, the fusion of point data and plane data (point-plane fusion) is an effective method to obtain quantitative products with high accuracy, large scope, and spatial continuity.

The basic idea of point-surface data fusion is to build a relational model and realize the mapping from plane data to point data. Fusion models can be divided into physical models and statistical models. The physical model mainly starts from the physical relationship of the data itself to establish the derivation process, while the statistical model mainly uses machine learning and other methods to build the statistical relationship between point and plane data. Statistical models are simple, efficient, and accurate, among which, benefiting from the development of computers, artificial intelligence, and other technologies, statistical models based on machine learning show superior performance and have been rapidly developed and widely used.

10.4.2 Point-surface data fusion machine learning model

The point-surface data fusion based on machine learning mainly includes three steps: spatio-temporal matching of point-surface multisource data, machine learning model construction and training, and quantitative inversion. Site data is often used as tag data in machine learning fusion models because of its high accuracy. There are various machine learning methods used for point-surface data fusion. In the early stage, linear regression (such as multiple linear regression, geographically weighted regression (GWR)), decision tree models (such as CART), support vector machines (SVM), and other methods were widely used. In recent years, with the improvement of computing power and the development of neural network technology, ensemble learning (such as Random Forest (RF), lightweight gradient lifting tree (LGBM), deep forest (DF)), neural networks (such as back propagation neural network (BPNN), generalized regression neural network (GRNN), Bayesian belief network (BBN)), deep learning (such as deep neural network (DNN), convolutional neural network (CNN), and long short-term memory neural network (LSTM) methods have been more widely used. The point tag data is used as the output, and the satellite observation and other auxiliary variables are used as the input to train the machine learning model. By inputting the surface data into the trained model, the quantitative product after point and surface fusion can be obtained. The product has the advantages of high precision of point data and wide coverage of plane data (Alhatami et al., 2022).

The point-surface data fusion model based on machine learning has the advantages of high precision and wide application range and is widely used in various fields. This paper lists some typical examples of point-surface fusion machine learning models in the hydrosphere, atmosphere, and biosphere to illustrate the application of point-surface fusion in quantitative inversion.

10.4.3 Water circle

Soil moisture plays a crucial role in the global water cycle. Accurate inversion of soil moisture can provide efficient decision-making reference for flood prediction, drought monitoring, etc. Machine learning can simplify ill-posed inversion problems with limited knowledge of physical processes, so it shows great advantages in soil moisture retrieval. Although the traditional ANN-based method of simulated brightness temperature or backscattering coefficient for soil moisture retrieval was the earliest, and its accuracy, stability, and effectiveness have been verified, it does not take into account the characteristics of site data. Focusing on the high-precision characteristics of site measurement in the small-scale range, the training dataset was expanded by using the method of training and calibrating ANN based on on-site observation.

However, the point-surface fusion mode of the site is limited by the spatial scale mismatch problem and the sparse distribution of the global soil moisture observation network. To solve this problem, the inversion method of adding the simulated data of the global surface model (LSM) to the training data of soil moisture inversion was proposed and developed. At the same time, focusing on the characteristics of high spatio-temporal revisit and low cost of GNSS-R data, the point-surface fusion inversion algorithm for soil moisture retrieval based on GNSS-R has been further expanded. The literature further proposed a GRNN inversion method based on reliable site screening, which can effectively weaken the space-ground spatial scale errors and obtain seamless soil moisture products with higher fidelity. In addition, XGBoost, RF, and other integrated learning methods also have excellent performance in soil moisture fusion retrieval tasks. Fig. 10.6 shows an example of soil moisture retrieval based on the GRNN point-surface fusion framework, which reflects the advantages of machine learning methods in retrieving soil moisture products by point-surface fusion.

Land snow cover is one of the key factors affecting climate change. The high reflectance of the snow surface and its low thermal conductivity will play a certain role in controlling the surface radiation balance and energy exchange. The two main factors in snow research are snow depth and snow water equivalent, respectively. In the inversion of these two factors, the machine learning model can integrate the satellite and earth observation data with point-surface fusion and achieve relatively good results. Some earlier studies have shown that Ann can achieve good results in the inversion of snow depth and snow water equivalent based on point-surface fusion. Later, some scholars proved that the inversion accuracy of neural networks can be further improved by adding planar microwave channel data and integrating multi-source data such as longitude, latitude, and terrain. Literature compares the effect of the empirical snow depth algorithm with BPNN, GRNN, and DBN neural networks in point-surface fusion and finds that the inversion accuracy of the neural network is significantly better than that of the traditional snow depth algorithm. Fig. 10.7 shows the network structure of DBN snow depth inversion, and Fig. 10.8 shows the interannual anomaly analysis of inversion results. Ensemble learning, as another kind of machine learning model, also performs well in snow parameter inversion based on point-surface fusion. Literature combined random forest and HUT snow physical models to improve the fusion inversion accuracy of the model in shallow snow and deep snow.

To retrieve large range, long time series, and high spatial resolution snow parameters, the support vector regression method was used to generate daily snow depth data in the Northern Hemisphere for nearly 25 years by fusing ground station data, MODIS land cover data, and passive microwave data. Some scholars have also considered further integrating snow assimilation data into machine learning input variables to obtain snow depth datasets with higher accuracy in the northern hemisphere.

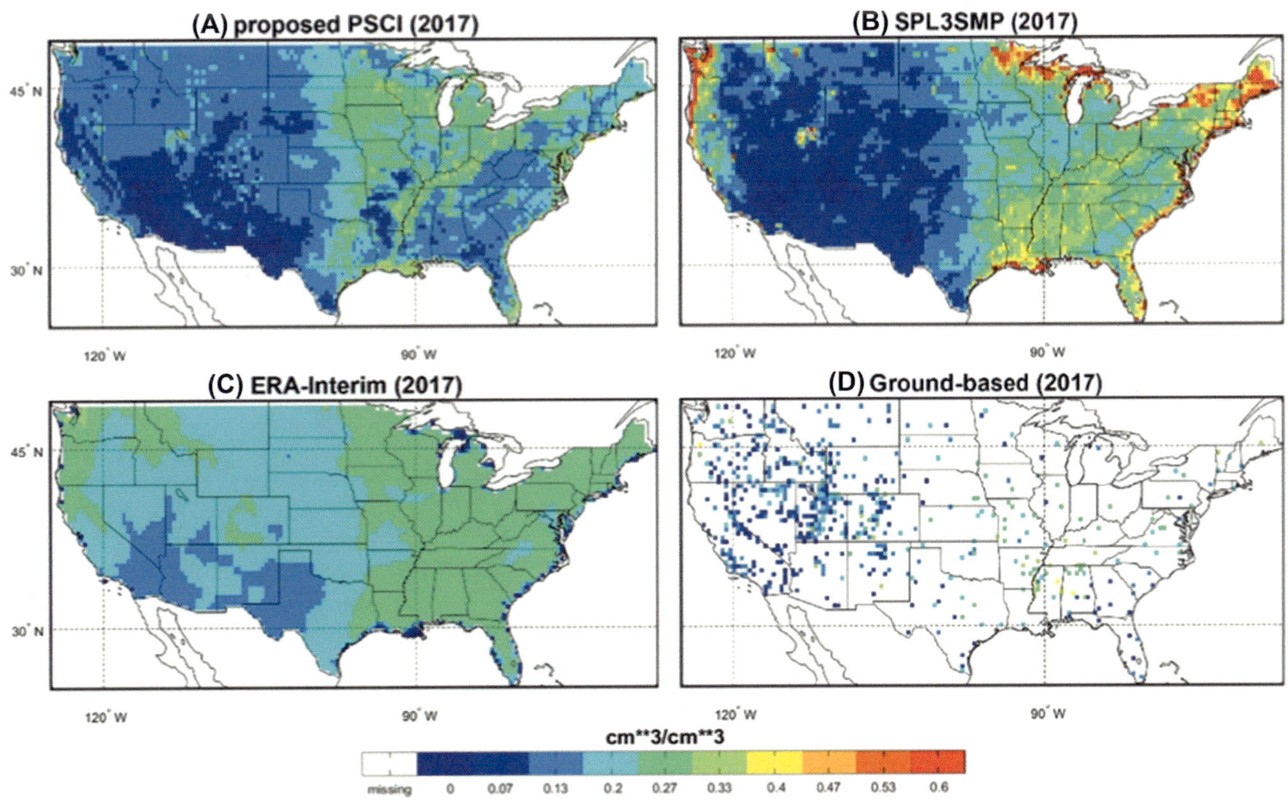

FIGURE 10.6 Example of soil moisture retrieval based on machine learning.

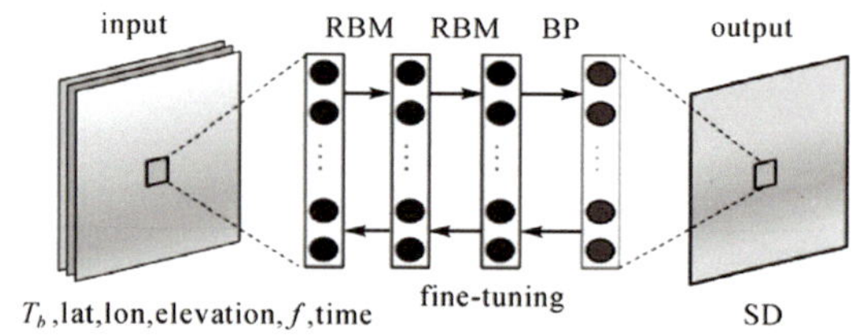

FIGURE 10.7 Snow depth retrieval of Alaska using deep belief network.

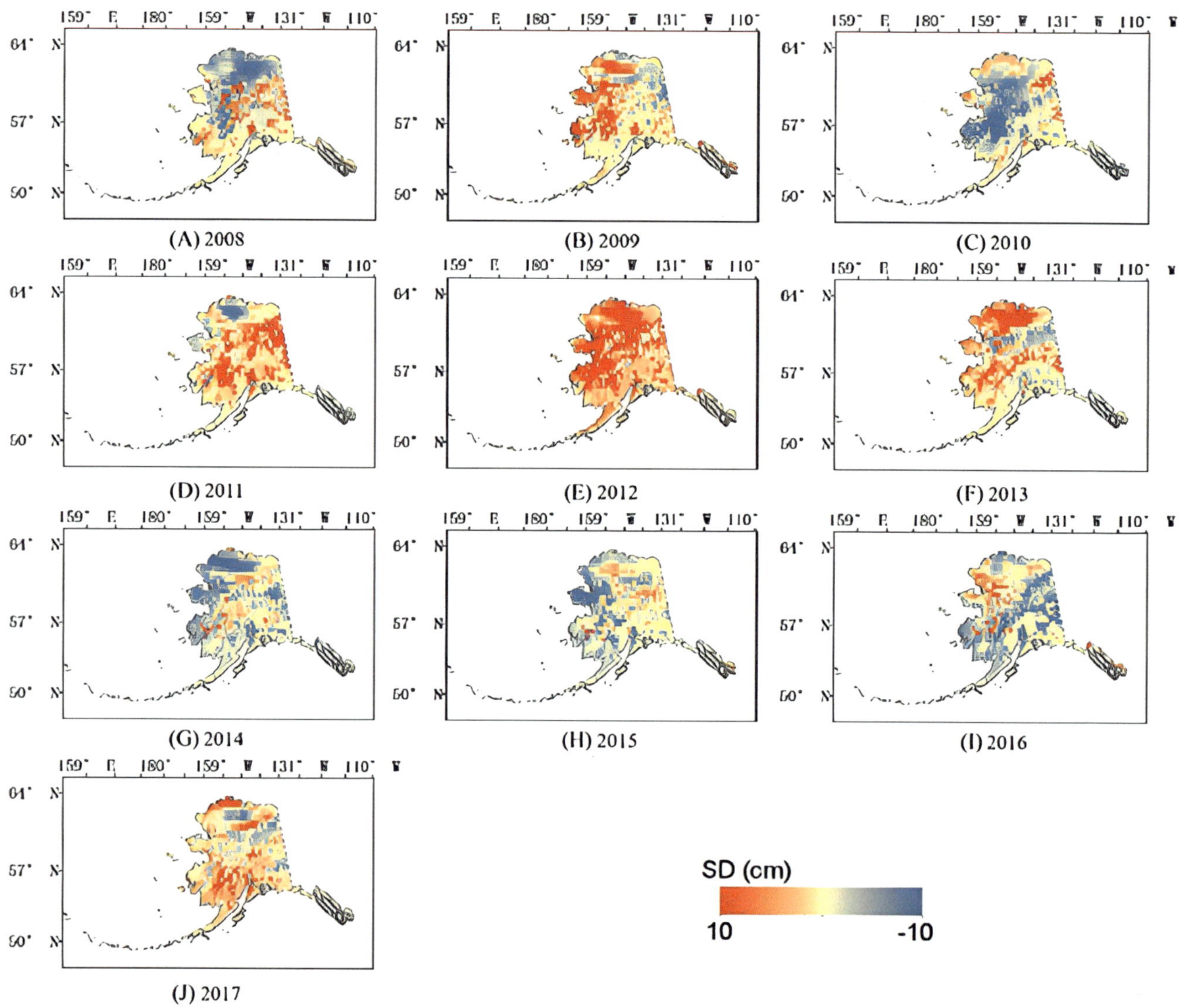

FIGURE 10.8 Anomaly analysis of snow depth by deep belief network.

In addition, precipitation is also an important part of the water cycle, and the precipitation process is characterized by complexity, nonlinearity, and high variability. Reliable and accurate surface precipitation data can be obtained from traditional rain gauges and ground-based radars. However, their distribution is not uniform, especially for most marine areas and mountainous areas with complex topography. Compared with ground stations, satellite observation can achieve large-scale and time-continuous observation without restrictions of geographical conditions, which can exactly make up for the defects of ground observation. According to the different fusion of satellite data, the precipitation inversion method based on fusion point-surface can be roughly classified into three broad categories: based on visible light/

infrared, based on active/passive microwave, and multisensor joint inversion method. The inversion method of infrared data fusion is to estimate precipitation by constructing the relationship between cloud top temperature and surface rainfall rate. Literature integrated infrared information and numerical weather forecast data in Himawari-8 to construct a precipitation inversion model based on random forest. Compared with infrared, microwaves are more sensitive to precipitation particles, so they can often achieve relatively high accuracy. However, there are still some limitations of a single sensor. Researchers have begun to try to combine multiple sensors to obtain high-quality global precipitation products, such as IMERG products, GSMaP products, and CMORPH products. In addition, with the RF-MEP model of the existing product site and the terrain elements with the ground, you can get more accurate local area rainfall estimation results.

10.4.4 Atmosphere

The atmosphere is mainly composed of a variety of gases, such as trace gases and CO_2. In recent years, the content and types of pollution components released into the atmosphere by human beings have been increasing year by year, causing serious harm to human health and the ecological system. The movement of the atmosphere produces a range of atmospheric phenomena, such as hurricanes, wildfires, and dust. These atmospheric disasters can have adverse effects on human health and the economy. These gas parameters and atmospheric phenomena can be retrieved from remote-sensing data because there are differences in spectral absorption characteristics of different gas observations, and remote-sensing images can identify atmospheric disasters. Traditional machine learning and deep learning have achieved great success in retrieving atmospheric parameters from satellite and earth data fusion, which is significantly better than traditional methods. This section focuses on the progress of data-driven point-surface fusion technology in the monitoring and mapping of air pollutants.

Aerosols are usually expressed by aerosol optical thickness (AOD). Data-driven aerosol remote sensing inversion incorporates satellite radiometric images of multiple bands and other auxiliary variables, such as angular information, terrain, and meteorological conditions. In reference, a neural network model for AOD inversion was constructed, and brightness temperature and AERONET site values were used. A similar attempt was also made in literature, where a deep neural network was proposed to fuse the estimation of AOD with better accuracy than physical models and traditional machine learning models. Since aerosols contain particulate matter (PM), AOD is a widely used parameter in near-surface PM inversion of satellite remote sensing. Literature integrates satellite and meteorological data, geographic data, and ground measurements and uses random forest, extreme gradient enhancement, and deep neural networks to predict PM2.5, respectively. The performance of the three methods is similar. Literature used a deep confidence network to accurately simulate the relationship between PM2.5 and AOD, meteorological conditions, and vegetation distribution considering the strong correlation of PM2.5 observation in time and space, and finally realized high-precision and high-resolution PM2.5 inversion. Literature used the LSTM model to integrate the spatio-temporal characteristics among Sunflower 8 TOA, meteorological data, NDVI, and PM2.5 to retrieve PM2.5 with the high temporal resolution, as shown in Fig. 10.9.

In the study of data-driven atmospheric trace gas monitoring, many studies integrate and use the atmospheric column or profile data observed by satellites to retrieve trace gas concentration. In addition, the column product of Sentinel 5 TROPOMI was introduced in literature, and combined with GEOS-FP data, the light gradient enhancement model (LightGBM) was used to estimate the O_3, CO, and NO_2 with full high-resolution coverage in China. In literature, RF was used to estimate ground SO_2 concentration by coupling TROPOMI column data and numerical model output. In addition, satellite remote sensing can effectively observe tropospheric ozone precursors, and these data can be used to estimate surface ozone concentrations. Literature proposes a spatio-temporal embedded deep residual learning model (STE-ResNet) to obtain daily high-resolution surface O_3 concentration data by integrating ground station measurements, satellite O_3 precursor data, reanalysis data, and emission data.

In addition, data-driven methods are also widely used in atmospheric disaster monitoring. Time series prediction models such as RNN and ConvLSTM can fuse site climate data and planar satellite observations to analyze hurricane behavior and predict hurricane paths. Deep learning can combine reanalysis data and satellite images to assess hurricane damage. A growing number of machine learning methods are being used to improve fire detection and prediction, classify and map wildfire severity, automatically detect wildfires on satellite images, and detect the source, transport, and wind erosion sensitivity of various dust types.

In general, the point-surface fusion machine learning model has achieved certain achievements in retrieving atmospheric parameters, and it is gradually trying to estimate high-resolution products and more optimally considering spatio-temporal characteristics and spatio-temporal heterogeneity. However, its application in the vertical structure of

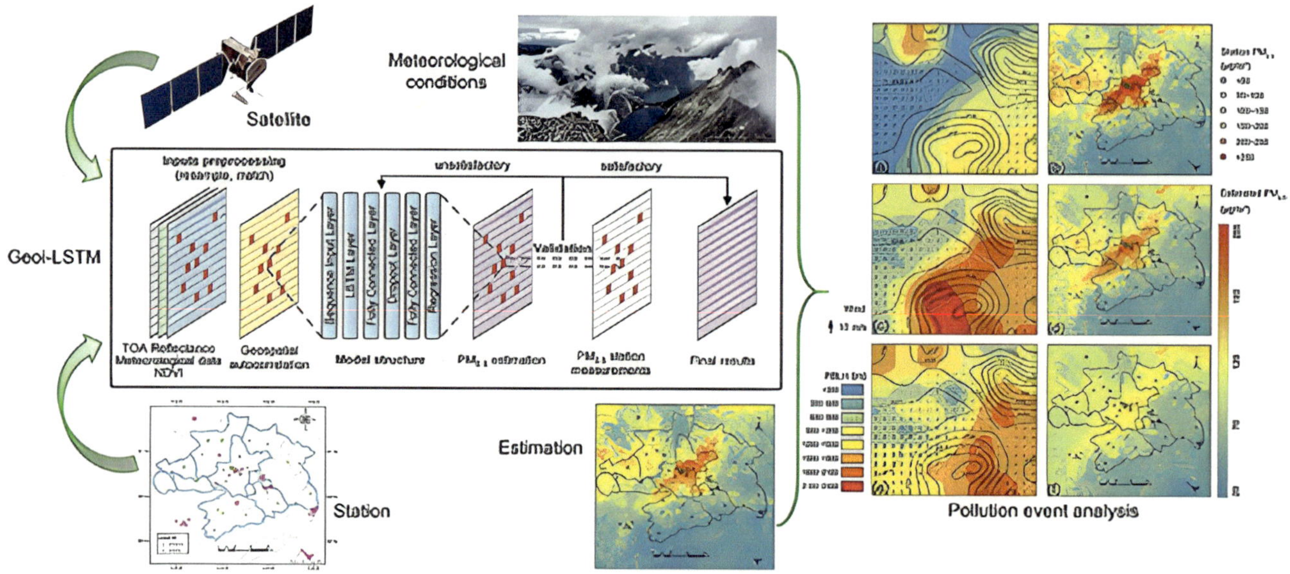

FIGURE 10.9 Hourly $PM_{2.5}$ concentration retrieval from long short-term memory neural network model.

the atmosphere is rare. In addition, the data-driven approach ignores the internal mechanism, weakens the causal relationship, and has poor interpretability, which is the bottleneck problem hindering the development of this direction.

10.4.5 Biosphere

Biomass, leaf area index, vegetation cover, and chlorophyll are important parameters for evaluating biosphere ecological balance. Only using remote-sensing data to calculate vegetation parameters (such as normalized vegetation index) can obtain limited information, and there is still a large room for improvement in accuracy. More abundant and accurate vegetation parameters can be obtained by using point-surface fusion quantitative inversion. At present, there have been many studies using support vector regression, Gaussian process regression, kernel ridge regression, neural networks, random forests, Bayesian networks, and deep learning, such as the machine learning method, combining multispectral and hyperspectral remote-sensing data and the measured data of vegetation, that have realized the accurate estimation of the vegetation of various biochemical parameters. The research results show that the point-surface fusion inversion based on machine learning greatly avoids the shortcomings of traditional field measurement methods, which are costly and destructive. At the same time, compared with the inversion results of single planar data, machine learning has a strong nonlinear fitting ability and achieves higher inversion accuracy.

In addition to the inversion of vegetation biochemical parameters, quantitative inversion of point-surface fusion based on machine learning has also been used to monitor surface phenology. For example, data from MODIS and the National Phenology Monitoring Network of the United States were used in literature to realize phenological inversion through neural networks and random forest models. The results show that the machine learning method achieves better results than the traditional method. In addition to optical data, point-surface fusion inversion using radar data is also used for phenological inversion. Literature used random forest algorithm, Radarsat-2 satellite data, and field experimental measurements during the SMAPVEX16-MB program to retrieve the phenology of oilseed rape, corn, soybean, and wheat in Manitoba, Canada.

10.5 Problems and challenges of multisource remote-sensing data fusion

A comprehensive, complete, and continuous Earth observation system is needed to deeply understand the complex natural and human phenomena on the Earth's surface. Although the spatial, temporal, and spectral resolution of remote-sensing earth observation systems has been continuously improved, they still cannot fully meet people's needs for in-depth understanding and cognition of complex geological processes and their internal mechanisms. Therefore, the new satellite sensors are still in continuous innovation, and how to further improve and develop the theory and method of Spatio-temporal spectrum integration for the emerging new earth observation data is an important direction for the research to continue to advance. In addition, with more and more remote-sensing observation satellite, the data quantity is becoming bigger and bigger, and some geological applications such as disaster emergency, such as environmental monitoring of timeliness has high requirements, how to excavate the existing capacity of computing resources, form a huge amount of data real-time transmission, processing, analysis, and application, is an effective approach to the elevation of potential applications in the project research.

Overall, in improving the data sheet, spatial, spectral, and time dimensions, the type of diversity and complexity of imaging for remote-sensing images, the model-driven method has stronger inclusiveness, can depict more data relations between time, space, and spectrum degradation, but is difficult to precise processing of empty, empty spectrum, and the nonlinear characteristics of an anisotropic relation of time and space. It limits the fusion accuracy in the case of spatial nonoverlap, spectral noncoverage, and spatiotemporal heterogeneity. Based on a data-driven approach, especially deep learning by building complex, multilayered networks characteristic of the relationship between data, which has strong nonlinear processing ability but is too dependent on data quality and quantity, a lack of imaging physical degradation processes leads to the result being unstable, less sensitive to sparse, and having the characteristics of a low score in the data. There are still some difficult problems to be further studied. Therefore, it is a necessary new approach to improve the fusion accuracy to establish a framework that can integrate data degradation mechanisms and statistical information and accurately extract effective features from multidimensional information. For the data-driven remote-sensing information fusion model, the main research directions in the future include:

Research lightweight networks to improve integration efficiency (Lertrattanapanich & Bose, 2002).

The significant improvement in the fusion effect of the existing data-driven algorithms comes at the cost of a large number of model parameters and computation, which greatly limits the fusion efficiency of multisource remote-sensing data. Therefore, how to propose a lightweight network with both model complexity and fusion effect is a major difficulty in the development of multisource remote-sensing information fusion.

Combination of data-driven and model-driven (Irani & Peleg, 1991).

Data-driven has the advantages of high efficiency and no manual prior, while model-driven has the advantages of strong generalization, inclusiveness, and physical interpretability. How to break the limitations, build a bridge between the two, and realize the combination of data-driven and model-driven multisource remote-sensing information fusion algorithms is also the top priority.

Generalization ability of the model (Stark & Oskoui, 1989).

The premise that the data-driven fusion algorithm can quickly adapt to sensor changes is to build a complete training set of different sensors and retrain them for the problem. The reason for this phenomenon is that the generalization ability of data-driven algorithms is weak, and it is difficult to take into account the degradation characteristics of different sensors at the same time. It is urgent to propose a unified framework for multisource remote-sensing information fusion to improve the model generalization ability of data-driven algorithms.

On-orbit real-time fusion processing (Farsiu et al., 2004).

The current data-driven algorithms need powerful computing power and hardware as support. However, with the increasing perfection of the space-space remote-sensing monitoring network, the on-orbit optimization of satellite image data is very important. How to develop a multisource remote-sensing information fusion algorithm that can operate on the satellite platform with limited payload is also an important direction in the future.

10.6 Conclusions

The data-driven multisource remote-sensing information fusion technology can realize the effective coupling of complementary information of multisource observation data by mining data features and learning feature relations in massive remote-sensing data. This paper summarizes the iteration process of data-driven multisource remote-sensing information fusion technology from three fields: isotropic remote-sensing data fusion, anisotropic remote-sensing data fusion, and point-surface fusion. From simple component replacement and multiresolution analysis to deep learning, graph convolution, and Transformer; From single time phase, single mode to multitime phase, multimode. Although the current data-driven fusion algorithm has achieved a good lead in the fusion effect, quantitative indicators, and other results, the data-driven algorithm ignores the physical meaning and has a strong dependence on the quality of training data. Although the model-driven algorithm relies on manual parameter adjustment and has limited prior expression, it has strong generalization and transfer ability. Therefore, in the future, the coupling of data-driven and model-driven can be realized by model derivation to help network construction, or network before optimizing model prior representation. This, of course, raises a new challenge: which models combine most efficiently with each other? Which combination route is the most convenient? In addition, only considering the improvement of data-driven algorithm architecture, how to achieve model lightweight and improve the generalization ability of model migration is also an important direction of future development.

References

Alhatami, E., Bhatti, U.A., Huang, M., & Feng, S.L. (2022). Image fusion based on discrete cosine transform with high compression. *7th international conference on signal and image processing*, ICSIP 2022, (pp. 606–610), https://doi.org/10.1109/ICSIP55141.2022.9887300, http://ieeexplore.ieee.org/xpl/mostRecentIssue.jsp?punumber = 9886003.

Bhatti, U. A., Bazai, S. U., Hussain, S., Fakhar, S., Ku, C. S., Marjan, S., Yee, P. L., & Jing, L. (2023). Deep learning-based trees disease recognition and classification using hyperspectral data. *Computers, Materials and Continua*, *77*(1), 681–697. Available from https://doi.org/10.32604/cmc.2023.037958, http://www.techscience.com/cmc/index.html.

Bhatti, U. A., Huang, M., Neira-Molina, H., Marjan, S., Baryalai, M., Tang, H., Wu, G., & Bazai, S. U. (2023). MFFCG – Multi feature fusion for hyperspectral image classification using graph attention network. *Expert Systems with Applications*, *229*. Available from https://doi.org/10.1016/j.eswa.2023.120496, https://www.journals.elsevier.com/expert-systems-with-applications.

Bhatti, U. A., Yu, Z., Yuan, L., Nawaz, S. A., Aamir, M., & Bhatti, M. A. (2022). A robust remote sensing image watermarking algorithm based on region-specific SURF. *Lecture Notes in Networks and Systems*, *350*, 75–85. Available from https://doi.org/10.1007/978-981-16-7618-5_7, http://springer.com/series/15179.

Bhatti, U. A., Yu, Z., Yuan, L., Zeeshan, Z., Nawaz, S. A., Bhatti, M., Mehmood, A., Ain, Q. U., & Wen, L. (2020). Geometric algebra applications in geospatial artificial intelligence and remote sensing image processing. *IEEE Access*, *8*, 155783–155796. Available from https://doi.org/10.1109/ACCESS.2020.3018544, http://ieeexplore.ieee.org/xpl/RecentIssue.jsp?punumber = 6287639.

Cheng, M., Tang, H., Khan, A., Melethil Sethumadhavan, S., Assam, M., Li, D., Ghadi, Y. Y., Mohamed, H. G., & Bhatti, U. A. (2023). A digital twin-based visual servoing with extreme learning machine and differential evolution. *International Journal of Intelligent Systems*, *2023*. Available from https://doi.org/10.1155/2023/6639963, https://www.hindawi.com/journals/ijis/.

Cliche, G., Bonn, F., & Teillet, P. (1985). Integration of the SPO panchromatic channel into its multispectral mode for image sharpness enhancement. *Photogrammetric Engineering & Remote Sensing*, *51*(3), 311–316.

Farsiu, S., Robinson, M. D., Elad, M., & Milanfar, P. (2004). Fast and robust multiframe super resolution. *IEEE Transactions on Image Processing*, *13*(10), 1327–1344. Available from https://doi.org/10.1109/tip.2004.834669.

Hallada, W. A., & Cox, S. (1983). Image sharpening for mixed spatial and spectral resolution satellite systems. *Environmental Research Institute of Michigan undefined Proceedings of the International Symposium on Remote Sensing of Environment*, *3*, 1023–1032.

Irani, M., & Peleg, S. (1991). Improving resolution by image registration. *CVGIP: Graphical Models and Image Processing*, *53*(3), 231–239. Available from https://doi.org/10.1016/1049-9652(91)90045-l.

Lertrattanapanich, S., & Bose, N. K. (2002). High resolution image formation from low resolution frames using delaunay triangulation. *IEEE Transactions on Image Processing*, *11*(12), 1427–1441. Available from https://doi.org/10.1109/tip.2002.806234.

Li, D., Li, Y., Yang, W., Ge, Y., Han, Q., Ma, L., Chen, Y., & Li, X. (2018). An enhanced single-pair learning-based reflectance fusion algorithm with spatiotemporally extended training samples. *Remote Sensing*, *10*(8), 1207. Available from https://doi.org/10.3390/rs10081207.

Liu, J., Li, J., Ma, J., Sadiq, N., Bhatti, U. A., & Ai, Y. (2019). A robust multi-watermarking algorithm for medical images based on DTCWT-DCT and Henon map. *Applied Sciences*, *9*(4), 700. Available from https://doi.org/10.3390/app9040700.

Nawaz, S. A., Li, J., Bhatti, U. A., Shoukat, M. U., & Ahmad, R. M. (2022). AI-based object detection latest trends in remote sensing, multimedia and agriculture applications. *Frontiers in Plant Science*, *13*. Available from https://doi.org/10.3389/fpls.2022.1041514, https://www.frontiersin.org/journals/plant-science.

Niu, Z. (2012). Use of MODIS and landsat time series data to generate high-resolution temporal synthetic landsat data using a spatial and temporal reflectance fusion model. *Journal of Applied Remote Sensing*, *6*(1), 063507. Available from https://doi.org/10.1117/1.jrs.6.063507.

Shen, H., Wu, P., Liu, Y., Ai, T., Wang, Y., & Liu, X. (2013). A spatial and temporal reflectance fusion model considering sensor observation differences. *International Journal of Remote Sensing*, *34*(12), 4367–4383. Available from https://doi.org/10.1080/01431161.2013.777488.

Stark, H., & Oskoui, P. (1989). High-resolution image recovery from image-plane arrays, using convex projections. *Journal of the Optical Society of America A*, *6*(11), 1715. Available from https://doi.org/10.1364/JOSAA.6.001715.

Xue, J., Leung, Y., & Fung, T. (2017). A Bayesian data fusion approach to spatio-temporal fusion of remotely sensed images. *Remote Sensing*, *9*(12), 1310. Available from https://doi.org/10.3390/rs9121310.

Yang, J., Wright, J., Huang, T. S., & Ma, Y. (2010). Image super-resolution via sparse representation. *IEEE Transactions on Image Processing*, *19*(11), 2861–2873. Available from https://doi.org/10.1109/tip.2010.2050625.

Section IV

Deep learning for lulc

Chapter 11

Climate change air quality monitoring using sentimental 2 dataset

Mughair Aslam Bhatti, Hao Tang, Uzair Aslam Bhatti, Dekai Li and Mingtian Yu
School of Information and Communication Engineering, Hainan University, Haikou, P.R. China

11.1 Introduction

11.1.1 Background and significance of the topic

We live on a blue planet. Although this planet has created a place for people to live, it faces many threats. Natural disasters are one of the many threats that people must deal with. From ancient times to the present, natural disasters have had a widespread and far-reaching effect on people, and even affect the direction of people's development to some extent, such as: catastrophic earthquakes, floods, forest fires, typhoons, or hurricanes, etc. Earthquakes are one of the most common natural disasters. They can destroy cities, damage buildings and infrastructure, and cause large-scale casualties and property losses. The 2010 Haiti earthquake killed more than 220,000 people and left millions homeless. Home can be returned. Flooding inundated cities and farmland, damaged homes and roads, and disrupted water and power facilities. The flooding caused by Hurricane Katrina in New Orleans in 2005 caused massive destruction and casualties. Bushfires have burned large areas of woodland, villages, and cities, causing severe impacts on air quality and ecosystems. Australia's bushfires in 2019 destroyed more than 12,000 structures and killed at least 33 people.

Typhoons are tropical cyclones with strong winds and storm surges. They can destroy buildings and infrastructure, flood cities and farmland, and damage water and power facilities. Super Typhoon Haiyan in the Philippines in 2013 killed more than 6000 people and left millions homeless. The impact and destructive power of natural disasters is so great that we must take appropriate measures to mitigate their effects and ensure the safety and well-being of our people.

With the development and progress of human science and technology, artificial satellites have been sent into space one by one. Humanity's first remote sensing satellite was the TIROS-1 satellite launched by the United States in 1960. Since then, mankind has had its first pair of eyes in space. The satellite is mainly used for earth meteorology research and can obtain the earth's meteorological information in real time, such as cloud images, temperature distribution, etc. (e.g., see Fig. 11.1). The successful launch of the TIROS-1 satellite also marked the beginning of remote sensing technology. In the following decades, remote sensing technology continued to develop and its application scope continued to expand, becoming an important means of modern earth observation and resource survey. Remote sensing satellites have important application value in natural disasters and can improve disaster monitoring, assessment, rescue and prediction capabilities. Using remote sensing technology, we can monitor disasters, assess disaster losses and assist with rescue operations. Remote sensing satellites can provide high-resolution, high-frequency images, monitor the occurrence and scope of natural disasters such as earthquakes, floods, and storms, and grasp disaster information in a timely manner.

Remote sensing satellites can also provide high-precision geographical information data such as terrain, land cover, vegetation, etc., to help rescuers understand the disaster situation, determine the location of personnel, materials, equipment, and other rescue resources, guide rescue operations, improve rescue efficiency, and can conduct disaster-stricken areas accurate damage assessment. Through remote sensing technology, the impact of disasters on life, property, and the environment can be quickly and accurately estimated, providing scientific basis for post-disaster reconstruction.

Deep Learning for Earth Observation and Climate Monitoring. DOI: https://doi.org/10.1016/B978-0-443-24712-5.00012-9

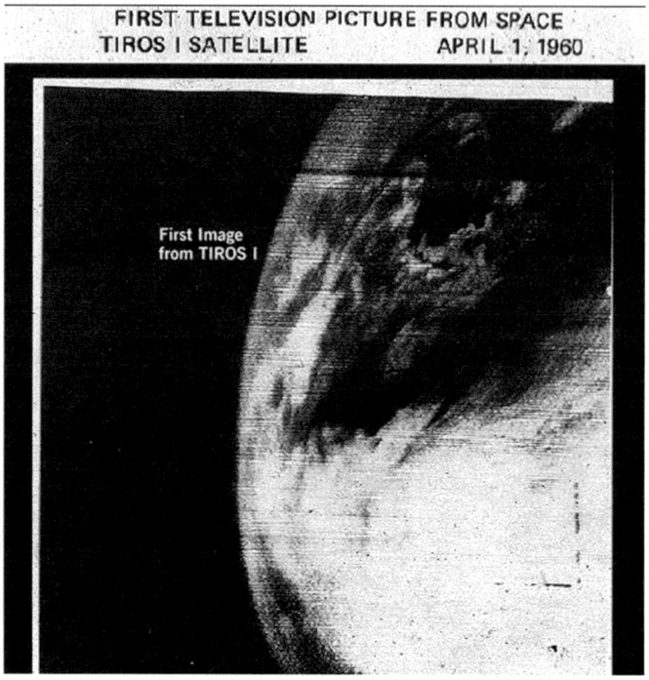

FIGURE 11.1 First television picture from space in human history. The first television picture from space in human history, taken by the weather satellite "Tyros 1" on April 1, 1960.

11.1.2 Research status at home and abroad

11.1.2.1 Traditional detection methods

Before the rise of deep learning, target detection algorithms were widely used. Most of these algorithms are based on traditional machine learning methods and computer vision technology, and mainly include the following:

- Haar features and cascade classifier: The main advantage is that images can be processed quickly and are suitable for real-time applications. However, the shortcomings of this algorithm are also obvious. For example, it requires manual design of Haar features and hyperparameters such as the number of stages of the cascade classifier and the number of weak classifiers. It is not very robust to the rotation, scaling, and illumination changes of the target object. With the development of deep learning algorithms, Haar features and cascade classifiers have been gradually replaced by deep learning algorithms.

  Traditional target detection algorithms perform well in some scenarios. However, because these algorithms lack the powerful ability of deep learning algorithms to express features, the application of traditional target detection algorithms is now gradually replaced by deep learning algorithms.
- Detection method based on deep learning: Domestically, domestic scholars have conducted extensive research on satellite image damage assessment in recent years, mainly focusing on disaster scenarios such as earthquakes, floods, and wildfires. For example, researchers from the University of Electronic Science and Technology of China, proposed a study on seismic image processing and evaluation methods based on deep learning (He & Bhatti, 2024), using CNNs to segment seismic remote sensing images, achieving automatic extraction of damaged areas, and evaluation of damage levels. In addition, researchers from Wuhan University also proposed the application of remote sensing technology in building damage detection after earthquakes. The article proposed the application method of remote sensing technology in building damage detection. If you have pre-disaster remote sensing data for the area, you can apply remote sensing change image detection technology and shoot post-disaster remote sensing data for comparison. This method is the most commonly used method when there is sufficient data. Image change technology can determine the following: The target, whether changes have occurred, the scope of the target changes and the type of changes (Alhatami et al., 2024). However, there are still some problems with this method. For example, there is no image data from the previous time in a certain place, or there is an image data and the images are captured by different sensors. Furthermore, the same sensor captures images at different times. The same building may be photographed from different angles or in different lighting conditions. Meteorological factors such as clouds and weather are also one of the unfavourable factors for comparative analysis of before and after images. In other words,

this method has higher requirements for data quality. Detection based only on post-disaster images of buildings does not have the above problems. After a natural disaster occurs, it is relatively easy to obtain post-disaster image remote sensing satellite resources. Therefore, damage detection based on post-disaster images is better than detection of changes in images before and after the disaster, the research is more practical. There are also other researchers who use neural networks to detect house collapse after earthquakes. The training speed and accuracy are superior to traditional methods and traditional neural networks (Bhatti et al., 2024). Abroad, scientists from the United States, Europe, and other countries have also conducted a series of research in this field. For example, NASA uses disaster observations to reduce the risk of natural disasters, improve rescue response times, facilitate disaster recovery, and improve resilience to natural and technological disasters at local and global scales.

- Ability (NASA, 2021): By sponsoring applied science and promoting domestic and international partnerships, the program hopes to prepare for natural disasters before they occur and use Earth observation data to aid decision-making when disasters occur.

11.2 Overall plan design

11.2.1 Requirements analysis

In emergency satellite image damage assessment, the first task is to identify damaged houses from satellite images. The reason is that, if people take shelter in buildings when encountering general natural disasters and if the building collapses at this time, then it is not only unable to protect the residents, it will also cause more serious consequences. If there is a disaster such as an earthquake that cannot be avoided in the building, many people are often still stranded in the building due to the sudden nature of the disaster.

Through the above analysis, the preferred technology should be instance segmentation of the image, which can not only detect the target in the image, but also segment the target (e.g., see Fig. 11.2). In instance segmentation tasks, deep learning models have become mainstream, mainly including two types of models: two-stage models based on region extraction and single-stage models. Both models have their own advantages in different scenarios (Nizamani et al., 2023).

Two-stage model:

Advantage:

1. High accuracy: Dual-stage models usually use a two-stage method. The first stage extracts regions from the image, and the second stage classifies and regresses candidate regions, so it can achieve high accuracy.
2. Strong ability to handle complex scenes: The two-stage model is usually able to handle complex scenes well, such as multiple targets, occlusions, and targets of different scales.
3. Good interpretability: The two-stage model can output the score and position information of each detection frame, so it has good interpretability and is beneficial to subsequent decision-making and analysis.

Shortcoming:

1. Slow speed: The two-stage model usually requires two stages of operations, so it is slow and not suitable for real-time applications.

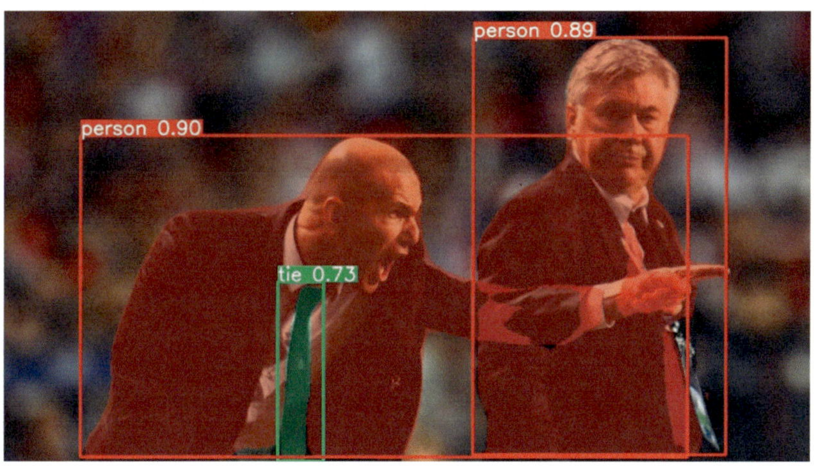

FIGURE 11.2 Segmentation algorithm. Example of instance segmentation algorithm.

2. Hyperparameter adjustment is complex: Two-stage models usually have multiple hyperparameters, which are difficult to adjust.
3. The training process is complex: The two-stage model usually requires multiple trainings, and the training process is relatively complicated.

Single stage model:
Advantage:

1. Fast: Detection methods based on single-stage models usually require only one stage of operation, so they are fast and suitable for real-time applications.
2. The training process is simple: The training process based on the single-stage model is relatively simple and only requires one training session.
3. Strong robustness: Detection methods based on single-stage models usually have good robustness to changes in the position and scale of the target.

Shortcomings:

1. The accuracy rate is relatively low: The detection method based on the single-stage model usually divides the detection frame into several anchor points, which has a certain error, so the accuracy rate is relatively low.
2. Poor interpretability: Detection methods based on single-stage models usually can only output the classification results and location information of each anchor point, and have poor interpretability.
3. Not suitable for processing complex scenes: Detection methods based on single-stage models are usually not suitable for processing complex scenes, such as multiple targets, occlusions, and targets of different scales.

In summary, the choice of a two-stage model or a detection method based on a single-stage model should be based on the specific application.

11.2.2 Solution selection

Commonly used instance segmentation models include Mask R-CNN, YOLACT, you only look once (YOLO), etc. Mask R-CNN: Mask R-CNN works by inserting a feature pyramid network between the detection branch and the segmentation branch, which can perform two tasks: object detection and instance segmentation. Mask R-CNN performs very well in the field of instance segmentation, but the corresponding computational overhead is also large. It is suitable for scenarios that require high accuracy and low speed requirements (He et al., 2017).

YOLACT: YOLACT is a single-stage instance segmentation model that uses a method called YOLO-style to combine instance segmentation and target detection. YOLACT can achieve instance segmentation tasks more efficiently by combining the prediction process of detection boxes and segmentation masks. Compared with Mask R-CNN, YOLACT has better performance in speed, but the accuracy is slightly lower. It is suitable for scenarios that require high speed and low accuracy requirements (Etten, 2018).

YOLO: In 2016, the YOLO (Redmon et al., 2016) algorithm proposed by Redmon was the first one-stage algorithm. Its characteristic is that only one network is used to complete target detection. The operation speed is very fast while ensuring detection accuracy, but the accuracy is slightly lower than Faster R-CNN. In 2017, Redmon proposed the YOLOv2 (Redmon & Farhadi, 2017) network, adding structures such as the a priori anchor box mechanism in Faster R-CNN and the global draw pooling in Network in Network (NIN) to the YOLOv1 network. It no longer directly predicts the target coordinates but predicts the deviation between the target and the anchor box, which reduces the training difficulty and improves the recall rate of the network. The feature pyramid and residual block structure were added to YOLOv3 proposed in 2018 (Redmon & Farhadi, 2018), which increased the detection effect of small targets. In 2020, Alexey Bochkovskiy, a participant in the YOLO community, took over the YOLO project and published YOLOv4. In YOLOv4 (Bochkovskiy et al., 2020), the author introduced a large number of training techniques and structural improvements, such as CSP structure, CIoU loss, Mish activation function, etc., which greatly improved the performance of the network. Soon after, YOLOv5 was released.

In emergencies, the first requirement is speed. In the transfer learning of satellite image damage assessment in emergencies, a faster model should be selected. The single-stage model can not only meet the detection needs of satellite images, but also, in view of the detection needs of real-time videos such as drones, this article chooses the YOLO algorithm for operation.

11.3 Principle of you can only look once algorithm

YOLO is a single-stage detection algorithm. Since it only requires one forward propagation, it is fast and can achieve real-time target detection.

The basic idea of the YOLO algorithm is to divide the entire image into multiple grids. The YOLO algorithm uses the cross-entropy loss function to optimize the model during training. For each bounding box, the loss function consists of two parts: classification loss and localization loss. The classification loss represents the classification error, and the localization loss represents the coordinate error between the predicted bounding box and the true bounding box. The training result is that each grid is responsible for predicting one or more bounding boxes and the category and confidence of the corresponding object.

11.3.1 Convolutional neural network

CNN is a deep learning model specially used for visual tasks such as image recognition, image classification, object detection, etc. Its basic components are convolution layer, pooling layer, and fully connected layer (Yu et al., 2024). The advantages of CNNs include good feature extraction capabilities and the ability to automatically learn features. Therefore, CNNs have been widely used in image recognition, image classification, object detection and other fields (Bhatti et al., 2024).

11.3.1.1 Convolutional layer

The convolutional layer is one of the most basic components of the CNN. It is used to perform convolution operations on input data to achieve feature extraction and abstraction (Li et al., 2024).

The input of the convolutional layer is usually a multichannel two-dimensional matrix (for color images, it is a three-dimensional tensor). The convolutional layer performs a convolution operation on the input data and a set of convolution kernels (convolution kernel) to generate A new feature map (feature map) (Bhatti et al., 2023).

The convolution operation can be regarded as a sliding window operation. The convolution kernel starts from the upper left corner of the input data and slides to the right through a certain stride. Each slide will perform a convolution operation to obtain the corresponding position. output. Specifically, for an input matrix and a convolution kernel, the output of the convolution operation can be calculated by the following formula:

$$O_{(i,j)} = (K * I)(i,j) = \sum_m \sum_n I(i-m, j-n)K(m,n) \tag{11.1}$$

Convolutional layers use different convolution kernels to achieve different features. Each set of convolution kernels can learn different features. Use the backpropagation algorithm to update the parameters of the convolution kernel to continuously optimize the performance of the model.

11.3.1.2 Activation function

Activation function: The activation function is an important part of the neural network. It is usually used to generate nonlinear mapping relationships between neurons, so that the neural network has stronger adaptive capabilities. The following are several common activation functions:

1. Sigmoid function: The Sigmoid function is a commonly used activation function, which maps the input to the (0,1) interval. The formula is:

$$f(x) = \frac{1}{1+e^{-x}} \tag{11.2}$$

   The Sigmoid function is differentiable and monotonic, but it is prone to the problem of gradient disappearance. For larger or smaller input values, its gradient is close to 0, making training difficult.
2. Rectified linear unit (ReLU) function: The function is a very popular activation function that maps all negative inputs to 0, leaving nonnegative values unchanged. The formula is:

$$f(x) = max(0,x) \tag{11.3}$$

   The ReLU function has the advantages of being simple, easy to calculate, and solving the problem of gradient disappearance. However, in some cases, the neuron "death" phenomenon may occur, that is, the output of the neuron is always 0 and cannot be recovered (Peng et al., 2024).

3. Leaky ReLU function: The Leaky ReLU function is a variant of the ReLU function. It no longer takes the value 0 on negative inputs, but is multiplied by a small slope, usually 0.01. The formula is:

$$f(x) = max(ax, x) \tag{11.4}$$

The Leaky ReLU function can solve the problem of neuron "death" to a certain extent, while still maintaining simplicity and ease of calculation.

4. Tanh function: The Tanh function is the hyperbolic tangent function, and the formula is:

$$f(x) = \frac{e^x - e^{-x}}{e^x + e^{-x}} \tag{11.5}$$

The Tanh function has differentiability, monotonicity and symmetry, and its value range is $(-1,1)$, but like the Sigmoid function, it is also prone to the problem of gradient disappearance.

In addition to the above-mentioned common activation functions, there are many other activation functions, such as Softmax function, ELU function, Swish function, etc. Different activation functions are suitable for different tasks and scenarios, and need to be selected according to specific problems.

11.3.1.3 Polling layer

The pooling layer is a level in the CNN. It is mainly used to downsample the output of the convolutional layer, thereby reducing the size of the feature map and reducing the amount of calculation. It can also prevent model overfitting. Commonly used pooling methods include max pooling and average pooling.

The pooling layer usually includes hyperparameters such as pooling window size, step size, and padding, which can be adjusted according to different application scenarios. In some deep learning models, pooling layers usually alternate with convolutional layers to form a sequence of convolutional pooling layers, which are used to extract multilevel feature representations in the input data.

11.3.1.4 Fully connected layer

Fully connected layer is a common layer in CNNs, also called dense layer. However, fully connected layers are not used in Yolov5, but are replaced by convolutional layers.

11.3.2 You can only look once network overview

11.3.2.1 YOLOv5 network structure

11.3.2.1.1 Conv

YoloV5l network structure Yolo's Conv module includes a convolution layer, a batch normalization (BN) and a SiLU activation function (Table 11.1). The calculation of the convolution has been introduced above. The BN normalizes the output of the previous convolution layer. Normalizing to a range with a mean of zero and a variance of one can narrow the range of data and reduce the sensitivity of the network to input data, thus improving the generalization ability of the network and preventing overfitting. Activation functions such as SiLU have been introduced above, which introduce nonlinear functions, giving the neural network stronger fitting capabilities.

11.3.2.1.2 C3

The C3 module consists of three convolutional layers. The first convolutional layer has a convolution kernel size of 1×1, and the second and third convolutional layers have a convolutional kernel size of 3×3. This design can help the network learn. to richer and higher-level feature information.

By introducing the C3 module in YOLOv5, the network can be made deeper and more complex, and the feature extraction capabilities of the network can be enhanced, thereby improving the accuracy of target detection and instance segmentation. At the same time, the design of the C3 module also conforms to the general design principles of convolutional layers in CNNs, that is, building a deep network with small convolution kernels can effectively extract more complex and high-level feature information (Zheng et al., 2022).

The structure diagram of the C3 module is as follows.

TABLE 11.1 Shows the structure of the Yolo network, where the from column indicates which layer the input of this layer comes from, −1 indicates the previous layer, and the number indicates the layer it comes from. Column *N* indicates the depth of the layer, that is, there are several identical modules. The module column indicates the name of the module, which will be introduced in detail below.

Layer	From	*N*	Module
0	−1	1	Conv
1	−1	1	Conv
2	−1	3	C3
3	−1	1	Conv
4	−1	6	C3
5	−1	1	Conv
6	−1	9	C3
7	−1	1	Conv
8	−1	3	C3
9	−1	1	SPPF
10	−1	1	Conv
11	−1	1	Upsampling
12	[−1,6]	1	Concat
13	−1	3	C3
14	−1	1	Conv
15	−1	1	Upsampling
16	[−1,4]	1	Concat
17	−1	3	C3
18	−1	1	Conv
19	[−1,14]	1	Concat
20	−1	3	C3
21	−1	1	Conv
22	[−1,10]	1	Concat
23	−1	3	C3
24	[17,20,23]	1	Segment

11.3.2.1.3 SPPF

The SPP structure is shown in the figure below. The input is passed through multiple MaxPools of different sizes in parallel, and then further fused, which can solve the target multiscale problem to a certain extent (e.g., see Fig. 11.3).

11.3.2.1.4 Others

Up sampling is an up sampling module that is mainly used to up sample low-resolution feature maps to high resolution, thereby improving the network's positioning capability and accuracy.

Concat is a feature fusion module, mainly used to splice feature maps from different layers to improve the detection and recognition capabilities of the network.

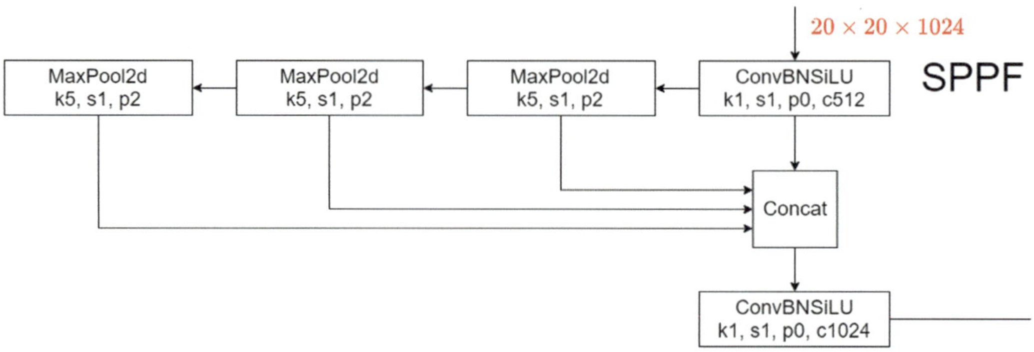

FIGURE 11.3 Spatial pyramid pooling fast (SPPF) module structure diagram.

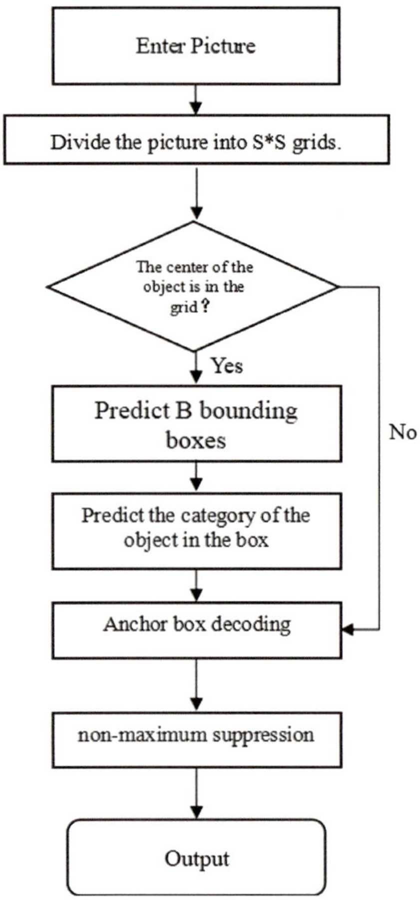

FIGURE 11.4 Yolo detection flow chart.

11.3.2.2 YOLO detection process

First, the input image is scaled to a fixed size network input size, and then the image is divided into S*S regions, with each grid responsible for predicting the object centered in it. Predict B bounding boxes for each style and give confidence:

$$confidence = P_r(Object) * IoU_{pred}^{truth} \tag{11.6}$$

This formula reflects that the confidence is equal to the predicted probability multiplied by the coincidence between the true and predicted values.

The final tensor shape output by the YOLO network is as follows:

$$filters = S \times S \times (B * 5 + C) \tag{11.7}$$

FIGURE 11.5 xBD data set.


```
⊟{
        "features":⊟{
                "lng_lat":⊟[
                        ⊞Object{...},
                        ⊞Object{...},
                        ⊞Object{...}
                ],
                "xy":⊟[
                        ⊟{
                                "properties":⊟{
                                        "feature_type":"building",
                                        "subtype":"minor-damage",
                                        "uid":"88703461-a33d-4327-9244-a0d4e2242ede"
                                },
                                "wkt":"POLYGON ((452.9629646916362 86.81005767060303, 490.8200366918895 9
0.17953528341397, 489.4326047321368 117.9281744461754, 483.8828768985862 117.9281744466502, 483.0900
586383122 135.3701762054355, 462.4767838315423 135.3701762049607, 462.2785792676607 130.415062067803
7, 452.3683509950063 129.424039241037, 452.5665555593626 117.5317653150889, 434.7281446704362 116.3
425379223041, 435.7191674981527 98.30592246641427, 448.9988733845685 99.29694529460532, 448.99887338
19571 94.73824028853485, 452.1701464292256 94.73824028853485, 452.9629646916362 86.81005767060303))"
                        },
```


FIGURE 11.6 xBD dataset labels.

```
0 0.841553 0.895222 0.848827 0.895213 0.848833 0.900548 0.841517 0.900545 0.841553 0.895222
0 0.433912 0.865404 0.435790 0.863317 0.441007 0.867491 0.437460 0.871352 0.433286 0.868013 0.433912 0.865404
0 0.423539 0.869536 0.432195 0.874863 0.429132 0.879790 0.420476 0.874330 0.420609 0.870868 0.423539 0.869536
0 0.769740 0.899090 0.773838 0.898515 0.774364 0.903825 0.771335 0.904734 0.769740 0.899090
0 0.297563 0.549710 0.302639 0.550882 0.303029 0.555566 0.300296 0.558299 0.295221 0.555957 0.294831 0.550491
```

FIGURE 11.7 Yolo tag.

In this formula, B represents the number of bounding boxes predicted by each cell, and C represents the number of object categories. Finally, according to the threshold and non-maximum suppression algorithm, low-confidence bounding boxes are filtered out, and highly overlapping bounding boxes are removed to obtain the final detection result. The flow chart is as follows (Fig. 11.4).

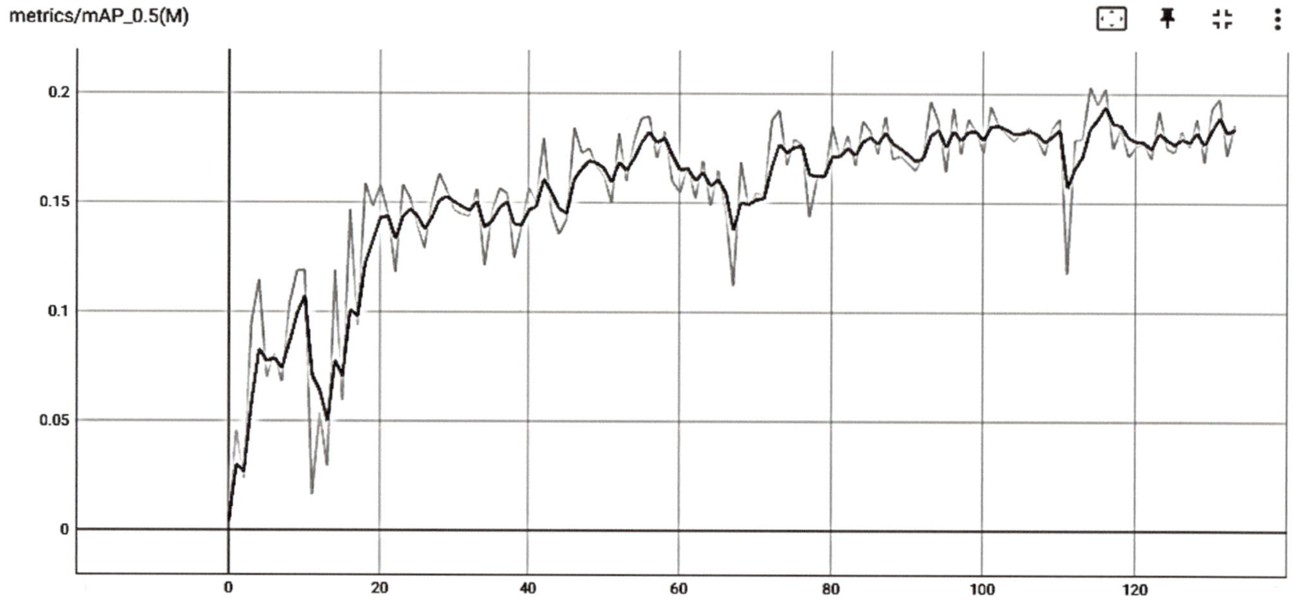

FIGURE 11.8 Original data set mAP0.5 curve.

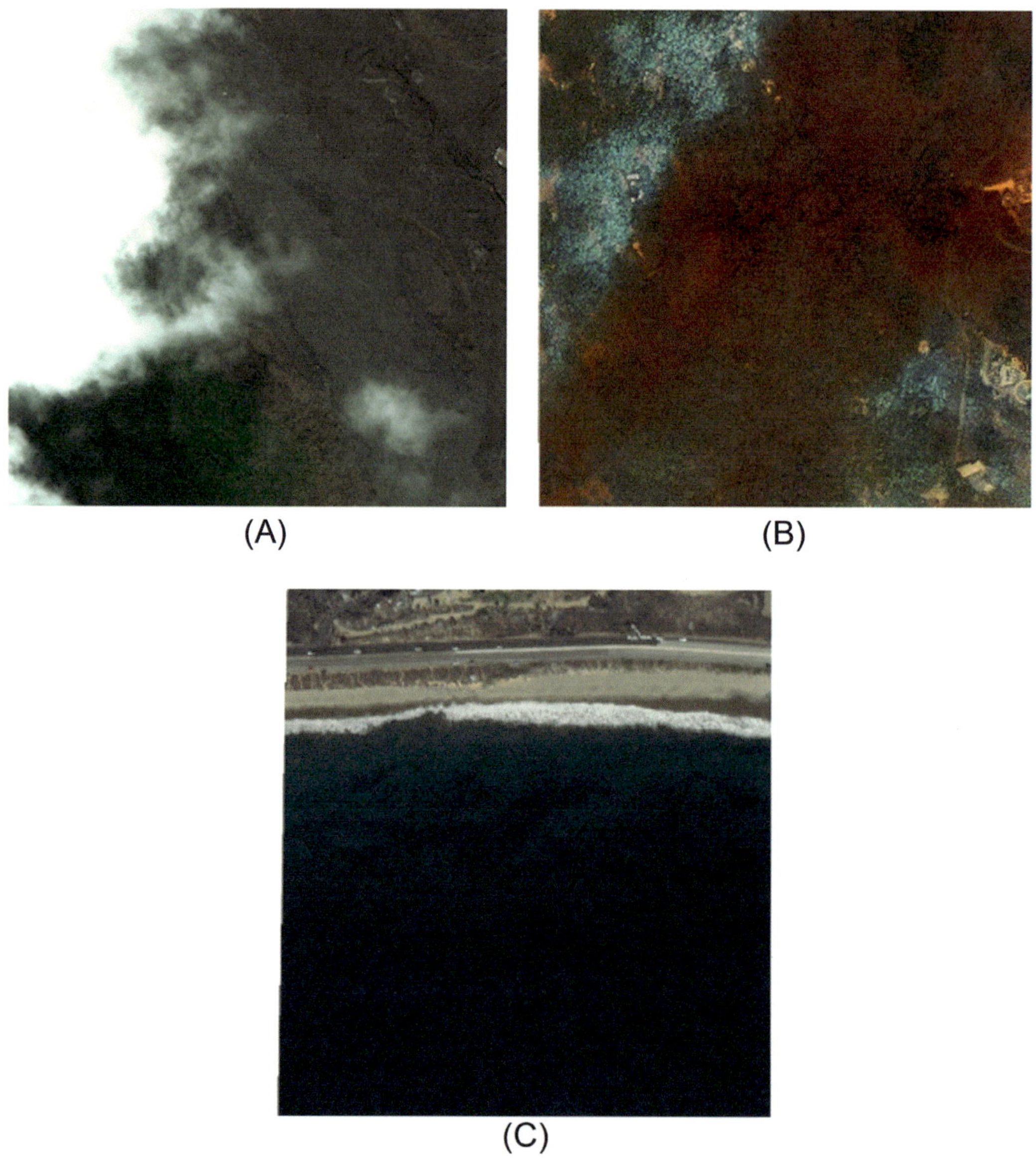

FIGURE 11.9 Problem picture (A) Remote sensing images with cloud, (B) Remote sensing images with hot weather, (C) remote sensing images with oceans.

11.4 Experiment and analysis

PyTorch is an open source Python-based machine learning library. It provides a flexible deep learning framework that can support a variety of neural network architectures. In this article, we will use PyTorch to implement the YOLOv5 model.

11.4.1 xBD dataset

The xBD dataset is a dataset used for building damage assessment. For now, natural disaster rescue requires an accurate understanding of damaged buildings in the disaster area, and the use of aerial imagery combined with computer vision algorithms for assessment can reduce potential personal risks. The xBD dataset provides satellite imagery of a variety of disaster events and provides data on building damage (e.g., see Fig. 11.5). Additionally, the dataset contains bounding boxes and labels for environmental factors such as fire, flood, smoke, etc. (Bhatti et al., 2024).

11.4.2 Experiment content

11.4.2.1 Experimental hardware platform

All data in this experiment were obtained by running on the following hardware platforms:

CPU: 12th Gen Intel CoreTM i5-12400 @ 2.50 GHz

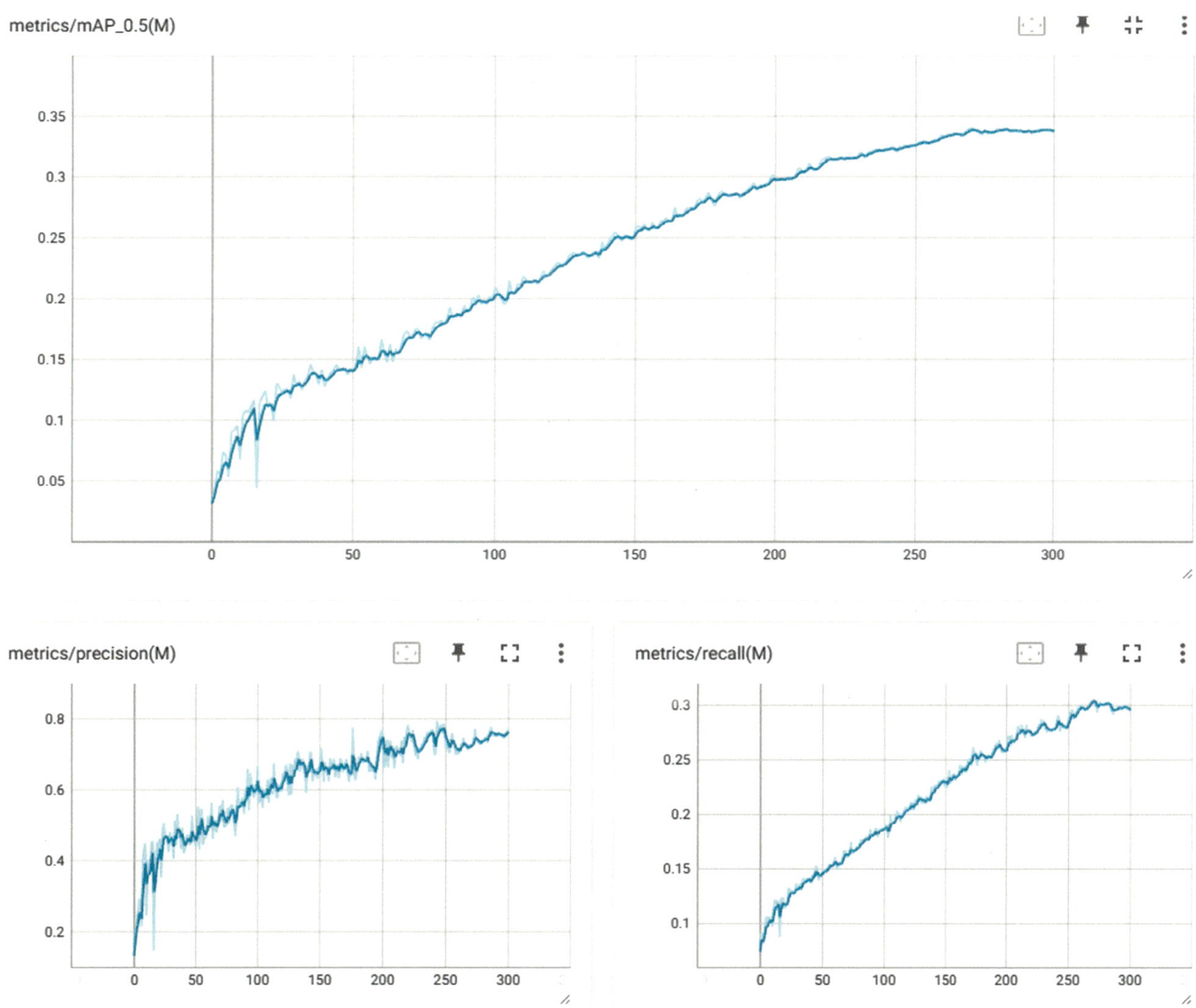

FIGURE 11.10 Training results after removing problematic images.

Memory: 32GB
GPU: NVIDIA GeForce RTXTM 3060, 12GB VRAM
Development and runtime environment:
Operating system: Ubuntu 22.04 LTS
PyTorch version: 2.0.1
CUDA version: 11.7

11.4.2.2 Data set processing

The label format given in the xBD data set cannot be directly sent to Yolo for training. The label must first be changed to a label in Yolo format (Nawaz et al., 2022). The labels of the xBD data set are stored in a Json file, which stores the target information and whether it is damaged in the form of longitude, latitude and coordinates in the picture(Fig. 11.6). Labels in the original data set:

The label format is as follows (Fig. 11.7):

$$<\text{class}> <x1> <y1> <x2> <y2> <x3> <y3> <x4> <y4> \ldots$$

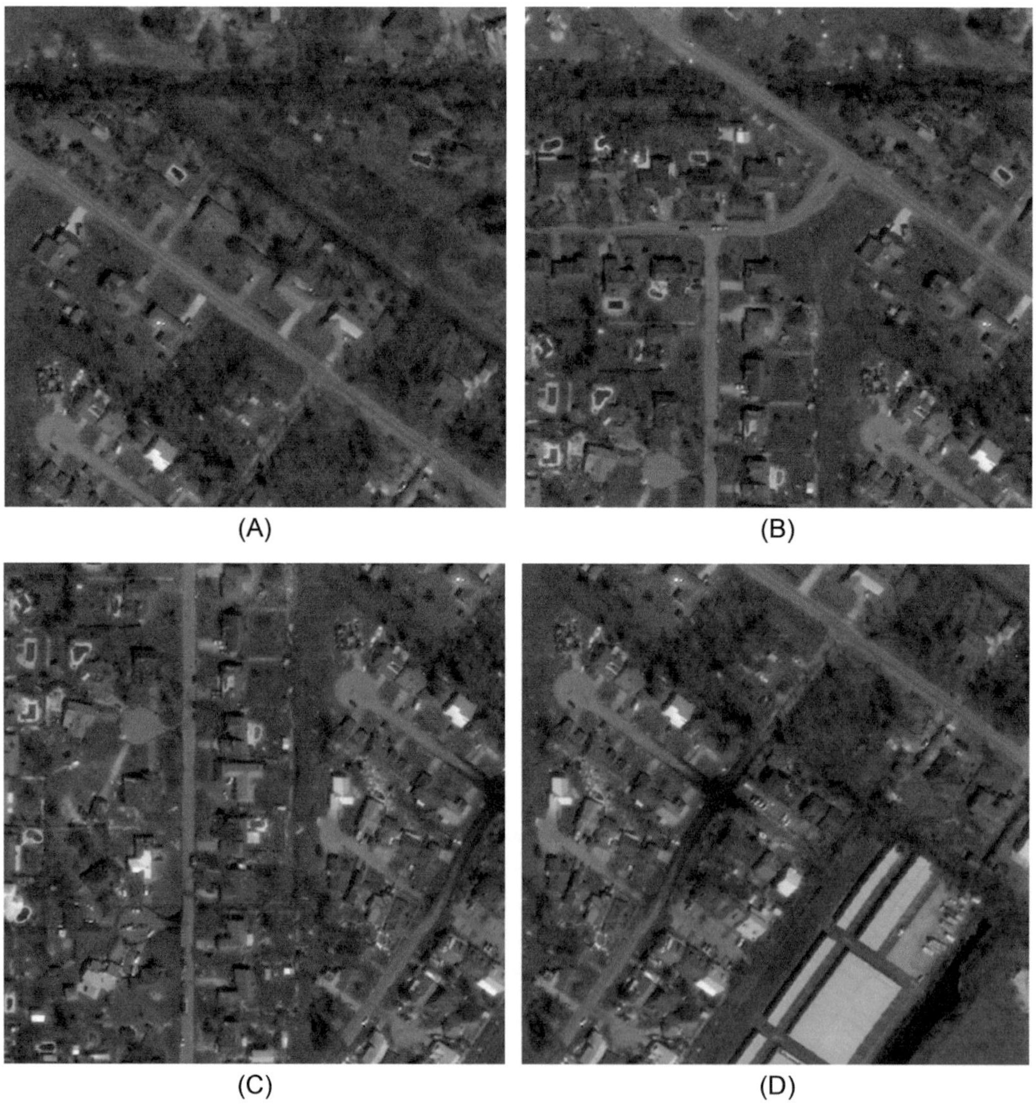

FIGURE 11.11 Image after segmentation (A-D) shows different images after segmentation.

The data set processing for this experiment first needs to convert the Json file into txt format. I wrote a Python script to implement this function. The script only reads the parts of the Json file that describe the damage type of the target and the image coordinates. Since Yolo uses Normalized coordinates, so the image coordinates in the xBD data set must be converted accordingly.

11.4.2.3 Analysis of results

This experiment is based on the Yolo model, 300 Epochs are trained, and evaluated.

11.4.2.3.1 Direct training on the original data set

By directly training the data set, we can find that the training results are very bad, with a precision rate of 0.54, a recall rate of 0.189, and mAP0.5 of 0.337. Almost impossible to complete the task. After checking the data set, it was found that there may be problems with the data set (Fig. 11.8).

11.4.2.3.2 Enhanced data set

After viewing and analyzing the data set, it was found that there are many problems in the data set, such as the target being blocked by clouds, the color of the picture being abnormal, the target being too small, etc. The picture is as follows (Fig. 11.9).

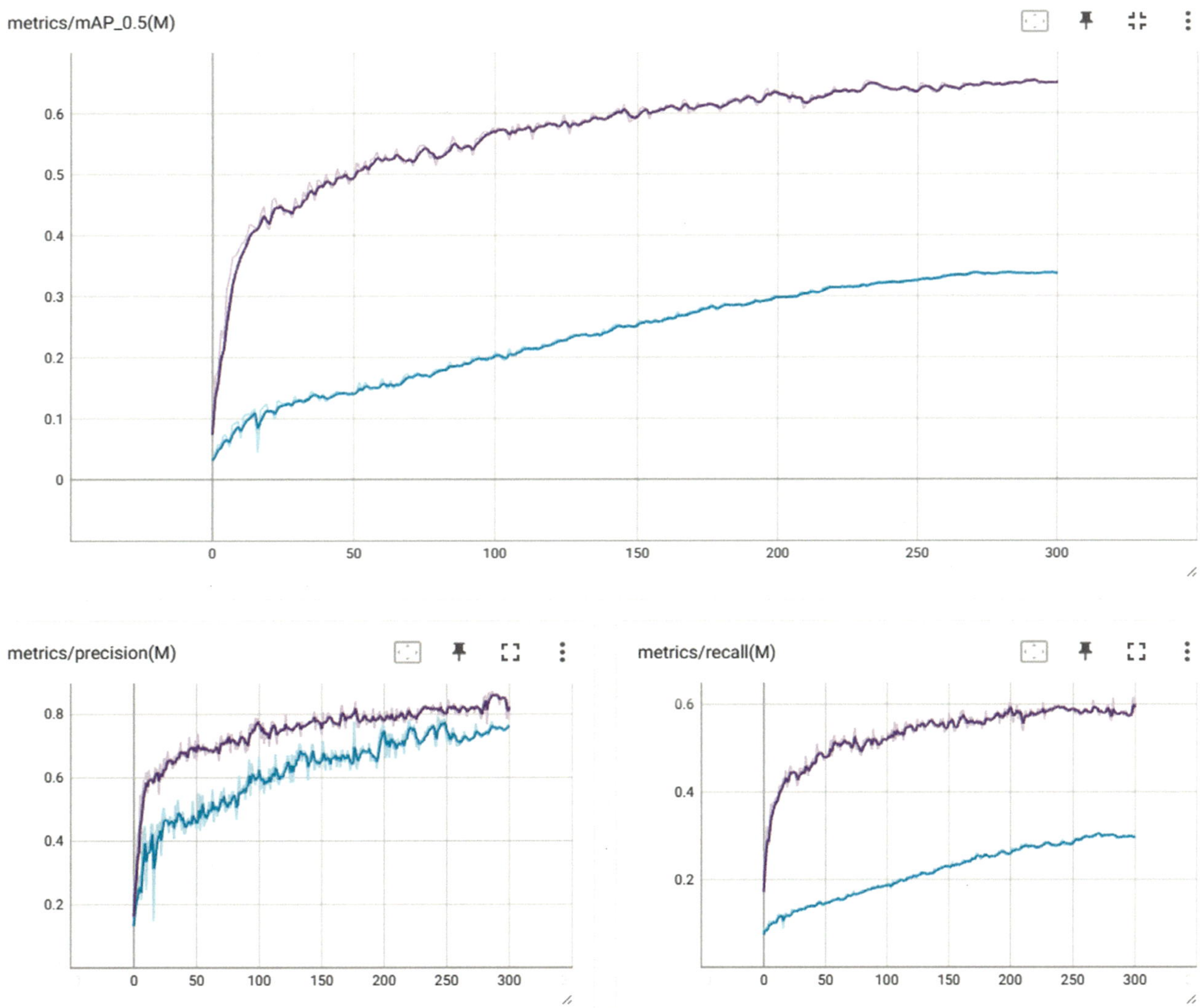

FIGURE 11.12 Freezing backbone network training results after segmentation (purple line).

In the picture above, the left half of the picture on the left is blocked by white clouds and casts a shadow on the ground, making it difficult to find the target in the picture. There is a large area of brown color in the middle picture. The reason is unknown, but if it is input into the network, the learning effect will be reduced a lot. Although there are no occlusions or color abnormalities in the picture on the right, a large area of the image is occupied by the ocean, and the target is too small for the human eye to distinguish. The learning effect of inputting it into the network will not be very good. Training is performed after removing the problematic images. The results are as follows (Fig. 11.10).

It can be found from the figure that mAP0.5 has increased to 0.338, but the recall rate is still low and the training effect is still unsatisfactory. A low recall rate means that the network failed to recognize the target, and the next step should be to improve the recognition rate.

11.4.2.3.3 Training after cutting pictures

After research, it was found that for Yolo, due to multiple downsampling of the image, if the width or height of the target in the original image is less than eight pixels, it is difficult for the network to learn the feature information of the target. The image input to YOLO in this experiment has a resolution of 1024×1024, and Yolo will resize it to a 640×640 image, making the target smaller and more difficult to learn. To solve this problem, first of all, the size of the target must remain unchanged when inputting into the network, so I divided the 1024×1024 picture into four 640×640 pictures, so that each picture of the target partially overlaps when inputting into the network (Fig. 11.11),

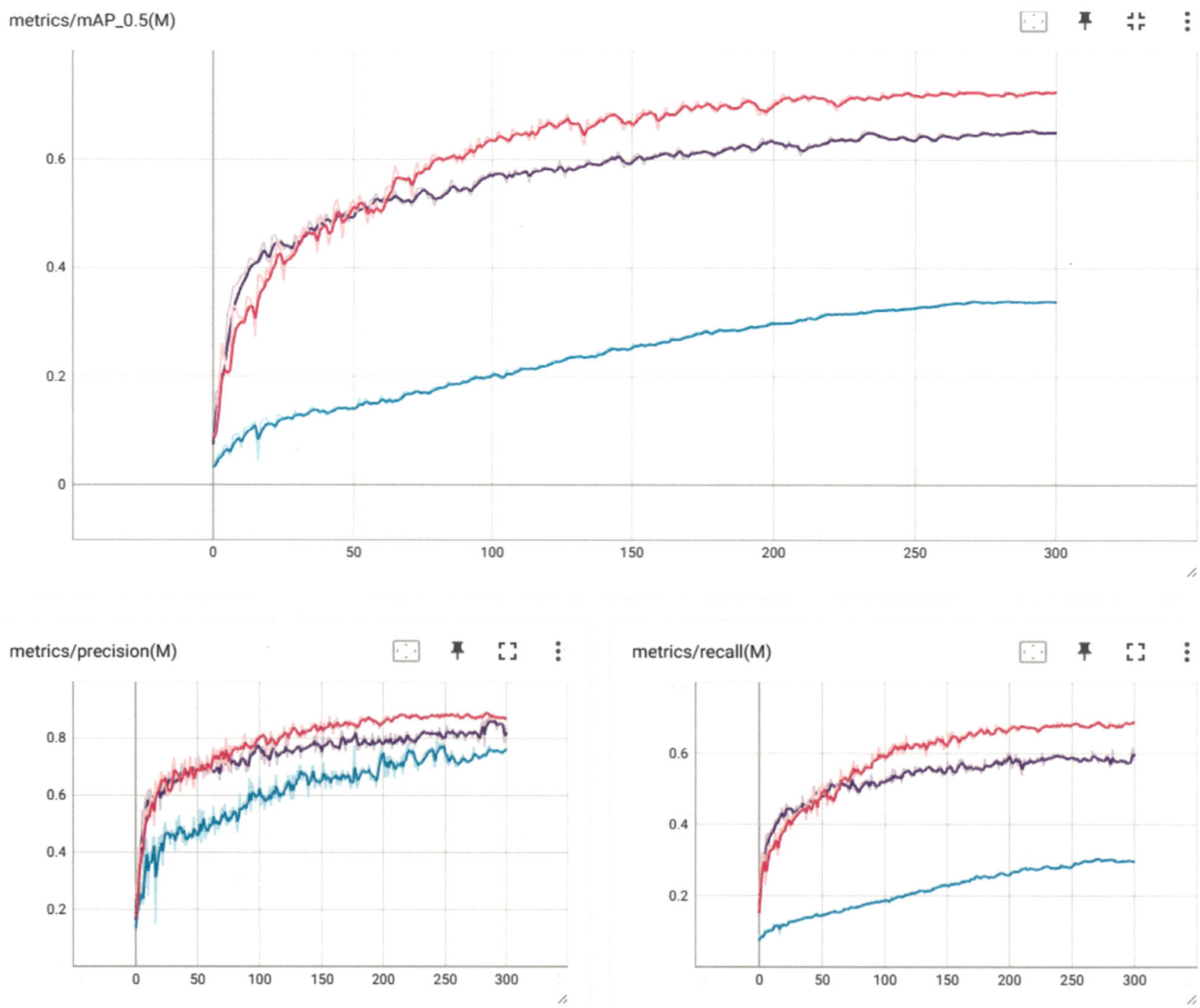

FIGURE 11.13 Training results using pretrained weights after segmentation (red line).

FIGURE 11.14 Original image and predicted image.

since 640 × 2 is larger than 1024, the images will overlap after segmentation. Rotating the segmented images can also enhance the performance of the network.

The images processed above are trained and the backbone network weights are frozen for transfer learning. The results are as follows (Fig. 11.12).

It can be seen that the recall rate has been greatly improved, from 0.296 to 0.592, and mAP0.5 has also increased from 0.338 to 0.65. This shows that at the resolution of the original data set, the network has been down sampled many times and cannot extract targets that are too small, and the rotated image also enhances the performance of the network. If the backbone network is not frozen and the pretrained weights are completely used for transfer learning, the results are as follows (Fig. 11.13).

It can be seen that during training, in the early stages of training, the rising speed of various indicators trained using pretraining weights is slower than that of the frozen backbone network, but at the end of the final training, all indicators are better than those of the frozen backbone network. trained model. The recall rate increased from 0.592 to 0.687, and mAP0.5 also increased from 0.65 to 0.724, which can basically achieve the goal (Fig. 11.14). The prediction chart is as follows.

The left side of the picture above is the original picture, and the right side is the predicted picture. It can be seen that although the target to be identified is small, the model using pretrained weights for transfer learning can still better identify collapsed houses from the picture.

References

Alhatami, E., Bhatti, U. A., MengXing, H., & Feng, S. (2024). Advanced fuzzy denoising technique for agricultural remote sensing: Modified partition filter for suppressing impulsive. *IEEE Access*, *40*, 1229–1251.

Bhatti, U. A., Tang, H., Wu, G., Marjan, S., & Hussain, A. (2023). Deep learning with graph convolutional networks: An overview and latest applications in computational intelligence. *International Journal of Intelligent Systems, 2023.* Available from https://doi.org/10.1155/2023/8342104, https://www.hindawi.com/journals/ijis/.

Bhatti, M. A., Zeeshan, Z., Syam, M. S., Bhatti, U. A., Khan, A., Ghadi, Y. Y., & Afzal, T. (2024). Advanced plant disease segmentation in precision agriculture using optimal dimensionality reduction with fuzzy c-means clustering and deep learning. *IEEE Journal of Selected Topics in Applied Earth Observations and Remote Sensing*, *1*, 74–81. Available from https://doi.org/10.20096/j.xhxb.1008-9233.2023.01.013.

Bochkovskiy, A., Wang, C. Y., & Liao, H. Y. M. (2020). YOLOv4: Optimal speed and accuracy of object detection. *arXiv*. Available from https://doi.org/10.48550/arxiv.2004.10934, https://arxiv.org.

Etten. (2018). You only look twice: Rapid multi-scale object detection in satellite imagery.

He, B., & Bhatti, U. A. (2024). Smart cities and smart networks: AI applications in urban geography and industrial communication. *International Journal of High Speed Electronics and Systems*, 2440122.

He, K., Gkioxari, G., Dollar, P., Girshick, R. (2017). Proceedings of the IEEE international conference on computer vision. Institute of Electrical and Electronics Engineers Inc., United States. Available from https://doi.org/10.1109/ICCV.2017.322 2980–2988. http://ieeexplore.ieee.org/xpl/conhome.jsp?punumber = 1000149.

Li, X., Hou, B., Tang, H., Talpur, B. A., Zeeshan, Z., Bhatti, U. A., & Al Naimi, I. S. (2024). Abnormal crops image data acquisition strategy by exploiting edge intelligence and dynamic-static synergy in smart agriculture. *IEEE Journal of Selected Topics in Applied Earth Observations and Remote Sensing*, 1008–1019.

NASA. (2021). Using Earth observations and applied research for disaster risk reduction, recovery and resilience.

Nawaz, S. A., Li, J., Bhatti, U. A., Shoukat, M. U., & Ahmad, R. M. (2022). AI-based object detection latest trends in remote sensing, multimedia and agriculture applications. *Frontiers in Plant Science*, *13*. Available from https://doi.org/10.3389/fpls.2022.1041514, https://www.frontiersin.org/journals/plant-science.

Nizamani, A. H., Chen, Z., Nizamani, A. A., & Bhatti, U. A. (2023). Advance brain tumor segmentation using feature fusion methods with deep U-Net model with CNN for MRI data. *Journal of King Saud University – Computer and Information Sciences*, *35*(9). Available from https://doi.org/10.1016/j.jksuci.2023.101793, http://www.journals.elsevier.com/journal-of-king-saud-university-computer-and-information-sciences/.

Peng, M., Liu, Y., Qadri, I. A., Bhatti, U. A., Ahmed, B., Sarhan, N. M., & Awwad, E. M. (2024). Advanced image segmentation for precision agriculture using CNN-GAT fusion and fuzzy C-means clustering. *Computers and Electronics in Agriculture*, *226*, 109431.

Redmon, J., Farhadi, A. (2017). Proceedings – 30th IEEE conference on computer vision and pattern recognition, CVPR 2017, 10.1109/CVPR.2017.690, 6517–6525, Institute of Electrical and Electronics Engineers Inc. United States YOLO9000: Better, faster, stronger 2017.

Redmon, J., & Farhadi, A. (2018). YOLOv3: An incremental improvement. *arXiv*. Available from https://doi.org/10.48550/arxiv.1804.02767, https://arxiv.org.

Redmon, J., Divvala, S., Girshick, R., & Farhadi, A. (2016). You only look once: Unified, real-time object detection. *Proceedings of the IEEE computer society conference on computer vision and pattern recognition* (pp. 779–788). United States: IEEE Computer Society. Available from https://doi.org/10.1109/CVPR.2016.91.

Yu, Z., Li, D., Du, P., Luo, W., Kou, K. I., Bhatti, U. A. . . . & Yuan. L. (2024). Hypercomplex Signal Processing in digital twin of the ocean: Theory and application [Hypercomplex signal and image processing]. IEEE Signal Processing Magazine, 41(3), 33–48.

Zheng, J., Li, J., Liu, J., Huang, M., Chen, Y. W., & Bhatti, U. A. (2022). A novel convolution kernel-based robust watermarking scheme applied in medical image. *International Journal of Wireless and Mobile Computing*, *22*(3-4), 290–299. Available from https://doi.org/10.1504/IJWMC.2022.124820, http://www.inderscience.com/ijwmc.

Further reading

Bhatti M. A., Song, Z., Bhatti U. A. (2024). AIoT-driven multi-source sensor emission monitoring and forecasting using multi-source sensor integration with reduced noise series decomposition. *Journal of Cloud Computing*, *13*(1), 65.

Peng, M., Liu, Y., Khan, A., Ahmed, B., Ghadi, S. K. . . . & Ali. Y. A. (2024). Crop monitoring using remote sensing land use and land change data: Comparative analysis of deep learning methods using pre-trained CNN models. *Big Data Research*, *36*, 100448.

Chapter 12

Latest trends in land use and land cover monitoring using deep learning

Ahsan Ahmed Nizamani[1], Yonis Gulzar[2], Hao Tang[3], Uzair Aslam Bhatti[3] and Abdul Haseeb Nizamani[1]

[1]Central South University, Computer Science and Engineering, Changsha, P.R. China, [2]Department of Management Information Systems, College of Business Administration, King Faisal University, Al-Ahsa, Saudi Arabia, [3]School of Information and Communication Engineering, Hainan University, Haikou, P.R. China

12.1 Introduction

Land use and land cover (LULC) monitoring is an essential practice that entails tracking alterations in how land is utilized and the variations in vegetation or structures covering it. It holds paramount importance in environmental management, urban planning, and gauging the repercussions of human activities on ecosystems. Remote sensing (RS) technologies, such as satellites and drones, serve as indispensable tools in LULC monitoring. By furnishing high-resolution imagery, these technologies facilitate the identification of changes in land patterns over time, enabling informed decision-making for sustainable land use and environmental preservation. The data obtained through RS not only aids in recognizing alterations but also contributes to the broader understanding of ecological shifts and the intricate dynamics between human activities and the environment. The emergence of RS technology has created new technological opportunities for monitoring LULC, capitalizing on its wide coverage and large amount of data it can access. RS imagery has emerged as the main source of data for the investigation and tracking of LULC changes (Bhatti et al., 2022). Utilizing big data effectively required a variety of approaches for the processing and analysis of RS data (Hermosilla et al., 2022; Talukdar et al., 2020). In addition, data from RS is widely used in many fields, including LULC categorization (Dong et al., 2022), flood prevention (Puno et al., 2021), fire mitigation (Ricotta et al., 2018), forest conservation (Duran et al., 2021), emission reduction (Zhu, 1970), and disaster prediction (Vali et al., 2020). Recently, deep learning (DL) and machine learning (ML) techniques have had a major influence on the processing and interpretation of hyperspectral data (Alzubaidi et al., 2021) (ML techniques, such as random forest (RF) (Zhang et al., 2022) and support vector machine (SVM) (Zhu et al., 2017), are very effective in classifying hyperspectral images. However, DL techniques have shown exceptional performance in LULC classification, particularly when sufficient training data are available (Gislason et al., 2006; Sefrin et al., n.d.). DL presents a distinctive advantage in LULC research by leveraging discriminative spectral-spatial characteristics, significantly enhancing classification accuracy. This study delves into the current landscape, potential future trajectories, and critical issues within LULC research through the lens of DL models. By exploring both object-level and pixel-level LULC RS classification datasets, the study scrutinizes the application of traditional DL methods evident in prior research endeavors. A comprehensive analysis unfolds, shedding light on the inherent challenges linked to addressing the limited training sample predicament. The study critically examines the merits and demerits of popular DL approaches and benchmark datasets, providing a nuanced understanding of their applicability in LULC classification. The chapter extends beyond retrospective analysis, incorporating a forward-looking discussion on recent advancements in LULC classification and illuminating potential future opportunities. By navigating through the intricacies of DL methodologies in the context of LULC, this study contributes not only to the understanding of current challenges but also serves as a compass for researchers charting their course into unexplored territories within this dynamic field.

12.2 Literature review

DL has grown rapidly since released of ground-breaking neural network model AlexNet in 2012 and went on to win the ImageNet image recognition competition by Krizhevsky et al. (2012). DL was previously listed as one of the top ten

Deep Learning for Earth Observation and Climate Monitoring. DOI: https://doi.org/10.1016/B978-0-443-24712-5.00013-0

revolutionary technologies (Zhu et al., 2017). Significant advancements in DL were made in the years that followed, especially in the areas of speech recognition, picture recognition, and natural language processing. When it comes to applying DL technology to the categorization of LULC, its quick progress has created both new opportunities and obstacles (Zhang et al., 2022). As a result, many models that use DL and ML for LULC analysis have been developed. Based on the input data, these models may be divided into three categories: sequence approaches, spatial methods, and pixel-based methods (Gislason et al., 2006). Using matching spectral data, pixel-based techniques like RF (Keller et al., 2016), SVM (Melgani & Bruzzone, 2004), and self-organizing maps (SOM) (Riese et al., n.d.) categorize each pixel separately. Using the google earth engine platform, Pande (2022) presented an RF learning algorithm for mapping LULC in India and performing change detection mapping using SAGA GIS software. The SOM framework for unsupervised land cover type categorization of hyperspectral data was introduced by Riese (2020). Beyond a single pixel, spatial classification techniques take into account a two-dimensional (2D) spatial neighborhood. Typically, hierarchical learning of low-level features in the first layer and higher-level features in the last layer is achieved using 2D convolutional neural networks (CNNs) with filter layers. Aamir et al. (2022) and Zhang et al. (2019) used a joint DL model to simulate LULC changes and classifications in Southampton and Manchester, UK. The model included a multilayer perceptron within CNNs. To mimic urban growth in China's Pearl River Delta, He et al. (2018) created an integrated CNN model with a cellular automaton (CA). Recurrent neural networks (RNNs), long short-term memory (LSTM) networks, and 3DCNNs are examples of sequence approaches (Bhatti, Ming-Quan, et al., 2021). A deep Siamese convolutional multilayer RNN was proposed by Chen et al. (2020) for the purpose of detecting changes in consecutive high-resolution photographs. The ST-CA was presented by Geng et al. (2022). It uses a spatial allocation module that uses a patch-based CA to simulate future LULC change and a potential generation module that uses a 3DCNN to compute the development potential of each cell.

12.3 Datasets for land use and land cover monitoring

In the recent past, researchers from various countries have actively contributed to the creation and release of sample datasets specifically designed for LULC classification. These datasets exhibit a diverse range of characteristics, including different sizes, sensor types, time intervals, spatial resolutions, and spectral resolutions. This diversity allows for a comprehensive exploration of LULC classification techniques and methods. These datasets exhibit a diverse range of characteristics, including different sizes, sensor types, time intervals, spatial resolutions, and spectral resolutions. This diversity allows for a comprehensive exploration of LULC classification techniques and methods. The primary purpose of these released datasets is to provide foundational or basic data for researchers working in the field of LULC classification. These datasets act as building blocks for conducting related research, enabling scientists to develop and test various algorithms and models. The available public datasets in the LULC sector are categorized into two main types: pixel samples and object samples.

12.3.1 Object-samples land use and land cover dataset

Essentially classifying photos into different LULC classes based on their major content is the process of scene-based classification, also known as patch-based classification. Typically, scene-based classification methods begin by extracting many patches for model training from a larger picture. The LULC map is then produced by the model after it has classified the scene category for every sampled patch. Scene-based CNNs are useful, but they have several drawbacks when it comes to LULC categorization. Determining appropriate patch sizes can be difficult, especially when there are large differences in the sizes of the ground targets. Moreover, throughout the classification process, contextual information is ignored in the separate processing of sampling patches within the same overall picture. While pixel-based techniques are better at capturing minute details, scene-based techniques are frequently used to recognize huge objects. RS object identification datasets are similar to the image-block-based LULC sample dataset. For this kind of dataset, DL algorithms are usually based on CNN or RNN image classification models. The benefit of this technique is that labeling is made simple, but it has the disadvantage of not offering boundary information for individual items. A list of popular and significant image-block-level sample datasets, together with pertinent details about each dataset, may be found in Table 12.1 and Fig. 12.1 shows the example of object-level dataset.

12.3.2 Pixel-samples land use and land cover dataset

The goal of pixel-level categorization methods is to give each pixel in a picture a category name. One type of pixel-level categorization called "semantic segmentation" entails categorizing each pixel by using contextual information.

TABLE 12.1 Object-sample land use and land cover datasets.

Name	Scene classes	Dimensions	Resolution	Size	Source
UC merced dataset	21	0.3 m	3	256 × 256	Aerial images
WHU-RS19	19	0.5 m	3	600 × 600	Google earth
RSSCN7	7	0.25 ~ 2 m	3	400 × 400	Google earth
EuroSAT	10	10/20/60 m	13	64 × 64	Sentinel-2
NWPU-RESISC45	45	0.2 ~ 30 m	3	256 × 256	Google earth

FIGURE 12.1 NWPU-RESISC45 example sample data.

Recently, there has been a lot of interest in fully convolutional networks (FCNs) (Dai et al., 2016) because to their remarkable results in semantic segmentation tasks. In contrast to conventional classifiers that classify pixels according to particular spectral information, FCNs use many fully convolutional layers to extract high-order context characteristics that are inherent in the pixels, allowing for accurate annotations at the pixel level.

As shown in Fig. 12.2, pixel-based land use/cover sample sets are comparable to RS semantic segmentation datasets in that they involve the annotation of all pixels covering a single land item. This sample set type has the benefit of being able to establish accurate borders for land objects, but it has the disadvantage of having a somewhat big labeling burden. Table 12.2 illustrates common pixel-level sample sets, indicating that the majority are bound by spectral resolution, resulting in lower spatial resolutions. Each dataset typically consists of a single picture plus annotations for a specified research region, with only a few newer datasets, such as DeepGlobe (Demir et al., 2018), containing larger sample numbers and geographic resolutions. It's worth mentioning, however, that these datasets are mostly made up of conventional RGB or RGBNIR photos with lesser spectral resolutions.

The division into pixel and object samples allows researchers to approach LULC classification in a holistic manner. Pixel-level datasets are suitable for fine-grained analysis of individual pixels, while object-level datasets provide a more encompassing view of entire scenes or regions. Researchers can use these datasets to test and improve algorithms and models under various conditions, taking into account factors like different sensor characteristics, temporal changes over intervals, and variations in spatial and spectral resolutions.

FIGURE 12.2 DeepGlobe example of sample data.

TABLE 12.2 Pixel-level land use and land cover datasets.

Name	Scene classes	Dimensions	Resolution	Size	Source
University of Pavia Baumgardner et al. (2015)	9	103	1.3 m	610 × 610	ROSIS
Salinas Baumgardner et al. (2015)	16	224	3.7 m	521 × 127	AVIRIS
Indian Pines Baumgardner et al. (2015)	16	224	20 m	145 × 145	AVIRIS
DeepGlobe Demir et al. (2018)	7	3	0.5 m	2448 × 2448	Mix

12.3.3 Deep learning models for land use and land cover

DL is a specialized branch of ML characterized by intricate layers of consecutive learning, enhancing the computational significance for addressing complex problems (Ahmad et al., 2019). As the neural network goes deeper, it generates more precise results, ensuring satisfactory performance. DL also plays a vital role in managing and processing extensive datasets (Uba, 2016). Notably, it has demonstrated significant progress in addressing various challenges, including LULC classification using hyperspectral images. However, DL faces challenges, particularly in terms of prolonged processing time and computational complexity.

The availability of training samples presents a notable challenge in LULC DL classification, necessitating a substantial quantity of samples for effective learning (Zhang et al., 2020). Despite previous efforts to overcome the limitation of training samples, the application of DL in processing hyperspectral images encounters ongoing challenges, such as the high dimensionality of hyperspectral imaging (HSI) (Tang et al., 2023), band noise, and the pronounced similarity between spectral bands.

The design of employing DL models for LULC research follows a structured framework encompassing four crucial steps (Zhu et al., 2021).

Data preprocessing: This foundational step involves the application of techniques to image data, including denoising, fusion, dimension reduction, and resampling. By refining and enhancing the quality of the input data, the subsequent stages of model training and evaluation benefit from improved accuracy and reliability.

Model training: With the preprocessed image data at hand, DL models undergo training. This phase is pivotal, as the models learn to discern intricate patterns and features within the data, enabling them to make accurate classifications of land use and cover.

Validation and evaluation: To ensure the functionality and reliability of the trained models, validation and evaluation processes are implemented. These steps rigorously assess the accuracy and effectiveness of the models, providing insights into their performance and suitability for LULC mapping.

LULC mapping: The final step involves leveraging the trained DL models to forecast maps of land use and cover. These predictive maps serve as invaluable tools for land resource managers and urban planners, assisting them in making informed decisions for sustainable resource management and urban development. Fig. 12.3 illustrates the

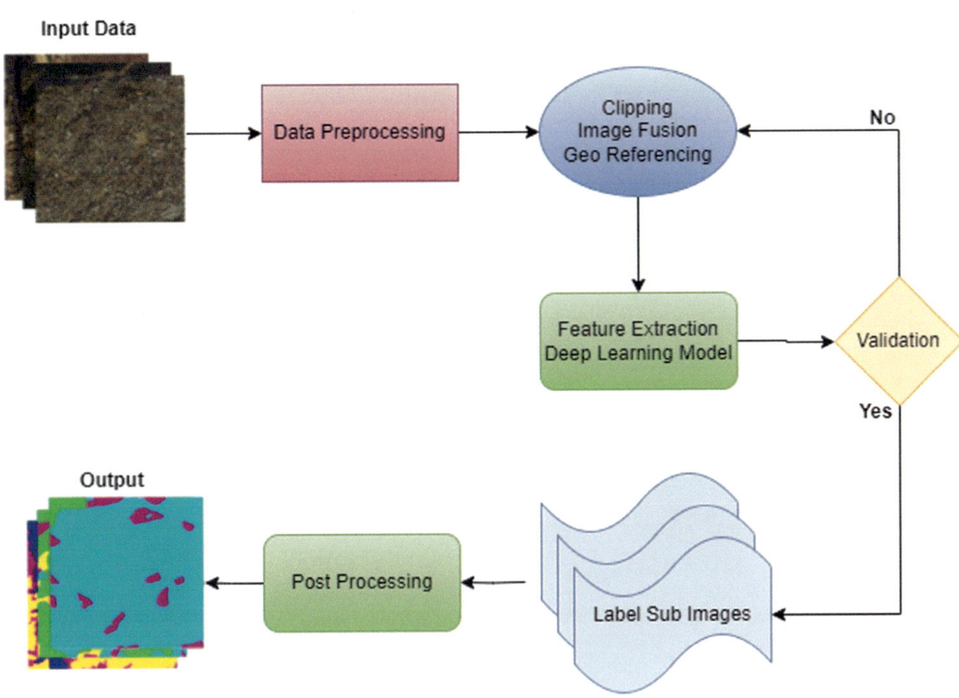

FIGURE 12.3 Deep learning model framework for land use and land cover monitoring.

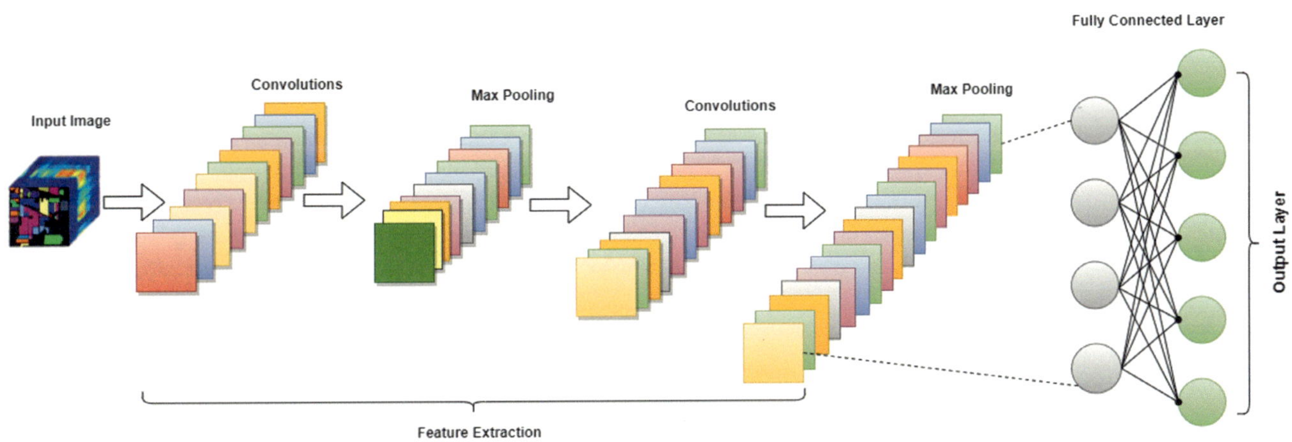

FIGURE 12.4 Convolution neural networks model.

interconnected nature of these steps, showcasing the seamless flow from data preprocessing to the generation of actionable LULC maps, facilitating informed decision-making in land management and urban planning endeavors.

12.3.4 Convolution neural networks model

CNNs are a potent DL technique that have shown promise in tasks including semantic segmentation, object identification, and picture categorization. CNNs are extensively utilized in computer vision and image classification applications (Zhao et al., 2017), and their strong feature extraction performance has drawn notice.

Numerous research have investigated the application of CNNs in LULC classification, with notable successes as a result of their efficacy in feature extraction (Bhatti, Tang et al., 2023; Verma & Jana, 2019). Fig. 12.4 shows how CNNs are used for LULC process. Feature representation or learning was the main emphasis of early DL-based LULC research, and simpler classifiers were frequently used for the final classification (Tuia et al., 2015). By using a CNN and a 10-band Sentinel-2B dataset, Verma and Jana (2019) provided a novel method that produced classification maps with a 10 m spatial resolution. In order to improve the accuracy of LULC mapping, Marcos et al. (2018) created a

CNN architecture that included rotation equivariance and was used to sub-decimeter land cover semantic labeling benchmarks. Zhu et al. (2018) added generative adversarial networks (GANs) as a regularization approach to the CNN model for hyperspectral RS image categorization in order to solve overfitting during model training. All of these methods demonstrate how CNNs have had a significant and positive influence on the advancement of LULC research.

A convolution equation can be described as

$$C_{a,b}(x, y) = \sum_i \sum_d h_{ij}(s, t) e_d \tag{12.1}$$

$$C_{i,j} = [c_{a,b}(1, 1), ., c_{a,b}(x, y), ., c_{a,b}(X, Y)] \tag{12.2}$$

Following feature extraction, a pooling or down sampling process is applied to provide a combined set of features that are resilient to small distortions and translational shifts.

$$Q_{i,j} = g_p(C_{i,j}) \tag{12.3}$$

Here, g_p stands for the pooling operation, and $Q_{i,j}$ is the pooling feature-map of the i-th layer for the j-th input feature-map. CNNs use a variety of pooling techniques, including as L2, overlapping, max, average, and spatial pyramid pooling. As a decision function for a convolved feature-map, activation functions are essential for speeding up learning.

$$x_{i,j} = g_a\left(C_{i,j}\right) \tag{12.4}$$

In the equation mentioned above, g_a represents the activation function, $C_{i,j}$ denotes the convolution output, and $x_{i,j}$ signifies the transformed output.

12.3.5 Recurrent neural network

The capacity of RNNs to capture temporal relationships and sequential patterns in data makes them a desirable tool for LULC monitoring. RNNs are not like conventional feedforward neural networks in that they can remember past inputs because of their directed cycle connections (Nawaz et al., 2022). In an effort to efficiently investigate spatial-spectral discriminating characteristics Lu et al. (2022) presents a unique technique that combines hyperspectral and multispectral pictures using an RNN model. In Singh and Kasana (2022), an additional model for evaluating soil attributes is put forward. After dimension reduction, this method incorporates both local and global information to create hybrid features using principal component analysis and locality preserving projections. After that, LSTM Networks are used to analyze the hybrid information to create an effective prediction model. Additionally, a paradigm for hyperspectral image categorization is presented in Liang et al. (2021). This model uses a hybrid technique that combines a bi-directional recurrent neural network for continuous spectrum information with a multi-scale densely connected convolutional network for extracting complex multi-scale spatial features. A visual illustration of the RNN model is shown in Fig. 12.5.

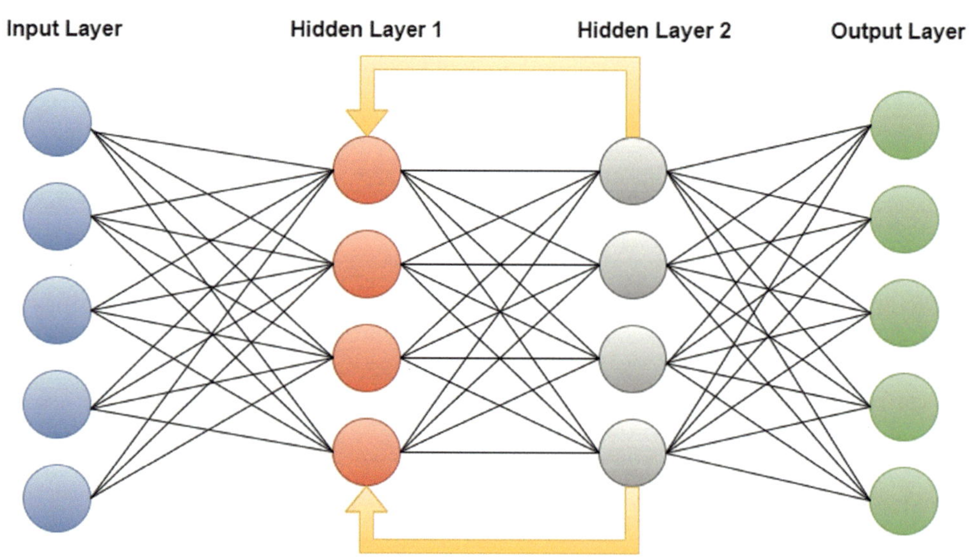

FIGURE 12.5 Recurrent neural networks model.

12.3.6 Fully convolution networks

FCNs-based semantic segmentation stands out as a critical DL approach in LULC applications. Using this method, each pixel in a picture is given a unique land cover label (Long et al., 2015). The encoder-decoder architecture used by the FCNs structure provides an end-to-end solution that is especially useful for the processing of RS data. In Fig. 12.6, the process of using FCNs for LULC is shown. The encoder uses a convolution technique, which can be stated mathematically as follows, to extract features.

$$(L * C)(i,j) = \sum_{m}\sum_{n} L(i-m, j-n)K(m,n) \quad (12.5)$$

The input is denoted by L, the kernel by C, and the spatial dimensions are indicated by (i, j) in the equation above. Zero-padding is used to maintain the input's height and breadth. The pooling function, which can be symbolically stated as follows, is applied after the convolution process to reduce the dimensionality of the input picture.

$$Q(i,j) = \max_{(x,y)\in O(i,j)} I(x,y) \quad (12.6)$$

In this case, Q stands for the pooled output, $Q(i, j)$ for the pixel's neighborhood, and $I(x, y)$ for the pixel's intensity at (x, y). Through this method, less important characteristics are removed while the important ones are kept.

12.3.7 Generative adversarial network

GANs are a family of DL models that are specifically engineered for data synthesis. They find usage in RS tasks such as super-resolution and LULC categorization. Fig. 12.7 illustrates the GANs method. Singh and Bruzzone (2022) introduced a data augmentation method that trains deep CNNs via spectral index GANs. This technique uses the spectrum characteristics of multispectral pictures to get around the issue of insufficient LULC sample data. Another addition is He et al. (n.d.)'s end-to-end GAN with a conditional random field for semantic segmentation of RS images.

12.3.8 Hybrid deep learning models

Hybrid DL models for RS combine traditional RS techniques with DL approaches to enhance the analysis of satellite or aerial imagery. These models leverage the strengths of both methods to overcome limitations and improve the accuracy

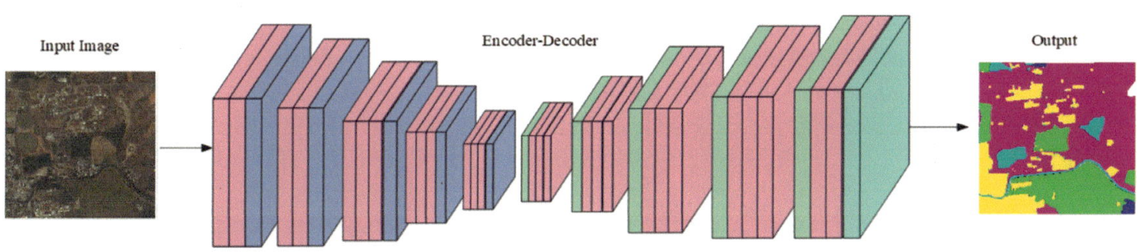

FIGURE 12.6 Fully convolutional neural networks model.

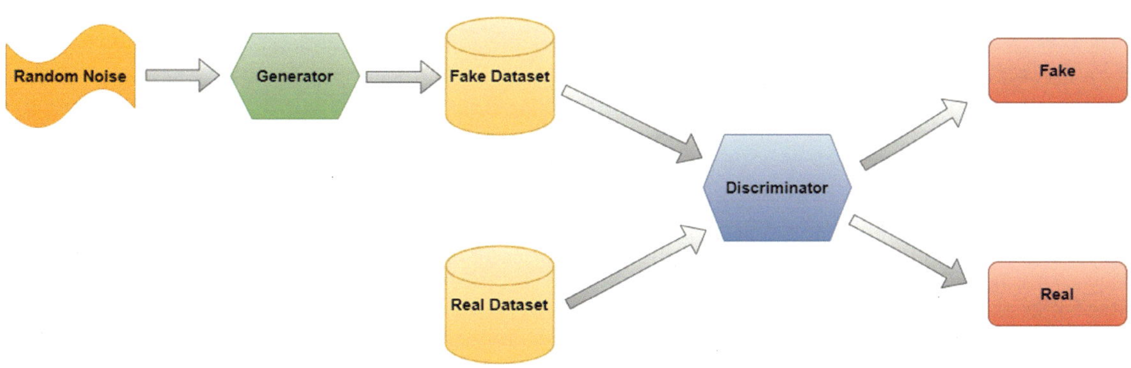

FIGURE 12.7 Generative adversarial networks model.

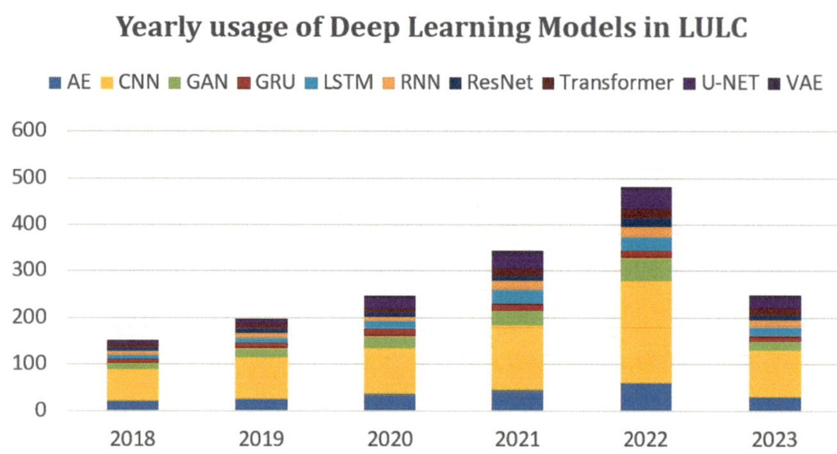

FIGURE 12.8 Distribution of deep learning models used in land use and land cover within 5 years.

of tasks such as classification, object detection, and semantic segmentation in RS applications (Bhatti, Yuan, et al., 2021). The advantage of hybrid models is that they can benefit from the interpretability of traditional RS techniques while harnessing the powerful feature learning capabilities of DL (Nizamani et al., 2021). This can be especially valuable in scenarios where labeled training data is limited, as DL models often require large amounts of annotated data to perform well. Researchers and practitioners working in RS applications often customize hybrid models based on the specific task at hand, the characteristics of the data, and the available computational resources. A multi-feature fusion of 3D-CNN and graph attention network MFFCG was proposed by (Bhatti, Huang et al., 2023). The two components of the algorithm are the 3D-CNN, which generates accurate classification for 3D HSI cube data, and the GAT-based encoder and decoder modules, which aid in enhancing the 3D-CNN's classification accuracy. Reducing dimensionalities in HSIs is especially crucial for extracting deeper characteristics that enhance classification in both the spectral and spatial domains. Local similarity projection (LSP) Gabor filtering, a spatial-spectral HSI classification technique, was proposed (Uzair et al., 2021). It combines a 2-D Gabor filtering approach with a reduced dimensional CNN based on LSP (Bhatti et al., 2022).

Through the examination of data from the literature from 2018 to 2023, we were able to discern trends in the application of various DL models for LULC classification tasks. The frequency of model usage during these years is represented by the appropriate bar chart in Fig. 12.8. Over the course of the specified period, CNNs became the most often used model. CNN's competence in image classification tasks, a crucial component of LULC analysis, might be linked to this tendency. A CNN variation designed for semantic segmentation tasks, called FCNs, also showed notable representation. Their suitability for LULC classification stems from their capacity to handle spatially ordered picture data. Less often, RNNs and their derivatives, LSTM and GRU, were employed. Although their use in conjunction with CNNs for spatiotemporal data was seen, this is probably because they are more suited for time-series data than spatial data. In conclusion, during the past 5 years, the literature data shows an expanding and varied application of DL models in LULC tasks.

12.4 Land use and land cover monitoring dataset evaluation parameters

Four common assessment criteria are used to test LULC categorization using hyperspectral images. These standards comprise individual class accuracies (ICA), average accuracy (AA), kappa coefficient (KC), and overall accuracy (OA).

The KC provides a quantitative estimate of the number of correctly identified pixels by illustrating the relationship between the chance agreement and the actual agreement (Zhu, 1970). The total number of pixels in column j is represented by Y_j. Furthermore, KC values fall into the range of [0, 1], where a value of "0" indicates a chance agreement in the connection between the categorized picture and the reference data, and a value of "1" indicates a true agreement. Consequently, the accuracy of categorization increases as the KC value approaches 1.

$$KC = \frac{M\sum_{ij=1}^{n} Cij - \sum_{ij=1}^{n} X_i.Y_J}{M^2 - \sum_{ij=1}^{n} X_i.Y_J} \tag{12.7}$$

Equation (Vali et al., 2020) provides the average of the proportion of properly categorized pixels for each class, which is known as AA. The total number of classes is denoted by n.

$$AA = \frac{\sum_{1}^{n} ICA_n}{n} \tag{12.8}$$

The percentage of pixels that are correctly recognized for each class is known as ICA, and it is represented by equation (Alzubaidi et al., 2021), where X_i is the total number of pixels in row i and C_{ij} is the total number of correctly categorized pixels in row i and column j.

$$ICA = \frac{C_{ij}}{X_i} \tag{12.9}$$

The proportion of all pixels that are correctly identified is known as OA, and it is represented by equation (Zhu et al., 2017). P is the total number of pixels in this case.

$$OA = \frac{\sum C_{ij}}{P} \tag{12.10}$$

12.5 Land use and land cover challenges and future directions

12.5.1 Challenges for land use and land cover monitoring

A significant obstacle in LULC classification is the limited quantity of labeled training samples. Moreover, getting the labeled training samples has several drawbacks, such as being expensive, time-consuming, and challenging. The pretrained model decreases processing time and enhances HSI classification ability. However, several studies have not included pretrained models in the classification of HSI data. Furthermore, because HSI include hundreds of narrow, redundant, and irrelevant spectral bands, processing HSI involves a large amount of computing time and complexity. HSI systems have several challenges, two of which are the high cost of the devices required to acquire the high spectral resolution pictures and the lack of requirements for hardware and software. Furthermore, the hyperspectral pictures contain a range of useful properties, such as discriminative spectral-spatial features. The region of interest will be effectively recognized and the HSI classification will be significantly improved with better results by merging the spectral-spatial information. However, spectral-spatial properties are not extensively utilized for LULC classification in many previous studies. Moreover, the extraction of the deep discriminative spectral-spatial characteristics is a tedious operation.

The low training sample problem is often addressed using one of four conventional approaches: weakly supervised techniques, data augmentation, active learning, or transfer learning.

Transfer learning (TL) stands out as a powerful technique to address the challenge of inadequate training samples in LULC classification. This approach proves essential for enhancing processing time and boosting classification accuracy by transferring knowledge and skills from one previously learned task to another. Leveraging preexisting knowledge helps overcome the limitations of insufficient training samples and contributes to the robustness of the classification model.

Active learning (AL) introduces an iterative procedure designed to reduce the number of required training samples by selecting the most relevant instances from a subset of unlabeled samples. By utilizing the structure of unlabeled instances and employing uncertainty measures, AL identifies informative examples with minimal redundancy. This not only accelerates the learning process but also enhances training effectiveness using fewer training samples. AL employs a query function to select the most significant samples from the unlabeled set, which are then labeled by a supervisor and added to the labeled sample set. The supervised classifier is subsequently retrained using these carefully selected samples, refining the model's performance.

Data augmentation emerges as a crucial solution to the restricted training sample problem in hyperspectral image classification within DL frameworks. This technique enhances DL performance by generating additional data through transformations like flip, rotation, and translation. Data augmentation proves instrumental in mitigating the challenges posed by limited training samples, successfully capturing the unique characteristics of each scene and enriching the original dataset with diversity. This augmentation process contributes to a more comprehensive understanding of the spectral-spatial features, resulting in improved classification accuracy.

In the realm of **weakly supervised learning**, DL exhibits remarkable effectiveness, particularly in scenarios where an extensive labeled training dataset is available. However, acquiring sufficient labeled training examples remains a challenge due to the time-consuming and expensive nature of the labeling process. Weakly supervised learning offers a solution to this predicament by leveraging partially labeled images. It eliminates the need for extensive manual picture annotation, reducing both time and costs associated with the labeling process. This approach proves particularly beneficial in situations where overfitting is a significant concern when training DL models with a limited amount of labeled training data.

These techniques, transfer learning, active learning, data augmentation, and weakly supervised learning, constitute a comprehensive arsenal of strategies to overcome challenges related to limited training samples in LULC classification. The integration of these methodologies not only enhances classification accuracy but also addresses practical constraints associated with resource-intensive processes, making DL models more robust and applicable in real-world scenarios.

12.5.2 Future directions

To enhance the reliability and effectiveness of LULC classification models, several critical factors need to be considered:

Integration of spectral-spatial features: Traditional LULC classification relied primarily on spectral information. However, incorporating both spectral and spatial features can significantly improve performance and accuracy. Technologies like transformers, which are adept at handling long-range dependencies, are particularly effective in extracting these combined features from hyperspectral images. They excel in processing sequential data and capturing global features, making them a more robust tool for hyperspectral image classification.

Advancements in technology: The rapid development of technology necessitates advanced hardware and software tools for HSI. With these advancements, it's now more feasible to efficiently extract informative features from hyperspectral images. The adoption of these technologies can greatly enhance the capability to process and analyze LULC data.

DL methods in the context of limited training samples: DL methods play a crucial role in LULC classification, especially when ample training data is available. However, one of the main challenges in this field is the limited availability of training samples. Addressing this issue requires dedicated efforts to develop methodologies or algorithms that can perform effectively even with fewer training data points.

Extensive use of DL techniques in LULC classification: DL approaches are highly effective in analyzing and extracting discriminative features from high-resolution hyperspectral images. These methods are capable of fully utilizing spectral-spatial data during the classification process, distinguishing them from traditional ML algorithms. The extensive use of DL techniques can lead to more accurate and detailed LULC classification, leveraging their ability to process and interpret complex patterns in hyperspectral data.

By improving LULC classification models hinges on leveraging advanced DL techniques that can utilize both spectral and spatial data, adopting the latest technological tools for HSI, and finding solutions to overcome the challenge of limited training samples. These factors collectively contribute to the development of more accurate, efficient, and robust LULC classification systems.

12.6 Conclusion

This chapter delves into the realm of knowledge graph analytics to scrutinize academic publications from 2018 to 2023, focusing on the application of DL technology in LULC research. Through this analytical lens, we have explored various dimensions of LULC monitoring and RS, assessing the methodologies and trends that have shaped this field over the past few years. We begin by examining the image sensing datasets pivotal for LULC monitoring. These datasets are the backbone of LULC research, providing the raw material for analysis and interpretation. Our discussion extends to the assessment processes involved in evaluating the effectiveness and accuracy of these datasets, offering insights into their strengths and limitations. The core of this chapter lies in our detailed exploration of the primary DL approaches and the emergence of hybrid DL models in LULC monitoring. These advanced methodologies have revolutionized the way LULC data is processed and interpreted, offering more nuanced and accurate insights than ever before. We delve into the specifics of these models, examining how they work and why they are effective in the context of LULC research. As the chapter progresses, we address significant challenges in LULC classification. These challenges include the perennial issue of limited training examples, which hampers the effectiveness of DL models. We discuss potential solutions to this issue and highlight the need for innovative approaches that can work with limited data while still providing reliable results. The chapter culminates by looking toward the future, outlining the practical applications of LULC classification and motivating future research directions. We underscore the importance of this field, not just in academic circles but also in its practical implications for environmental monitoring, urban planning, and policy-making. This work contributes to the field of LULC in three significant ways. First, it aims to enrich the understanding of the evolution, current state, and future prospects of DL technology in LULC research, guiding both academics and policymakers. Second, our analysis of high-yield publishing themes and sources offers strategic insights for scholars seeking publication avenues. Lastly, the utilization of word clouds and thematic analysis provides an innovative and visual means of highlighting the central questions and trends within the LULC domain, equipping researchers with crucial insights into the field's trajectory and focal points.

References

Aamir, M., Bhatti, M. A., Bazai, S. U., Marjan, S., Mirza, A. M., Wahid, A., Hasnain, A., & Bhatti, U. A. (2022). Predicting the environmental change of carbon emission patterns in South Asia: A deep learning approach using BiLSTM. *Atmosphere*, *13*(12). Available from https://doi.org/10.3390/atmos13122011, http://www.mdpi.com/journal/atmosphere.

Ahmad, J., Farman, H., & Jan, Z. (2019). *Deep learning methods and applications springerbriefs in computer science* (pp. 31–42). Springer. Available from http://www.springer.com/series/10028, 10.1007/978-981-13-3459-7_3.

Alzubaidi, L., Zhang, J., Humaidi, A. J., Al-Dujaili, A., Duan, Y., Al-Shamma, O., Santamaría, J., Fadhel, M. A., Al-Amidie, M., & Farhan, L. (2021). Review of deep learning: Concepts, CNN architectures, challenges, applications, future directions. *Journal of Big Data*, *8*(1). Available from https://doi.org/10.1186/s40537-021-00444-8, http://journalofbigdata.springeropen.com/.

M.F. Baumgardner, L.L. Biehl, D.A. Landgrebe, 220 band aviris hyperspectral image data set: 12 June 1992 indian pine test site 3. PURR – Purdue University Research Repository, 10 (2015), 2015.

Bhatti, U., Huang, M., Neira-Molina, H., Marjan, S., Baryalai, M., Tang, H., Wu, G., & Ullah Bazai, S. (2023). MFFCG – Multi feature fusion for hyperspectral image classification using graph attention network. *Expert Systems With Applications*, *229*.

Bhatti, U. A., Ming-Quan, Z., Huo, Q., Ali, S., Hussain, A., Yuhuan, Y., Yu, Z., Yuan, L., & Ali Nawaz, S. (2021). Advanced color edge detection using clifford algebra in satellite images. *IEEE Photonics Journal*, *13*(2), 1–20. Available from https://doi.org/10.1109/jphot.2021.3059703.

Bhatti, U. A., Tang, H., Wu, G., Marjan, S., Hussain, A., & Sarker, S. K. (2023). Deep learning with graph convolutional networks: An overview and latest applications in computational intelligence. *International Journal of Intelligent Systems*, *2023*, 1–28. Available from https://doi.org/10.1155/2023/8342104.

Bhatti, U. A., Yuan, L., Yu, Z., Nawaz, S. A., Mehmood, A., Bhatti, M. A., Zeeshan, Q. ul A., Nizamani, M. M., Zeeshan., & Xiao, S. (2021). Predictive data modeling using sp-kNN for Risk factor evaluation in urban demographical healthcare data. *Journal of Medical Imaging and Health Informatics*, *11*(1), 7–14. Available from https://doi.org/10.1166/jmihi.2021.3313.

Bhatti, U. A., Yu, Z., Yuan, L., Nawaz, S. A., Aamir, M., & Bhatti, M. A. (2022). A robust remote sensing image watermarking algorithm based on region-specific SURF. *Lecture Notes in Networks and Systems*, *350*, 75–85. Available from https://doi.org/10.1007/978-981-16-7618-5_7, http://springer.com/series/15179.

Chen, H., Wu, C., Du, B., Zhang, L., & Wang, L. (2020). Change detection in multisource VHR images via deep siamese convolutional multiple-layers recurrent neural network. *IEEE Transactions on Geoscience and Remote Sensing*, *58*(4), 2848–2864. Available from https://doi.org/10.1109/tgrs.2019.2956756.

Dai, J., Li, Y., He, K., & Sun, J. (2016). Neural information processing systems foundation United States R-FCN: Object detection via region-based fully convolutional networks. *Advances in Neural Information Processing Systems*, 379–387.

Demir, I., Koperski, K., Lindenbaum, D., Pang, G., Huang, J., Basu, S., Hughes, F., Tuia, D., & Raska, R. (2018). DeepGlobe 2018: A challenge to parse the earth through satellite images, *IEEE computer society conference on computer vision and pattern recognition workshops* (pp. 172–181). http://ieeexplore.ieee.org/xpl/conferences.jsp, doi: 10.1109/CVPRW.2018.00031.

Dong, S., Guo, H., Chen, Z., Pan, Y., & Gao, B. (2022). Spatial stratification method for the sampling design of LULC classification accuracy assessment: A case study in Beijing, China. *Remote Sensing*, *14*(4), 865. Available from https://doi.org/10.3390/rs14040865.

Duran, E., Mas, J.-F., & Velázquez, A. (2021). *Land use/cover change in community-based forest management regions and protected areas in Mexico, the community forests of México* (2021, pp. 215–238). University of Texas Press.

Geng, J., Shen, S., Cheng, C., & Dai, K. (2022). A hybrid spatiotemporal convolution-based cellular automata model (ST-CA) for landuse/cover change simulation. *International Journal of Applied Earth Observation and Geoinformation*, *2022*.

Gislason, P. O., Benediktsson, J. A., & Sveinsson, J. R. (2006). Random forests for land cover classification. *Pattern Recognition Letters*, *27*(4), 294–300. Available from https://doi.org/10.1016/j.patrec.2005.08.011.

Hermosilla, T., Wulder, M. A., White, J. C., & Coops, N. C. (2022). Land cover classification in an era of big and open data: Optimizing localized implementation and training data selection to improve mapping outcomes. *Remote Sensing of Environment*, *268*. Available from https://doi.org/10.1016/j.rse.2021.112780, http://www.elsevier.com/inca/publications/store/5/0/5/7/3/3.

He, C., Fang, P., Zhang, Z., Xiong, D., & Liao, M. (n.d.). An end-to-end conditional random fields and skip-connected generative adversarial segmentation network for remote sensing images.

He, J., Li, X., Yao, Y., Hong, Y., & Jinbao, Z. (2018). Mining transition rules of cellular automata for simulating urban expansion by using the deep learning techniques. *International Journal of Geographical Information Science*, *32*(10), 2076–2097. Available from https://doi.org/10.1080/13658816.2018.1480783, http://www.tandf.co.uk/journals/titles/13658816.asp.

Keller, S., Braun, A. C., Hinz, S., & Weinmann, M. (2016). Investigation of the impact of dimensionality reduction and feature selection on the classification of hyperspectral EnMAP data. *Workshop on Hyperspectral Image and Signal Processing, Evolution in Remote Sensing*. Available from https://doi.org/10.1109/WHISPERS.2016.8071759, http://ieeexplore.ieee.org/xpl/conferences.jsp.

Liang, L., Li, J., & Zhang, S. (2021). Hyperspectral image classification method based on multi-scale densenet and Bi-RNN joint network. *IOP conference series: Earth and environmental science* (783 (1), 012087), doi: 10.1088/1755-1315/783/1/012087.

Long, J., Shelhamer, E., & Darrell, T. (2015). Fully convolutional networks for semantic segmentation. *Proceedings of the IEEE computer society conference on computer vision and pattern recognition* (pp. 431--440). IEEE Computer Society United States. Available from https://doi.org/10.1109/CVPR.2015.7298965.

Krizhevsky, A., Sutskever, I., & Hinton, G. E. (2012). ImageNet Classification with Deep Convolutional Neural Networks. *Advances in Neural Information Processing Systems*.

Lu, R., Chen, B., Sun, J., Chen, W., Wang, P., Chen, Y., Liu, H., & Varshney, P. K. (2022). Heterogeneity-aware recurrent neural network for hyperspectral and multispectral image fusion. *IEEE Journal on Selected Topics in Signal Processing*, *16*(4), 649–665. Available from https://doi.org/10.1109/JSTSP.2022.3180896, http://www.ieee.org/products/onlinepubs/news/0806_01.html.

Marcos, D., Volpi, M., Kellenberger, B., & Tuia, D. (2018). Land cover mapping at very high resolution with rotation equivariant CNNs: Towards small yet accurate models. *ISPRS Journal of Photogrammetry and Remote Sensing*, *145*, 96–107. Available from https://doi.org/10.1016/j.isprsjprs.2018.01.021.

Melgani, F., & Bruzzone, L. (2004). Classification of hyperspectral remote sensing images with support vector machines. *IEEE Transactions on Geoscience and Remote Sensing*, *42*(8), 1778–1790. Available from https://doi.org/10.1109/TGRS.2004.831865.

Nawaz, S. A., Li, J., Bhatti, U. A., Shoukat, M. U., & Ahmad, R. M. (2022). AI-based object detection latest trends in remote sensing, multimedia and agriculture applications. *Frontiers in Plant Science*, *13*. Available from https://doi.org/10.3389/fpls.2022.1041514, https://www.frontiersin.org/journals/plant-science.

Nizamani, M. M., Bhatti, U. A., Cheng, X. L., Nizamani, F. G., Rind, R. A., Khokhar, A. A., Ma, C. W., Zeeshan, Z., Bahadur, S., & Yang, D. M. (2021). The connections between above-ground biomass and plant diversity of roadside trees, density and diversity on different types of roads in karachi. *Polish Journal of Environmental Studies*, *30*(3), 2691–2700. Available from https://doi.org/10.15244/pjoes/128581, http://www.pjoes.com/pdf-128581-64007?filename = The%20Connections%20between.pdf.

Pande, C. B. (2022). Land use/land cover and change detection mapping in Rahuri watershed area (MS), India using the google earth engine and machine learning approach. *Geocarto International*, *37*(26), 13860–13880. Available from https://doi.org/10.1080/10106049.2022.2086622, http://www.tandfonline.com/toc/tgei20/current.

Puno, G. R., Puno, R. C. C., & Maghuyop, I. V. (2021). Two-dimensional food model for risk exposure analysis of land use/land cover in a watershed. *Global Journal of Environmental Science and Management*, *7*(2), 225–238. Available from https://doi.org/10.22034/gjesm.2021.02.06, http://www.gjesm.net/.

Ricotta, C., Bajocco, S., Guglietta, D., & Conedera, M. (2018). Assessing the influence of roads on fire ignition: Does land cover matter? *Fire*, *1*(2), 1–9. Available from https://doi.org/10.3390/fire1020024, https://www.mdpi.com/2571-6255/1/2/24/pdf.

Riese, F. (2020). Karlsruher Institut für Technologie (KIT) development and applications of machine learning methods for hyperspectral data.

Riese, F.M., Keller, & Hinz, S. (n.d.). Supervised and semi-supervised self-organizing maps for regression and classification focusing on hyperspectral data.

Sefrin, O., Riese, F.M., & Keller (n.d.). Deep learning for land cover change detection.

Singh, A., & Bruzzone, L. (2022). Data augmentation through spectrally controlled adversarial networks for classification of multispectral remote sensing images. *International Geoscience and Remote Sensing Symposium (IGARSS)*, 651-–654. Available from https://doi.org/10.1109/IGARSS46834.2022.9884928.

Singh, S., & Kasana, S. S. (2022). Quantitative estimation of soil properties using hybrid features and RNN variants. *Chemosphere*, *287*. Available from https://doi.org/10.1016/j.chemosphere.2021.131889, http://www.elsevier.com/locate/chemosphere.

Talukdar, S., Singha, P., Mahato, S., Shahfahad., Pal, S., Liou, Y.-A., & Rahman, A. (2020). Land-use land-cover classification by machine learning classifiers for satellite observations—A review. *Remote Sensing*, *12*(7), 1135. Available from https://doi.org/10.3390/rs12071135.

Tang, H., Bhatti, U. A., Li, J., Marjan, S., Baryalai, M., Assam, M., Ghadi, Y. Y., Mohamed, H. G., & Gastaldo, P. (2023). A new hybrid forecasting model based on dual series decomposition with long-term short-term memory. *International Journal of Intelligent Systems*, *2023*, 1–15. Available from https://doi.org/10.1155/2023/9407104.

Tuia, D., Flamary, R., & Courty, N. (2015). Multiclass feature learning for hyperspectral image classification: Sparse and hierarchical solutions. *ISPRS Journal of Photogrammetry and Remote Sensing*, *105*, 272–285. Available from https://doi.org/10.1016/j.isprsjprs.2015.01.006.

Uba, N. K. (2016). *Land use and land cover classification using deep learning techniques*.

Uzair, A., Bhatti, Z., Yu, J., Chanussot., Zeeshan., Zeeshan, L., Yuan, W., Luo, A. S., Nawaz, M. A., & Bhatti. (2021). Local similarity-based spatial—-spectral fusion hyperspectral image classification with deep CNN and gabor filtering. *IEEE Transactions on Geoscience and Remote Sensing*, *60*.

Vali, A., Comai, S., & Matteucci, M. (2020). Deep learning for land use and land cover classification based on hyperspectral and multispectral earth observation data: A review. *Remote Sensing*, *12*(15), 2495. Available from https://doi.org/10.3390/rs12152495.

Verma, D., & Jana, A. (2019). LULC classification methodology based on simple convolutional neural network to map complex urban forms at finer scale: Evidence from Mumbai. *arXiv*, India arXiv. https://arxiv.org. doi: 10.48550/arxiv.1909.09774.

Zhang, C., Harrison, P. A., Pan, X., Li, H., Sargent, I., & Atkinson, P. M. (2020). Scale sequence joint deep learning (SS-JDL) for land use and land cover classification. *Remote Sensing of Environment*, *237*. Available from https://doi.org/10.1016/j.rse.2019.111593, http://www.elsevier.com/inca/publications/store/5/0/5/7/3/3.

Zhang, C., Sargent, I., Pan, X., Li, H., Gardiner, A., Hare, J., & Atkinson, P. M. (2019). Joint deep learning for land cover and land use classification. *Remote Sensing of Environment*, *221*, 173–187. Available from https://doi.org/10.1016/j.rse.2018.11.014, http://www.elsevier.com/inca/publications/store/5/0/5/7/3/3.

Zhang, X., Zhou, Y., & Luo, J. (2022). Deep learning for processing and analysis of remote sensing big data: A technical review. *Big Earth Data*, *6*(4), 527–560. Available from https://doi.org/10.1080/20964471.2021.

Zhao, B., Feng, J., Wu, X., & Yan, S. (2017). A survey on deep learning-based fine-grained object classification and semantic segmentation. *International Journal of Automation and Computing*, *14*(2), 119–135. Available from https://doi.org/10.1007/s11633-017-1053-3.

Zhu, E. (1970). Carbon emissions induced by land-use and land-cover change from China. *The Science of the Total Environment*, *646*(1970), 930.

Zhu, L., Chen, Y., Ghamisi, P., & Benediktsson, J. A. (2018). Generative adversarial networks for hyperspectral image classification. *IEEE Transactions on Geoscience and Remote Sensing*, *56*(9), 5046–5063. Available from https://doi.org/10.1109/TGRS.2018.2805286.

Zhu, Y., Geis, C., So, E., & Jin, Y. (2021). Multitemporal relearning with convolutional LSTM models for land use classification. *IEEE Journal of Selected Topics in Applied Earth Observations and Remote Sensing*, *14*, 3251–3265. Available from https://doi.org/10.1109/JSTARS.2021.3055784, http://ieeexplore.ieee.org/xpl/RecentIssue.jsp?punumber = 4609443.

Zhu, X. X., Tuia, D., Mou, L., Xia, G. S., Zhang, L., Xu, F., & Fraundorfer, F. (2017). Deep learning in remote sensing: A comprehensive review and list of resources. *IEEE Geoscience and Remote Sensing Magazine*, *5*(4), 8–36. Available from https://doi.org/10.1109/MGRS.2017.2762307, http://ieeexplore.ieee.org/xpl/RecentIssue.jsp?punumber = 6245518.

Chapter 13

Oceanic biometric recognition algorithm based on generalized zero shot learning

Hao Tang, Dekai Li, Ahsan Ahmed Nizamani, Mughair Aslam Bhatti, Jinru Liu, Yonghui Zhang and Uzair Aslam Bhatti
School of Information and Communication Engineering, Hainan University, Haikou, P.R. China

13.1 Introduction

The ocean is the cradle of life on earth, accounting for approximately 71% of the earth's surface area. The ocean nurtures a wide variety of marine life, and its biomass accounts for approximately 87% of the earth's total biomass. It is a treasure house of biodiversity (Kadi et al., 2023). The physical system of nature is a process of increasing entropy, and the system always automatically tends to chaos and disorder when isolated. On the contrary, organisms evolve in an orderly manner, tending to decrease in entropy and toward chaos, which is also the basis of biological diversity. As the global demand for the development and protection of marine resources continues to grow, the importance of accurately identifying marine organisms has become increasingly apparent. Against this background, although machine learning technology has made significant progress in the field of natural image recognition (Bhatti et al., 2019, 2021), its application in marine biological identification faces unique challenges. Traditional recognition models require a large number of manually annotated training samples to achieve high prediction accuracy. However, the difficulty in obtaining marine biological samples, low quality and limited data, these factors together limit the effectiveness of the recognition model (Bhatti et al., 2023; Nawaz et al., 2021). Generally, thousands of labeled samples are required for each object class. For marine life, training samples are often taken by divers and underwater robots. Due to the complex and changeable seabed conditions, the use of imaging equipment varies from person to person, and the lighting conditions are also uneven, resulting in the final imaging effect including Issues such as contrast, low brightness, color deviation, and blurred details (Gao et al., 2024). In addition, marine life is highly mobile and hidden, often appearing in a short period of time and disappearing quickly. These factors combine to make it difficult to obtain marine life.

If you directly select underwater images to create a data set without filtering or processing the images, you will get a low-quality data set. Models trained based on low-quality data sets will cause severe information loss and fail to achieve the expected prediction results. If the model is trained based on high-resolution images, then in real business scenarios, due to the relatively low quality of the data source and the large resolution gap between the actual image and the training image (Huang et al., 2020), the model effect will be greatly reduced, and it will have a significant impact on information acquisition and processing in advanced vision tasks. Exploitation produces effects that increase the difficulty of the identification task (Bhatti et al., 2020).

The marine creature recognition algorithm based on generalized zero-shot learning can solve the problem of traditional sample image classification. This article has the following research significance (Kadi et al., 2023).

Improve the applicability of marine life identification. Traditional marine biological recognition methods are usually limited to samples of known categories and cannot accurately identify unseen categories. However, methods based on generalized zero-shot learning can make use of the semantic relationships between categories without any direct correlation with the category to be identified. Accurate identification can be carried out under the conditions of training samples, thereby improving the applicability and scope of marine biological identification (Gao et al., 2024).

Solve the problem of scarcity of marine biological samples. It provides an important reference for in-depth understanding of marine life and its ecological environment and promotes the progress of marine life research. By accurately identifying marine life types, more accurate and comprehensive marine life data can be obtained. Marine biological identification is of great significance for scientists to explore marine biodiversity and formulate effective protection

Deep Learning for Earth Observation and Climate Monitoring. DOI: https://doi.org/10.1016/B978-0-443-24712-5.00014-2

measures. By accurately identifying different marine biological species, scientists can have an in-depth understanding of the structure and function of marine ecosystems, discover endangered species in a timely manner, and take corresponding protective measures (Marani et al., 2023).

Marine resource management. The identification of marine organisms is of great significance for monitoring the status of fishery resources and formulating sustainable fisheries management policies. By accurately identifying different marine life species, the abundance and distribution of fishery resources can be effectively assessed, providing important data support for managers.

13.2 Research background

Zero-shot learning is similar to how humans learn, inferring unseen things from prior knowledge. For example, humans can recognize a zebra when encountering it through prior knowledge of the shape of a horse and its black and white stripes, but have not seen a zebra before. The concept diagram of zero-shot learning is shown in Fig. 13.1. Zero samples were first proposed in 2010, providing a solution to the practical problem of only identifying specific types in image classification tasks, and also providing the possibility for invisible class data (Marani et al., 2023). After continuous development, a large number of zero-shot learning methods have been spawned, which can be roughly divided into three categories according to current research directions: attribute-based zero-shot learning, embedding-based zero-shot learning, and generative network-based zero-shot learning.

13.2.1 Research status at home and abroad on attribute-based zero-shot learning

Marani et al. studied the problem of object classification when training and test categories do not overlap (Marani et al., 2023). In order to solve this problem, the author proposed an attribute-based classification method. This method performs object detection via manually specified high-level target object descriptions without using training images. The description of the target object can include arbitrary semantic attributes such as shape, color, and even geographical information. Because these properties transcend the specific learning task, they can be pre-learned on image datasets that are irrelevant to the current task. New categories can then be detected based on their attribute representations

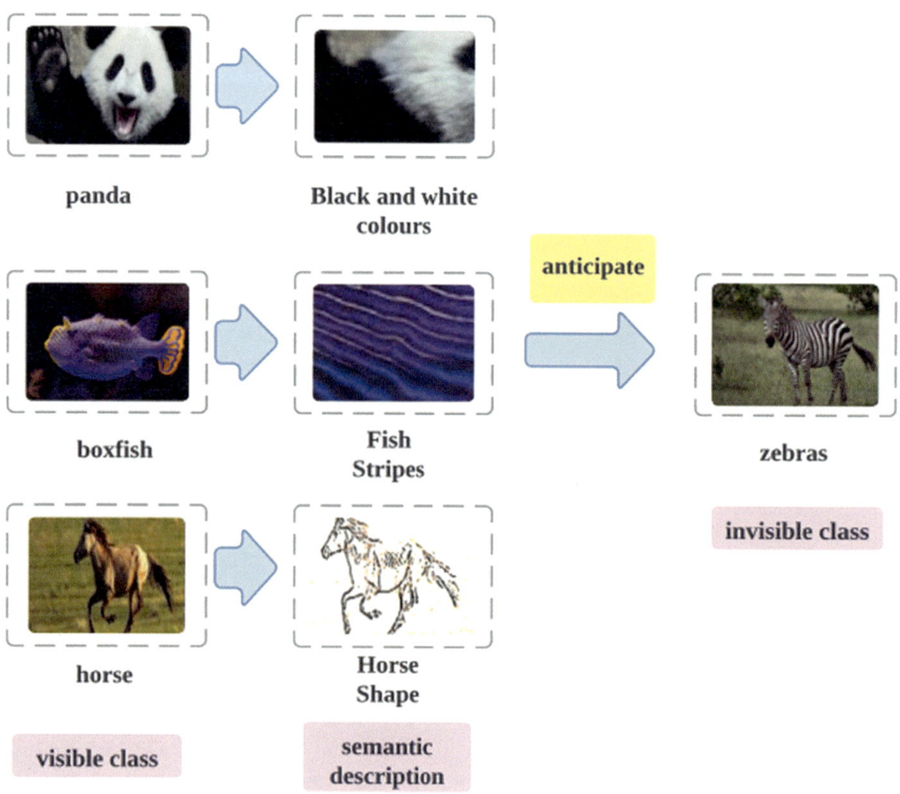

FIGURE 13.1 Conceptual diagram of zero sample learning.

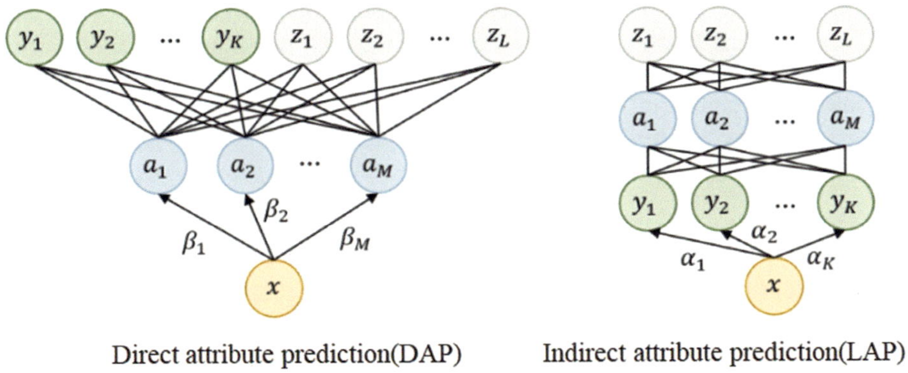

FIGURE 13.2 Attribute-based prediction.

without the need for a new training phase. The author proposed two generalization methods derived from attribute classification methods: direct attribute prediction (DAP) and indirect attribute prediction (IAP), as well as the concept of multi-layer or cascade classifiers. As shown in Fig. 13.2. The yellow node X can always be observed, the green node Y can only be observed during training, the gray node Z has never been observed, and the blue node a is the parameter to be trained.

Liang et al. proposed a method to solve the hubness problem prevalent in zero-shot learning (Liang & Liang, 2023). By using unsupervised methods to extract vector-based word representations from text corpora, and then utilizing these representations for mapping. But the mapping process will cause the hubness problem, that is, pushing the correct word label to a lower position in the neighbor list. Dinu et al. proposed a simple correction method to solve the hubness problem by re-ranking the mapped hub elements to correct the hubness problem.

Almomani et al. proposed a new bimodal attribute prediction (BAP) model (Almomani et al., 2023). BAP combines the advantages of DAP and IAP. It includes an attribute classifier pooling process to generate a large number of basic classifiers, and uses a combination strategy to integrate these classifiers to improve the performance of attribute prediction.

Sharma et al. introduced a new attribute learning framework called hypergraph-based attribute predictor (HAP) (Sharma et al., 2023). In HAP, by using a hypergraph to describe attribute relationships in data, the attribute prediction problem is transformed into a regularized hypergraph cutting problem. HAP simultaneously learns a set of attribute projections from feature space to hypergraph embedding space, which directly act as attribute classifiers. By treating the model as a multigraph cutting task, the HAP framework is able to flexibly integrate other available information, especially category labels.

Xayitov & Mirzakarimov et al. introduced a simple and effective zero-shot learning method based on a more general framework in which the relationship between features, attributes and categories is modeled as a network containing two linear layers (Xayitov & Mirzakarimov, 2023). Among them, the weight of the bottom layer is obtained through learning, while the weight of the top layer depends on the environment. Furthermore, by treating zero-shot learning methods as domain-adaptive methods, learning bounds on the generalization error of such methods are provided.

Current methods often fail to distinguish small visual differences between images due to the lack of fine-grained annotations, as well as issues such as attribute imbalance and co-occurrence. Delgado-Santos et al. proposed an end-to-end zero-shot learning method DUET based on Transformer, which integrates latent semantic knowledge from pretrained language models through a self-supervised multimodal learning paradigm. model (Delgado-Santos et al., 2023). Specifically, (1) developed a cross-modal semantic guidance network to study the model's ability to parse semantic attributes from images; (2) applied an attribute-level contrastive learning strategy to further enhance the model's ability to distinguish fine-grained visual features and overcome the attribute co-occurrence and imbalance problems; (3) a multitask learning strategy is proposed to consider multi-model objectives.

Pereira et al. proposed the objective function and training method of deep symmetric structured joint embedding (SJE). By minimizing the empirical risk, the embedding function of images and texts is learned, so that the prediction results can be conditionally predicted based on images and texts at the same time (Pereira et al., 2023).

Schlett et al. proposed a new model that can classify unseen categories based on text descriptions (Schlett et al., 2024), using text features to predict the convolutional layers and sums in deep convolutional neural networks (CNN). The output weights of the fully connected layer. Different from existing methods, the architecture of CNN is used to learn features at different layers instead of just learning the embedding space of two modalities. The model can also automatically generate a list of pseudo-attributes for each visual category, derived from the vocabulary of Wikipedia articles.

Attribute-based zero-shot learning is limited to accurately available attribute information, which requires the use of professional domain knowledge and a lot of manual labor. Although attribute-based zero-shot learning has achieved some success, further exploration is needed when dealing with more complex and ambiguous attributes or when no obvious attributes are available. Facing the huge number of marine organisms, clearly classifying their attributes is undoubtedly a huge challenge, which requires multifaceted research and manual annotation.

13.2.2 Research status at home and abroad on embedding-based zero-shot learning

Deep visual-semantic embedding (DeViSE) Fig. 13.3 is a method using a deep visual-semantic embedding (DeViSE) model, which aims to solve the limitations of modern visual recognition systems on large-scale target categories (Abou Elazm et al., 2023). The training of the DeViSE model is divided into three stages: language model pretraining, visual model pretraining and deep visual-semantic model construction. In the language model pretraining stage, the skip-gram (Attar & Lyashenko, 2023; Rozema, 2023) model is used to pretrain unlabeled text data and learn to generate semantically rich word vector representations. In the visual model pretraining stage, the core visual model is pretrained using labeled image data. Finally, in the deep visual-semantic model construction stage, the image is mapped into a rich semantic space by embedding visual and semantic information. Combining visual and semantic information, it provides more accurate and semantically consistent visual object recognition predictions, demonstrates the ability to predict image labels for unseen classes, and has the potential to be applied in zero-shot learning tasks.

Kaur et al. introduced a new zero-shot learning framework, transducing multi-view embedding, which solves the projection domain offset and prototype sparse problems in zero-shot learning through perspective multi-view embedding (Kaur et al., 2023). This method utilizes the complementary information of multiple views and combines heterogeneous hypergraph annotation propagation to enhance the robustness of the model and improve performance in the process of combining different popular structures across views.

Liu et al. proposed a framework called label embedding to solve the zero-shot learning problem in image classification. This framework utilizes attributes as intermediate representations to achieve parameter sharing between categories to cope with the scarcity of training data (Liu et al., 2024). By embedding each class into an attribute vector space and learning a function that measures the compatibility between image and label embeddings, it is guaranteed that the correct class will be ranked higher than the wrong class given an image. Compared with traditional attribute prediction methods, the label embedding framework exhibits better performance in zero-shot learning scenarios. In addition, the tag embedding framework also has the ability to leverage other information sources such as category hierarchies or textual descriptions. According to this framework, a method based on the compatibility function, SJE, is proposed to achieve classification by learning the scores of matching and mismatching embedding vectors (Prakash et al., 2023). The authors used state-of-the-art image features and combined supervised attributes and unsupervised output embedding vectors for classification. Unsupervised output embedding uses methods such as manually annotated attributes, unsupervised word embedding, and hierarchical structure embedding.

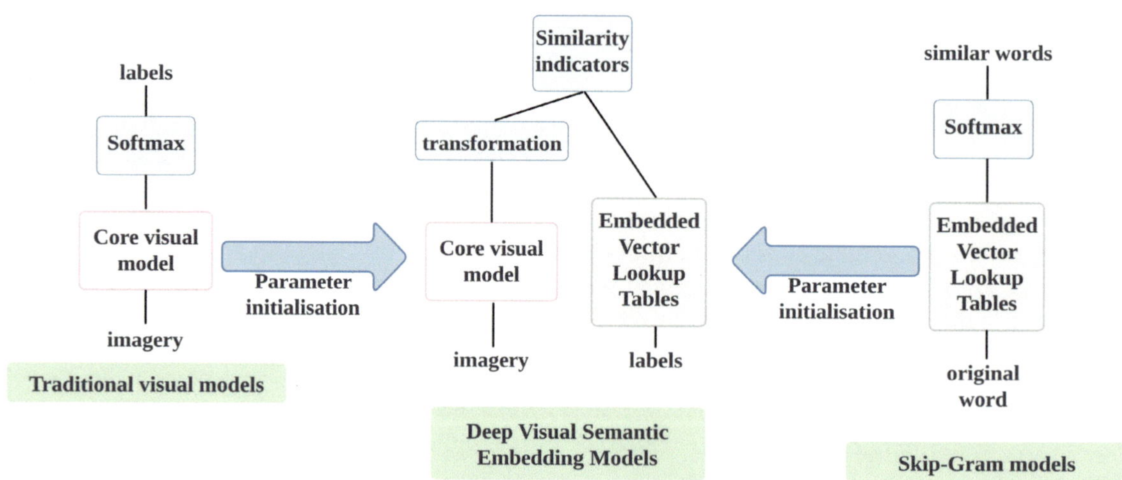

FIGURE 13.3 DeViSE framework.

Rukhiran et al. proposed zero-shot learning through semantic similarity embedding, treating each source or target data as a mixture of seen category proportions. It is assumed that if two instances belong to the same unseen category, then the blending modes must be similar (Rukhiran et al., 2023). This perspective is able to learn source/target embedding functions that map arbitrary source/target domain data into the same semantic space, allowing easy measurement of similarity.

Zhang et al. pointed out that choosing an appropriate embedding space is a key factor in making deep zero-shot learning models successful (Zhang et al., 2023). Instead of embedding textual descriptions of objects into semantic space or intermediate spaces, it is proposed to embed image representations of objects into visual space. Because in visual space, subsequent nearest neighbor searches will not suffer from the "hub problem" and can effectively classify.

Tataar et al. proposed a zero-shot learning solution based on learning semantic autoencoder (Tatar, 2023). The authors adopt an encoder-decoder paradigm, where the encoder's goal is to project visual feature vectors into semantic space, and the decoder imposes an additional constraint, that is, the projection/encoding must be able to reconstruct the original visual features.

Minaee et al. proposed a maximum margin framework based on semantic manifolds, which incorporates distance constraints of (supervised and unsupervised) lexical atoms to ensure that labeled samples are closest to their correct prototypes in the embedding space and not others. Prototype (Minaee et al., 2023).

By introducing the concept of semantic descriptors, this framework unifies multi-domain learning (MDL) and multi-task learning (MTL) and combines various classic and recent MTL. The MDL algorithm is explained as a different way of building semantic descriptors (Khan et al., 2023), providing an alternative process to zero-shot learning, where a model for a new category can be built without training data.

Embedding-based zero-shot learning is extremely sensitive to the quality of embeddings. If embeddings cannot accurately and inferentially capture the semantic relationships between categories, then the performance of zero-shot learning will be greatly affected. At the same time, there are too many types of unseen classes, causing the distribution in the embedded space to be more dispersed than the training classes, resulting in performance degradation.

13.2.3 Current status of domestic and international research on zero-sample learning based on generative models

Current status of domestic and international research on zero-sample learning based on generative models, Parashar et al. proposed an innovative generative adversarial network (GAN), named f-CLSWGAN, which can synthesize CNN features based on category-level semantic information and directly go from the semantic description of the category to category-specific. The characteristic distribution provides a shortcut (Parashar et al., 2023). Using Wasserstein GAN and classification loss, the author is able to generate sufficiently discriminative CNN features for training Softmax classifiers or other multimodal embedding methods. The process is shown in Fig. 13.4.

In response to the difficulties of regularization training related to zero samples, as well as the limitations of generation quality and the problem of seen–unseen bias, Parashar et al. proposed the concept of invertible zero-shot flow, using reversible flow. The forward pass of the network learns decomposed data embeddings, such as semantic factors and non-semantic factors (Parashar et al., 2024). Backward pass is used to generate data samples. This process theoretically extends the traditional generative flow model to the conditional scheme of decomposition.

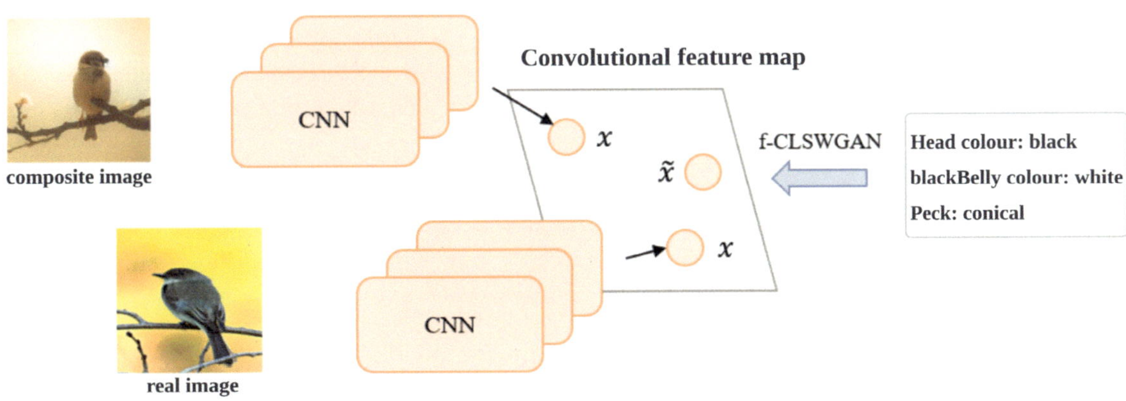

FIGURE 13.4 f-CLSWGAN block diagram.

Yin Haohao proposed an attention-based f-VAEGAN-D2 model method (Mandal & Pattnaik, 2023). In the f-VAEGAN-D2 model, virtual invisible class samples are generated through the VAE-GAN model, and two discriminators are used to identify the generated visible class samples and invisible class samples respectively. In order to selectively pay attention to visual features, emphasize discriminative local features and ensure the quality of samples generated by the model. This method adds a visual attention module to the model, inputs the selectively attended visual features into the discriminator, improves the discriminator's distinguishing ability, and then optimizes the generator during the confrontation process to improve the quality of the generated samples.

Verma et al. (2024) proposed a generative framework for generalized zero-shot learning, where training and test categories are not necessarily disjoint (Shaheed et al., 2024). The model is based on a variational autoencoder architecture and contains a probabilistic encoder and a probabilistic conditional decoder, capable of generating new instances from seen and unseen categories based on their respective category attributes. These instances can then be used to train any off-the-shelf classification model. A key aspect of the encoder-decoder architecture is the feedback-driven mechanism, where a discriminator (multiple regressor) learns to map generated instances to corresponding category attribute vectors, thereby improving the performance of the generator.

Zero-shot learning based on generative networks usually requires a large amount of computing resources and time to train the generative model. At the same time, during the training process, it may face the problem of mode collapse. And the quality of the generated model directly affects the performance of zero-shot learning, which has strong uncertainties.

13.3 Method

Currently available marine biological data sets face the challenge of small size and limited species diversity, which largely limits the possibility of in-depth research on marine biodiversity. In order to overcome these limitations and promote research progress in related fields, we chose to collect data extensively from the Internet and successfully constructed a new marine life data set, named it halobios-image. This data set covers 10,240 species of marine organisms, each of which is equipped with a detailed description file, aiming to provide researchers with a rich information resource. The data set was designed to reflect the complex distribution of marine life in the real world as closely as possible, thereby helping researchers better understand and simulate the dynamics of marine ecosystems. The species image information in the data set shows significant differences, which is consistent with the long-tail distribution characteristics of organisms in real environments. This design not only increases the practical value of the data set, but also provides a more realistic and comprehensive perspective for studying marine life. Through the use of this dataset, researchers can more deeply explore the diversity of marine life, as well as their distribution and interactions in the natural environment.

13.3.1 Dataset

The commonly used data sets in ZSL are mainly animal with attribute 2 (AWA2) (Gowda, 2023), Caltech-UCSD Birds-200−2011 (CUB) (Liu et al., 2023) data set. The AWA2 data set is used to evaluate transfer learning, such as attribute classification and zero The benchmark data set for sample learning is an expanded version of AWA, which increases the number of images and removes images with copyright issues. It contains 50 animal categories, 37,322 images, and provides an 85-dimensional attribute information for each category. attribute matrix. The CUB dataset is widely used for fine-grained visual classification and contains 11,788 images of 200 bird subcategories, providing detailed annotations, including 15 feature point locations, 312-dimensional attributes, and target detection boxes.

Marine ecosystems are famous for their wide variety of organisms, including all levels of biological taxonomy: phylum, class, order, family, genus, and species. To support generalized zero-shot classification research, halobios-image focuses on collecting a series of images of common marine organisms that represent a portion of marine biodiversity and are suitable for use as samples in visible categories. This includes many species of fish, marine mammals, invertebrates, shellfish, reptiles, birds, and molluscs. Some of the image resources of halobios-image come from the iNaturalist (Wang et al., 2023) community, which is a global community of nature lovers whose members share their photos of various plants and animals. The photos not only cover species from around the globe, but also capture their diverse scenes in their natural environments. The images were captured with various equipment, resulting in differences in image quality, but all images have been verified by community members to ensure the accuracy and reliability of their observation records.

Since 2017, iNaturalist has started releasing its own datasets, with the initial dataset covering images of 5000 species. In 2018, 2019, and 2021, iNaturalist expanded the data set to include 10,000 different species and approximately 2.7 million images. This provides a valuable resource for machine learning model training, especially species identification and classification. However, although the iNaturalist data set has grown in size and diversity, there is relatively little marine life data, containing only 369 species of marine life and 89,936 images, accounting for a small part of the entire data set, showing that research on marine life identification and classification needs to be done for further development.

The halobios-image dataset significantly enhances its content and value by integrating information from the Wildfish+ + (Li et al., 2023) dataset. Wildfish+ + focuses on fish diversity, including 2348 fish categories from five orders, 49 families, and 931 genera, with a total of 103,034 wild images. In addition to images, Wildfish+ + also contains a large number of fish description files, which are mainly crawled from professional websites such as FishBase and Fishes of Australia. They provide detailed information about various types of fish, including morphological characteristics and living habits, and provide data analysis a reliable knowledge base.

One of the distinctive features of Wildfish+ + is its scale, which is in a leading position among similar fish databases. Compared with the fish data of Fish4Knowledge (Zhang & Cao, 2023), QUT (Huang et al., 2023) and iNaturalist, Wildfish+ + has a larger amount of data, which is important for Improving the accuracy and robustness of the model is extremely beneficial. In addition, as a fine-grained dataset, Wildfish+ + not only covers a large number of species, but also has detailed classification and description of each species, requiring the model to be able to identify and distinguish subtle differences. The dataset also contains a large number of hand-drawn images depicting the outlines of fish, providing another form of data for studying fish morphological characteristics.

As an effective data organization and management method, knowledge graph serves as an important resource in the process of building the halobios-image database. The marine fish encyclopedia knowledge graph (Su et al., 2023) contributed by Xu Yajing and others is a data set that stores a wealth of basic information on fish, covering basic information such as Chinese and English names of fish, picture addresses, fish characteristics, etc. The amount of data reaches There were more than 30,000 entries. In addition, this data set also contains habitat and geographical distribution information of marine fish, providing extremely valuable data support for the halobios-image database.

The ocean biodiversity information system (OBIS) (Yang et al., 2023) is a global data and information exchange platform. The OBIS database contains more than 45 million observation records, covering nearly 120,000 marine species. Bacteria to whales, from sea level to 10,900 meters deep, and across a wide geographical range from the tropics to the polar regions. These records not only include species distribution information, but also detailed information on species habits, ecological status, and environmental preferences, providing rich basic information for the study of marine biodiversity. To ensure data accuracy and consistency, OBIS adopts standardized scientific names from the world register of marine species (WoRMS). WoRMS is an authoritative marine species database that provides detailed species descriptions, taxonomic information and references. The cooperation between OBIS and WoRMS ensures the accuracy of species names in the data set, providing users with reliable scientific information. In the process of constructing the marine biological image dataset, the data selected from OBIS provided an important supplement to the dataset, including key information such as species morphological characteristics and living habits, which increased the richness and scientific nature of the marine biological image dataset and improved the accuracy of the dataset.

In the process of constructing the halobios-image data set, in addition to using network resources and public platforms, the introduction of books and materials significantly enriched the content and depth of the data set. These books provide detailed species information, including morphological characteristics, geographical distribution and living habits, etc., which are crucial to building a comprehensive and accurate image database of marine life. "Field Identification Manual of Common Seaside Animals" (Verma et al., 2024): This manual records the identification methods of common seaside animals, including morphological characteristics, living habits, etc., and provides a practical guide for the observation and research of seaside animals. "Ecological Atlas of Coral Reef Fishes in the Nansha Islands" (Kurzendörfer et al., 2024): Focuses on the ecological characteristics of coral reef fish in the Nansha Islands, including habitat habits, eating habits and endangerment levels. Fish are divided into different functional groups according to their feeding habits, and the morphological characteristics and ecological habits of 301 fish species are described in detail. "Illustrated Chronicle of China's Offshore Molluscs" (Cai et al., 2023): Reorganized and identified marine shellfish samples collected along the country's coast, and combined with the latest research progress, provided detailed descriptions of 1040 species of marine shellfish. This illustrated book not only revised previously incorrectly identified species, but also included descriptions of new species. "Beibu Gulf Fish Illustrated Book" (Zhai et al., 2023): Detailed records of 424 species of marine bony fish collected in the Beibu Gulf of Guangxi, including Chinese names, Latin scientific names, morphological and quantitative characteristics, ecology, distribution, living conditions and utilization methods.

When building a data set for image recognition tasks, especially when the task involves fine-grained recognition, we are faced with higher data quality and accurate annotation requirements. Fine-grained recognition tasks require high-quality images and very precise annotation information so that the model can learn and recognize these subtle differences. For most applications, due to resource and technology constraints, it is preferable to perform a broad class of recognition tasks. These tasks do not require high-precision data like fine-grained recognition, so the requirements on data collection and processing are relatively low, and they have wider applicability in practical applications. The introduction of the Wikipedia-based image text (WIT) (Haas et al., 2023) data set provides valuable resources for a wide range of recognition tasks. WIT is a multimodal data set based on Wikipedia, which contains a large amount of image and text information. This information is not only rich, but also edited, verified and corrected by the Wikipedia community to ensure high quality and accuracy. For the halobios-image dataset, leveraging the WIT dataset for broad category descriptions can significantly improve the quality and usefulness of the dataset. At the same time, for some common fine-grained marine life descriptions, although the WIT dataset may not be sufficient to provide all necessary details, it can still serve as a valuable starting point to provide directions for further data collection and refinement.

To enrich the dataset, halobios-image introduces image resources from Flickr. Flickr is a widely used online image and video sharing platform, but the image tagging system on Flickr cannot meet the needs of accurately segmenting marine life. Its tagging system tends to broadly classify rather than accurately identify specific species. To solve this problem, the strategy of classifying images into broader categories, such as clownfish, turtles, starfish, and sea anemones, is adopted to compensate for the lack of label accuracy and facilitate data management and application. Through this method, 88 images of different marine life categories were successfully collected, with the number reaching 207,261. Although this strategy sacrifices a certain degree of accuracy, it effectively expands the coverage of the data set.

13.3.2 Marine biological data cleaning

Research by Thao Kim et al. pointed out that simply collecting large amounts of data from the Internet is not the most effective way to construct a pretraining data set suitable for robust generalization, and further research on data set design is needed (Kim et al., 2023). Data cleaning is a critical step in processing raw data in machine learning projects and is extremely important to ensure the accuracy and reliability of the model. Raw data often contains errors and inconsistencies, and unprocessed data may seriously affect model performance and prediction accuracy. Therefore, data cleaning is an important step in improving model prediction capabilities. Its main steps include correcting or deleting errors and incomplete data, removing duplicate information, removing irrelevant data, unifying formats, handling missing values and spelling errors, aiming to improve data quality and consistency and build efficient and accurate machine learning. The model lays the foundation.

13.3.2.1 Image cleaning

There are some pictures in the data set that interfere with the marine life recognition model. These problems mainly include: (1) Obvious text information. Some pictures contain text information that interferes with recognition. (2) The tags do not match the content. The tags of some pictures do not match the actual content. (3) Motion blur or poor lighting conditions cause blur, making it difficult to extract special features. These problematic images are collectively called noise data and have a negative impact on the performance of the model. Therefore, these noisy data need to be cleaned. However, the cleaning work does not need to be too thorough to ensure that the model can adapt to and handle a certain degree of noise, thereby ensuring the robustness of the model.

The cleaning effect directly affects the actual performance of subsequent tasks. After using various methods to clean the image, it is finally stored in the database. Please see Fig. 13.5 for the specific process. In the experimental environment, thanks to the support of multicore processors, the processing of large-scale images can be accelerated through multi-threaded parallel computing.

Image resolution has a significant impact on the performance of the model, and larger resolution means higher computational complexity. To deal with this problem, models typically compress the image to a specific size, and during this compression process, information loss may occur. For the classification of large targets, the impact of this loss is less, and the running efficiency after compression is actually higher. However, for small targets, compression may cause distortion. Furthermore, the details provided by high-resolution images may be redundant for classification models and may even reduce the model's generalization ability. When collecting images, a resolution of 1920*720 was selected as the baseline. Images beyond this resolution were scaled, while images below this resolution were kept unchanged in their original size, but if the image resolution If it is too small, discard it. Reducing the resolution not only speeds up image processing, but also reduces the size of the image matrix that needs to be calculated, thereby improving computational efficiency.

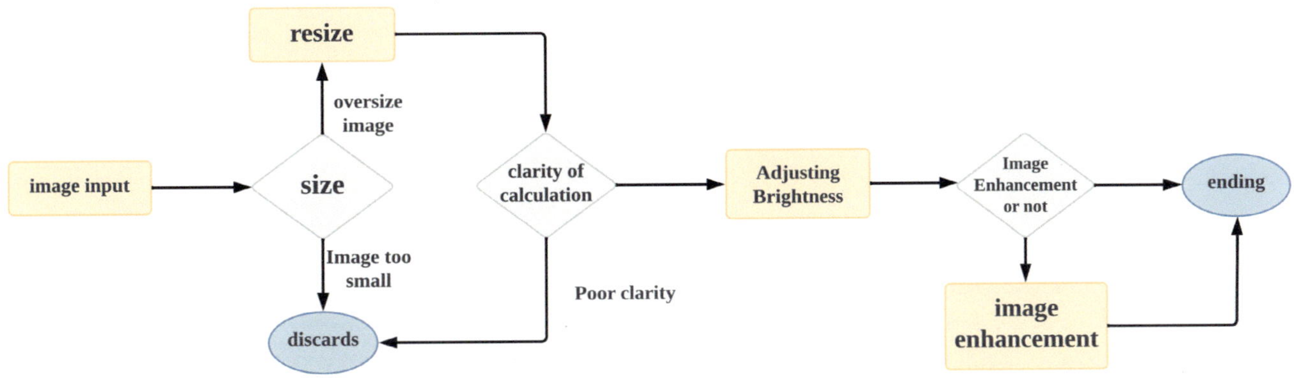

FIGURE 13.5 Image cleaning process.

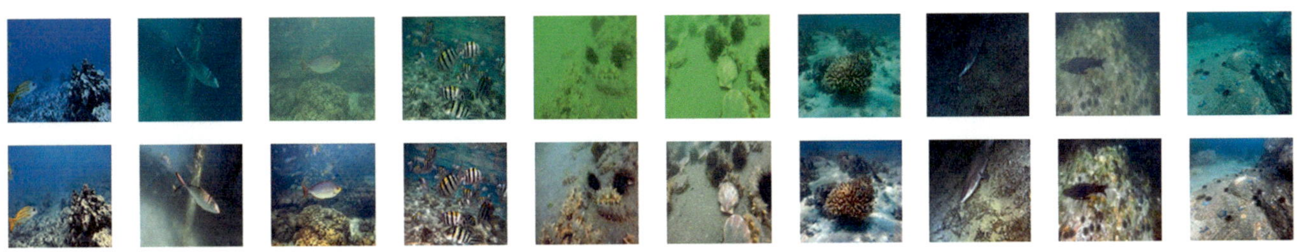

FIGURE 13.6 Comparison of underwater image enhancement effects.

Sharpness is evaluated by calculating the Laplacian variance of the image. The Laplacian operator is a second-order derivative operator that can highlight the edges of the image. The clearer the image, the richer the edge information, the greater the variance of the Laplacian operator. On the contrary, if the image is blurred, the edge information will be reduced, resulting in a smaller variance of the Laplacian operator. Set the sharpness threshold, discard low-definition images caused by motion blur or focus failure, and automatically average the image histogram to ensure that the image clarity reaches the standard (Fig. 13.6).

Use underwater image enhancement algorithms to enhance underwater images. The (UICE^2-Net) (Rao et al., 2023) algorithm is adopted to efficiently integrate RGB color space and HSV color space in a single CNN. This algorithm can process many unclear underwater photos. This algorithm has good performance and generalization, and the enhanced comparison chart is shown in Fig. 13.6.

13.3.2.2 Text cleaning

In NLP, text cleaning is a very important step, which has the meaning of improving data quality, reducing computational complexity, improving model performance and promoting feature extraction. The main process of text cleaning is to remove abnormal text and merge and organize WIT information and English translation, as shown in Fig. 13.7.

Original text data often contains a large amount of noise information, such as spelling errors, special symbols, and irrelevant information. The text crawled from the Internet must be thoroughly cleaned to ensure the accuracy and availability of the information. This process includes removing HTML tags and other non-text elements to improve the quality of the data. Specifically, the cleaning step involves removing unusual characters, redundant characters, supplementary content within brackets, URL links, email addresses, and phone numbers.

In response to the problem of mixed Chinese and English in the data set, the field of NLP has weaker processing capabilities for Chinese than English, so the Chinese content was chosen to be translated into English. This study uses google translation API for efficient and unified translation processing. Google cloud translation API is based on Google's advanced neural machine translation technology and provides two modes: using Google's pretrained model for translation, or using a model specially designed for translation tasks. large language model. In this way, users can choose the most appropriate translation service based on their needs. At the same time, the API provides a certain amount of free translation characters per month (500,000 characters per month), which is usually sufficient for general data set text translation needs.

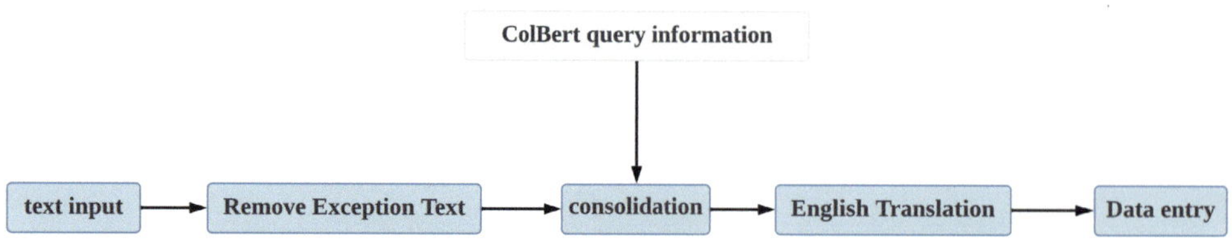

FIGURE 13.7 Text cleaning process.

TABLE 13.1 Description of each name of the point field.

Name	Hidden meaning
Option_harm	Potential harm
Option_shape	Morphological characteristic
Option_imminent	Endangered level
Option_distributing	Distributions
Option_functional_taxa	Functional group
Option_ecological_habits	Ecological habitat

13.3.3 Marine life data storage

Data warehousing can convert temporary or volatile data into persistent storage to ensure data security and stability and avoid data loss. Through the database management system, operations such as adding, deleting, modifying, and checking data can be effectively performed to improve the efficiency and accuracy of data management. Among multiple data sources and different types of data, data warehousing helps the integration and unified management of data, and facilitates the integrated analysis of data.

For image information, use the marine life management platform introduced below for storage, and use the MySQL database to record the image storage location information, including file name and image md5 value, to ensure the uniqueness of the image.

For the integrated species description data, use MySQL database for storage, use structured storage, and set the field type to varchar (1000) to accommodate more description information. At the same time, for some more detailed species descriptions, use the point field. Storage, the storage format is JSON format, see Table 13.1 for details.

13.3.4 Marine life data analysis

The halobios-image dataset successfully collects photos of 10,241 species of marine life, bringing the total number of images to 1,048,462. This significantly exceeds many conventional datasets in terms of data size, although it is still insufficient compared with some very large datasets, mainly due to the current difficulty of collecting photos of marine life. A detailed statistical analysis of the halobios-image dataset clearly shows the imbalance in the data distribution, where the difference between the species with the most images and the fewest is as much as 285 times greater. This uneven distribution is 47 times greater than the wildfish++ dataset and 13 times greater than the large-scale noisy WebVision (Cao et al., 2023) dataset. The data set comparison chart is shown in Fig. 13.8.

A traditional image classification experiment was conducted on the halobios-image data set to obtain its benchmark data. Split the data set into training set, validation set and test set using a ratio of 7:3:1. The models selected are ResNet (Sheng et al., 2023; Wu et al., 2023), ViT, DenseNet Iman et al., 2023), ECAResNet (Zhu et al., 2023), etc. The experimental environment is unified as NVIDIA V100x4 16G. The pretraining weights are from ImageNet (Salehi et al., 2023). The number of training times is uniformly set to 300. The experimental results are shown in Table 13.2.

Conduct separate experiments on the 200 types with the largest number of images in the data set, using the same experimental environment, but without using any image enhancement method. The experimental results are shown in Table 13.3.

It is easy to see that the accuracy is significantly higher than the unselected data set. The fundamental reason for this difference is that the random data set contains too many variable types and the traditional model parameter number

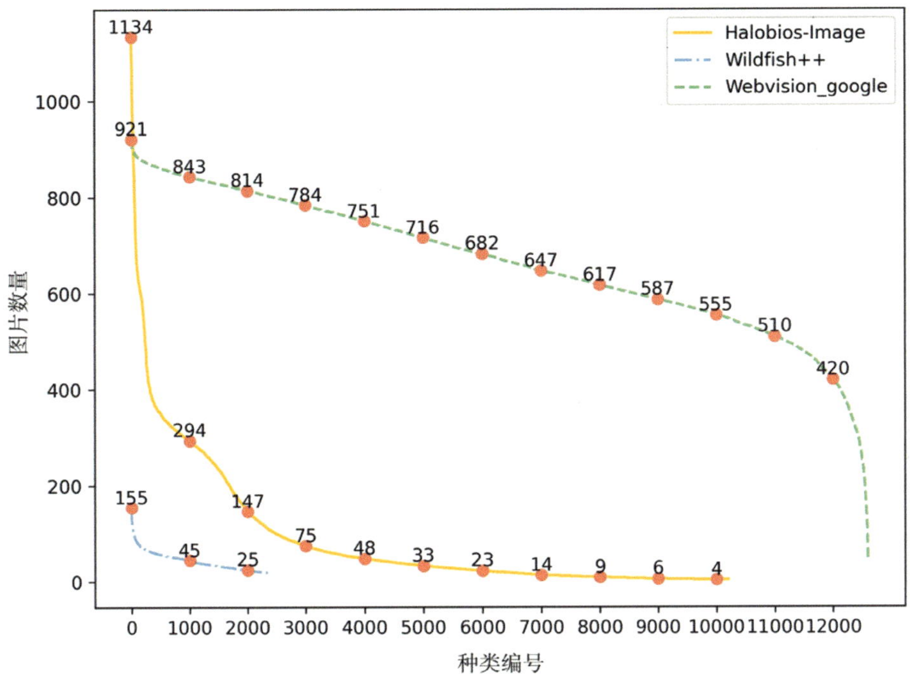

FIGURE 13.8 Dataset comparison

TABLE 13.2 Benchmarking results.

Modeling	Batch-size	Data enhancement	Maximum accuracy
Densenet121	128	RandAugment	58.96%
Densenet201	64	CutMix	63.42%
ecaresnet26t	64	None	60.94%
ecaresnet269d	32	None	63.45%
vit_base_patch16_224	8	None	61.59%
vit_base_patch8_224	8	None	69.88%
vit_base_patch16_384	16	None	68.48%
Resnet50	256	RandAugment, Chen et al. (2024)	62.24%
Resnet101	128	Mixup, Theodoris et al. (2023)	64.13%
Resnet50	256	Mixup, Gulzar (2023)	63.17%
Resnet101	128	CutMix	62.78%
resnetv2_152	128	Mixup	65.72%
Resnetv2_152d	128	CutMix	67.93%

is not enough to handle this size of data. In addition, the long tail effect in the data set is also very significant, making the model more biased toward those categories with a larger proportion. For long-tail distribution problems in data sets, traditional solutions are resampling and reweighting. These methods may require fine tuning of parameters or the design of complex loss functions, which increases the complexity of implementation. Secondly, traditional methods mostly jointly learn representations and classifiers, which may lead to difficulties in dealing with class imbalance problems and require more optimization skills and adjustments (Chen et al., 2016).

TABLE 13.3 Baseline for the first 200 types of quantities.

Modeling	Batch-size	Maximum accuracy
Resnet152	64	84.20%
Resnetv2_152	64	75.72%
Resnetv2_152d	64	77.93%
efficientnet_b2	32	83.39%

13.4 Conclusion

This chapter first provides a comprehensive overview of the halobios-image data set, detailing the data collection process and data cleaning methods. Then, this chapter deeply explores the diversity of data sets and existing data imbalance problems, and proposes corresponding solution strategies. In addition, this chapter evaluates the difficulty of data collection and presents a series of benchmark results through detailed data analysis. These tests not only show the specific details of the models and training used, but also include metrics to evaluate the performance of these models. Through the introduction of these contents, we aim to provide an in-depth and comprehensive understanding of the halibios-image data set and how to effectively use these data for model training and evaluation.

References

Almomani, I., El-Shafai, W., AlKhayer, A., Alsumayt, A., S. Aljameel, S., & Alissa, K. (2023). Proposed biometric security system based on deep learning and chaos algorithms. *Computers, Materials & Continua*, *74*(2), 3515–3537. Available from https://doi.org/10.32604/cmc.2023.033765.

Attar, A.-J. A. T., & Lyashenko, H. (2023). Access control to robotic systems based on biometric: The generalized model and its practical implementation. *International Journal of Intelligent Engineering & Systems* (5).

Bhatti, U. A., Huang, M., Wu, D., Zhang, Y., Mehmood, A., & Han, H. (2019). Recommendation system using feature extraction and pattern recognition in clinical care systems. *Enterprise Information Systems*, *13*(3), 329–351. Available from https://doi.org/10.1080/17517575.2018.1557256, http://www.tandf.co.uk/journals/titles/17517575.asp.

Bhatti, U. A., Tang, H., Wu, G., Marjan, S., Hussain, A., & Sarker, S. K. (2023). Deep Learning with graph convolutional networks: An overview and latest applications in computational intelligence. *International Journal of Intelligent Systems*, *2023*, 1–28. Available from https://doi.org/10.1155/2023/8342104.

Bhatti, U. A., Yu, Z., Chanussot, J., Zeeshan, Z., & Yuan, L. (2021). Local similarity-based spatial–spectral fusion hyperspectral image classification with deep CNN and Gabor filtering. *IEEE Transactions on Geoscience and Remote Sensing*, *60*, 1–15.

Bhatti, U. A., Yu, Z., Yuan, L., Zeeshan, Z., Nawaz, S. A., Bhatti, M., Mehmood, A., Ain, Q. U., & Wen, L. (2020). Geometric algebra applications in geospatial artificial intelligence and remote sensing image processing. *IEEE Access*, *8*, 155783–155796. Available from https://doi.org/10.1109/access.2020.3018544.

Cai, J., Wu, L., & Wu, D. (2023). Multi-dimensional information alignment in different modalities for generalized zero-shot and few-shot learning. *J. Information* (3).

Cao, W., Wu, Y., Sun, Y., Zhang, H., Ren, J., Gu, D., & Wang, X. (2023). A review on multimodal zero-shot learning. *Wiley Interdisciplinary Reviews: Data Mining and Knowledge Discovery*, *13*(2). Available from https://doi.org/10.1002/widm.1488, http://onlinelibrary.wiley.com/journal/10.1002/(ISSN)1942-4795.

Chen, H., Luo, H., Huang, B., Jiang, B., & Kaynak, O. (2024). Transfer learning-motivated intelligent fault diagnosis designs: A survey, insights, and perspectives. *IEEE Transactions on Neural Networks and Learning Systems*, *35*(3), 2969–2983. Available from https://doi.org/10.1109/tnnls.2023.3290974.

Chen, X., Yang, R., & Xue, Y. (2016). Deep transfer learning for bearing fault diagnosis: A systematic review since. *IEEE Transactions on Instrumentation and Measurement*, 2016.

Delgado-Santos, P., Tolosana, R., & Guest, R. (2023). Mobile biometric gait verification using transformers. *Engineering Applications of Artificial Intelligence*.

Elazm, A., El-Shafai, W., Ibrahim, S., et al. (2023). Efficient hardware design of a secure cancellable biometric cryptosystem. *IASC-Intelligent Automation & Soft Computing. 2023 36*(1), 929–955.

Gao, J., Zhang, Y., Geng, X., Tang, H., & Bhatti, U. A. (2024). PE-transformer: Path-enhanced transformer for improving underwater object detection. *Expert Systems with Applications*, *246*, 2024.

Gowda, S. N. (2023). Synthetic sample selection for generalized zero-shot learning, *IEEE computer society conference on computer vision and pattern recognition workshops* (pp. 58–67). Available from http://ieeexplore.ieee.org/xpl/conferences.jsp, https://doi.org/10.1109/CVPRW59228.2023.00011.

Gulzar, Y. (2023). Fruit image classification model based on MobileNetV2 with deep transfer learning technique. *Sustainability, 15*(3), 1906. Available from https://doi.org/10.3390/su15031906.

Haas, L., Alberti, S., & Skreta, M. (2023). Learning generalized zero-shot learners for open-domain image geolocalization. *arXiv*, United States arXiv. Available from https://arxiv.org, https://doi.org/10.48550/arXiv.2302.00275.

Huang, S., Huang, M., Zhang, Y., Chen, J., & Bhatti, U. (2020). Medical image segmentation using deep learning with feature enhancement. *IET Image Processing, 14*(14), 3324–3332. Available from https://doi.org/10.1049/iet-ipr.2019.0772.

Huang, J., Li, Z., & Zhou, Z. (2023). A simple framework to generalized zero-shot learning for fault diagnosis of industrial processes. *IEEE/CAA Journal of Automatica Sinica, 10*(6), 1504–1506. Available from https://doi.org/10.1109/jas.2023.123426.

Iman, M., Arabnia, H. R., & Rasheed, K. (2023). A review of deep transfer learning and recent advancements. *Technologies, 11*(2), 40.

Kadi, A., Oubelaid, A., & Towfek, S. K. (2023). Electrocardiogram comparison as a biometric identifier: A review. *Journal of Artificial Intelligence and Metaheuristics* (2), 31.

Kaur, P., Kumar, N., & Singh, M. (2023). Biometric cryptosystems: a comprehensive survey. *Multimedia Tools and Applications, 82*(11), 16635–16690. Available from https://doi.org/10.1007/s11042-022-13817-9.

Khan, S. U., Khan, N., Hussain, T., Muhammad, K., Hijji, M., Del Ser, J., & Baik, S. W. (2023). Visual appearance and soft biometrics fusion for person re-identification using deep learning. *IEEE Journal on Selected Topics in Signal Processing, 17*(3), 575–586. Available from https://doi.org/10.1109/JSTSP.2023.3260627, http://www.ieee.org/products/onlinepubs/news/0806_01.html.

Kim, J., Shim, K., & Kim, J. (2023). Vision transformer-based feature extraction for generalized zero-shot learning. *IEEE International Conference on Acoustics, Speech and Signal Processing (ICASSP)* (pp. 1–5).

Kurzendörfer, D., Mercea, O. B., & Koepke. (2024). *Audio-visual generalized zero-shot learning using pre-trained large multi-modal models, 2024.*

Liang, Y., & Liang, W. (2023). *ResWCAE, biometric pattern image denoising using residual wavelet-conditioned autoencoder.*

Liu., Ashbaugh., & Chimitt. (2024). Farsight: A physics-driven whole-body biometric system at large distance and altitude. *Applications of Computer Vision*, 6227–6236.

Liu, M., Li, F., Zhang, C., Wei, Y., Bai, H., & Zhao, Y. (2023). Progressive semantic-visual mutual adaption for generalized zero-shot learning. *Proceedings of the IEEE computer society conference on computer vision and pattern recognition* (pp. 15337–15346), Available from https://doi.org/10.1109/CVPR52729.2023.01472.

Li, X., Zhang, Y., Bian, S., Qu, Y., Xie, Y., Shi, Z., & Fan, J. (2023). VS-boost: Boosting visual-semantic association for generalized zero-shot learning. In *IJCAI international joint conference on artificial intelligence*, (pp. 1107–1115). Available from http://www.ijcai.org/.

Mandal, D., & Pattnaik, S. S. (2023). *Machine learning and deep learning for multimodal biometrics. Multimodal biometric and machine learning technologies: Applications for computer vision* (pp. 163–172). Wiley, https://www.wiley.com/en-us/Multimodal + Biometric + and + Machine + Learning + Technologies%3A + Applications + for + Computer + Vision-p-9781119785477.

Marani, M., Soltani, M., & Bahadori, M. (2023). The role of biometric in banking: A review. *EAI Endorsed Transactions on AI and Robotics.*

Minaee, S., Abdolrashidi, A., Su, H., Bennamoun, M., & Zhang, D. (2023). Biometrics recognition using deep learning: A survey. *Artificial Intelligence Review, 56*(8), 8647–8695. Available from https://doi.org/10.1007/s10462-022-10237-x.

Nawaz, S. A., Li, J., Bhatti, U. A., Bazai, S. U., Zafar, A., Bhatti, M. A., Mehmood, A., Ain, Q. ul, Shoukat, M. U., & Zeng, Q. (2021). A hybrid approach to forecast the COVID-19 epidemic trend. *PLoS One, 16*(10), e0256971. Available from https://doi.org/10.1371/journal.pone.0256971.

Parashar, A., Parashar, A., Abate, A. F., Shekhawat, R. S., & Rida, I. (2023). Real-time gait biometrics for surveillance applications: A review. *Image and Vision Computing, 138*, 104784. Available from https://doi.org/10.1016/j.imavis.2023.104784.

Parashar, A., Parashar, A., & Shabaz, M. (2024). Advancements in artificial intelligence for biometrics: A deep dive into model-based gait recognition techniques. *Journal of Engineering Applications of Artificial Intelligence.*

Pereira, T. M. C., Conceição, R. C., Sencadas, V., & Sebastião, R. (2023). Biometric recognition: A systematic review on electrocardiogram data acquisition methods. *Sensors, 23*(3). Available from https://doi.org/10.3390/s23031507, http://www.mdpi.com/journal/sensors.

Prakash, A. J., Patro, K. K., & Samantray, S. (2023). A deep learning technique for biometric authentication using ECG beat template matching. *Journal of Information* (2).

Rao, Z., Guo, J., Lu, X., Zhou, Q., Zhang, J., Wei, K., Li, C., & Guo, S. (2023). Attribute-aware representation rectification for generalized zero-shot learning. *arXiv*, Hong Kong arXiv. Available from https://arxiv.org, https://doi.org/10.48550/arXiv.2311.14750.

Rozema, J. J. (2023). Refractive development I: Biometric changes during emmetropisation. *Ophthalmic and Physiological Optics, 43*(3), 347–367. Available from https://doi.org/10.1111/opo.13094, http://onlinelibrary.wiley.com/journal/10.1111/(ISSN)1475-1313.

Rukhiran, M., Wong-In, S., & Netinant, P. (2023). IoT-based biometric recognition systems in education for identity verification services: Quality assessment approach. *IEEE Access, 11*, 22767–22787. Available from https://doi.org/10.1109/access.2023.3253024.

Salehi, A. W., Khan, S., Gupta, G., Alabduallah, B. I., Almjally, A., Alsolai, H., Siddiqui, T., & Mellit, A. (2023). A study of CNN and transfer learning in medical imaging: Advantages, challenges, future scope. *Sustainability, 15*(7), 5930. Available from https://doi.org/10.3390/su15075930.

Schlett, T., Rathgeb, C., Tapia, J., & Busch, C. (2024). Considerations on the evaluation of biometric quality assessment algorithms. *IEEE Transactions on Biometrics, Behavior, and Identity Science, 6*(1), 54–67. Available from https://doi.org/10.1109/tbiom.2023.3336513.

Shaheed, K., Szczuko, P., & Kumar, M. (2024). Deep learning techniques for biometric security: A systematic review of presentation attack detection systems. *Journal of Engineering Applications of Artificial Intelligence.*

Sharma, S., Saini, A., & Chaudhury, S. (2023). A survey on biometric cryptosystems and their applications. *Computers & Security, 134*, 103458. Available from https://doi.org/10.1016/j.cose.2023.103458.

Sheng, M., Li, J., Bhatti, U. A., Liu, J., Huang, M., & Chen, Y. W. (2023). Zero watermarking algorithm for medical image based on resnet50-DCT. *Computers, Materials and Continua*, *75*(1), 293–309. Available from https://doi.org/10.32604/cmc.2023.036438, https://www.techscience.com/cmc/v75n1/51538.

Su, H., Li, J., Lu, K., Zhu, L., & Tao Shen, H. (2023). Dual-aligned feature confusion alleviation for generalized zero-shot learning. *IEEE Transactions on Circuits and Systems for Video Technology*, *33*(8), 3774–3785. Available from https://doi.org/10.1109/tcsvt.2023.3239390.

Tatar, A. B. (2023). Biometric identification system using EEG signals. *Neural Computing and Applications*, *35*(1), 1009–1023. Available from https://doi.org/10.1007/s00521-022-07795-0, https://www.springer.com/journal/521.

Theodoris, C. V., Xiao, L., Chopra, A., Chaffin, M. D., Al Sayed, Z. R., Hill, M. C., Mantineo, H., Brydon, E. M., Zeng, Z., Liu, X. S., & Ellinor, P. T. (2023). Transfer learning enables predictions in network biology. *Nature*, *618*(7965), 616–624. Available from https://doi.org/10.1038/s41586-023-06139-9, https://www.nature.com/nature/.

Verma., Mehta., & Liang, K. J. (2024). Meta-learned attribute self-interaction network for continual and generalized zero-shot learning. *Applications of Computer Vision*, 2721–2731.

Wang, X., Liu, A., Wu, L., Li, C., Liu, Y., & Chen, X. (2023). A generalized zero-shot learning scheme for SSVEP-based BCI system. *IEEE Transactions on Neural Systems and Rehabilitation Engineering*, *31*, 863–874. Available from https://doi.org/10.1109/tnsre.2023.3235804.

Wu, Y., Liang, T., Feng, S., Jin, Y., Lyu, G., Fei, H., & Wang, Y. (2023). MetaZSCIL: A meta-learning approach for generalized zero-shot class incremental learning. *Proceedings of the AAAI Conference on Artificial Intelligence*, *37*(9), 10408–10416. Available from https://doi.org/10.1609/aaai.v37i9.26238.

Xayitov, A., & Mirzakarimov, B. (2023). The use of biometric authentication techniques for safeguarding data in computer systems against unauthorized access or breaches. *J. Ротомки Аль-Фаргани* (2), 33–36.

Yang, F. E., Lee, Y. H., Lin, C. C., & Wang, Y. C. F. (2023). Semantics-guided intra-category knowledge transfer for generalized zero-shot learning. *International Journal of Computer Vision*, *131*(6), 1331–1345. Available from https://doi.org/10.1007/s11263-023-01767-0, https://www.springer.com/journal/11263.

Zhai, Z., Li, X., & Chang, Z. (2023). Center-VAE with discriminative and semantic-relevant fine-tuning features for generalized zero-shot learning. *Signal Processing: Image Communication*, *111*, 116897. Available from https://doi.org/10.1016/j.image.2022.116897.

Zhang, Z., & Cao, W. (2023). Visual-semantic consistency matching network for generalized zero-shot learning. *Neurocomputing*, *536*, 30–39. Available from https://doi.org/10.1016/j.neucom.2023.03.007.

Zhang, G., Liu, J., Chen, Y., Zheng, Y., & Zhang, H. (2023). Multi-biometric unified network for cloth-changing person re-identification. *IEEE Transactions on Image Processing*, *32*, 4555–4566. Available from https://doi.org/10.1109/tip.2023.3279673.

Zhu, Z., Lin, K., Jain, A. K., & Zhou, J. (2023). Transfer learning in deep reinforcement learning: A survey. *IEEE Transactions on Pattern Analysis and Machine Intelligence*, *45*(11), 13344–13362. Available from https://doi.org/10.1109/TPAMI.2023.3292075, https://ieeexplore.ieee.org/servlet/opac?punumber = 34.

Section V

Deep learning for oceans

Chapter 14

Remote sensing image fusion based on deep learning and convolutional neural network technique

Emadalden Alhatami, Mengxing Huang and Uzair Aslam Bhatti
School of Information and Communication Engineering, Hainan University, Haikou, P.R. China

14.1 Introduction

Remote sensing image fusion plays a crucial role in extracting valuable information from multiple sources of remotely sensed data. By combining images acquired from different sensors or platforms, the fused image provides enhanced spatial, spectral, and temporal details, enabling improved analysis and interpretation. In recent years, convolutional neural networks (CNNs) and deep learning techniques have emerged as powerful tools for remote sensing image fusion (Bhatti et al., 2020, 2023), revolutionizing the field with their ability to automatically learn and extract complex features from the input data (Zhang et al., 2021).

Remote sensing encompasses various domains, including satellite imagery, aerial photography, and hyperspectral imaging, among others. Each domain presents unique challenges due to differences in sensor properties, resolutions, and spectral characteristics (Nawaz et al., 2022). Traditional image fusion methods often rely on handcrafted feature extraction and fusion rules, which may not fully exploit the inherent complexities of remote sensing data (Saxena et al., 2021). This limitation led to the exploration of CNNs and deep learning techniques, which have shown promising results in capturing intricate patterns and relationships present in the images. CNNs are deep learning models specifically designed to process gridlike structured data, such as images. They employ learnable filters and pooling layers to automatically extract hierarchical features from the input image. This hierarchical feature extraction allows CNNs to capture both local details and global contextual information, making them well-suited for remote sensing image fusion tasks (Tan et al., 2019).

The utilization of CNNs in remote sensing image fusion involves the design of specialized network architectures tailored to the specific fusion task at hand. These architectures often adopt an encoder-decoder structure, where the encoder captures the spatial or spectral information from the input images, and the decoder generates the fused output image. The network learns to optimize a loss function by comparing the fused image with ground truth data, enabling it to gradually improve its performance through training (Pahlevan et al., 2022).

Deep learning techniques, including CNNs, offer several advantages in remote sensing image fusion. They can automatically learn discriminative features from largescale datasets, leveraging the power of transfer learning by finetuning pretrained models on remote sensing data. This transfer of knowledge enhances the fusion network's ability to generalize and extract relevant information from the input images. Furthermore, deep learning techniques allow for multi-scale fusion, combining information from different scales and resolutions. This capability is particularly valuable in remote sensing, where images often contain details at various scales, ranging from fine structures to broader spatial contexts. By incorporating multiscale information, CNNs can generate fused images that preserve both local details and global patterns, leading to more accurate and comprehensive analysis (Yang et al., 2017).

The field for image fusion they can see considerable work has been done in the image fusion perspective, starting from image fusion method can be broadly classified into two groups from the temporal domain to a spatial realm. Research conducted by Wan et al. (2018) utilized CNN based multi-sensor fusion techniques for change detection in remote sensing images. The study demonstrated how CNN architectures could effectively capture and identify changes

Deep Learning for Earth Observation and Climate Monitoring. DOI: https://doi.org/10.1016/B978-0-443-24712-5.00016-6

in the environment. A case study by Tan et al. (2019) utilized CNN architectures for fusing multi-sensor data to classify forest cover in remote sensing images. The study revealed significant improvements in accuracy and resolution, aiding in better forest cover classification. A study by Dong et al. (2020) presented a methodology using a CNN-based framework for fusing hyperspectral and multispectral remote sensing images. The work demonstrated significant improvements in preserving spectral information and spatial details. Research conducted by Liu et al. (2020) employed a CNN based fusion method for combining multispectral remote sensing images. The study utilized an end-to-end learning approach, achieving high accuracy in fusing multispectral data from different sensors. A study by Li et al. (2020) showcased the use of CNN based image fusion for urban development mapping from aerial and satellite images. The research emphasized the accuracy and efficiency of CNN models in extracting and classifying urban structures. A case study by Pahlevan et al. (2022) explored the fusion of Landsat and Sentinel remote sensing imagery using CNN architectures. The study focused on optimizing data integration and improving classification accuracy for land cover mapping. Research by Saxena et al. (2021) applied deep learning techniques to fuse multimodal remote sensing data for agricultural monitoring and crop classification. The study demonstrated the robustness of CNN based fusion methods in accurately identifying and classifying crop types. These case studies illustrate how CNN and deep learning techniques are effectively used in diverse remote sensing applications.

The evaluation of remote sensing image fusion using CNNs, and deep learning techniques involves quantitative and qualitative assessments. Quantitative metrics, such as peak signal-to-noise ratio (PSNR), structural similarity index (SSIM), or entropy, provide objective measures of the fused image quality. Additionally, visual inspection and comparison with ground truth data allow for qualitative evaluation, ensuring that the fused image retains the essential characteristics of the input images while enhancing their interpretability (Wu et al., 2022).

The CNNs and deep learning techniques have revolutionized remote sensing image fusion by enabling automatic feature extraction, multi-scale fusion, and end-to-end learning. Their ability to capture complex patterns and relationships in the input data has significantly enhanced the spatial, spectral, and temporal information integration in fused images. Ongoing research and advancements continue to push the boundaries of CNN based fusion methods, paving the way for more accurate and robust analysis of remotely sensed data.

14.2 Remote sensing image fusion

14.2.1 Image fusion systems

There are two types of imaging fusion systems models provided in Fig. 14.1.

14.2.1.1 Single sensor fusion system

A single sensor fusion system (SSFS) is a system that combines data from one sensor to produce an output. In applications like remote sensing or medical imaging, multiple measurements may come from one sensor, and the aim is to merge these measurements for a more accurate depiction. SSFS enhances the quality, reliability, and interpretability of the information. The process involves data acquisition, preprocessing, feature extraction, fusion method, and fusion output. The selection of fusion technique depends on specific requirements and the nature of the acquired data. The efficacy of an SSFS can be assessed based on criteria like accuracy and computational efficiency. Overall, SSFS integrates and harnesses information from a single sensor for improved data interpretation and decision-making (Saxena et al., 2021).

14.2.1.2 Multi-sensor fusion system

The multi-sensor fusion system (MSFS) is a technological solution that merges data collected from multiple sensors to produce a cohesive result. This system finds utility in a range of fields, including autonomous vehicles, robotics, surveillance, and military systems. The primary objective of MSFS is to improve situational awareness, object detection and tracking, and decision-making processes. Through the integration of data from multiple sensors, the system can surpass limitations and deliver a more reliable depiction of the observed phenomenon. The process entails acquiring sensor

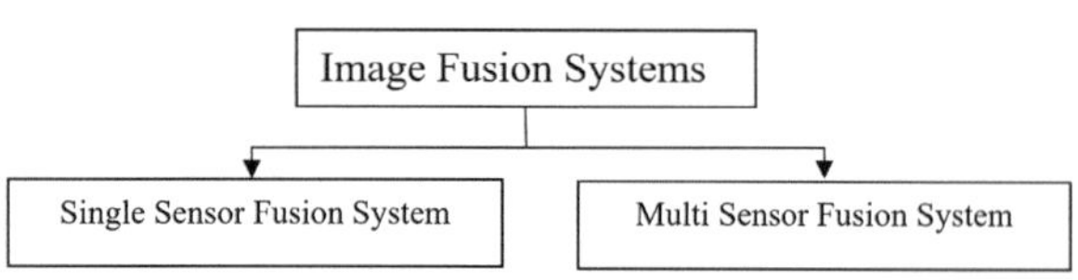

FIGURE 14.1 Image fusion systems.

data, registering it, preprocessing it, extracting relevant features, applying fusion methods, and producing the fused output. The selection of the fusion method relies on the particular needs of the application, the sensor properties, data quality, and the intended outcome. The assessment of MSFS can consider factors like precision, resilience, computational effectiveness, and its capacity to manage diverse sensor modalities and data circumstances. The MSFS integrates information from diverse sensors to enhance situational awareness, perception, and decision-making capabilities across various applications (Saxena et al., 2021).

14.2.2 Remote sensing image

Remote sensing images encompass images of the Earth's surface or celestial bodies that are obtained from a distance, often through the use of sensors on satellites, aircraft, drones, or other remote platforms. These images have extensive applications, including but not limited to environmental monitoring, agriculture, urban planning, disaster management, natural resource management, and scientific research. They offer valuable insights into the Earth's surface, its features, and temporal changes. Remote sensing images commonly include various spectral bands, including visible light, infrared, and microwave, offering a wide range of information about factors such as the well-being of vegetation, land utilization, and temperature. Remote sensing images exhibit a spectrum of spatial resolutions, spanning from finely detailed high-resolution images to broader low-resolution images that encompass larger geographic areas (Alhichri et al., 2021). The selection of resolution is contingent upon the particular application at hand. Certain remote sensing systems enable temporal monitoring through the acquisition of images of the same region at distinct time points, facilitating the monitoring of seasonal variations, deforestation, and urban expansion. Images can be obtained through a variety of platforms, including orbiting satellites, aircraft, drones, and sensors positioned on the ground. The choice of platform is influenced by factors such as the desired coverage area, resolution capabilities, and the frequency at which the area can be revisited. Specialized software and algorithms are utilized to process and analyze remote sensing images, performing various tasks such as image classification, detecting changes, and extracting relevant information (Alkhelaiwi et al., 2021). Remote sensing plays a crucial role in a wide range of disciplines, such as environmental science, agriculture, forestry, geology, and disaster response. It enables the collection of data across vast regions and facilitates the monitoring of environmental changes and trends, providing valuable insights for decision-making and policy development.

14.2.3 Image fusion techniques

Image fusion techniques involve combining information from multiple images to create a single, more informative image (Alhatami et al., 2022). These techniques are used in various fields, including remote sensing, medical imaging, computer vision, and more. These are some of the common image fusion techniques show in Table 14.1.

The choice of image fusion technique depends on the specific application, the characteristics of the input images, and the desired properties of the fused image.

14.2.4 Image fusion application

Image fusion has a wide range of applications across various fields, including remote sensing, medical imaging, computer vision, and more. These are some common applications of image fusion shown in Fig. 14.2.

14.2.4.1 Remote sensing

Image fusion is an indispensable tool in remote sensing, amplifying the quality and interpretability of acquired data. By integrating multiple images from various sensors and sources, image fusion aids in creating a single, comprehensive image or dataset that enhances the overall information content. This technique significantly contributes to accurate land cover classification, change detection, disaster management, and environmental monitoring. It allows for a more detailed and informative interpretation of the land surface, enabling analysts to extract more precise and reliable information for various applications. Remote sensing image fusion assists in monitoring environmental changes, such as deforestation, estimating soil moisture, and drought assessment. Moreover, in disaster response efforts, it provides high-resolution imagery crucial for accurate disaster assessment and management (Dong et al., 2020).

TABLE 14.1 Some of common techniques used in remote sensing image fusion.

Technique	Description
Pixel-level fusion	Combines pixel values of corresponding locations in multiple images using mathematical operations such as averaging, weighted summation, or maximum/minimum selection. While this approach is uncomplicated and computationally efficient, it might not fully harness the spatial and spectral information present in the images (Zhang et al., 2023).
Wavelet transform	Decomposes images into different frequency bands using wavelet transforms and combines the corresponding sub-bands to create a fused image (Nawaz et al., 2023). This technique can capture spatial details at different scales and has been widely used in remote sensing image fusion (Huyan et al., 2021).
Principal component analysis (PCA)	Applies PCA to the input images to transform them into a new set of uncorrelated variables called principal components. The principal components are then fused to create the final image. PCA-based fusion can effectively capture the spectral information of the input images (Sarker, 2021).
Intensity-hue-saturation (IHS) transform	Converts the input images from the RGB color space to the IHS color space, where intensity represents the spatial details, and hue and saturation represent the spectral information. The transformed components are fused, and the fused image is then converted back to the RGB color space. IHS fusion balances the spatial and spectral information in the fused image (Sarker, 2021).
Multi-resolution analysis	Utilizes multi-resolution representations, such as pyramids or scale-space analysis, to extract features at different levels of detail. The corresponding features are then combined to create the fused image. Multi-resolution analysis is effective in preserving spatial details while integrating spectral information from multiple images (Joudah Mounir et al., 2021).
Sparse representation-based fusion	Represents the input images using overcomplete dictionaries or sparse representations and combines the sparse coefficients to generate the fused image. This technique can effectively promote sparsity and preserve both spatial and spectral information in the fused image (Huyan et al., 2021).
Deep learning	Extensive training data is utilized to employ deep learning techniques, such as convolutional neural networks (CNNs) and other architectures, in order to obtain the fusion strategy deep learning-based fusion has the ability to capture intricate relationships and patterns within the data and has exhibited encouraging outcomes in the context of remote sensing image fusion tasks (Dong et al., 2020).
Model-based fusion	the fusion process incorporates statistical models or physical principles to provide guidance. To optimize the fusion outcome, these models consider sensor characteristics, imaging conditions, and prior knowledge about the scene. Model-based fusion approaches typically involve optimization algorithms or Bayesian inference to achieve the desired fusion result (Wan et al., 2018).

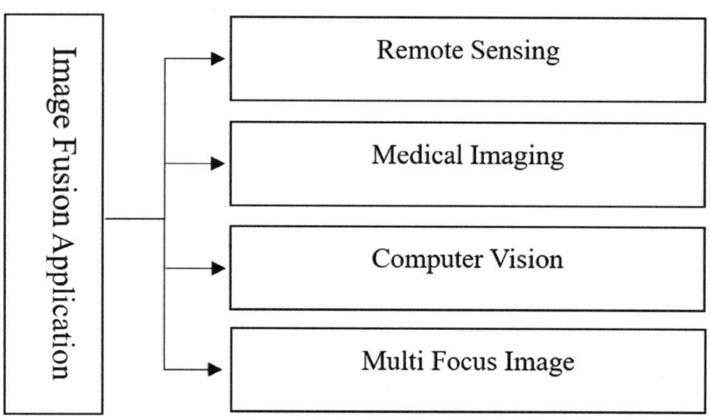

FIGURE 14.2 Image fusion application.

14.2.4.2 Medical image

Image fusion is a critical process in the field of medical imaging, revolutionizing diagnostic and treatment capabilities (Dong et al., 2023; Zhang et al., 2024). By combining data from various imaging modalities like Magnetic Resonance Imaging (MRI), Computed Tomography (CT), Positron Emission Tomography (PET), and ultrasound, image fusion enables a comprehensive view of anatomical structures and functional information (Nizamani et al., 2023). This technique assists in accurate disease diagnosis, tumor delineation, and treatment planning. It offers a more comprehensive understanding of complex conditions by integrating different imaging perspectives into a single, detailed representation. In medical applications image fusion helps in precise localization of abnormalities, offering a more thorough and cohesive view of patient conditions (Li et al., 2024; Xia et al., 2019). This allows healthcare professionals to make more informed decisions regarding patient care, treatment strategies, and surgical planning (Wu et al., 2023).

14.2.4.3 Computer vision

Image fusion serves as a critical element in the realm of computer vision, offering a powerful methodology to amalgamate and enhance multiple images for improved analysis and interpretation. Its application extends across various domains within computer vision, from aiding in object recognition and detection to bolstering surveillance systems, the integration of image fusion techniques has proven its efficacy. By consolidating data from diverse sources and sensors, it facilitates more accurate and detailed image analysis. Within the domain of computer vision, image fusion significantly contributes to more precise feature extraction, improved object localization, and heightened image clarity. It empowers the deciphering of complex scenes, extracting critical information, and offering a more nuanced understanding of visual data (Wang, Chen, et al., 2020). Ultimately, the integration of image fusion within computer vision methodologies elevates the accuracy and depth of image analysis, benefiting a wide array of fields and applications.

14.2.4.4 Multi focus image

Image fusion serves as a valuable technique in addressing the challenges posed by multi-focus images. Multi-focus imaging often involves capturing a scene with varying focal points, resulting in distinct areas of sharpness across multiple images. Image fusion techniques work to integrate these diverse images, creating a single composite image that captures the in-focus elements from each, resulting in a final output with an extended depth of field. This process is invaluable in scenarios such as macro photography or microscopic imaging, where different parts of an object or scene are at varying focal lengths. By merging and combining these multi-focus images, Image Fusion assists in producing a final composite image with an enhanced overall depth of field, ensuring that all areas of the scene are captured sharply and coherently (Zhang et al., 2021).

14.2.5 Advantages of image fusion

Image fusion is a powerful technique that offers several advantages across various fields. These advantages make image fusion a valuable tool in a wide range of applications, from scientific research to practical fields like agriculture, defense, and healthcare. These are some of the advantages of image fusion image fusion combines data from multiple sources or modalities, providing a more comprehensive and informative representation of a scene. This results in increased detail, clarity, and richness in the fused image.

1. Fused images often exhibit higher quality and reduced noise, making them more visually appealing and useful for various applications.
2. Fusion can increase the spatial resolution of images, which is especially valuable in remote sensing, where high-resolution imagery is crucial for detailed analysis.
3. Image fusion allows the integration of data from different sensors, such as optical, infrared, radar, or LiDAR, enabling a broader spectrum of information to be captured.
4. Fusion can improve the accuracy of feature detection, object recognition, and classification, making it invaluable for applications like object tracking and target detection.
5. Fused images are often more robust in various lighting and environmental conditions, making them suitable for applications where data quality may vary.
6. Image fusion eliminates redundant information, resulting in a more compact representation of the data without loss of important details.
7. It enables the integration of data from multiple modalities, such as color, thermal, and spectral, which is beneficial in applications like medical imaging and remote sensing.
8. Fused images are easier to interpret, making them beneficial in applications like medical diagnostics, where detailed visualization of anatomical and functional information is critical.
9. Fused images can simplify data interpretation, reducing the cognitive load on human analysts and aiding in decision-making processes.

10. Image fusion techniques can be implemented in real time, making them suitable for applications like surveillance, robotics, and autonomous vehicles.
11. In scientific research, image fusion aids in the analysis of complex data, enabling the extraction of valuable insights.
12. It is essential for monitoring and managing environmental changes, such as deforestation, agriculture, and natural disasters.
13. In astronomy, image fusion is used to enhance the visibility of celestial objects and phenomena, allowing for more detailed observations.
14. In medical imaging, it contributes to accurate diagnosis, treatment planning, and surgical navigation by fusing data from various medical imaging modalities.
15. Image fusion is a key component of remote sensing, enabling the creation of high-quality maps and land cover classification, among other applications.
16. It supports precision agriculture by providing detailed information for crop health monitoring and yield prediction.
17. Fused images aid in urban planning, helping assess changes in urban landscapes and infrastructure.
18. Image fusion is used to reveal hidden archeological features and preserve cultural heritage sites.
19. It is crucial for improving situational awareness, target detection, and reconnaissance in military and defense applications.

14.3 Convolutional neural network and deep learning technique for remote sensing image fusion

CNN and deep learning techniques have greatly impacted remote sensing image fusion. These methodologies enable the integration of multi-sensor data, facilitating the combination of information from various sources such as satellites and aerial platforms. They excel in capturing complex features from diverse data sets, enhancing the accuracy and precision of image fusion processes (Fig. 14.3). In this section we explore the application of CNNs and deep learning in remote sensing image fusion, then describe the different architectures, methodologies, algorithms, and explore the also application of CNNs and deep learning in remote sensing image fusion.

14.3.1 Convolutional neural network technique for remote sensing image fusion

Different network models or architectures can be designed to take advantage of the capabilities of CNNs for capturing spatial and spectral features in remote sensing image fusion. Below are some commonly used architectures:

1. U-Net: The U-Net architecture is widely employed in image fusion tasks. It consists of an encoder-decoder structure with skip connections. The encoder down samples the input image to extract hierarchical features, while the decoder up samples the features to generate the fused output. The skip connections enable the fusion of low-level and high-level features, preserving local details and global contextual information.
2. Residual network (Res Net): Res Net is a deep CNN architecture that addresses the vanishing gradient problem by introducing residual connections. These connections allow direct information flow from earlier layers to subsequent layers, enabling the network to learn residual mappings. Res Net architectures, such as ResNet-18 or ResNet-50, can be adapted for image fusion tasks, incorporating fusion operations at specific stages in the network.

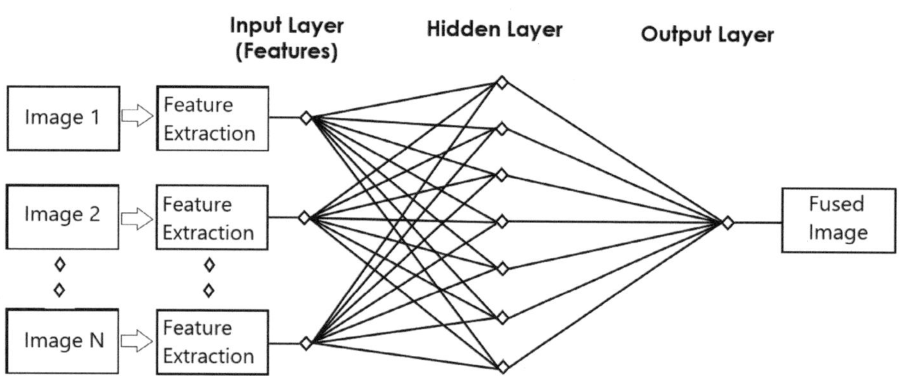

FIGURE 14.3 Convolutional neural network and deep learning technique for image fusion.

3. Dense net: Dense net is another CNN architecture that promotes feature reuse and information flow across layers. It introduces dense connections, where each layer receives input from all preceding layers. Dense net architectures, such as DenseNet-121 or DenseNet-169, can be modified for image fusion by incorporating fusion operations between dense blocks or at specific transition layers.
4. Inception network: The Inception network, also known as GoogLeNet, utilizes inception modules for multi-scale feature extraction. Inception modules consist of parallel convolutional layers with different kernel sizes, capturing features at multiple scales. These modules can be adapted for image fusion by incorporating fusion operations at different scales within the network.
5. Attention mechanisms: Attention mechanisms have gained attention in image fusion tasks. They allow the network to focus on informative regions and selectively fuse features. Self-attention mechanisms, such as the transformer architecture, use attention maps to assign weights to spatial locations. These mechanisms can be integrated into fusion networks to enhance the fusion process.
6. Encoder-decoder networks: Apart from U-Net, other encoder-decoder architectures like the fully convolutional network or SegNet can be adapted for image fusion. These networks encode the input image into a feature representation and decode it to generate the fused output. Fusion operations can be integrated at various stages in the encoder or decoder to facilitate information integration.
7. Hybrid architectures: Hybrid architectures combine multiple network components or architectures to leverage their complementary strengths. For example, a fusion network can incorporate both U-Net and ResNet components, where U-Net captures fine spatial details and Res Net captures high-level spectral features. These hybrid architectures aim to achieve a more comprehensive fusion process by combining different network characteristics.

It's important to consider the specific requirements of the image fusion task, the nature of the input data, and the available computational resources when choosing a network architecture. Researchers often experiment with various architectures and tailor them based on the characteristics of the remote sensing data and desired fusion outcomes. Ongoing research continues to explore new network architectures and fusion strategies to further enhance the performance of image fusion in remote sensing.

14.3.2 Convolutional neural network

The architecture of a CNN for fusing remote sensing images through deep learning techniques depends on the specific requirements of the fusion task and the characteristics of the input images. Fig. 14.4 shows a general outline of a CNN architecture for image fusion:

Input images: In image fusion typically have two or more input images from different sensors or modalities. Each input image is treated as a separate channel, and the number of channels in the input data is equal to the number of input images.

Convolutional layers: The initial convolutional layers extract features from each input image independently. These layers can have different numbers of filters and kernel sizes, depending on the information you want to extract from each image.

Feature concatenation: After the initial convolutional layers, the feature maps from different input images are concatenated. This merging of features allows the network to consider cross-channel information and relationships between the input images.

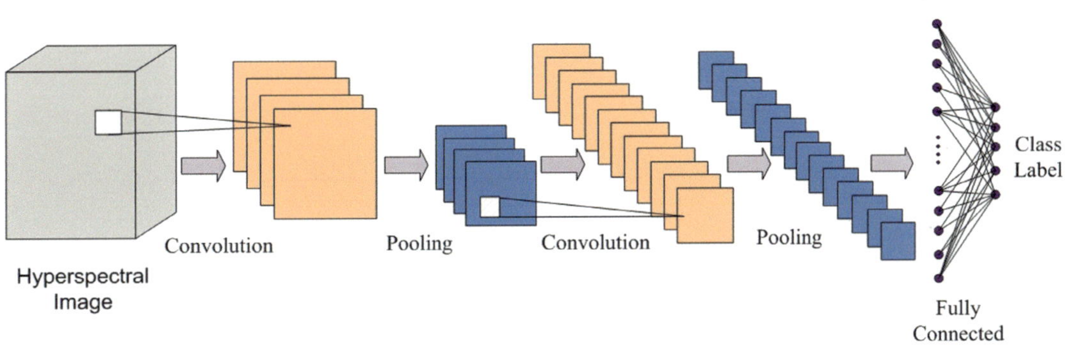

FIGURE 14.4 Architecture of convolutional neural network.

Convolution and fusion layers: Additional convolutional layers are applied to the concatenated feature maps to further capture interchannel interactions. These layers help in fusing the information from different channels.

Activation function: Apply an activation function, such as ReLU, to introduce nonlinearity in the network.

Pooling layers (Optional): incorporate pooling layers to down sample the feature maps and reduce computational complexity.

Fully connected layers: After processing the fused feature maps, you can add fully connected layers to the network. The number of neurons in the output layer depends on the specific fusion task.

Output layer: The output layer produces the final fused image or feature representation. The architecture of this layer may vary depending on the fusion goal. For instance, it could be a single channel (if you're fusing two images) or a multichannel image, when fusing more than two images.

Loss function: Define an appropriate loss function that quantifies the error between the predicted fused image and the ground truth. The choice of loss function depends on the fusion objective, such as mean squared error for image quality, or cross-entropy for classification tasks.

Optimization algorithm: Use an optimization algorithm (e.g., Adam or SGD) to train the network by adjusting the weights to minimize the loss function.

Hyperparameters: Set hyperparameters such as learning rate, batch size, and the number of epochs to control the training process.

The specific architecture details, including the number of layers, filter sizes, and hyperparameter values, will vary based on the complexity of the fusion task and the amount of available data. Customizing the architecture and hyperparameters is often necessary to optimize performance. Additionally, the choice of loss function and the evaluation metrics used depend on the specific objectives of the image fusion task, which can range from enhancing image quality to enabling advanced analyses such as land cover classification or change detection.

14.3.3 Commonly used remote sensing satellites

Remote sensing satellites are essential for capturing the data used in image processing applications. They provide a valuable source of imagery and data for various purposes, including land monitoring, environmental assessment, disaster management and more. Here are some sample of sensing images shown in Fig. 14.5.

Some of the commonly used remote sensing satellites that play a significant role in image processing.

FIGURE 14.5 Sample of sensing images.

Landsat satellites capture multispectral and thermal infrared data, which are often fused to produce high-quality, information-rich images. They have been pivotal in land cover classification, change detection, and environmental monitoring.

The Sentinel satellites, part of the European Space Agency's Copernicus program, provide both optical and radar imagery. Their data can be fused to create more detailed and versatile images for applications like agriculture, forestry, and disaster response.

Moderate resolution imaging spectroradiometer on NASA's Terra and Aqua satellites collects data across various spectral bands. This data is often fused to create images with enhanced information for climate modeling, land cover mapping, and environmental monitoring.

High-resolution optical images from Worldview satellites can be fused to create detailed, information-rich images for urban planning, disaster management, and land use analysis.

Advanced spaceborne thermal emission and reflection radiometer on NASA's Terra satellite captures multispectral data, including thermal infrared. These datasets are fused to create images for land cover classification, geology, and environmental monitoring.

These German synthetic aperture radar (SAR) satellites offer high-resolution radar data, which is often fused with optical data to create all-weather, detailed images for applications like agriculture and topographic mapping.

RADARSAT series: Canadian RADARSAT satellites provide SAR data suitable for fusion with optical imagery. This fusion enhances disaster monitoring, ship tracking, and agriculture applications.

Suomi national polar-orbiting partnership collects data for atmospheric and environmental research. Fusion of this data with other sources is valuable for weather forecasting, air quality assessment, and climate monitoring.

Quick Bird, IKONOS, and commercial satellites, the high-resolution optical images from commercial satellites can be fused for applications like urban planning, land use classification, and forestry management.

COSMIC-2 (constellation observing system for meteorology, ionosphere, and climate) satellites are used for atmospheric and ionospheric research and are suitable for fusion with other sources for climate studies and weather forecasting.

Image fusion involving remote sensing satellites is critical for improving the quality and in formativeness of images used in diverse applications. The selection of satellites and sensors for fusion depends on the specific requirements of the application.

14.3.4 Supervised learning for spatial spectrum remote sensing image fusion

Supervised learning for spatial spectrum remote sensing image fusion involves using labeled datasets to train models. By providing explicit data labels, this method enables models to learn and identify features, patterns, and relationships within the spectral and spatial data of remote sensing imagery. It offers enhanced accuracy and interpretability due to the labeled information but is constrained by the need for comprehensive, high-quality labeled datasets. This approach contributes to precise feature extraction and effective fusion, particularly when distinct features in the spectral and spatial domains are clearly identified and labeled for the models (Kang et al., 2021; Liu, Ding, et al., 2020).

14.3.5 Unsupervised learning for remote sensing image fusion

Unsupervised learning in remote sensing image fusion involves the utilization of algorithms that process data without explicit labeling or supervision. This technique aims to identify patterns, features, or relationships within the data autonomously. By doing so, it discovers inherent structures within multi-sensor images without prior knowledge or labeled datasets (Wang, Tan, et al., 2020). However, challenges arise concerning the interpretability and validation of the results, as well as potential limitations in accuracy when compared to supervised methods. Unsupervised learning offers an exploratory and data-driven approach to image fusion in remote sensing, allowing for the discovery of inherent patterns and structures within diverse data sources (Kang et al., 2021; Wang, Chen, et al., 2020).

14.3.6 Semi-supervised learning for remote sensing image fusion

Semi-supervised learning in remote sensing image fusion involves a hybrid approach that utilizes both labeled and unlabeled data to train models. This method capitalizes on the limited availability of labeled data by incorporating a more extensive set of unlabeled data (Kang et al., 2021). By harnessing the combined information, semi-supervised learning enhances the model's performance and robustness. However, challenges such as ensuring effective model generalization and maintaining accuracy with partial labeling persist within this approach. Nonetheless, the integration of both labeled and unlabeled data provides a valuable solution to enhance the efficiency of image fusion techniques in remote sensing applications (Kang et al., 2021).

14.4 Evaluation metrics used to assess the performance of convolutional neural network and deep learning techniques in image fusion

Several evaluation metrics are commonly used to assess the performance of CNN and deep learning techniques in image fusion (Alhatami et al., 2023). These metrics help measure the quality, accuracy, and similarity of the fused images compared to the original or input images. Some of the key evaluation metrics include Measures the similarity between the fused image and the original image based on luminance, contrast, and structure. Higher SSIM values indicate greater similarity. Assesses the quality of the fused image by measuring the ratio of the maximum possible power of a signal to the power of corrupting noise. Higher PSNR values indicate better quality. Measures the shared information content between the fused image and the source images. Higher mutual information values imply greater shared information. Evaluates the linear relationship between the fused image and the source images. Higher correlation coefficients indicate a stronger linear relationship. Measures the average difference between pixel values in the fused image and the source images. Lower RMSE values indicate higher similarity. An index that combines multiple criteria to assess the overall quality of the fused image, considering spatial and spectral information. Measures the amount of information or randomness in the fused image. Higher entropy values indicate greater diversity in pixel values. Measures the dispersion or variation of pixel values in the fused image. Higher standard deviation values indicate more diversity in pixel values. Subjective assessments by experts, focusing on visual quality, clarity, and semantic fidelity of the fused images compared to the original or reference images (Sarker, 2021).

These evaluation metrics collectively provide a comprehensive understanding of the quality, and accuracy of the fused images produced by CNN and deep learning techniques in image fusion. They help in quantifying and analyzing the effectiveness of the fusion algorithms in remote sensing and other domains.

14.5 Application of convolutional neural networks and deep learning in remote sensing image fusion

CNNs and deep learning techniques are extensively applied in remote sensing image fusion, enabling accurate, automated, and data-driven fusion processes. CNNs are utilized to fuse multimodal remote sensing data, such as optical, radar, LiDAR, and thermal imagery. This integration offers a comprehensive understanding of the Earth's surface by combining information from various sensors. Deep learning techniques, especially CNNs, automatically extract features from remote sensing data, reducing the need for manual feature engineering. This process assists in identifying relevant patterns and characteristics from different sensor inputs. End-to-end learning in CNNs facilitates direct fusion from raw input data to the production of fused images, simplifying the fusion process and eliminating the need for intermediate feature extraction steps (Wu et al., 2022). CNN based fusion techniques focus on preserving the semantic content of the input images. This approach ensures that important information is retained during the fusion process, maintaining the quality and interpretability of the fused imagery. Deep learning methods improve the accuracy of remote sensing image fusion by capturing complex relationships within the data. This leads to more precise and detailed fused images, aiding in better interpretation and analysis. CNN based fusion approaches are optimized for real-time or near-real-time applications, making them suitable for scenarios like autonomous vehicles, disaster monitoring, and surveillance systems, where rapid data processing is crucial. Transfer learning with pretrained CNN models (often from natural image datasets) reduces the need for extensive labeled training data. These pretrained models are fine-tuned for remote sensing tasks, enhancing the fusion process's efficiency. Generative adversarial networks are employed to generate synthetic remote sensing data, expanding the training dataset and enhancing the robustness of CNN based fusion models. CNN based fusion techniques bring significant quantitative and qualitative improvements to the fused imagery, leading to higher resolution, increased accuracy, and better spectral fidelity. The applications of CNN based fusion techniques span a wide range of sectors, including agriculture, environmental monitoring, disaster management, urban planning, surveillance, and military operations, providing more detailed and comprehensive information for decision-making (Sarker, 2021; Zhang et al., 2021).

These applications demonstrate the versatility and effectiveness of CNNs and deep learning techniques in improving the fusion of multimodal remote sensing data, leading to more accurate, informative, and actionable insights across various domains.

14.6 Challenges faced by convolutional neural network and deep learning techniques in remote sensing image fusion

CNN and deep learning techniques in remote sensing image fusion face several challenges, hindering their seamless implementation and performance. Some of the primary challenges include:

1. Data heterogeneity: Remote sensing datasets are often diverse in terms of sensor types, resolutions, and spectral characteristics. Integrating heterogeneous data sources requires adaptable and robust fusion techniques.
2. Limited labeled data: Training CNNs and deep learning models in remote sensing often suffer from a scarcity of labeled data, hindering their learning capacity. Supervised methods require extensive, high-quality labeled datasets for effective training.
3. Scale and resolution discrepancies: Handling multi-scale and multi-resolution data fusion is challenging. CNNs need to effectively handle differences in spatial resolutions between various sensors to avoid loss of information.
4. Complex feature extraction: Extracting relevant features from multi-sensor data is complex due to the high dimensionality and diverse characteristics of remote sensing data. Designing architectures that effectively extract such complex features is challenging.
5. Overfitting and generalization: Overfitting, where the model learns specific details from the training data, leading to poor generalization, remains a challenge. Models need to generalize well to unseen data, particularly in varied environmental conditions.
6. Interpretability and explain ability: CNNs are often considered as black-box models, lacking explain ability. In remote sensing, understanding how and why the model makes specific fusion decisions is crucial, especially in critical decision-making scenarios.
7. Computational demands: Deep learning models often require significant computational resources, including high-end GPUs and considerable training time, making them resource-intensive for some applications.
8. Robustness to environmental conditions: Models need to be robust against environmental variations, such as changes in weather conditions, illumination, and seasonal alterations, to ensure consistent performance.
9. Optimal fusion parameters: Selecting the optimal fusion parameters, network architectures, and hyperparameters to ensure the best fusion performance across diverse applications remains a challenge. Addressing these challenges is essential for the effective deployment of CNN and deep learning techniques in remote sensing image fusion. Researchers and practitioners continue to work toward developing more robust, adaptable, and efficient fusion methods to overcome these hurdles.

14.7 Conclusion

Deep learning and CNN techniques have revolutionized remote sensing image fusion by providing powerful tools to extract and fuse information from multiple sources. With further advancements and research, these techniques hold great promise for addressing the complex challenges in remote sensing applications and unlocking new possibilities in information extraction and analysis by improving fusion quality and adaptability. These methods excel at feature extraction and can learn fusion strategies directly from data. End-to-end fusion simplifies the workflow and enhances efficiency. Transfer learning enables leveraging pretrained models for improved performance with limited training data. Deep learning models can handle the fusion of heterogeneous data modalities. Applications span various domains, including land cover classification, urban monitoring, and disaster assessment. Ongoing research focuses on advancing architectures, attention mechanisms, and interpretability. Overall, deep learning and CNN techniques have significantly impacted remote sensing image fusion, offering valuable insights and benefits for analysis and decision-making.

References

Alhatami, E., Bhatti, U.A., Huang, M., & Feng, S.L. (2022). Image fusion based on discrete cosine transform with high compression. In *Seventh international conference on signal and image processing, ICSIP 2022* (pp. 606–610). China: Institute of Electrical and Electronics Engineers Inc. Available from http://ieeexplore.ieee.org/xpl/mostRecentIssue.jsp?punumber = 9886003, https://doi.org/10.1109/ICSIP55141.2022.9887300.

Alhatami, E., Bhatti, U.A., Huang, M.X., & Feng, S.L. (2023). Review and enhancement of discrete cosine transform (DCT) for medical image fusion. *Lecture notes in networks and systems 614* (pp. 89–97). Springer Science and Business Media Deutschland GmbH. Available from https://www.springer.com/series/15179, https://doi.org/10.1007/978-981-19-9331-2_8.

Alhichri, H., Alswayed, A. S., Bazi, Y., Ammour, N., & Alajlan, N. A. (2021). Classification of remote sensing images using efficientNet-B3 CNN model with attention. *IEEE Access*, *9*, 14078–14094. Available from https://doi.org/10.1109/ACCESS.2021.3051085, http://ieeexplore.ieee.org/xpl/RecentIssue.jsp?punumber = 6287639.

Alkhelaiwi, M., Boulila, W., Ahmad, J., Koubaa, A., & Driss, M. (2021). An efficient approach based on privacy-preserving deep learning for satellite image classification. *Remote Sensing*, *13*(11), 2221. Available from https://doi.org/10.3390/rs13112221.

Bhatti, U. A., Huang, M., & Neira-Molina, H. (2023). MFFCG—multi feature fusion for hyperspectral image classification using graph attention network. *Expert Systems with Applications*, *229*, 120496.

Bhatti, U. A., Yu, Z., Yuan, L., Zeeshan, Z., Nawaz, S. A., Bhatti, M., Mehmood, A., Ain, Q. U., & Wen, L. (2020). Geometric algebra applications in geospatial artificial intelligence and remote sensing image processing. *IEEE Access*, *8*, 155783–1155796. Available from https://doi.org/10.1109/access.2020.3018544.

Dong, F., Li, J., Bhatti, U. A., Liu, J., Chen, Y. W., & Li, D. (2023). Robust zero watermarking algorithm for medical images based on improved NasNet-mobile and DCT. *Electronics*, *12*(16). Available from https://doi.org/10.3390/electronics12163444, http://www.mdpi.com/journal/electronics.

Dong, S., Zhuang, Y., Yang, Z., Pang, L., Chen, H., & Long, T. (2020). Land cover classification from VHR optical remote sensing images by feature ensemble deep learning network. *IEEE Geoscience and Remote Sensing Letters*, *17*(8), 1396–1400. Available from https://doi.org/10.1109/LGRS.2019.2947022, http://ieeexplore.ieee.org/xpl/RecentIssue.jsp?punumber = 8859.

Huyan, L., Bai, Y., Li, Y., Jiang, D., Zhang, Y., Zhou, Q., Wei, J., Liu, J., Zhang, Y., & Cui, T. (2021). A lightweight object detection framework for remote sensing images. *Remote Sensing*, *13*(4), 683. Available from https://doi.org/10.3390/rs13040683.

Joudah Mounir, A., Mallat, S., & Zrigui, M. (2021). Analyzing satellite images by apply deep learning instance segmentation of agricultural fields. *Periodicals of Engineering and Natural Sciences (PEN)*, *9*(4), 1056. Available from https://doi.org/10.21533/pen.v9i4.2569.

Kang, J., Fernandez-Beltran, R., Duan, P., Liu, S., & Plaza, A. J. (2021). Deep unsupervised embedding for remotely sensed images based on spatially augmented momentum contrast. *IEEE Transactions on Geoscience and Remote Sensing*, *59*(3), 2598–2610. Available from https://doi.org/10.1109/TGRS.2020.3007029, https://ieeexplore.ieee.org/servlet/opac?punumber = 36.

Li, J., Zhang, Q., Yang, M., & Yuan, Y. (2024). *Robust zero watermarking algorithm for encrypted medical images based on SUSAN-DCT deep learning for multimedia processing applications: Volume one: Image security and intelligent systems for multimedia processing* (pp. 18–30). CRC Press. Available from https://www.taylorfrancis.com/books/edit/10.1201/9781003427674/deep-learning-multimedia-processing-applications-uzair-aslam-bhatti-jingbing-li-sibghat-ullah-bazai-huang-mengxing-muhammad-aamir, https://doi.org/10.1201/9781003427674-2.

Li, X., Gong, P., Zhou, Y., Wang, J., Bai, Y., Chen, B., Hu, T., Xiao, Y., Xu, B., Yang, J., Liu, X., Cai, W., Huang, H., Wu, T., Wang, X., Lin, P., Li, X., Chen, J., He, C., … Zhu, Z. (2020). Mapping global urban boundaries from the global artificial impervious area (GAIA) data. *Environmental Research Letters*, *15*(9), 094044. Available from https://doi.org/10.1088/1748-9326/ab9be3.

Liu, X., Liu, Q., & Wang, Y. (2020). Remote sensing image fusion based on two-stream fusion network. *Information Fusion*, *55* (, 1–15. Available from https://doi.org/10.1016/j.inffus.2019.07.010, http://www.elsevier.com/inca/publications/store/6/2/0/8/6/2/index.htt.

Liu, Y., Ding, L., Chen, C., & Liu, Y. (2020). Similarity-based unsupervised deep transfer learning for remote sensing image retrieval. *IEEE Transactions on Geoscience and Remote Sensing*, *58*(11), 7872–7889. Available from https://doi.org/10.1109/TGRS.2020.2984703, https://ieeexplore.ieee.org/servlet/opac?punumber = 36.

Nawaz, S. A., Li, J., Bhatti, U. A., Shoukat, M. U., & Ahmad, R. M. (2022). AI-based object detection latest trends in remote sensing, multimedia and agriculture applications. *Frontiers in Plant Science*, *13*. Available from https://doi.org/10.3389/fpls.2022.1041514, https://www.frontiersin.org/journals/plant-science.

Nawaz, S. A., Li, J., Shoukat, M. U., Bhatti, U. A., & Raza, M. A. (2023). Hybrid medical image zero watermarking via discrete wavelet transform-ResNet101 and discrete cosine transform. *Computers and Electrical Engineering*, *112*. Available from https://doi.org/10.1016/j.compeleceng.2023.108985, https://www.journals.elsevier.com/computers-and-electrical-engineering.

Nizamani, A. H., Chen, Z., Nizamani, A. A., et al. (2023). Feature-enhanced fusion of U-NET-based improved brain tumor images segmentation. *Journal of Cloud Computing*, *12*, 170. Available from https://doi.org/10.1186/s13677-023-00554-z.

Pahlevan, N., Smith, B., Alikas, K., Anstee, J., Barbosa, C., Binding, C., Bresciani, M., Cremella, B., Giardino, C., Gurlin, D., Fernandez, V., Jamet, C., Kangro, K., Lehmann, M. K., Loisel, H., Matsushita, B., Hà, N., Olmanson, L., Potvin, G., … Ruiz-Verdù, A. (2022). Simultaneous retrieval of selected optical water quality indicators from Landsat-8, Sentinel-2, and Sentinel-3. *Remote Sensing of Environment*, *270*, 112860. Available from https://doi.org/10.1016/j.rse.2021.112860.

Sarker, I. H. (2021). Deep learning: A comprehensive overview on techniques, taxonomy, applications and research directions. *SN Computer Science*, *2*(6). Available from https://doi.org/10.1007/s42979-021-00815-1, https://www.springer.com/journal/42979.

Saxena, V., Dwivedi, R.K., & Kumar, A. (2021). Analysis of machine learning algorithms for crop mapping on satellite image data. *Proceedings of the 2021 10th international conference on system modeling and advancement in research trends, SMART 2021* (pp. 23–28). India: Institute of Electrical and Electronics Engineers Inc. Available from http://ieeexplore.ieee.org/xpl/mostRecentIssue.jsp?punumber = 9676173, https://doi.org/10.1109/SMART52563.2021.9676320.

Tan, Z., Di, L., Zhang, M., Guo, L., & Gao, M. (2019). An enhanced deep convolutional model for spatiotemporal image fusion. *Remote Sensing*, *11* (24). Available from https://doi.org/10.3390/rs11242898, https://res.mdpi.com/d_attachment/remotesensing/remotesensing-11-02898/article_deploy/remotesensing-11-02898.pdf.

Wan, L., Zhang, T., & You, H. J. (2018). Multi-sensor remote sensing image change detection based on sorted histograms. *International Journal of Remote Sensing*, *39*(11), 3753–3775. Available from https://doi.org/10.1080/01431161.2018.1448481, https://www.tandfonline.com/loi/tres20.

Wang, M., Tan, K., Jia, X., Wang, X., & Chen, Y. (2020). A deep Siamese network with hybrid convolutional feature extraction module for change detection based on multi-sensor remote sensing images. *Remote Sensing*, *12*(2), 205. Available from https://doi.org/10.3390/rs12020205.

Wang, Z., Chen, B., Lu, R., Zhang, H., Liu, H., & Varshney, P. K. (2020). FusionNet: An unsupervised convolutional variational network for hyperspectral and multispectral image fusion. *IEEE Transactions on Image Processing*, *29*, 7565–7577. Available from https://doi.org/10.1109/TIP.2020.3004261, https://ieeexplore.ieee.org/xpl/mostRecentIssue.jsp?punumber = 83.

Wu, X., Hong, D., & Chanussot, J. (2022). Convolutional neural networks for multimodal remote sensing data classification. *IEEE Transactions on Geoscience and Remote Sensing*, *60* (, 1–10. Available from https://doi.org/10.1109/tgrs.2021.3124913.

Wu, Y., Zhang, L., Bhatti, U. A., & Huang, M. (2023). Interpretable machine learning for personalized medical recommendations: A LIME-based approach. *Diagnostics*, *13*(16), 2681. Available from https://doi.org/10.3390/diagnostics13162681.

Xia, Kj, Yin, Hs, & Wang, Jq (2019). A novel improved deep convolutional neural network model for medical image fusion. *Cluster Computing*, *22*, 1515–1527. Available from https://doi.org/10.1007/s10586-018-2026-1, http://www.kluweronline.com/issn/1386-7857.

Yang, J., Fu, X., Hu, Y., Huang, Y., Ding, X., & Paisley, J. (2017). PanNet: A deep network architecture for pan-sharpening. *Proceedings of the IEEE international conference on computer vision* (pp. 1753–1761). China: Institute of Electrical and Electronics Engineers Inc. Available from http://ieeexplore.ieee.org/xpl/conhome.jsp?punumber = 1000149, https://doi.org/10.1109/ICCV.2017.193.

Zhang, H. K., Roy, D. P., & Luo, D. (2023). Demonstration of large area land cover classification with a one dimensional convolutional neural network applied to single pixel temporal metric percentiles. *Remote Sensing of Environment*, *295*113653. Available from https://doi.org/10.1016/j.rse.2023.113653.

Zhang, H., Xu, H., Tian, X., Jiang, J., & Ma, J. (2021). Image fusion meets deep learning: A survey and perspective. *Information Fusion*, *76*, 323–3336. Available from https://doi.org/10.1016/j.inffus.2021.06.008, http://www.elsevier.com/inca/publications/store/6/2/0/8/6/2/index.htt.

Zhang, Y., Chen, J., Ma, X., Wang, G., Bhatti, U. A., & Huang, M. (2024). Interactive medical image annotation using improved attention U-net with compound geodesic distance. *Expert Systems with Applications*, *237*121282. Available from https://doi.org/10.1016/j.eswa.2023.121282.

Chapter 15

Oil spills and the ripple effect: exploring climate and environmental impacts through a deep learning lens

Yonis Gulzar[1], **Faheem Ahmad Reegu**[2] **and Shahnawaz Ayoub**[3]
[1]*Department of Management Information Systems, College of Business Administration, King Faisal University, Al-Ahsa, Saudi Arabia,* [2]*Department of Electrical and Electronics Engineering, College of Engineering and Computer Science, Jazan, Arabia Arabia,* [3]*Glocal School of Science and Technology, Glocal University, Saharanpur, Uttar Pradesh, India*

15.1 Introduction

Oil spills pose a significant environmental hazard, impacting both water and land ecosystems. They lead to marine and coastal pollution, endangering birds, fish, and various aquatic organisms (Frate et al., 2000). Primarily stemming from accidents involving oil tankers, ships, and ruptured pipelines transporting petroleum products, these spills also result from illegal oil discharges and residue from vessel cleaning (Brekke & Solberg, 2005; Solberg et al., 2007). With the continuous growth of maritime transportation in recent decades, oil spill incidents have surged. Between 1970 and 2010, an alarming 5.71 million tons of oil leaked from vessels (Dave, 2011). Suspicion often falls on oil tankers and maritime vessels for clandestine oil dumping activities. Hence, prioritizing swift and accurate detection of such incidents is crucial for mitigating environmental harm, preserving cleanliness, and safeguarding marine and coastal life (Frate et al., 2000).

Currently, a range of tools including ships, aircraft, and satellites are employed for the detection and surveillance of oil spills. Traditional methods such as aerial photography and on-site inspections have been conventionally utilized for this purpose, albeit requiring skilled personnel, specialized equipment, and substantial time investments, rendering them economically burdensome (Brekke & Solberg, 2005; Solberg et al., 2007). Vessels, particularly those outfitted with specialized radar systems, offer a means to identify oil slicks on the sea surface; however, their effectiveness is limited by factors such as restricted visibility and the inability to cover expansive areas efficiently. Consequently, satellite-based observations, notably leveraging synthetic aperture radar (SAR) technology, emerge as the preeminent approach for monitoring and detecting oil contamination in marine ecosystems (Fingas & Brown, 2018). SAR, functioning as an active microwave sensor, captures backscattered signals across various frequencies and polarizations, generating detailed two-dimensional images. Widely favored for its independence from sunlight, resilience to cloudy conditions, extensive coverage capabilities, and cost-effectiveness compared to SAR-equipped aircraft, SAR sensors represent a pivotal tool in oil spill detection efforts at sea (Ciappa, 2023).

Furthermore, satellite-based SAR technology offers unparalleled advantages in scalability and accessibility. Unlike aerial surveys, which are constrained by factors like weather conditions and operational costs, satellite observations provide consistent coverage regardless of environmental variables. This consistency allows for continuous monitoring of vast oceanic regions, enabling early detection and rapid response to oil spill incidents. Additionally, the integration of computer vision techniques into SAR data analysis holds immense potential for enhancing oil spill detection efforts (Basit et al., 2022). Algorithms based on machine learning and image processing can automatically identify and classify oil slicks in SAR imagery, streamlining the detection process and enabling real-time monitoring. By leveraging computer vision capabilities, stakeholders can improve the efficiency and accuracy of oil spill detection, thereby mitigating environmental impacts and contributing to climate resilience and conservation efforts (Dehghani-Dehcheshmeh et al., 2023;

Deep Learning for Earth Observation and Climate Monitoring. DOI: https://doi.org/10.1016/B978-0-443-24712-5.00017-8

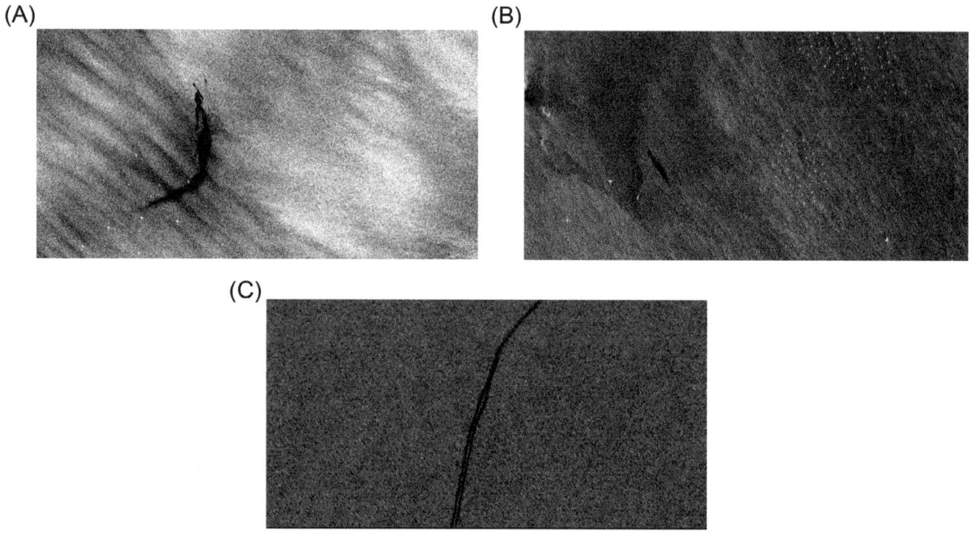

FIGURE 15.1 Showcases a selection of images depicting oil spill incidents.

Malik et al., 2023). Fig. 15.1 presents a series of image samples portraying the aftermath of oil spills, showcasing the visible effects of these incidents on the environment.

15.1.1 Deep leaning

Deep learning, a subset of artificial intelligence inspired by the structure and function of the human brain's neural networks, has found widespread applications across various domains. In healthcare, it enables the analysis of medical images for disease diagnosis and prognosis, revolutionizing patient care (Alam et al., 2022, 2023; Anand et al., 2023; Gulzar & Khan, 2022; Khan et al., 2021, 2023; Khan, Ayoub, et al., 2023; Majid et al., 2023; Mehmood et al., 2023). In education, deep learning facilitates personalized learning experiences through adaptive tutoring systems and intelligent educational platforms (Sahlan et al., 2021). In agriculture, it aids in crop monitoring, yield prediction, and pest detection, enhancing agricultural productivity and sustainability (Dhiman et al., 2023; Gulzar, 2023; Gulzar et al., 2023; Mamat et al., 2023; Albarrak et al., 2022; Alkanan & Gulzar, 2024; Amri et al., 2024; Gulzar et al., 2020; Gulzar, Alwan, et al., 2023; Gulzar, Ünal, Ayoub, Reegu, 2024; Hamid et al., 2022). In finance, deep learning powers algorithmic trading strategies, fraud detection systems, and customer relationship management tools, optimizing financial decision-making processes (Ayoub et al., 2022; Gulzar, Ünal, Ayoub, Reegu & Altulihan, 2024; Hamid et al., 2023).

When applied to the detection of oil spills, deep learning holds significant promise for improving detection accuracy and efficiency. By analyzing satellite imagery and SAR data, deep learning algorithms can automatically identify and classify oil slicks with high precision, reducing the reliance on manual inspection and human intervention. This not only streamlines the detection process but also minimizes labor costs and accelerates response times to oil spill incidents. Furthermore, deep learning techniques can enhance the understanding of oil spill dynamics and environmental impacts through predictive modeling and data analytics, aiding in the development of more effective mitigation strategies. Overall, leveraging deep learning in oil spill detection offers a cost-effective and sustainable approach to safeguarding marine ecosystems and mitigating the adverse effects of oil pollution on the environment and climate.

15.1.2 Objectives

The objective of this critical review is to explore the multifaceted impacts of oil spills on climate and the environment, with a particular focus on leveraging deep learning methodologies for comprehensive analysis. This study aims to:

1. Literature review: Conducting a comprehensive literature review to synthesize existing knowledge on the environmental and climate impacts of oil spills, emphasizing the role of deep learning methodologies.
2. Data extraction and analysis: Extracting and analyzing relevant data sources, including satellite imagery, SAR data, and environmental datasets, to assess the applicability of deep learning techniques in oil spill detection and monitoring.

3. Performance evaluation of deep learning models: Assessing the performance of deep learning models in detecting oil spills and predicting their environmental consequences, evaluating metrics such as accuracy, precision, and recall.
4. Identification of challenges and limitations: Identifying challenges and limitations associated with the application of deep learning in oil spill research, including issues related to data availability and quality, model interpretability, and generalization, and Integration with existing monitoring systems.
5. Exploration of future directions: Exploring future directions and opportunities for advancing deep learning-based approaches in addressing the complexities of oil spills and their impacts on climate resilience, environmental conservation, and sustainable development.

15.2 Material and method

In this critical review, we undertook a comprehensive exploration of the Scopus database to uncover relevant literature. Employing a carefully crafted search string, we aimed to retrieve data pertinent using following search string:

("deep learning" OR "neural networks" OR cnn OR "convolutional neural network") AND ("image classification" OR "object detection" OR "image detection") AND ("oil spill" OR "oil slick")

Notably, these keywords were strategically chosen to encompass a broad spectrum of research encompassing various fields, including the title, abstract, author keywords, and Keywords Plus.

In our pursuit of retrieving pertinent research articles, we identified a total of 64 studies as of March 10, 2024. Among these, we observed a diverse distribution comprising 34 research articles, 25 conference papers, and 5 book series entries. To refine our focus and ensure relevance to our objectives, we subsequently narrowed our attention to research articles specifically addressing oil spill detection or classification. This meticulous curation process led us to select 22 articles for in-depth examination and analysis. Through this methodological approach, we aimed to capture a comprehensive understanding of the current landscape surrounding the utilization of deep learning methodologies in addressing the challenges posed by oil spills.

15.3 Findings and discussion

In the pursuit of enhancing our understanding and capabilities in detecting and classifying oil spills, the studies outlined in Table 15.1 provide a diverse array of insights into various facets of this complex problem. In this section, we embark on a comprehensive exploration of the methodologies, findings, and implications encapsulated within the dataset. We begin by delving into the intricate world of model architecture, where researchers have ingeniously crafted deep learning frameworks tailored to the specific challenges posed by oil spill detection. Subsequently, we scrutinize the study types employed, ranging from classification to detection methodologies, each offering unique perspectives and approaches to tackling the problem at hand. Following this, we dissect the datasets utilized in these endeavors, ranging from synthetic data to real-world SAR imagery, illuminating the crucial role of data in driving advancements in this field. Moreover, we delve into the nuanced interplay between accuracy achieved and model tuning techniques, elucidating how fine-tuning parameters and architectures have propelled the field forward in achieving ever-higher accuracies. Finally, we contextualize these findings within the geographical landscapes where these studies were conducted, shedding light on the diverse environmental contexts and the relevance of oil spill detection in safeguarding these ecosystems.

15.3.1 Model architecture

The studies presented in the table employ a wide array of model architectures to tackle the challenges inherent in oil spill detection and classification. Among these architectures, convolutional neural networks (CNNs) stand out as a cornerstone methodology, given their adeptness at automatically learning hierarchical representations of image data. Utilized across various tasks such as classification and detection, CNNs offer a versatile framework for analyzing SAR images, UV images, RGB images, and hyperspectral images. Ranging from shallow CNNs to deeper variants like VGG16 and ResNet, these architectures strike a balance between computational efficiency and performance. Moreover, generative adversarial networks (GANs) find their place in image enhancement and dehazing tasks, significantly improving the quality of SAR images by generating synthetic counterparts that closely mimic real-world scenarios. In certain studies, recurrent neural networks (RNNs) make an appearance, leveraging their sequential modeling capabilities to capture temporal dependencies in SAR image sequences, thereby facilitating the detection and monitoring of oil spill events over time. Fully convolutional networks, on the other hand, are tailored for semantic segmentation tasks, proving instrumental in delineating oil spill boundaries within SAR images. Meanwhile, object detection architectures like

TABLE 15.1 Summary of different research works about detecting/classifying oil spill using deep learning.

Paper	Model trained	Study type	Dataset type	Accuracy	Model tuning techniques	Area of oil spill
Fan and Liu (2023)	MTGANs	Classification	SAR images	97.22%	GAN, fully convolutional symmetric structure, multiple convolution blocks	Not specified
Blondeau-Patissier et al. (2023)	Deep learning, empirical approach	Detection	SAR images	>98%	Empirical approach, deep learning model	Great barrier reef marine park
Topouzelis et al. (2009)	Genetic algorithms	Detection	SAR images	85.3% (oil spills), 84.4% (look-alikes)	Genetic algorithms	Not specified
Baek and Jung (2021)	SVM, random forest, DNN	Classification	SAR images	~0.889 (DNN, dual-pol)	Support vector machine (SVM), random forest (RF), deep neural network (DNN)	Kerch strait oil spill event
Lang et al. (2022)	Shallow CNN	Classification	SAR images	84.67%	Shallow CNN, pre-training on task-specific dataset, feature refinement	Not specified
Yuan et al. (2018)	Not specified	Detection	Synthetic data	100% (Level 2 classification)	Computer vision, natural scene statistics, white balance preprocessing, background modeling, multi-dimensional feature parameters	Not specified
Zheng et al. (2022)	Dehaze-AGGAN	Dehazing	SAR images	Not specified	Enhanced attention-guide generative adversarial networks (AGGAN), total variation loss, cycle consistency loss	Not specified
de Moura et al. (2022)	Deep learning	Detection	SAR images	98%	Semantic segmentation architectures (U-net, DeepLabv3 + , LinkNet), backbones (ResNet-101, ResNet-50, efficient-net-B0, efficient-net-B3)	Brazilian territory
Garcia-Pineda et al. (2009)	TCNNA	Detection	SAR images	Not specified	Edge-detection filters, textural descriptors, neural network algorithm	Gulf of Mexico
Li et al. (2017)	Genetic neural network	Detection	SAR images	91.42%	Graph-based visual saliency model, spectral similarity match model, genetic neural network	Not specified
Passah and Kandar (2023)	Lightweight CNN	Classification	SAR images	97%	Single-unit kernel for feature mapping, depth-wise convolutions	Not specified
Mehdi et al. (2022)	YOLOv3 (Modified)	Detection	UV and RGB images	86.89	MobileNetv2 backbone architecture, generalized intersection over union (GIoU) loss function	Not provided
Das et al. (2023)	Convolutional neural network (CNN)	Classification	SAR images	99.06	Not provided	Not provided
Ghorbani and Behzadan (2021)	VGG16, mask R-CNN, PSPNet, YOLOv3	Detection, segmentation, classification	Visual dataset of oil spills	92 (Classification), 49–68 (Segmentation), ~71 (Classification)	Transfer learning	Not provided

Dasari et al. (2022)	Supervised SVM and neural network classifiers	Classification	SAR images	98.13	Preprocessing, supervised classification	Chennai, India
Scardigli et al. (2023)	Neural network architectures	Detection	Synthetic aperture radar (SAR) images	82–92	Nonlinear time-based block layers	Aegean Sea
Sudha and Vijendran (2024)	Faster R-CNN with enhanced mobileNetV2	Detection	Synthetic and real-world oil spill photos	97%	Preprocessing (non-adaptive threshold with CLAHE), Fused UNet segmentation, CNN (AlexNet architecture) for feature extraction	Not provided
Huang et al. (2020)	Faster R-CNN	Identification	UV images at 365 nm, blue channel images, RGB images	88.8%	Faster R-CNN-based methods for segmentation	Not provided
Xu et al. (2014)	Support vector machine, artificial neural network, tree-based ensemble classifiers (bagging, bundling, boosting), generalized additive model, penalized linear discriminant analysis	Identification	RADARSAT-1 imagery	N/A	Feature construction techniques	Canada
Bianchi et al. (2020)	Deep learning framework	Detection and categorization	SAR images	State-of-the-art performance	N/A	Not provided
Han et al. (2022)	LCSE-ResNet	Classification	Optical remote sensing imagery	N/A	Laplacian of Gaussian (LoG) operator, connected domain controller	Not provided
Yang et al. (2022)	Deep learning object detector	Detection	Sentinel-1 SAR imagery	Average precision: 69.10% (validation), 68.69% (test)	N/A	Eastern Mediterranean Sea

Faster R-CNN, YOLOv3, and Mask R-CNN excel in localizing and classifying oil spill instances within images, thanks to their region proposal and classification heads coupled with feature extraction backbones. Furthermore, hybrid architectures emerge, combining elements from different methodologies, such as CNNs for feature extraction and RNNs for temporal modeling, to address the multifaceted nature of oil spill detection and monitoring comprehensively. This diverse repertoire of model architectures underscores the importance of selecting appropriate methodologies tailored to the specific requirements of oil spill detection tasks, aiming to develop robust and adaptable models capable of accurately identifying and characterizing oil spill events across various environmental conditions and geographical regions.

15.3.2 Study type

The studies presented in the Table 15.1 primarily focus on two key study types: classification and detection, both integral to effectively addressing oil spill events. In classification studies, researchers employ various machine learning algorithms to categorize image pixels or regions, distinguishing between those containing oil spills and those representing other elements. These studies utilize diverse imagery types, including SAR, UV, RGB, and hyperspectral images, achieving remarkable accuracy rates often exceeding 90%. Genetic algorithms are also utilized to optimize feature selection and improve classification performance. On the other hand, detection studies concentrate on automatically identifying and localizing oil spill instances within images. Researchers deploy sophisticated object detection architectures like Faster R-CNN, YOLOv3, and Mask R-CNN, showcasing high precision and recall rates, with accuracies surpassing 98% in some instances, particularly in SAR image-based detection. Additionally, hybrid approaches, blending deep learning with empirical methods or computational techniques such as GANs and genetic algorithms, are employed in several studies to bolster accuracy and robustness in oil spill detection and classification systems. These classification and detection methodologies underscore the pivotal role of machine learning and deep learning techniques in advancing environmental monitoring and safeguarding efforts against oil spill occurrences.

15.3.3 Dataset used

The datasets utilized across these studies encompass a wide array of sources and types, reflecting the diversity of oil spill monitoring applications. SAR images, known for their effectiveness in capturing oil spill features, are predominant, highlighting their significance in oil spill detection and classification research. SAR imagery offers valuable insights due to its ability to penetrate cloud cover and provide all-weather, day-and-night monitoring capabilities. Other studies leverage UV, RGB, and hyperspectral images, each offering unique advantages in detecting oil spills based on spectral characteristics and surface properties. Furthermore, some studies utilize synthetic datasets, allowing researchers to explore algorithm performance under controlled conditions before deployment in real-world scenarios (Ayoub et al., 2022). This diverse range of datasets enables researchers to develop and evaluate models across various environmental conditions, facilitating the creation of robust and adaptable oil spill monitoring systems.

15.3.4 Accuracy achieved and model tuning

Table 15.1 illustrates a remarkable range of accuracies achieved across different studies, showcasing the efficacy of various deep learning models and techniques in oil spill detection and classification. Several studies report high accuracies exceeding 90%, underscoring the potential of deep learning approaches in accurately identifying oil spills from different types of imagery. Model tuning emerges as a crucial factor contributing to the enhancement of accuracy in many cases. Techniques such as fine-tuning model architectures, adjusting hyperparameters, and optimizing loss functions have been instrumental in fine-tuning models to achieve higher accuracies. For instance, the utilization of genetic algorithms, deep learning models, and ensemble methods has allowed researchers to refine model performance, particularly in distinguishing oil spills from look-alike phenomena. Additionally, preprocessing techniques such as image dehazing, contrast enhancement, and background modeling have improved model robustness and accuracy by enhancing image quality and reducing noise. The successful implementation of these model tuning techniques underscores their significance in achieving higher accuracies and improving the reliability of oil spill detection and classification systems.

15.3.5 Geographical context

The studies encompass a diverse array of geographical locations, each presenting unique challenges and contexts for oil spill detection and mitigation. These areas include globally significant marine environments such as the Great Barrier

Reef Marine Park, the Kerch Strait Oil Spill Event, the Gulf of Mexico, and the Eastern Mediterranean Sea. Additionally, specific regions like the Brazilian territory, Chennai in India, and the Aegean Sea have also been focal points of investigation. The selection of these areas underscores the global impact of oil spills on both marine ecosystems and human livelihoods. Researchers have tailored their methodologies to address the specific characteristics and challenges posed by each location, ranging from the extensive use of satellite imagery in vast oceanic regions to the application of optical remote sensing techniques in coastal areas. By focusing on diverse geographical regions, these studies contribute to a comprehensive understanding of oil spill dynamics and facilitate the development of tailored detection and mitigation strategies suited to various environmental contexts.

15.4 Identification of challenges and limitations

Despite the advancements made in utilizing deep learning for oil spill detection and classification, several challenges and limitations persist, hindering the widespread adoption and effectiveness of these methodologies.

1. Data availability and quality
   a. Limited availability of labeled datasets poses a significant challenge for training and evaluating deep learning models. Acquiring sufficiently large and diverse datasets that encompass various environmental conditions and oil spill scenarios remains a challenge, particularly for regions with sparse or inaccessible data.
   b. Moreover, the quality of available datasets can vary significantly, leading to biases and inaccuracies in model training. Ensuring the accuracy and representativeness of training data is crucial for developing robust and generalizable models.
2. Generalization and transferability:
   c. Deep learning models trained on specific datasets may struggle to generalize to unseen data or different environmental conditions. Transferability of models across geographical regions, sensor platforms, and imaging modalities is often limited, necessitating retraining or fine-tuning for new scenarios.
   d. Variations in oil spill characteristics, such as size, shape, and spectral properties, further compound the challenge of developing universally applicable detection and classification algorithms.
3. Model interpretability:
   e. Deep learning models, particularly complex architectures like CNNs and RNNs, often lack interpretability, making it challenging to understand the underlying features driving model predictions. Interpretable models are essential for building trust and confidence in algorithmic outputs, especially in critical applications such as environmental monitoring.
   f. Addressing the black-box nature of deep learning models is crucial for providing actionable insights and facilitating decision-making by stakeholders and policymakers.
4. Integration with existing monitoring systems:
   g. Integrating deep learning-based solutions into existing oil spill monitoring systems and decision-support frameworks presents technical and logistical challenges. Compatibility with legacy systems, data interoperability issues, and ensuring real-time or near-real-time processing capabilities are key considerations for seamless integration.
   h. Collaboration among interdisciplinary teams, including environmental scientists, data scientists, engineers, and policymakers, is essential for bridging the gap between technological innovation and practical application in the field.

Addressing these challenges and limitations requires a concerted effort from the scientific community, industry stakeholders, policymakers, and civil society. By fostering collaboration, investing in research and development, and prioritizing ethical and equitable deployment practices, the potential of deep learning to mitigate the environmental impacts of oil spills can be realized effectively.

15.5 Exploration of future directions

As we reflect on the findings and discussions presented in this study, several avenues for future research and development emerge, offering promising opportunities to further enhance the application of deep learning in addressing the multifaceted challenges posed by oil spills and their broader environmental implications.

1. Integration of multimodal data sources: Future research endeavors should explore the integration of multimodal data sources, including satellite imagery, aerial surveys, in-situ sensor data, and environmental monitoring networks, to enrich the information available for oil spill detection and characterization. Leveraging diverse data modalities can

provide complementary insights into oil spill dynamics, environmental impacts, and response strategies, thereby enhancing the effectiveness and robustness of deep learning models.

2. Advancements in model interpretability and explainability: Efforts to enhance the interpretability and explainability of deep learning models represent a critical research frontier, enabling stakeholders to better understand model predictions and trust the reliability of model-driven decisions. Exploring techniques for model interpretability, such as attention mechanisms, feature visualization, and uncertainty quantification, can facilitate transparent communication of model outputs and foster informed decision-making in oil spill response and management.
3. Scaling deep learning approaches for global impact: Scaling deep learning approaches to address global-scale challenges requires collaborative efforts across interdisciplinary teams, leveraging cloud computing infrastructure, and distributed data repositories. By harnessing the power of distributed computing and federated learning techniques, researchers can develop scalable and adaptive deep learning models capable of addressing diverse environmental contexts and supporting decision-making at local, regional, and global scales.
4. Incorporation of climate resilience and adaptation strategies: Integrating climate resilience and adaptation strategies into deep learning-based approaches can enhance the resilience of coastal communities and ecosystems to oil spill events and associated environmental stressors. Future research should explore the incorporation of climate change projections, sea-level rise scenarios, and ecosystem vulnerability assessments into oil spill risk assessment and response planning, thereby fostering adaptive management strategies that mitigate the impacts of oil spills in a changing climate.
5. Empowering stakeholder engagement and capacity building: Promoting stakeholder engagement and capacity building is essential for ensuring the effective deployment and uptake of deep learning technologies in oil spill monitoring and management. Future research initiatives should prioritize the codesign and co-development of deep learning solutions in collaboration with end-users, including government agencies, industry partners, nongovernmental organizations, and local communities. By empowering stakeholders with the necessary skills and knowledge, we can foster a culture of data-driven decision-making and collective action toward sustainable oil spill management.

15.6 Conclusion

This study has provided a comprehensive exploration of the application of deep learning in addressing the challenges of oil spill detection and classification, with a focus on understanding the broader environmental impacts and implications. Through an in-depth exploration of methodologies, model architectures, dataset types, accuracy achieved, and geographical contexts, we have identified key trends and challenges shaping the field of oil spill detection and monitoring. The diverse array of model architectures, from CNNs to RNNs and object detection frameworks like Faster R-CNN and YOLOv3, highlights the versatility and adaptability of deep learning approaches in addressing the complexities of oil spill detection. Moreover, the utilization of various study types, including classification and detection methodologies, underscores the importance of leveraging machine learning techniques to effectively identify and localize oil spill instances within imagery data. Dataset availability and quality emerge as critical challenges, with the need for large, diverse, and accurately labeled datasets to train and evaluate deep learning models effectively. Additionally, the generalization and transferability of models across different environmental conditions and geographical regions remain areas of concern, requiring further research and development efforts.

The integration of multimodal data sources, advancements in model interpretability, scaling deep learning approaches for global impact, incorporation of climate resilience strategies, and empowerment of stakeholder engagement and capacity building represent promising avenues for future research and development. By addressing these challenges and embracing these opportunities, we can enhance the effectiveness and sustainability of oil spill detection and management efforts, contributing to the protection of marine ecosystems and the promotion of environmental resilience worldwide.

Acknowledgments

This work was supported by the Deanship of Scientific Research, Vice Presidency for Graduate Studies and Scientific Research, King Faisal University, Saudi Arabia, under project KFU241962.

References

Alam, S., Raja, S. P., Gulzar, Y., & Mir, M. S. (2023). Enhancing autism severity classification: Integrating LSTM into CNNs for multisite meltdown grading. *International Journal of Advanced Computer Science and Applications*, *14*(12), 670–677. Available from https://doi.org/10.14569/IJACSA.2023.0141269, http://thesai.org/Publications/Archives?code = IJACSA.

Alam, S., Raja, P., & Gulzar, Y. (2022). Investigation of machine learning methods for early prediction of neurodevelopmental disorders in children. *Wireless Communications and Mobile Computing, 2022*. Available from https://doi.org/10.1155/2022/5766386, https://www.hindawi.com/journals/wcmc/.

Albarrak, K., Gulzar, Y., Hamid, Y., Mehmood, A., & Soomro, A.B. (2022.). A deep learning-based model for date fruit classification. *Sustainability, 14*(10), 6339. https://doi.org/10.3390/su14106339.

Alkanan, M., & Gulzar, Y. (2024). Enhanced corn seed disease classification: leveraging mobilenetV2 with feature augmentation and transfer learning. *Frontiers in Applied Mathematics and Statistics*. Available from https://doi.org/10.3389/FAMS.2023.1320177.

Amri, E., Gulzar, Y., Yeafi, A., Jendoubi, S., Dhawi, F., & Mir, M. S. (2024). Advancing automatic plant classification system in Saudi Arabia: Introducing a novel dataset and ensemble deep learning approach. *Model Earth Systems and Environment*, 1–17. Available from https://doi.org/10.1007/S40808-023-01918-9/METRICS.

Anand, V., Gupta, S., Gupta, D., Gulzar, Y., Xin, Q., Juneja, S., Shah, A., & Shaikh, A. (2023). Weighted average ensemble deep learning model for stratification of brain tumor in MRI images. *Diagnostics*, 1320. Available from https://doi.org/10.3390/DIAGNOSTICS13071320.

Ayoub, S., Gulzar, Y., Reegu, F. A., & Turaev, S. (2022). Generating image captions using Bahdanau attention mechanism and transfer learning. *Symmetry (Basel), 14*.

Ayoub, S., Gulzar, Y., Rustamov, J., Jabbari, A., Reegu, F. A., & Sherzod, T. (2022). Adversarial approaches to tackle imbalanced data in machine learning. *Sustainability, 15*(9), 7097. Available from https://doi.org/10.3390/su15097097.

Baek, W. K., & Jung, H. S. (2021). Performance comparison of oil spill and ship classification from x-band dual-and single-polarized SAR image using support vector machine, random forest, and deep neural network. *Remote Sensing, 13*(16). Available from https://doi.org/10.3390/rs13163203, https://www.mdpi.com/2072-4292/13/16/3203/pdf.

Basit, A., Siddique, M. A., Bhatti, M. K., & Sarfraz, M. S. (2022). Comparison of CNNs and vision transformers-based hybrid models using gradient profile loss for classification of oil spills in SAR images. *Remote Sensing (Basel), 14*. Available from https://doi.org/10.3390/RS14092085.

Bianchi, F. M., Espeseth, M. M., & Borch, N. (2020). Large-scale detection and categorization of oil spills from SAR images with deep learning. *Remote Sensing, 12*(14). Available from https://doi.org/10.3390/rs12142260, https://res.mdpi.com/d_attachment/remotesensing/remotesensing-12-02260/article_deploy/remotesensing-12-02260-v2.pdf.

Blondeau-Patissier, D., Schroeder, T., Suresh, G., Li, Z., Diakogiannis, F. I., Irving, P., Witte, C., & Steven, A. D. L. (2023). Detection of marine oil-like features in Sentinel-1 SAR images by supplementary use of deep learning and empirical methods: Performance assessment for the Great Barrier Reef marine park. *Marine Pollution Bulletin, 188*. Available from http://www.elsevier.com/locate/marpolbul, https://doi.org/10.1016/j.marpolbul.2023.114598.

Brekke, C., & Solberg, A. H. S. (2005). Oil spill detection by satellite remote sensing. *Remote Sensing of Environment, 95*(1), 1–13. Available from https://doi.org/10.1016/j.rse.2004.11.015.

Ciappa, A. C. (2023). Oil trajectory analysis for oil spill surveillance by SAR in the Mediterranean Sea. *Marine Pollution Bulletin, 190*. Available from https://doi.org/10.1016/j.marpolbul.2023.114825, http://www.elsevier.com/locate/marpolbul.

Dasari, K., Anjaneyulu, L., & Nadimikeri, J. (2022). Application of C-band sentinel-1A SAR data as proxies for detecting oil spills of Chennai, East Coast of India. *Marine Pollution Bulletin, 174*, 113182. Available from https://doi.org/10.1016/j.marpolbul.2021.113182.

Das, K., Janardhan, P., & Narayana, H. (2023). Application of CNN based image classification technique for oil spill detection. *Indian Journal of Geo-Marine Sciences, 52*(1), 5–14. Available from https://doi.org/10.56042/ijms.v52i01.5438, http://doi.org/10.56042/ijms.v52i01.5438.

Dave. (2011). Remediation technologies for marine oil spills: A critical review and comparative analysis. *American Journal of Environmental Science, 7*, 423–440. Available from https://doi.org/10.3844/AJESSP.2011.423.440.

de Moura, N. V. A., de Carvalho, O. L. F., Gomes, R. A. T., Guimarães, R. F., & de Carvalho Júnior, O. A. (2022). Deep-water oil-spill monitoring and recurrence analysis in the Brazilian territory using Sentinel-1 time series and deep learning. *International Journal of Applied Earth Observation and Geoinformation, 107*. Available from https://doi.org/10.1016/j.jag.2022.102695, http://www.elsevier.com/locate/jag.

Dehghani-Dehcheshmeh, S., Akhoondzadeh, M., & Homayouni, S. (2023). Oil spills detection from SAR Earth observations based on a hybrid CNN transformer networks. *Marine Pollution Bulletin, 190*, 114834. Available from https://doi.org/10.1016/j.marpolbul.2023.114834.

Dhiman, P., Kaur, A., Balasaraswathi, V. R., Gulzar, Y., Alwan, A. A., & Hamid, Y. (2023). Image acquisition, preprocessing and classification of citrus fruit diseases: A systematic literature review. *Sustainability*. Available from https://doi.org/10.3390/SU15129643.

Fan, J., & Liu, C. (2023). Multitask GANs for oil spill classification and semantic segmentation based on SAR images. *IEEE Journal of Selected Topics in Applied Earth Observations and Remote Sensing, 16*, 2532–2546. Available from https://doi.org/10.1109/JSTARS.2023.3249680, http://ieeexplore.ieee.org/xpl/RecentIssue.jsp?punumber = 4609443.

Fingas, M., & Brown, C. (2018). A review of oil spill remote sensing. *Sensors, 18*(1), 91. Available from https://doi.org/10.3390/s18010091.

Frate, F. D., Petrocchi, A., Lichtenegger, J., & Calabresi, G. (2000). Neural networks for oil spill detection using ERS-SAR data. *IEEE Transactions on Geoscience and Remote Sensing, 38*(5), 2282–2287. Available from https://doi.org/10.1109/36.868885.

Garcia-Pineda, O., Zimmer, B., Howard, M., Pichel, W., Li, X., & MacDonald, I. R. (2009). Using SAR images to delineate ocean oil slicks with a texture-classifying neural network algorithm (TCNNA). *Canadian Journal of Remote Sensing, 35*(5), 411–421. Available from https://doi.org/10.5589/m09-035.

Ghorbani, Z., & Behzadan, A. H. (2021). Monitoring offshore oil pollution using multi-class convolutional neural networks. *Environmental Pollution, 289*. Available from https://doi.org/10.1016/j.envpol.2021.117884, https://www.journals.elsevier.com/environmental-pollution.

Gulzar, Y. (2023). Fruit image classification model based on mobilenetV2 with deep transfer learning technique. *Sustainability, 15*(3), 1906. Available from https://doi.org/10.3390/su15031906.

Gulzar, Y., Alwan, A. A., Abdullah, R. M., Abualkishik, A. Z., & Oumrani, M. (2023). OCA: Ordered clustering-based algorithm for E-commerce recommendation system. *Sustainability*, 2947. Available from https://doi.org/10.3390/SU15042947.

Gulzar, Y., Ünal, Z., Aktąs, H. A., & Mir, M. S. (2023). Harnessing the power of transfer learning in sunflower disease detection: A comparative study. *Agriculture*. Available from https://doi.org/10.3390/AGRICULTURE13081479.

Gulzar, Y., Hamid, Y., Soomro, A. B., Alwan, A. A., & Journaux, L. (2020). A convolution neural network-based seed classification system. *Symmetry*, *12*, 2018.

Gulzar, Y., & Khan, S. A. (2022). Skin lesion segmentation based on vision transformers and convolutional neural networks—A comparative study. *Applied Sciences*. Available from https://doi.org/10.3390/APP12125990.

Gulzar, Y., Ünal, Z., Ayoub, S., Reegu, F. A., & Altulihan, A. (2024). Adaptability of deep learning: Datasets and strategies in fruit classification. *BIO Web of Conferences*, *85*. Available from https://doi.org/10.1051/bioconf/20248501020, https://www.bio-conferences.org/.

Gulzar, Y., Ünal, Z., Ayoub, S., & Reegu, F. A. (2024). Exploring transfer learning for enhanced seed classification: Pre-trained xception model. *Lecture Notes in Civil Engineering*, *458*, 137–147. Available from https://doi.org/10.1007/978-3-031-51579-8_14, https://www.springer.com/series/15087.

Hamid, Y., Elyassami, S., Gulzar, Y., Balasaraswathi, V. R., Habuza, T., & Wani, S. (2023). An improvised CNN model for fake image detection. *International Journal of Information Technology*, *15*(1), 5–15. Available from https://doi.org/10.1007/s41870-022-01130-5.

Hamid, Y., Wani, S., Soomro, A.B., Alwan, A.A., & Gulzar, Y. (2022). Smart seed classification system based on mobilenetV2 architecture. *Proceedings of 2022 second international conference on computing and information technology, ICCIT 2022* (pp. 217–222). https://doi.org/10.1109/ICCIT52419.2022.9711662, http://ieeexplore.ieee.org/xpl/mostRecentIssue.jsp?punumber = 9711508.

Han, Z., Xing, J., Wang, X., Xue, F., & Fan, J. (2022). A robust LCSE-ResNet for marine man-made target classification based on optical remote sensing imagery. *International Journal on Artificial Intelligence Tools*, *31*(6). Available from https://doi.org/10.1142/S021821302240022X, http://www.worldscinet.com/ijait/ijait.shtml.

Huang, H., Wang, C., Liu, S., Sun, Z., Zhang, D., Liu, C., Jiang, Y., Zhan, S., Zhang, H., & Xu, R. (2020). Single spectral imagery and faster R-CNN to identify hazardous and noxious substances spills. *Environmental Pollution*, *258*, 113688. Available from https://doi.org/10.1016/j.envpol.2019.113688.

Khan, F., Ayoub, S., Gulzar, Y., Majid, M., Reegu, F. A., Mir, M. S., Soomro, A. B., & Elwasila, O. (2023). MRI-based effective ensemble frameworks for predicting human brain tumor. *Journal of Imaging*, *9*(8). Available from https://doi.org/10.3390/jimaging9080163, http://www.mdpi.com/journal/jimaging/.

Khan, F., Gulzar, Y., Ayoub, S., Majid, M., Mir, M. S., & Soomro, A. B. (2023). Least square-support vector machine based brain tumor classification system with multi model texture features. *Frontiers in Applied Mathematics and Statistics*. Available from https://doi.org/10.3389/FAMS.2023.1324054.

Khan, S. A., Gulzar, Y., Turaev, S., & Peng, Y. S. (2021). A modified HSIFT descriptor for medical image classification of anatomy objects. *Symmetry*, *13*, 1987.

Lang, H., Wang, R., Zheng, S., Wu, S., & Li, J. (2022). Ship classification in SAR imagery by shallow CNN pre-trained on task-specific dataset with feature refinement. *Remote Sensing*, *14*(23), 5986. Available from https://doi.org/10.3390/rs14235986.

Li, Y., Cui, C., Liu, Z., Liu, B., Xu, J., Zhu, X., & Hou, Y. (2017). Detection and monitoring of oil spills using moderate/high-resolution remote sensing images. *Archives of Environmental Contamination and Toxicology*, *73*(1), 154–169. Available from https://doi.org/10.1007/s00244-016-0358-5.

Majid, M., Gulzar, Y., Ayoub, S., Khan, F., Reegu, F. A., Mir, M. S., Jaziri, W., & Soomro, A. B. (2023). Enhanced transfer learning strategies for effective kidney tumor classification with CT imaging. *International Journal of Advanced Computer Science and Applications*, *14*(8), 421–432. Available from https://doi.org/10.14569/IJACSA.2023.0140847, http://thesai.org/Publications/Archives?code = IJACSA.

Malik, I., Ahmed, M., Gulzar, Y., Baba, S. H., Mir, M. S., Soomro, A. B., Sultan, A., & Elwasila, O. (2023). Estimation of the extent of the vulnerability of agriculture to climate change using analytical and deep-learning methods: A case study in Jammu, Kashmir, and Ladakh. *Sustainability*, *15*(14), 11465. Available from https://doi.org/10.3390/su151411465.

Mamat, N., Othman, M. F., Abdulghafor, R., Alwan, A. A., & Gulzar, Y. (2023). Enhancing image annotation technique of fruit classification using a deep learning approach. *Sustainability*, *15*(2), 901. Available from https://doi.org/10.3390/su15020901.

Mehdi, S. R., Raza, K., Huang, H., Naqvi, R. A., Ali, A., & Song, H. (2022). Combining deep learning with single-spectrum UV imaging for rapid detection of HNSs spills. *Remote Sensing*, *14*(3). Available from https://doi.org/10.3390/rs14030576, https://www.mdpi.com/2072-4292/14/3/576/pdf.

Mehmood, A., Gulzar, Y., Ilyas, Q. M., Jabbari, A., Ahmad, M., & Iqbal, S. (2023). SBXception: A shallower and broader xception architecture for efficient classification of skin lesions. *Cancers*. Available from https://doi.org/10.3390/CANCERS15143604.

Passah, A., & Kandar, D. (2023). A lightweight deep learning model for classification of synthetic aperture radar images. *Ecological Informatics*, *77*, 102228. Available from https://doi.org/10.1016/j.ecoinf.2023.102228.

Sahlan, F., Hamidi, F., Misrat, M. Z., Adli, M. H., Wani, S., & Gulzar, Y. (2021). Prediction of mental health among university students. *International Journal on Perceptive and Cognitive Computing*, *7*, 85–91.

Scardigli, A., Risser, L., Haddouche, C., & Jatiault, R. (2023). Integrating unordered time frames in neural networks: Application to the detection of natural oil slicks in satellite images. *IEEE Transactions on Geoscience and Remote Sensing*, *61*, 1–14. Available from https://doi.org/10.1109/tgrs.2023.3241681.

Solberg, A. H. S., Brekke, C., & Husøy, P. O. (2007). Oil spill detection in Radarsat and Envisat SAR images. *IEEE Transactions on Geoscience and Remote Sensing*, *45*(3), 746–754. Available from https://doi.org/10.1109/TGRS.2006.887019.

Sudha, V., & Vijendran, A. S. (2024). Oil spill detection and recognition utilizing faster R-CNN with enhanced mobilenetv2 architecture. *International Journal of Intelligent Systems and Applications in Engineering*, *12*(17s), 304–321. Available from https://ijisae.org/index.php/IJISAE/article/download/4876/3564.

Topouzelis, K., Stathakis, D., & Karathanassi, V. (2009). Investigation of genetic algorithms contribution to feature selection for oil spill detection. *International Journal of Remote Sensing*, *30*(3), 611–625. Available from https://doi.org/10.1080/01431160802339456.

Xu, L., Li, J., & Brenning, A. (2014). A comparative study of different classification techniques for marine oil spill identification using RADARSAT-1 imagery. *Remote Sensing of Environment*, *141*, 14–23. Available from https://doi.org/10.1016/j.rse.2013.10.012.

Yang, Y. J., Singha, S., & Mayerle, R. (2022). A deep learning based oil spill detector using Sentinel-1 SAR imagery. *International Journal of Remote Sensing*, *43*(11), 4287–4314. Available from https://doi.org/10.1080/01431161.2022.2109445, https://www.tandfonline.com/loi/tres20.

Yuan, F., Huang, Y., Chen, X., & Cheng, E. (2018). A biological sensor system using computer vision for water quality monitoring. *IEEE Access*, *6*, 61535–61546. Available from https://doi.org/10.1109/ACCESS.2018.2876336, http://ieeexplore.ieee.org/xpl/RecentIssue.jsp?punumber = 6287639.

Zheng, Y., Su, J., Zhang, S., Tao, M., & Wang, L. (2022). Dehaze-AGGAN: Unpaired remote sensing image dehazing using enhanced attention-guide generative adversarial networks. *IEEE Transactions on Geoscience and Remote Sensing*, *60*. Available from https://doi.org/10.1109/TGRS.2022.3204890, https://ieeexplore.ieee.org/servlet/opac?punumber = 36.

Index

Note: Page numbers followed by "*f*" and "t" refer to figures and tables, respectively.

F

G

H

I

S

T

U

V

W

X

Y

Z

Made in the USA
Monee, IL
12 September 2026

62979090R10188